Method	Type of Change	Time pattern	Pros	Cons
Time series	Movement or change in character	Trend cycle before and after	Strong visual impact if change is substantial; shows conditions at each data/time	Readers have to visually compare maps to see where, and how much change occurred
Tracking map	Movement	Trend cycle before and after	Easier to see movement and rate of change than with time series, especially if change is subtle	Can be difficult to read if more than a few features
Measuring change	Change in character	Trend before and after	Shows actual difference in amounts or values	Doesn't show actual conditions at each time; change is calculated between two times only

FUNDAMENTALS OF GEOGRAPHIC INFORMATION SYSTEMS

FOURTH EDITION

Michael N. DeMers
New Mexico State University

JOHN WILEY & SONS, INC.

VICE PRESIDENT AND PUBLISHER Jay O'Callaghan

EXECUTIVE EDITOR Ryan Flahive

ASSISTANT EDITOR Courtney Nelson

SENIOR PRODUCTION EDITOR Nicole Repasky

MARKETING MANAGER Danielle Torio

DESIGNER Michael St. Martine

PRODUCTION MANAGEMENT SERVICES Katie Boilard/Pine Tree Composition, Inc.

PHOTO EDITOR Hilary Newman

EDITORIAL ASSISTANT Erin Grattan

MEDIA EDITOR Lynn Pearlman

Image provided by USGS EROS Data Center Satellite Systems Branch. Courtesy NASA.

This book was set in 10/12 Cheltenham by Laserwords Private Limited, Chennai and printed and bound by Hamilton Printing. The cover was printed by Phoenix Color.

This book is printed on acid free paper. ♾

To order books or for customer service please, call 1-800-CALL WILEY (225-5945).

Library of Congress Cataloging-in-Publication Data

DeMers, Michael N.
Fundamentals of geographic information systems / Michael N. DeMers.—4th ed.
p. cm.
Includes bibliographical references.
ISBN 978-0-470-12906-7 (alk. paper)
1. Geographic information systems—Textbooks. I. Title.

G70.212.D46 2009
910.285—dc22

2007038332

Printed in the United States of America

10 9 8 7 6 5 4 3 2 1

PREFACE

GIS once meant only geographic information systems. Today it is evolving into a more comprehensive discipline in its own right and the term now represents not only the system itself, but the underlying science and technology foundations. It has even acquiring a new acronym: GIS&T, meaning Geographic Information Science and Technology. Just this year we have completed a major undertaking with the publication of the first edition of what is hoped will be an ongoing set of editions of the GIS&T Body of Knowledge (Dibiase, et. al. 2006). Afer the culmination of almost seven years of work involving geographers, computer scientists, researchers, academics, and industry participants, this document represents a first approximation definition of the discipline as it stands today. Beyond setting the standards and outlining some 1,660 learning objectives within ten knowledge areas, it suggests what a student completing a four-year university degree emphasizing GIS&T would be expected to know. This also provided both an incentive and a set of guidelines for the 4*th* edition of the text you are now reading.

When I wrote the first edition of this text over ten years ago there were few guidelines for content or learning objectives. I was forced to glean what I could from personal mentors, my own class notes and teaching experiences, discussions with other GIS faculty teaching GIS, student responses and a gut reaction. As I look at the GIS&T Body of Knowledge I must say that I was not too far off the mark. But, the discipline is changing constantly, and the many people involved in creating the disciplinary standard represent an interesting mix of the disciplinary historical roots and the future advances in cutting edge research. As a result of working with this awesome set of participants and advisors I have rethought some major portions of this text to reflect the new standard.

At first blush the structure of the text doesn't look all that different from its initial inception, but the content, particularly with regard to the computer science, cartographic visualization, and design are considerably more robust. These are meant to reflect both the Body of Knowledge Standard and the changing level of computer expertise of the students. The percentage of tech savvy students that I have seen over the past decade has grown logarithmically. This is fortunate because the industry itself is adapting to acknowledge both a need for new concepts and geographical constructs to include in the toolkit and new ways to implement them. This edition of Fundamentals of Geographic Information Systems is meant to target students who are comfortable with e-mail and text messaging, digital file formats (mp3, mp4, jpeg), computer games and visualizations, and a host of technologies that did not exist ten years ago. Some educators are now referring to this new generation of students as the neomillenials, not because of when they were born, but because of their daily immersion in a technology-rich environment.

At the same time I have added a pre-GIS chapter (Chapter 0) called "A Spatial Learner's Permit." This chapter is a reaction to the ever increasing diversity of student who enters the field of GIS, often with little or no knowledge of geographic terminology, explicit training in geographic principals, or experience in spatial cognition and spatial analysis. The concept here is that of a permitting process where those students who are new to the field of GIS, or those who wish to renew their background, have an opportunity to do so without being thrown in head first and watching to see who survives and who doesn't. There is no reason why everyone, no matter what their discipline, cannot fully appreciate and successfully employ the tools and techniques of GIS as long as they have a chance to get this necessary background.

Beyond their growing technological sophistication, my own students are increasingly entering my GIS classes from disciplines outside of geography. They are coming from departments you might expect such as criminal justice, geology, biology, surveying, wildlife, and anthropology. More recently they have started arriving from economics, history, engineering, public health, hospitality, agronomy, and a host of others. This has made the teaching of GIS and the production of this supporting text both exciting and difficult. It is stimulating getting the often wildly different perspectives these students bring to the classroom but it also highlights a fundamental difficulty with the current state of GIS education. How do we teach students with a solid geography and mapping skills background with the same class, or the same textbook, as those with little or no background in the spatial sciences? This is an on-going topic of GIS educational research, but I have made a stab at a partial solution in this book. For geographers and others well versed in spatial tools I suggest they review Chapters 1–3 and begin studying in earnest with Chapter 4. For the neophyte I often recommend that they think of the first three chapters, especially 2 and 3, as a GIS Bootcamp. Many GIS software companies provide a set of pre-GIS exercises as a part of their software training. This has proven successful and might provide a model for academic instruction as well.

The pressure from students to get more software exposure and less theory and concepts continues to mount. And as I stated in the third edition, I continue to resist. It is getting even more difficult with the neomillenials because they are so inextricably linked to technology. So what do we instructors do? My approach has been to increase the complexity of the hands-on problems and force the students to ask questions about the underlying concepts that are required for them to answer them. My lab components are becoming less recipe-like and more open ended. For example, it is much more difficult to ask your students to solve a problem where there is no set recipe of GIS commands to follow than it is to just take them through the problem. Such an approach requires more than just an ability to read a lab manual, type the appropriate commands, and hand in the results. It requires problem solving. These are the skills the GIS industry constantly tells me they need! I've found it useful to focus on the learning objectives for each chapter and to ask some of these questions as required output for lab reports or as practical lab exam questions.

So what's new in this edition? As I mentioned before, for those entering GIS without a substantial background in the geographic underpinnings of the discipline, I have added a short, gentle introduction called A Spatial Learner's Permit. This is not only for non-geographers, but also for those whose geography is a little rusty and would like to brush up a little. Next, I've split the old GIS data models chapter (old Chapter 4) into two separate chapters: Chapter 4–GIS: Computer Structure Basics and Chapter 5–GIS Data models. This seems to have reduced the confusion for my students. It also allows for a more in-depth, although not comprehensive, examination of the underlying computer science behind GIS, without overwhelming the student with a large single chapter. Another chapter split came in the old Chapter 10 on statistical surfaces is now Chapter 11–Statistical Surfaces and Chapter 12–Terrain Analysis. This suggestion came from a faculty member who used the text and pointed out to me that there is now a pretty robust set of analytical techniques focusing just on the terrain type of statistical surface. As with the other split it allowed me to spend a little more time on techniques that had been neglected before. The reaction in my classes was positive regarding this as well.

Some of the most massive changes came at the end of the text where the old Chapters 16 and 17 dealing with applications and research were eliminated. I received feedback from instructors that they were unable to find time to get to these chapters, and that the selection of topics for both was limited. I was also unable to get to those chapters and unwilling to rush through the material to do so. As a result I have expanded and completely updated Chapter 16 (Chapter 14 in the 3rd edition) to reflect the increasingly robust literature on cartographic visualization. Given the graphical sophistication of the current student I thought this would be a better use of my time. I have also reworked Chapter 17 (old chapter 15) to reflect an equally improved body of literature on GIS design and management. Both of these major changes are also a direct response to the Body of Knowledge Document.

A cautionary note here: neither Fundamentals of Geographic Information Systems, nor any other introductory text I can envision, is capable of containing the sum total of the 1,660 learning objectives nor all of the topics and knowledge areas found in the Body of Knowledge with any reasonable level. An introductory text is meant to introduce, not overwhelm. Instead this text focuses on the ideas, concepts, and technologies that one might expect a student to encounter in an early exposure to GIS. More detailed topical coverage of the existing material and more selection of topics should be expected in advanced coursework and accompanying textbooks. Still, there is always room for improvement. I look forward to your experiences and feedback, both make my own courses better.

ACKNOWLEDGMENTS

Much of this revision came after a long stay in the hospital. There was some doubt about my chances for survival. That I beat the odds is due in large part to the actions and expertise of Renee Williams, Dr. Azhir Zahir, Dr. E. Rhett Jabour, the NMSU emergency services personnel, the doctors, nurses and other medical staff of Memorial Medical Center in Las Cruces (especially the 4th floor, and ICU). I am also grateful to all whose prayers, visits, gifts, and kind thoughts maintained me during that difficult time, especially Jack Wright, Carol and Jim Campbell, Chris Brown, Bob and Beth Czerniak, Dan and Karla Dugas, David and Francine McNeil, Jaffer Hanfoosh, my students, and all those whose names I have forgotten. Of course, I am extremely blessed to have had the constant love and prayers of my wife, Dolores, and my brothers James and Dennis, and all my family members. You are all responsible to some degree for this text.

Thanks to those who taught me and mentored me, especially Bill Dando, Lee Williams, Duane Marble, Kang Tsung Chang, Vince Robinson, Peter Fisher, and many others. In this regard I also want to thank all the members of the GIS curriculum project and my Body of Knowledge co-editors for the ideas, knowledge, and insights you provided. The students who had to suffer through all the previous editions deserve thanks for their tolerance, and, in some cases, their criticism. Finally, I want to thank the folks at Pine Tree Composition, Inc. for their editorial services, and at John Wiley and Sons for their continued faith in my work, especially Laura Spence Kelleher, and Ryan Flahive, the production staff, and graphic artists. A special note to all the following who took the time out of their busy schedules and their own projects to review this book: Richard Beck, University of Cincinnati; William Cooke, Mississippi State University; Alison Feeney, Shippensburg University; Todd Fritch, Northeastern University; David Fyfe, Penn State University; Alberto Giordano, Texas State University–San Marcos; Shunfu Hu, Southern Illinois University–Edwardsville; Dorleen Jenson, Salt Lake Community College; Nicholas Kohler, University of Oregon; Helmut Kraenzle, James Madison University; Debbie Kreitzer, Western Kentucky University; Dean Lambert, San Antonio College; Mark Leipnik, Sam Houston State University; Chor Pang Lo, University of Georgia; Jia Lu, University of Wisconsin–Stevens Point; Yongmei Lu, Texas State University–San Marcos; Robert Martin, Kutztown University; Darrel McDonald, Stephen F. Austin State University; Jaewan Yoon, Old Dominion University; and May Yuan, University of Oklahoma. You all share in this product.

CONTENTS

CHAPTER 0

Spatial Learner's Permit

The technological sophistication of modern geographic information systems (GIS) software brings with it a misconception that its use as a map-analysis and decision-making tool is easy—requiring little if any background in the disciplinary and conceptual background from which it was derived. As with word processing software, it is not the software that produces short stories or novels, but the user who employs the tool effectively. Without some basic understanding of plot and character, and without a proper vocabulary and the ability to employ that vocabulary to construct sentences and paragraphs that communicate the events of a story, the software itself produces nothing. The author envisions the setting and location of the story; creates the look of the characters, how they are dressed, how they move, how they sound, and how they interact with each other and their environment. Each element and each interaction is put together in a coherent storyline that carries the characters along from beginning to end. Each step is calculated for effect, each scene proceeds logically from the one before. Without the author's ability to piece this together in a logical sequence the story moves slowly, the characters are flat and unrealistic, and the entire work is unsuccessful and artificial.

So it is with GIS. Without some familiarity with the explicit terminology and underlying principles of geography, without an understanding of the spatial (i.e., geographic) nature of the problems the software was designed to address, and without an awareness of how the representation of geographic data in either analog or digital map form might impact its effectiveness and accuracy, GIS will inadequately serve the user. Much of the output produced with GIS software suffers from an inadequate conceptual foundation that results in extremely simplistic and often incorrect models. Some show a lack of understanding of how the system they are modeling works. Still others demonstrate a lack of understanding of how data at different measurement levels can properly interact.

Think of this chapter much like the experience many of you had when you took classroom training before you could get a driver's permit. The driver's permit was essentially a learner's permit to allow you to take behind-the-wheel driving instruction. It provided you with an appropriate vocabulary, understanding of the rules of the road, insight into traffic signs, and gave you a basic understanding of how the transportation system works. While many have obtained driver's licenses without this background, they tend to learn the necessary rules of the road through experience—much of which involves many close calls and an occasional accident. This chapter will provide you with some of the basic spatial terminology, geographic concepts, and spatial data relationships that you will encounter in the remainder of the text in considerably more detail. It lets you encounter this information gradually and helps you gain a fuller understanding of the foundations of the discipline. It should make you a better GIS practitioner, whatever your goals in that profession are.

This chapter differs somewhat from the remaining chapters in this book. In the first place it assumes you are beginning from scratch with regard to your spatial and geographic background. Second, its material is more overarching and more integrative than the other chapters. Finally, there is a structural change in that the questions at the end of the other chapters of the book are replaced here with suggested exercise and discussion topics that can be applied throughout your course and even continued throughout your professional years to keep your skills sharp. Like any skill, the more it is exercised the stronger it becomes.

SPATIAL TERMINOLOGY

Every discipline has its own terminology, and that terminology both reflects the content and directs it. Language instructors will tell you that to understand a culture you must understand the language because the culture itself is embedded in the language. Language and terminology is the intellectual filter within which the discipline operates. Historians have a time-based terminology, scientists have the scientific method, mathematicians employ a complex set of symbols, and logicians construct a set of formal logic rules. Geographers integrate many disciplines and incorporate the terminology of those disciplines, but what separates them is their in-depth comprehension and frequent use of a terminology that is explicitly spatial (Downs, et al. 2006). This robust spatial terminology is integral to the discipline and has been developing for over twenty-five hundred years. It is the terminology that has been explicitly incorporated into the tool about which you are learning—geographic information systems.

GIS software is designed to input, store, edit, retrieve, analyze, and output **geographically referenced** data (data tied explicitly to known points on the earth's surface) (Figure 0.1). In short, it is about spatially located objects, spatial distributions, and their spatial analysis. We might, for example, have a database that contains roads, counties, towns, rivers, and many other geographic features. Because we all grow up, develop, learn, experience, and are constantly immersed in a spatial world, we often take space for granted. We often fail to explicitly recognize its many aspects and the degree to which our lives are affected by it. Our vocabulary reflects this spatial dependency with a robust set of spatial terminology and related concepts and relationships. While we operate effortlessly and mostly unconsciously in a spatial world, our ability to translate this into an ability to explicitly define, evaluate, and manipulate geographic phenomena in GIS software requires that we have a thorough, explicit, and in-depth understanding of spatial terms in a wide variety of geographic settings

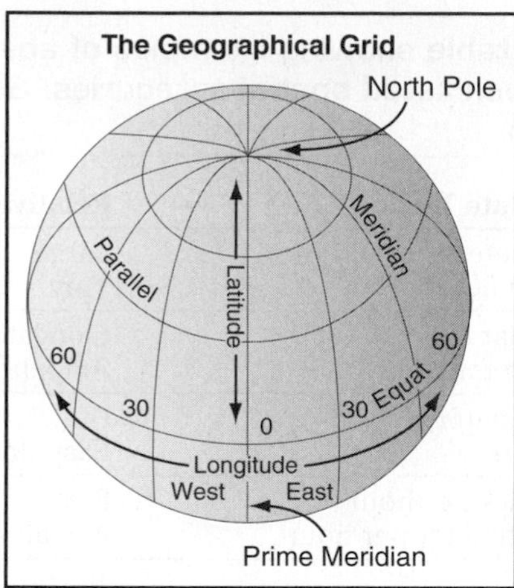

Figure 0.1 To find our way on the earth, all of the objects that we map must be linked to specific positions on the earth. This is known as geographical referencing and is shown here with a set of geographical reference lines.

and environments. The most successful GIS practitioners have a heightened sense of spatial awareness and spatial relationships.

Spatial terminology is everywhere and comprises a nearly endless list. To facilitate your learning it, this section includes a simple structure with some sample terms that allow you to get a feel for how the spatial terminology works. The following is a list of some basic forms that spatial terms can take:

- Size and scale
- Shape and dimension
- Distance
- Speed (movement)
- Direction (angle)
- Containment and adjacency
- Location and Distribution

Each of these categories can be defined through an exhaustive list of terms that characterize it as measurable absolutes (i.e., absolute terms). We can also characterize them in relative terms. By that I mean we can characterize them relative to other conditions, terms, or relationships. Here's an incomplete table (Table 0.1) that might help get you started as you compile your own list. Some spatial categories have lots of absolute terms and few relative terms while others are the reverse.

The ability to communicate in both absolute and relative terms is vital to operate in a GIS environment. For example, you might be trying to determine which fences are at least 200 meters long (absolute), or you might just want to find which are large compared to some other measurable object, like a farm or yard (relative). You might also want to know if they are adjacent to the farm or yard (absolute), partially or fully contained within them (relative), or are at least 75 percent contained within them (absolute). It might be important to know how far you have to travel to get to these farms or yards from your existing location (absolute), or you might just want to know if it is close enough to travel

TABLE 0.1 Sample portion of a table showing examples of absolute and relative terms for each of the seven listed spatial categories. See if you can add more terms to this table.

Spatial Category	Absolute Term	Relative Term
Size and Scale	200 meters 1 millimeter	Large Tiny
Shape and Dimension	Circular Perimeter/area = 0.5	Elongate Amoeboid
Distance	2 kilometers 2 hours	Far Easy drive
Speed	60 miles per hour 1 centimeter per hour	Fast A snail's pace
Direction	North 289°	Northish Mostly west
Containment and Adjacency	In 2 centimeters apart	Partly in, partly out Very near
Location and Distribution	42° N, 97° W 3,200 people per square mile	Central U.S. Dense population

by foot or by car (relative). Once you have that answer you might want to know how difficult a trip (relative) it might be or what speed you would need to travel (absolute) to get there in two hours. When you arrive you might also want to describe the farm or yard in terms of its size in acres or hectares (absolute), or describe it in relative terms (a large farm). You might also want to describe the shape to someone who wants to purchase it as a wedge with radius of 300 meters and a circumference of 270,000 square meters (absolute) or in general as a sort of wedge-shaped piece of land (relative).

Look at Table 0.1 to see how many terms you can think of that fit within the categories listed. Think about books and stories you have read that use explicitly spatial terms or whose themes might be strongly suggestive of spatial concepts. Think about how you might describe movie action scenes and the juxtaposition of characters and scene elements like movie script writers are expected to do. Imagine having to describe to a sightless person how you move around your house or apartment, avoiding objects and navigating inside and outside different rooms. Because we do this so automatically it is not something that is always easy to translate into words. Imagine, then, if you as a GIS programmer must translate such things and actions into computer code. Think what a GIS analyst must do to make the software define spatial concepts to describe and analyze the environment. Many people like to participate in scavenger hunts, or the GPS equivalent, **geocaching** (Figure 0.2). This is the reverse of our previous examples in that we must now move from verbal description to action within geographic space. Before you leave this part of the chapter take some time to think of as many spatial terms as you can. Think of places where spatial terms occur, how they are used, and how they are represented. Continue to think of terms for table 0.1. Keep a log. Stop to think about the terms and their meanings. This will help you think explicitly about

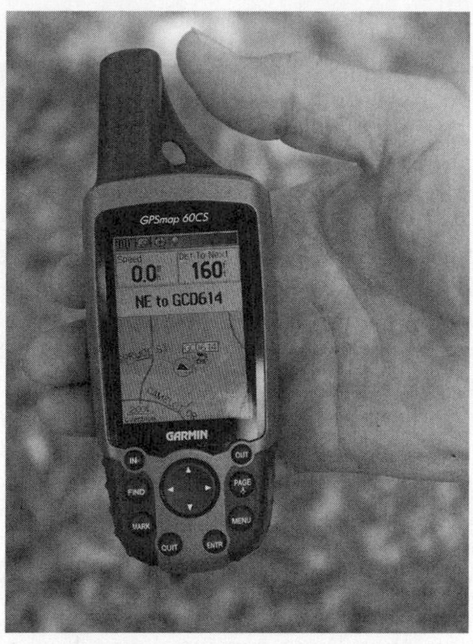

Figure 0.2 Geocaching. is an activity where people use GPS and maps to locate items hidden at specific locations. To do this requires a knowledge of grid and reference systems.

spatial terms, thus providing the first necessary steps in becoming effective in your GIS endeavors.

APPLYING SPATIAL TERMS

Now that you have begun to use spatial terminology explicitly, the next step is to evaluate how these terms are applied to geographic features that we encounter in our GIS analysis. Sometimes geographic representation, referred to as **ontology,** is important both in how geographic features are represented and quantified inside the software, and in how we conceptualize the problems that employ them (Kuipers, 1978; Mark, et al. 1999; Smith and Mark, 1998). A common starting point is to think about the term "mountain." Possibly without exception we inherently know what a mountain is. If we have not seen one in reality we've seen pictures, and if we have never hiked or climbed one it is no major intellectual leap to imagine what that experience would be like. Here's the trick: describe to the computer exactly what a mountain is both in relative and absolute spatial terms (Fisher, 2005).

Imagine we are trying to analyze the impact of both absolute and relative position on a mountain of bird populations using a GIS. What aspects of the mountain are explicitly contained in the software that will allow us to analyze the relationships between its environments and various bird species? Before we can even begin to separate out the aspects of mountain environments that might be important in our study, we must first incorporate a quantifiable digital description of a mountain. We must be very explicit in our definition. Is a hill a mountain? What about a plateau? A mesa? A butte? What spatial descriptors do we need to separate out each of these features? A shortcut might be to go to Wikipedia.com for answers. Here's what it says:

> A **mountain** is a landform that extends above the surrounding terrain in a limited area. A mountain is generally steeper than a hill, but there is no universally accepted standard definition for the height of a mountain or a hill although a mountain usually has an identifiable summit. [http://en.wikipedia.org/wiki/Mountain (last visited 5/19/07)].

Sounds pretty reasonable, doesn't it? But it still isn't very explicit. First, it uses terms that themselves are not well defined, such as *landform*, *terrain*, and *hill*. It also mentions things like *steeper*, which is not explicitly defined, and *summit*, which is also not defined. We not only need to define these terms, but we must do this using numbers and quantities that a computer can calculate.

In the movie *The Englishman Who Went Up a Hill but came Down a Mountain* the main character is tasked with determining whether a local landmark could be considered a mountain. The primary determinate it seems was its elevation. While not enough to separate out a mountain from other geographical formations, it's as good a place as any to start. So, to provide the computer with the necessary information upon which our definition will be based, we start by including primary elevational information. The details of how this is accomplished are covered later in the book.

Based just on that information, then, how can we define how a mountain differs from a hill or a plateau? We know, for example, that plateaus and mesas are flat on top. We also know that the top or "summit" of a mountain is not flat. But how do we define this inside the software? How is "flat" defined? Let's think about this for a moment. So far we have assumed that elevation is the only component because that is all we put into our computer. But now that it's there we can also assume that all the elevation data are also located somewhere in geographic space as well. This means that we can make comparisons among individual elevational values, and we can compare the change in elevation from one location to another.

What does this do for us? First, it allows us to compare rise over run (change in elevation over distance). This is what we refer to as **slope.** Slope, then, is a measure of steepness. So, if we have a mountain we might want to say that it has a steep slope. Does this exactly separate a mountain from a hill? Can you think of steep hills? What if we combined the steepness quality with an absolute elevation of, perhaps, 6,000 feet? While there is no agreement that this value of the **summit** (the maximum elevation) must be at least this high, we can choose to designate it so in this instance. Thus, the definition of mountain is problem- or situation-specific. We define the criteria that in turn quantitatively allow us to select the features and attributes we wish to extract from the computer for our application. So to define a mountain we must be able to define that it has a substantial rise in elevation with distance (slope), is over 6,000 feet in elevation at it highest point, and has no substantial flat portions. But we still must define our terms more explicitly. What is a "substantial" slope? We might decide that for our purposes and for our definition this value might need to be greater than 45° (calculated as elevation divided by horizontal distance). We can also stipulate that no more than 5 percent of the top of the mountain can have less than 2 percent slope to allow us to quantitatively define that it is not a tableland feature like a butte or mesa.

As you can see from this discussion, the naming of geographic features brings with it a substantial amount of specific knowledge about the features themselves. It requires that we know how the feature looks, how it acts, and its relationships to other features around it. Not only is this important so we can define what data go into the computer to allow us to define and extract these features, but what properties must be searched to do so.

The concept of a mountain is just the beginning. Imagine having to define some of the following terms: urban area, land use, agriculture, bluff, riparian zone, 100-year flood zone, prairie, forest, lake, pond, and many more. Let's compare them to our mountain example just briefly so you can understand the importance of ontology to GIS input and analysis. Does an area with dense population qualify as an urban area? What if it has a lake in it? How dense is dense? If it's outside the city is it still urban? Are there standards for this terminology?

Consider land use: Is forest a land use or is it a form of land cover? Is grassland a land use? What about rangeland? Is it a land use? Does it qualify as agriculture? How are you going to categorize these terms when you input them? What questions are you going to ask that require the data? How do the questions relate to the input? You can see the same difficulties with the other terms in the preceding paragraph. How big is a lake? How small is a pond? How many trees does a grassland have before it's called a savanna rather than a prairie? While this may seem rather philosophical navel gazing, many of the inconsistencies in GIS input and analysis occur because they are not well thought out before they are input. If the person inputting the data uses different terminology or does not provide enough primary information for features to be defined during query formulation, the GIS software will not be able to perform its function.

To be really expert at defining such terms requires that we spend some time thinking not only about the meanings of our terms, but about the implications of our definitions and the contexts within which they are embedded. A village may mean one thing in North America and another in Asia. Wetlands are defined one way by one agency and quite differently by another. And, of course, one person's mountain is another person's molehill. What is a meadow? A glen? A valley? (Figure 0.3) What is important is that such terms be dissected and their

Figure 0.3 Is this a valley, a meadow, or a glen? The definition of such features must be explicit so the software can store it correctly and the items can be retrieved.

properties defined so explicitly that their digital essence allows users to exactly specify the meaning within a search strategy or for analysis.

As with all things our knowledge is based on the degree to which we experience them. This is a good reason to experience the environment about which your GIS operations will be conducted. But beyond just characterizing individual spatial elements we must also be able to identify, quantify, and ultimately computerize collections of features (houses, trees, landfills, etc.), or **distributions.** Moreover, the ability to identify distributions (e.g., burrowing owls, auto thefts, and cancer mortality) must also allow us to identify the very spatial nature of our environment and the questions that are derived from that environment. That is the topic of our next section.

SPATIAL COGNITION

The term **spatial cognition** may seem pretty imposing, but all it really means is that you posses the ability to recognize spatial patterns and the spatial nature of problems. As with spatial terminology, we often take our ability to see patterns and recognize spatial problems for granted. And, as with spatial terminology, we should not. I remember an anecdote provided by Duane Marble (an early GIS pioneer) that describes a situation in which a gas station had paid significant amounts of money to a transportation engineering firm to help improve their customer base. After spending weeks counting traffic in different directions, their conclusion was that there was nothing wrong with the location and that there must be another factor besides location that was contributing to their lack of business. Dr. Marble was then asked to take a look at the problem from a geographer's perspective. After he arrived he quickly explained to the station manager that a road median was making it very difficult for people to turn into the station to get gas, thus contributing to their low sales. The problem was not numbers—it was totally spatial.

Many people living in large cities discuss the potential routes to get to new locations, such as a restaurant that they go to occasionally, or to the movies, a play, and so on. Their day-to-day travels are pretty much a function of habit, but when new travel scenarios are required, the first thing needed is to determine the spatial nature of their problem by examining alternative routes for distance, speed limit, traffic flows, number of turns, and such. They do this in a rather ad hoc fashion, but GIS might be able to solve these problems. In short, they have recognized the spatial components they need to determine and can now employ some form of routing GIS operations that examine, quantify, and provide solutions for these problems. Everything from simple online routing software to modern in-vehicle navigation systems might be employed. But before they were available, someone had to acknowledge that the problem was a geographic one.

The placement of facilities such as factories, retail stores, schools, emergency services, and the like are often made without much consideration for the impact that geographic space might have on their success. Factories should be located where raw materials are *nearby*, but should also be *far* from *dense* populations, *downwind* of urban areas, and so forth. Notice how our newly acquired vocabulary assists us in defining the geographic nature of the placement problem. Think about the way you travel to places in your own area. How often have you said things like, "I wish there were a grocery store near our home." You might tell your realtor that you want to own a home within a particular school district and close to the school. A wildlife expert might want to know what places and environments are likely to foster animal diversity, or a city might want to know the best place to put a civic center. Planners question the impact of new developments on the patterns of traffic and population change. Police departments examine maps to decide on the placement of peace officers during different shifts. Military officers need to know the most effective locations for troop deployment and battle strategy.

The potential number of scenarios for spatial model building is often limited by our ability to recognize that there is a spatial component in the first place. Like the transportation engineer mentioned earlier, a failure to understand this often results in an unsatisfactory solution or no solution at all. Our task, then, is to focus on whatever context in which we intend to use GIS and ask ourselves explicitly what the spatial dimensions are that comprise the problem. Using our new vocabulary helps us do this, but we must also understand the environment itself. If possible, it is best to experience it firsthand as Dr. Marble did. Among the more interesting outcomes of such experiential spatial learning resulted when Dr. Peter Fisher (1991) was struggling with visualizing how the GIS software calculated what parts of an environment could and could not be seen from selected observation points. His field trip into the environment showed him that the way the software was calculating the visible areas was not realistic (Figure 0.4). His subsequent changes resulted in a much improved method of calculating these areas. In short, his skills as a GIS professional were enhanced by a heightened understanding of the spatial components of the visibility problem. There is no substitute for firsthand experience, but a concentration on the existing maps and satellite images of an area can improve one's sensitivity to the geography of an area and nearly always results in better GIS input, design, and modeling.

Figure 0.4 A view from a particular place allows some things to be seen and others to be hidden. The question is whether or not the software is capable of recreating the actual experience.

SPATIAL QUANTITIES

This chapter has frequently alluded to quantification of spatial entities. The ability to provide not just classifications, but to measure properties of spatial objects, is paramount to the ability of the software to compute spatial responses to spatial queries (searches performed by geographic location). There are no spirits in the computer that read our minds; we must explain things very explicitly to the computer.

Among the first comments I heard as a geography teaching assistant was, "I didn't know there was math in geography." Well, of course there is math everywhere. And a basic understanding of the properties of mathematics and the levels of data measurement (i.e., nominal, ordinal, interval, ratio, scalar) that we use in GIS is going to save us a lot of embarrassment when we employ GIS for analysis. The principles of mathematics drive the queries we make and the calculations that take place in the computer. They are inherent in the way the computer software stores maps and spatial data and computes the equations for navigation, map comparison, and more complex computations.

For us, this all begins with an understanding of the basic measurement levels we encountered in our youth. Most of us have at least a working knowledge of **nominal** scaled data (names and categories), **ordinal** (ranked) data, **interval** and **ratio** scaled data, but would have a difficult time defining some of these terms. Later on in this text we will revisit these terms, and another one—**scalar.** Because geographic data come in all of these different measurement levels, we need to be aware that their improper combinations might result in some pretty strange results. The computer stores all geographic data, whatever the data measurement level, as numbers. These numbers might be integers (whole numbers) or they might be rational numbers, but even categories, such as "lake" or "pond," are going to be stored as numbers. You wouldn't think of multiplying "wheatfields" by "elevation." If you did, what would the value mean? If, for example, your GIS is using the number 3 to represent the category "wheatfields" in a map and you wanted to use multiplication to compare it to your elevation data (let's just say the value is 2,000 feet), what would you get? The software would give a correct value of $3 \times 2{,}000 = 6{,}000$, but what would the category label be? What exactly is 6,000 wheatfeet? While you might think this is silly, I have seen even sillier examples of GIS misuse. Imagine if you will some of the following: "seweracres," "lowrentals," "dryfeet," and (laughter, please), even "bigfeet." It is funny now, but we won't laugh if it happens to us when presenting our results to a paying customer or in a public forum. When you approach the material on geographic data measurement levels, take the time to think about some of the silly labels you've seen here and the many more that we can create.

Another fundamental principle is that the software and mathematical queries used in GIS also conform to the basic properties of addition, subtraction, multiplication, and division. Such properties as the commutative $\{a + b = b + a$ or $a \times b = b \times a\}$, associative $\{a + (b + c) = (a + b) + c$ or $a \times (b \times c) = (a \times b) \times c)$, identity $\{a + 0 = a$ or $a \times 1 = a\}$, distributive $\{a(b + c) = a \times b + a \times c$ or $(b + c)a = b(a) + c(a)\}$, reciprocals $\{1/25 \times 25 = 1\}$, and zero property $(x \times 0 = 0)$ will all come into play when solving modeling equations, comparing maps, selecting maps and objects based on complex combinations of mathematical operations, and on nearly all major analytical operations of the GIS. Long after you learned these principles you might have either forgotten them or internalized them. In either case, it is time to recall them and externalize them. Some of us have chosen to forget them as much as possible. If you do so while using GIS, you do so at your own peril. GIS software is neither a math tutor nor the algebra police. It will do any combination of operations you choose in exactly the sequence regardless of how many rules you violate. You may want to revisit your basic algebra as you get into GIS analysis.

SPHERICAL EARTH

For the most part GIS examines data that are explicitly linked to geographic locations on the earth, although other planetary bodies are also studied with the software. The shape and size of the earth create some interesting and sometimes complex problems for GIS because we normally use flat (two-dimensional) maps, whether analog or digital, while the earth is roughly spherical (three-dimensional). As you read through this text, three major issues related to this are going to become important and will remain so throughout your GIS career. They are **projection, scale,** and **grid.**

Because we often input flat maps, analyze data from them, and often produce flat maps at the end of our analysis, it is important that we have a feel for the impact the transformation from sphere to plane has on what we do with GIS. Moving from a three-dimensional object to a two-dimensional object affects a number of basic properties such as distance, direction, angles, shape, and size (Figure 0.5). This process is called map **projection** because we are projecting a three-dimensional object onto a flat medium. Conceptually it's much like trying to take an orange peel and flatten it out. If you were to draw a figure or a drawing of a continent on an orange and then peel it, you will see that the figure is distorted from its original shape. Some lines are squashed and therefore shorter than they were. Others are stretched, meaning they are longer. The angles are moved and the spatial relationships are different.

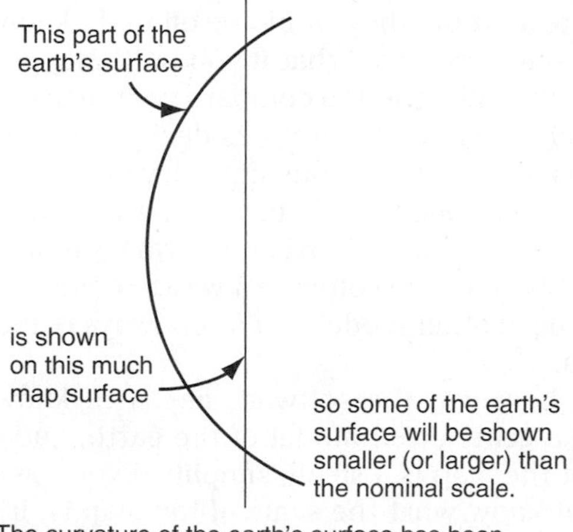
This part of the earth's surface
is shown on this much map surface
so some of the earth's surface will be shown smaller (or larger) than the nominal scale.
(The curvature of the earth's surface has been exaggerated in this diagram.)

Figure 0.5 When we project a three-dimensional object like the earth into a two-dimensional display, spatial properties are distorted. As an example, think about how the length of lines changes as the globe is projected.

While this concept seems pretty obvious, we tend to forget about it when we use sophisticated GIS software because the software does all the hard work for us. It transforms projected maps into a digital globe when we input our data. It knows the proper distances, angles, directions, shapes, and sizes of things when it does its calculations. Finally, the software converts the globe back into a flat, projected map when we are done with our work. So why do we need to think about this stuff?

First, because the process of map projection distorts the spatial relationships, cartographers (mapmakers) have developed a large number of methods of map projection. Each has its own characteristics and each either preserves some of the properties of the spherical object while sacrificing others in the process. We need to know which map projection was used as input to the GIS because the errors and distortions are still there and the computer can't fix them. So if we need to use a map to measure areas of distributions, such as changing distributions of plants or animals, our results will be slightly compromised if the original data were input from a map that does not preserve the property of the area. Likewise, if we want to measure distance, we would like to have had our maps input from a map whose projection preserves that property. Before you get to the section in Chapter 3 on projections, you might want to visit the U.S. Geological Survey site that contains a wealth of information on projections and their properties (http://erg.usgs.gov/isb/pubs/MapProjections/projections.html; last visited 5/26/2007).

The preserved properties of a map's projection are not enough to guarantee map accuracy, because each projection is also based on a particular way of measuring the earth and some generalized models are employed to do that. These generalized models have as a major factor a set of starting points for measurement, called a **datum.** There are many different datums available to us, and we must be able to move back and forth from one to another. One might wonder why this is an issue with today's modern GPS systems. Imagine if you had a GPS that was accurate to within a meter for any given point on the earth. Now keep in mind that the earth is almost 6,378,206 kilometers in circumference at the equator. Let's say that you were using your GPS to measure the earth's circumference and you made a one meter mistake every 1,000 kilometers. That's not much, is it? However, that gives you a potential total error of 6,378 meters (over 6 kilometers). If someone else was using a GPS that was only half as accurate as yours they would be off by 12 kilometers. And, keep in mind, that's only in one direction. What if you both mapped the entire surface of the earth this way and then tried to compare your maps? You can see that the measurement of such a large body forces us decide on a general model of its shape and dimensions so we aren't constantly trying to compare each other's measurements. When you begin a GIS project by inputting maps into the computer, you will be asked to provide basic information about the map projection and the datum. This gives the software a way to compare your maps with others that might have used other models (different ways of measuring the earth) to start their projects.

To fully compare maps, however, the software needs to know one more thing—**scale.** A map is just a generalized model of the earth, and a projected one at that. This means that the map is a small, simplified version of the earth itself. The software does not know what the scale of the map is. It knows how big your digitizer might be in digitizer inches or how large your monitor is in dots per inch, but it must be able to translate these into something close to real-world measurements. Beyond this, scale has another major impact on our GIS operations. When the cartographer produces a map, there is only so much space on which to produce symbols. Decisions are made during the mapping process to select some objects for inclusion on the map and to omit others. Much of this is based on scale. And even when some objects are selected for input they must be simplified, their symbols enlarged or reduced in size or even moved to make room for other symbols. In short, the size and scale of the map places limitations on the mapmaker, which affects the map user, and also affects the process of map input into the GIS, and the quality of the data inside the GIS. These concepts are covered in more detail later in this book, but you need to begin thinking about them now so you are comfortable with the concepts as you encounter them.

A final property related to our spherical earth is important to present here, again because of its importance to all aspects of GIS. Without some way to navigate, in other words, without a grid or coordinate system of some kind, we would be unable to record the locations of objects and distributions we expect to analyze. This appears in two basic forms. The first is a **geographic grid** based on the spherical globe model of the earth. You might remember it under the latitude-longitude moniker. Most of us have heard of this system, and many of us assume we know it. However, it is striking how easy it is to forget the differences between latitude and longitude. You might want to listen carefully to the made-for-TV movie pilot for *The 4400*. One of the characters gets this completely backwards, which means that the script writer got it wrong and that nobody caught it before it was filmed.

The second set of coordinate systems is based on the two-dimensional map after it is projected. There are many forms of coordinate systems—far too many to cover in a single GIS textbook. Even with the limited selection of coordinate systems included in this book, however, if you are not familiar with the term you might want to start looking at this information before the course begins. To limit the initial exposure to this topic you can focus on coordinate systems that are commonplace in your own nation. As your experience and need for alternative systems expand, look at other coordinate systems in other nations and regions of the world. Each coordinate system is different, is based on one or more projections, and has its own starting and ending points and method of calibration and measurement. The following Web sites provide some basic information on the British, U.S., and Australian systems:

http://en.wikipedia.org/wiki/British_national_grid_reference_system; last visited 5/26/2007

http://en.wikipedia.org/wiki/United_States_National_Grid; last visited 5/26/2007

www.ga.gov.au/geodesy/datums/ang.jsp; last visited 5/26/2007

CONCLUDING REMARKS

The material in this chapter constitutes a set of essential skills that every GIS practitioner needs and it is important to internalize and to practice them. As with learning anything, focus, repetition and application are fundamental to a thorough understanding. If you are reading this as part of a class, form study groups with your fellow students and work on these exercises together. Revisiting Table 0.1 when you encounter Spatial Terms later in the course will also prove useful. Above all, remember that the sooner you become comfortable with these concepts, the sooner you will be able to incorporate them into your GIS activities.

Terms

datum
distributions
geocaching
geographic grid
geographically referenced
grid
interval
mountain
nominal
ontology
ordinal
projection
ratio
scalar
scale
spatial cognition
summit

Practice Exercises

1. Recreate the spatial terms in Table 0.1 in a computer-based table or document so that you can expand it as needed. Using only the knowledge you already possess, fill in as much of the table as you can with as many terms as you can think of. Spend an hour or so on the first attempt. Now think about what you have written down and contemplate it from time to time over the next day or so. As you think of, hear, or see new terms that might fit in the table, go back and include them. After doing this for a few days get together with others in your class (perhaps a study group of four or five) and compare tables. Spend an hour or so discussing these and see if you can collectively come up with still more. See if you get more terms than other study groups in your class. Finally, compare your results with theirs and begin a dialog to see if you can increase the size of your table. For a more thorough workout try some of these:

   a. Create a travelogue using as many of the terms as you can.

   b. Read a chapter out of a book and note all the spatial descriptors and terms.

   c. Watch a TV or movie action scene and try to create a script that uses the terms you just discovered to describe the action as explicitly as you can.

   d. Use Google Earth (http://earth.google.com) or a similar program to examine a place you know or a place you have never been to. Look at the bottom of the screen and make a note of the estimated distance above the earth. Now describe, with as much spatial terminology as you can muster, what that scene looks like spatially. Pretend you are trying to do this for an alien creature that has no eyes.

2. Compile a list of terms that are commonly used to describe the geographic features of our world. You could group these terms into human and natural,

biological or landforms, whatever you like. For each term write as explicit a definition as you can, keeping in mind that the ultimate goal is to incorporate these terms into the computer and/or to use the definition to retrieve these features from the GIS. As before, spend time doing this on your own, then with your study group, and finally with the entire class.

3. Create four expandable tables each with two columns. Label the first table "Points," the second "Lines," the third "Areas," and the final one "Surfaces and Volumes." In the left column list the following: Nominal, Ordinal, Interval, Ratio, and Scalar. Now for each table and for each measurement level compile a list of appropriate examples of geographic objects. As before, do this first on your own for an hour or so, then take some time to think of others and include them, finally moving to your study group and then comparing your group's results to those of the other groups.

4. This chapter includes some pretty weird types of geographic labels resulting from inappropriate mathematical combinations. Using the procedures established in Questions 1–3, compile a list of other possible "silly" combinations. You might also want to try your hand at listing possible mistakes represented by misapplication of the mathematical principles defined in this chapter (i.e., commutative, associative, identity, distributive, reciprocals, and the zero property).

5. Find a number of maps of wildly different scales. With a ruler, measure as accurately as you can some of the point, line (width only), and area symbols found on the map in fractions of inches or millimeters. For each scale of map you select compile a list of the measurements you took. Now convert the measurements into the scale of the map. For example, if you measured a map symbol and it is 2 millimeters, and your map is at a scale of 1:100,000 or one inch equals 100,000 inches, your symbol will, in theory, occupy 200,000 millimeters or 200 meters. When you are finished, write out your reactions to these results and discuss their impact on the accuracy of the map, symbol placement, and map use.

6. This chapter suggests that there are many types of grid or coordinate systems, datums, and projections. Use whatever resources you have at hand or on the Internet to compile a list of each together with their properties. As with Questions 1 through 3, you should compare your results with your own study group and/or with other study groups to see how complete your list is. Write a description of at least two coordinate systems, two datums, and two projections as if you were to give a lecture to your study group. Each member of your study group should have at least one unique coordinate system and one unique projection, and, if possible, at least one unique datum. Share your write-ups so you each now have a more complete set of information from which you will later be able to refer when you begin your GIS work. In your write-ups you should not only include the information you found, but also document its source and some evaluation of your comfort level with the authority of that source. Discuss with your groups any differences in information you find and think about why these differences exist.

References

Downs, R.M., et al., 2006. *Learning to Think Spatially: GIS as a Support System in the K–12 Curriculum*. Washington, D.C.: National Academies Press.

Fisher, P.F., 1991. "First Experiments in Viewshed Uncertainty: The Accuracy of the Viewshed Area." *Photogrammetric Engineering and Remote Sensing*, 57(10):1321–1327.

Fisher, P.F., 2005. "Fuzziness and Ambiguity in Multi-Scale Analysis of Landscape Morphometry." In Petry, et al., *Fuzzy Modeling and Spatial Information for Geographic Problems*. Berlin: Springer-Verlag.

Kuipers, B.J., 1978. "Modeling Spatial Knowledge." *Cognitive Science*, 2:129–153.

Mark, D.M., B. Smith, and B. Tversky, 1999. "Ontology and Geographic Objects: An Empirical Study of Cognitive Categorization." In Freksa, C., and D.M. Mark, Eds. *Spatial Information Theory: A Theoretical Basis for GIS*. Lecture Notes in Computer Sciences. Berlin: Springer-Verlag, pp. 283–298.

Smith, B., and Mark, D.M., 1998. "Ontology and Geographic Kinds." In Poiker, T.K., and N. Chrisman, Eds. *Proceedings. of the 8th International Symposium on Spatial Data Handling (SDH'98), Vancouver*. International Geographical Union, pp. 308–320.

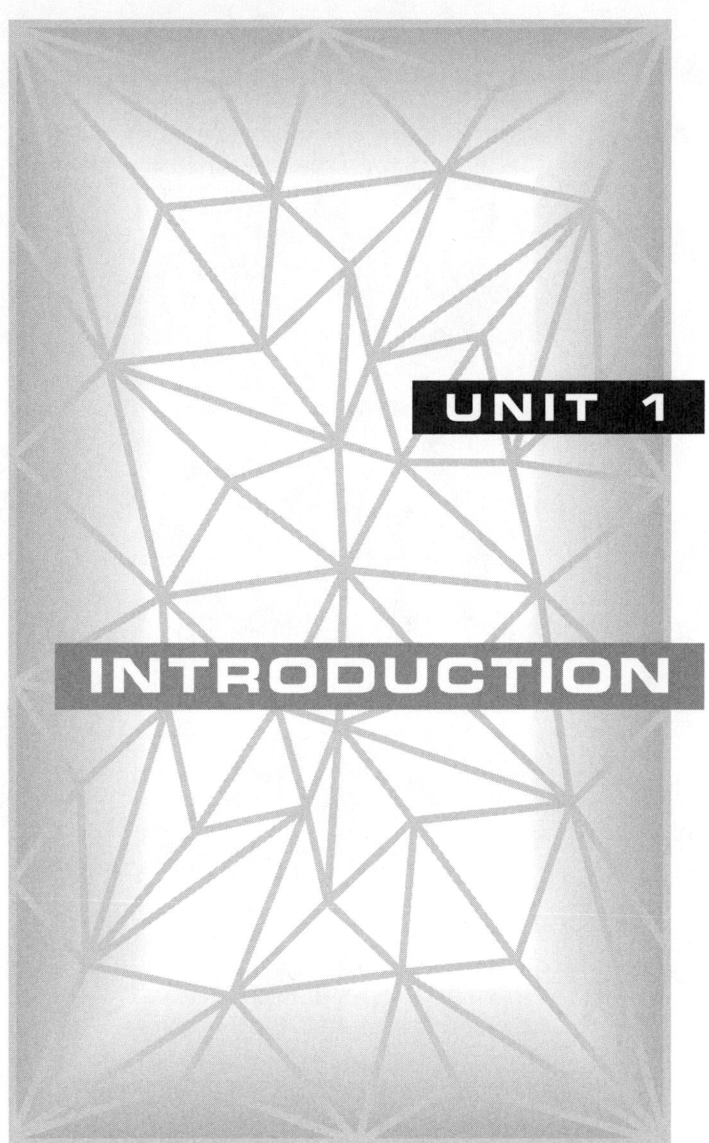

UNIT 1

INTRODUCTION

CHAPTER 1

Introduction to Digital Geography

LEARNING OBJECTIVES

When you are finished with this chapter you should be able to:

1. Provide a complete definition of geographic information systems (GIS).
2. Explain why the definition of GIS as merely a software system is incomplete.
3. Explain the parallels between the evolution of geographic thought and the advent of GIS.
4. Explain the initial impetus for the development of GIS.
5. Describe some of the difficulties encountered during the early development of GIS.
6. Describe the relationships among a GIS, computer-assisted cartography (CAC), and computer-assisted drafting (CAD).
7. Describe some basic analytical capabilities of a modern GIS.
8. Suggest possible users of a GIS and how it might benefit them.

GEOGRAPHIC INFORMATION SYSTEMS DEFINED

While many consider geographic information systems (GIS) to be software programs that manipulate **spatial** data, this definition is very restrictive. As the name implies, geographic information systems are systems designed to input, store, edit, retrieve, analyze, and output geographic data and information (DeMers 2005). Like all systems (e.g., ecosystems, digestive systems, ventilation systems, etc.), the GIS is composed of an orchestrated set of parts that allow it to perform its many interrelated tasks. These parts include computer hardware and software, space and organizations within which these reside, personnel who use the system in a number of levels and capacities, data and information upon which the system operates, clients who obtain and use the products, vendors who supply the hardware and software, and other systems (financial, institutional, and legal) within which the GIS functions (Figure 1.1). While the software component of GIS is most often what we think of when we hear the term, its scope is far bigger and more comprehensive than that. Initially we will focus our discussion on the computer programs themselves—the technology component—and revisit the other components later in this book as appropriate.

The primary task of a GIS is to analyze spatially referenced data and information. To perform meaningful analysis requires that the software be able to perform many other tasks, such as input, editing, retrieval, and output. Still, analysis is the strength of GIS. There are many ways of classifying the analytical and modeling capabilities of GIS because many of these capabilities interact. Ultimately, the software most certainly contains algorithms and computer code specifically designed to (1) organize geographic data within appropriate referencing systems, (2) selectively query those data and aggregate them for easy understanding, (3) count and measure both individual objects and collections of objects, (4) classify and reclassify objects based on user specified properties, (5) overlay related thematic map data, and ultimately (6) be able to combine these individual techniques into ordered sequences of operations designed to simulate some natural or anthropogenic activities for decision making. All of these tasks tend to involve, either directly or indirectly, some form of mapped data.

To understand how this works, imagine how you currently read, analyze, and interpret the analog equivalent of GIS—maps. Many first-time GIS students, even those who do not regularly refer to maps, are surprised when they discover that they are already GIS practitioners when using road maps to find routes from one place to another. This activity requires us to select portions of the road map for analysis (query), to find the shortest route from place to place (measurement), and to mark that route with a highlighter pen (classification). We use a special-purpose digital GIS when we use online map services such

Figure 1.1 The multiple nature of geographic information systems.

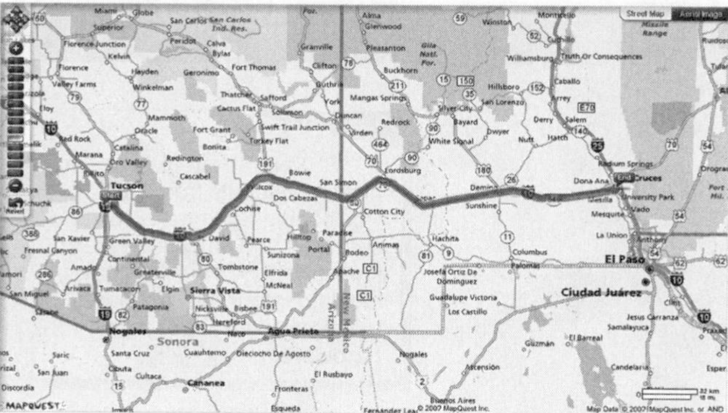

Figure 1.2 MapQuest output for a query showing directions from Las Cruces, NM to Tucson, AZ. Most people who use this service are unaware that they are actually using a rudimentary GIS.

as MapQuest to perform this same task for us (Figure 1.2). Some of us have onboard global positioning system (**GPS**)/GIS components in our automobiles such as OnStar (Figure 1.3) that not only tell us where we are and give us routing information, but also connect us to emergency services through wireless telecommunication. Both of these examples deal with the movement along networks and demonstrate relatively simple applications of the existing technology.

We become more sophisticated in our GIS skill set when we begin looking for places to buy or build a home. Whether we directly employ maps or not we most certainly employ what geographers call **mental maps** (mental perceptions of our spatial environment) when we do this. To perform this task we frequently develop spatial queries by defining the criteria we employ in selecting our candidate locations. For example, we may tell a real estate agent that we wish to buy a home that is in a new development (a query of our geographic data), costs less than $150,000 (another query), "near" our workplace (a measurement combined with a query), and within a particular school district (a form of overlay that we will later call *point-in-polygon*). These operations can readily be implemented by a digital GIS and are commonplace within commercial GIS software. Some real estate firms already employ GIS on a regular basis. So from these simple illustrations you can see that you already possess some of the skills necessary to perform GIS analysis. You only need to expand your concept of how space can be examined, measured, and compared, and then envision how the geographic objects you wish to examine might be encoded (input), stored, retrieved, and manipulated inside the computer.

The last example illustrates the analytical power of the GIS. Some people believe, for example, that there is no difference between **computer-assisted cartography (CAC)**, **computer aided drafting (CAD),** and GIS. Because the graphic display from these three systems can look identical to both casual and trained observers, it is easy to assume that they are, with minor differences, the same thing. Anyone attempting to analyze maps will discover, however, that CAC systems, computer systems designed to create maps from graphical objects combined with descriptive **attributes**, are excellent for display but

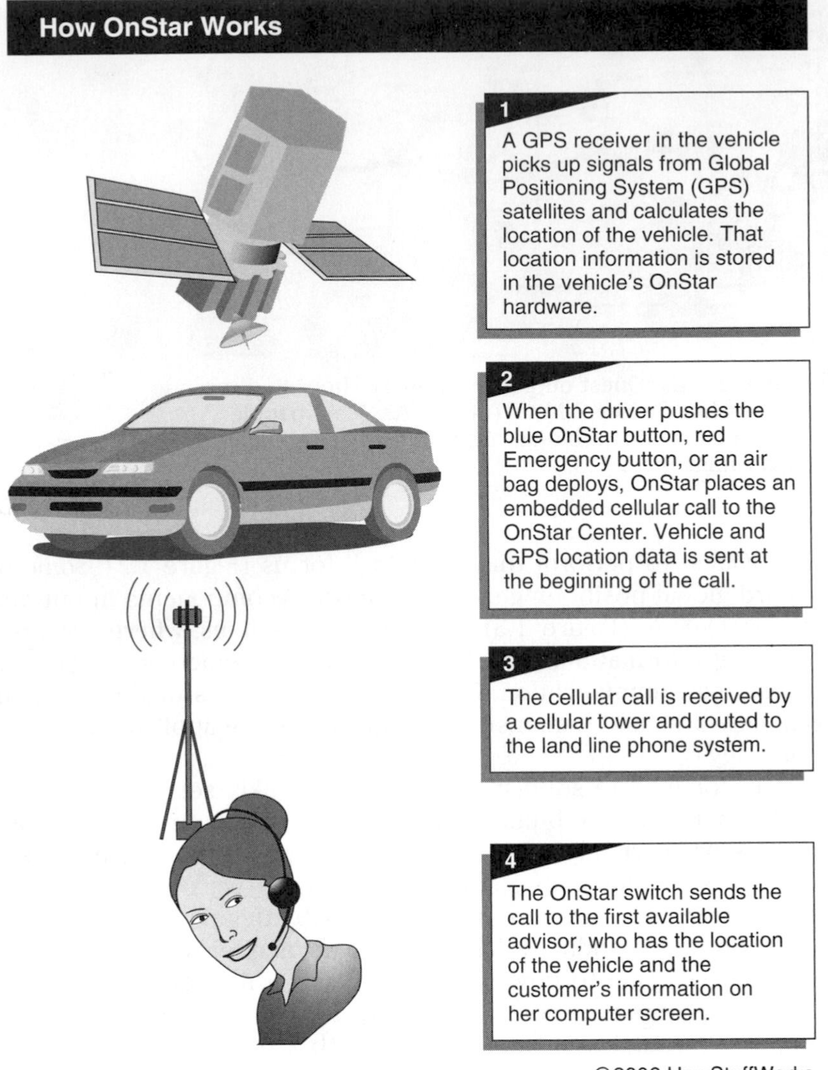

How OnStar Works
1
A GPS receiver in the vehicle picks up signals from Global Positioning System (GPS) satellites and calculates the location of the vehicle. That location information is stored in the vehicle's OnStar hardware.
2
When the driver pushes the blue OnStar button, red Emergency button, or an air bag deploys, OnStar places an embedded cellular call to the OnStar Center. Vehicle and GPS location data is sent at the beginning of the call.
3
The cellular call is received by a cellular tower and routed to the land line phone system.
4
The OnStar switch sends the call to the first available advisor, who has the location of the vehicle and the customer's information on her computer screen.
@2006 HowStuffWorks

Figure 1.3 OnStar system at work.

generally lack the analytical capabilities of a GIS. Likewise, for pure mapping purposes it is highly desirable to use a CAC system developed specifically for the input, design, and output of mappable data rather than working through the myriad analytics of the GIS to produce a simple map. CAD—a computer system developed to produce graphic images but not normally tied to external descriptive data files—is excellent software for architects, speeding the process of producing architectural drawings and simplifying the editing process. It would not be as easy to use for producing maps as would CAC, nor would it be capable of analyzing maps—generally the primary task assigned to the GIS (Cowen, 1988). As each of these three technologies matures, however, we are finding a large crossover of techniques and capabilities, thus blurring the definitional lines among them.

A BRIEF HISTORY OF GEOGRAPHIC INFORMATION SYSTEMS

Geographic information systems are the result of over twenty-five hundred years of geographic research and exploration. As such, they reflect the culmination of that evolutionary process from exploration to description to explanation and finally prediction. The earliest stage of geographic thought was relatively simple, focusing on *exploration* of unknown lands, their inhabitants, and other phenomena. Its primary outcome was one of discovery. Later, as writing and graphic tools became available, the geographer's craft began to systematically *describe* the observed lands, people, and other objects. Among the most powerful of such descriptions involved the preparation of maps showing not just locations but, eventually, distributions of phenomena. This phase led to the need to *explain* what these distributions might mean. Geographers began to look for both natural and anthropogenic explanations for different types of distributions. A natural outcome of pattern explanation was that of predicting new distributions based on those explanations and, ultimately, to exploit those predictions for planning. It is this need to exploit our knowledge of our resources for planning and management that led to the eventual development of what is now known as GIS. The following situation closely parallels the evolution of geographic thought and demonstrates how this process ultimately led to the development of the GIS toolkit.

In the early 1960s the Department of Forestry and Rural Development of Canada decided to pursue a large-scale project to manage the resources for much of their territory (Tomlinson, 1984). Among the initial tasks was the inventory *(exploration and discovery)* and mapping *(description)* of the available forest and mineral resources, wildlife habitat requirements, and water availability and quality to list just a few. The inventory was meant as a first step in a larger endeavor to compare the maps to each other to explain their patterns and finally to predict the longevity of the resource base. The goal of this proposal was to develop a management strategy to ensure the long-term availability of both renewable and nonrenewable resources without damaging the environment.

The mission just outlined was a daunting one, requiring enormous amounts of data gathering, compilation, evaluation, analysis, and modeling. Yet this is often the very task that faces natural resource managers. Immediately it became obvious that maps of the resources would permit viewing the extent, quality, and current rate of use at a single glance. The manual production of such very helpful maps covering an entire country, especially a large one, would call for the employment of perhaps hundreds of cartographers and considerable amounts of time and money. Many of you are painfully aware of the limited coverage and variable quality of topographic, vegetation, and soil maps for your own region or nation. Depending on the size of your area, it is entirely possible that long before such a task is completed the resources themselves will have disappeared, leaving the environment despoiled and the local residents up in arms.

A chance meeting between Roger Tomlinson, the director of the Department of Forestry and Rural Development, and an IBM executive resulted in a suggested application of the emerging computer technology to the problem. What was needed was a GIS for Canada. Thus the then newly instituted Regional Planning Information Systems Division, funded by the Canadian government, was assigned to produce what was to become the first fully operational geographic information system ever built—the **Canada Geographic Information System (CGIS).** Its initial task was to classify and map the land resources of Canada, but more advanced uses were envisioned as these maps became available for analysis.

Before any of this could be accomplished there were substantial hurdles that needed to be negotiated. Because Canada encompasses such a large area of land, the first problem the CGIS developers encountered was how to input such vast amounts of data into the computer without using clear plastic grids and cell-by-cell input. This required the refinement of the rather crude **digitizers** of the day to a large (48″ × 48″) pencil-following cartographic digitizing table to input point data. At the same time a large format (48″ × 48″) cartographic-quality drum scanner was also invented to replace manual line tracing. In tandem with the development of the digitizer, and because graphic displays were not available, Tomlinson's team was forced to developed computational versions of **topology**—a branch of mathematics dealing with spatial properties and their conditions upon deformation—to detect input errors (e.g., closure of polygons).

Related to the large data volume problem was the limitation of the computers available in the early 1960s. In 1962 neither laptop nor desktop computers existed. Moreover, even the largest mainframe computers had very little core memory (kilobytes, not megabytes nor gigabytes). The IBM 1401, for example, had 16K of memory and the 1964 IBM 360/65 had 512K of core memory. There was no such thing as random access memory (RAM), so all the data had to be kept on tape. Imagine trying to build a geographic information system with a computer that has less core memory than one of today's simplest hand calculators. A simple text file with the three characters *Hi.* would take up nearly 1/3 of the core memory of most computers of the time (Figure 1.4). One major solution to the memory limitation was the development of the **Morton Matrix,**

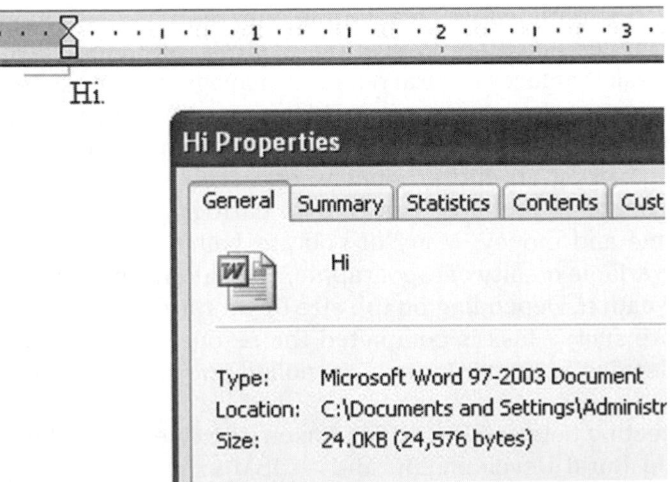

Figure 1.4 Screen capture shows how a tiny Microsoft Word document with only the three characters *Hi.* takes up 24K of memory—more than the total core memory of the IBM 1401 computer used to create the first GIS.

the first of today's facet or tile data structures. It provided a method of splitting up the database into small portions called *tiles* (Guy Morton's idea). This "**tiling**" approach to dividing large spatial databases into manageable portions is still used today.

A third major problem needing a solution was that of finding a way to store and to link the locational data (points, lines, and polygons) with their nonspatial descriptions—a storage and editing problem. On analog maps this is done through the legend, a graphical device interpreted by the map reader, but the computer required that these be explicitly defined and linked. The spatial database management system that was developed was called a **retrieval monitor** and consisted of two separate databases—one for the locational data and one for the descriptions—linked by software pointers. The retrieval monitor also contained a language structure that allowed users to reclassify attribute data, dissolve lines, merge polygons, measure cartographic objects, change scale, generate circles, generate new polygons, search attributes, create lists and reports, and carry out polygon overlay operations. This was the essence of the analytical portion of a GIS and is still the core of modern GIS software.

One additional problem to overcome was the need to build a complex spatial software package using (pre-PL1) computer languages that were not designed for graphics or for the complex nature of geographic data. Moreover, nearly all the programmers at that time either worked for computer companies or for very centralized governmental service bureaus. None of these programmers were trained in digital spatial data handling. This required rethinking the training of computer scientists to understand the nature of spatial (geographic) data and the training of geographers in computer science.

GIS AS A GROWTH INDUSTRY

As with the Regional Planning Information Systems Division in 1960s Canada, today there is an ever increasing recognition of the need to perform large-scale mapping and map analysis operations for a wide variety of traditionally manual tasks. Foresters wanting to keep an up-to-date inventory of their timber resources see GIS as an efficient management tool for their day-to-day operations. Fire departments need GIS to enhance their routing capabilities to ensure rapid response in emergencies. The military uses GIS to determine appropriate battle plans and to organize troop movements. Cellular phone companies, wanting to provide the best service for a mobile customer base, employ GIS to site their transmission towers to avoid conflicts with neighbors and while still allowing clear line-of-sight for signal transmission. Local governments employ GIS to develop growth and development plans and to modify zoning regulations to account for increasing population pressures. Businesses are using GIS to market products and even to develop mailing lists based on selected spatial criteria. Real estate companies are using GIS to isolate available housing on the basis of customer criteria such as proximity to schools, type of neighborhood, or access to highways. Police departments are currently using GIS to compile information to characterize the movements and operational settings of suspected serial killers. Academic disciplines such as geography, biology, geology, landscape architecture, range science, and wildlife management now have the capability to employ the technology to develop and test hypotheses concerning patterns of natural phenomena on the earth. Even social science and scientific researchers are increasingly adopting GIS technology, although their numbers are small by comparison with many other groups. The potential users of GIS are nearly limitless, and their types and numbers are growing at a logarithmic pace.

This growth is indicative of the nature of GIS as an empowering technology. It is not unlike the development of the printing press, the creation of the first telephone, the replacement of the horse and buggy with the automobile, or the introduction of the first computer. All these innovations had a profound impact on the way in which we communicated, the way we moved from place to place, and the way we solved problems—even on the nature of the problems we solved. Modern GIS enhanced the utility of the map by replacing a single map with a large number of interrelated thematic maps. These maps could be automatically analyzed, and their themes combined to give meaningful answers for decision makers. GIS has changed the way we do things with maps, the way we think about geographic information, even the way in which geographic data are collected and compiled. Tasks that were impossible with traditional maps are now commonplace.

A decade or more ago, market trends for GIS and related technology indicated that it would be a major growth industry, far outstripping many others, even during recession years (*U.S. News & World Report*, 1995). In early 2004, *Directions Magazine* estimated that there would be over $2 billion in GIS software sales alone by the end of that year, with additional billions spent for services, hardware, and related activities (www.directionsmag.com/press.releases/?duty=Show&id=10412; accessed 6/20/06). As more organizations become dependent on GIS, the need to become familiar with its basic principles will grow as well. There will also be an increase in the demand for people knowledgeable about the concepts behind the technology. We will examine these concepts here, with an aim toward understanding how spatial phenomena can be manipulated and how the technology can help us in an increasingly complex world.

SAMPLE APPLICATION AREAS OF GIS

Whether we think of GIS as software or not, its primary purpose is the organization and analysis of spatially referenced (usually geographical) data. The utility of GIS to solve real-world geographical problems and to provide long-term return on investment has contributed to its continued growth and increasing popularity. Some general areas of endeavor for which GIS is useful include natural resources management; city, regional, and environmental planning; transportation planning; crime analysis; emergency services; and site selection to name just a few.

You have already seen some examples of the specific types of problems for which the technology is currently applied. While at first it might seem unnecessary to identify specific examples of problems, I have found that the more examples I encounter, the more possibilities I can imagine. These possibilities

Figure 1.5 A satellite image from Operation Iraqi Freedom focusing on the tactical application of 3-D representation of imagery (www.geospatial-online.com/geospatialsolutions/article/articleDetail.jsp?id=56052; last accessed August 20, 2006). The satellite image was collected on April 1, 2003. It is a 0.9 meter resolution DigitalGlobe image and shows the parade grounds in central Baghdad with the Hands of Victory monument, in the form of two crossed swords (lower center), marking the entrances. The image reveals the billowing smoke plume from a burning oil trench. The structure on the right of the smoke is the Monument of the Unknown Soldier.

often result in pushing the technology even further so that the tool improves to meet these new demands.

Take, for example, the burgeoning use of GIS and remote sensing technologies in the defense industry. During "Operation Iraqi Freedom" in 2003–4, many of us saw three-dimensional images of urban areas in Baghdad and other Iraqi cities, both as static pictures (Figure 1.5) and as dynamic "flythroughs." Not only was the general public able to view these images in extreme detail on their television sets as the news media tried to explain tactical details of operations, but the U.S. Congress saw them as well. Within months of these images appearing on the evening news, the U.S. federal agency then known as the National Imagery and Mapping Agency (NIMA) was ordered to change its name to the National Geospatial-Intelligence Agency (NGA) to more closely reflect its increasing use of GIS and remote-sensing technologies in the defense and intelligence missions.

Prior to that, the terrorist attacks on September 11, 2001 on the World Trade Center, the Pentagon, and the thwarted attack on Washington D.C. brought about new applications of the technology. Within hours after the World Trade Center buildings collapsed, some entrepreneurial students at the City University of New York began creating a database that could, and ultimately would, be used to reroute traffic around the site. In fact, the GIS database was also used to plan for debris removal, to allocate and route service vehicles, and to perform a wide variety of planning strategies in and around the area. Figure 1.6 shows a 3-D image of the World Trade Center towers (www1.cuny.edu/events/cunymatters/2001_december/groundzero.html; last visited 7/27/2007). With the resulting development in the United States of a cabinet-level Office of Homeland Security (OHS), the utility of GIS was quickly recognized. It is now an integral part of OHS operations both at the state and the national levels. It may one day become an integrative tool to empower the various agencies to share vital security-related geospatial data and information.

The recognition of the utility of geospatial data and software has also been responsible for some radical changes in how police and federal crime agencies perform their missions. In September of 2000 a series of related sniper attacks on innocent civilians in the Virginia and Washington, D.C. area prompted the FBI to request the assistance of D. Kim Rossmo of the Vancouver Police Department to employ the newly developing geospatial toolkit now know as "geographic profiling" to identify suspects. By comparing geographic space used by these

Figure 1.6 A 3-D GIS view of the condition of the World Trade Center towers and surrounding area after the attack on September 11, 2001. (www1.cuny.edu/events/cunymatters/2001_december/groundzero.html; last accessed 7/27/2007).

perpetrators with that of their victims (sometimes called *activity space*), together with psychological profiles of similar types of criminals, the agencies hoped to narrow down the potential search area for their investigations. Although the results were not clearly responsible for the eventual apprehension of the suspects, it shows once again the high level of sophistication and increasing

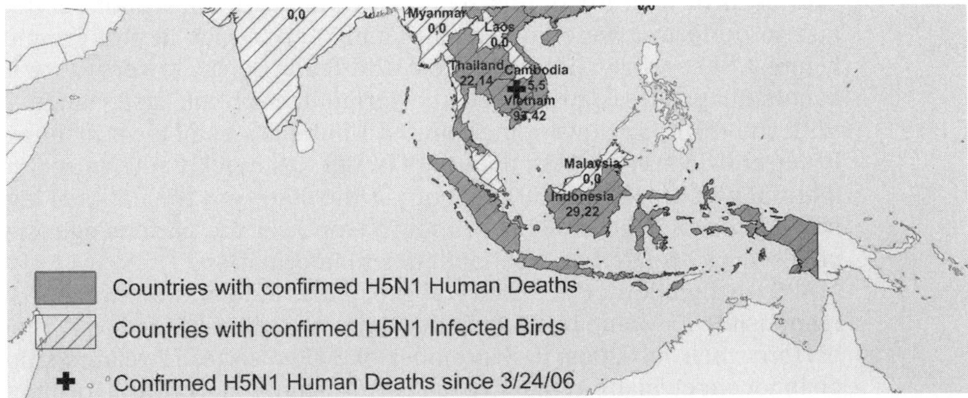

Figure 1.7 A portion of a map showing avian influenza around the world. The map is helpful in tracking the spread of the disease (www.passporthealthtampa.com/fluinfo.html; last accessed 7/27/2007).

application areas of GIS and related technologies. In fact, the U.S. Department of Justice (DOJ) created the Crime Mapping Research Center (CMRC) to encourage and enhance the utility of GIS and related technologies within the DOJ.

In recent years there has been a major concern about the potential pandemic effects of the avian flu. This has prompted a variety of international health organizations to begin tracking the occurrences of the disease worldwide (Figure 1.7) (www.passporthealthtampa.com/fluinfo.html; accessed 10/19/2007). While these maps may be relatively crude by cartographic standards, they represent an increasing use of GIS for such things as epidemiology and other forms of medical research. There is even an effort by the National Geographic Society to map the deep ancestry of people from all over the world.

THE STUDY OF GIS

GIS is an exciting, even glamorous, field with rapidly expanding opportunities for those who are familiar with the concepts and the technology (University Consortium for Geographic Information Science, 2007). A common fallacy is that because GIS has become readily available and is showing up in a wide variety of organizations, anyone can just sit down and start using the software. GIS software, however, is not like the personal computer word processing software we are so accustomed to. Although most of us know some basics about writing and are perhaps very familiar with computer word processing, few of us are comfortable with the analytical operations necessary to make decisions with maps. Just as word processing software assumes that you can organize your thoughts and ideas into coherent sentences and paragraphs, GIS assumes you are familiar with the vocabulary of maps.

When asked, most of us will say that we are fairly comfortable with maps. We use road maps routinely, and when necessary we consult a world atlas with its political, physical, and economic boundaries, and associated colors, graphic symbols, text, and, of course, north arrows. Most of us, however, don't often think about how much information a map contains. Nor do we give much thought to the generalization processes that take place to decide which details get included and which do not. Nobody wants to think about the problem of representing an essentially spherical surface onto a flat piece of paper. Because the map is such an elegantly designed document—so well thought out—we simply accept it at face value.

On occasion, however, the limitations of the cartographic craft begin to show through. How often have you wondered why a road that looks straight on a map really curves all over the place? The graphic limits imposed on the cartographer by available data quality, pen size, size of paper, and other conditions all require him or her to make conscious decisions about how much detail can and should be placed on a given map document. Much of this generalization is imposed by the map scale. The smaller the map scale (the larger the mapped area), the greater is the required generalization to produce the cartographic model.

This concept of the map as a model of reality is perhaps the most important concept that the future GIS professional must learn. Because the map has such a strong visual appeal, the viewer tends to accept it as reality. Those who work with maps, and especially those who work with the interactions of many maps, must constantly remind themselves of the limitations of the cartographic product. Here are just a few simple exercises you can do to become familiar with the cartographic model and some of its limitations.

Take a look at a number of world maps from different atlases you might find in your library. Pick out a country familiar to you. Notice how maps of it differ with respect to sizes, shapes, boundary configurations, numbers of cities, and the like. You might be surprised at the wide variation from one map to another. Consider, then, what you would have to do if you were to digitize a map of this country into a GIS. Which one would you select? Why? How does focusing on the purpose of your GIS project help you decide which map you want?

Obtain two or three adjacent topographic maps for your areas. What are the dates for each map? Are they the same? Different? Now the fun begins. With clear plastic tape (preferably removable), tape the maps together so that all the lines match. Be sure to turn on some relaxing music while you do this. What do you discover? The lines don't match exactly. Imagine how you are going to input 20 or 30 of these documents into a GIS if the lines don't line up.

Soil data might be nice to include in your GIS. Try the last experiment with your local soil map sheets. The match between sheets is even worse. If you are using soil maps from the U.S. Department of Agriculture Soil Conservation Service, you might be quite taken with the use of the aerial photography in the background. Admittedly, this feature is nice to have. But the addition comes at a cost. If you are going to input this map with other maps inside a geographic information system, you will have to co-register it with all the others so that the features match. This requires that the locational coordinates be specified on all maps. Try to find these on the soil survey maps. How do you solve this little dilemma?

If the foregoing examples haven't convinced you of the importance of understanding the vocabulary of maps before you begin speaking GIS, perhaps this one will. You need to create a map of presettlement vegetation for your state or region. It turns out that three very well-known vegetation mappers have compiled such maps for portions of your area of interest. Taking a trip to the library to obtain these maps, you discover that the first shows vegetation classified by its structural components (herbaceous, grasses, trees, shrubs, etc.), and the second map, which intersects with the first map, shows vegetation classed by floristics (based on species). You also note, to your annoyance, that the two systems seem to have only limited map areas that correspond. Hoping for help from the third map, you discover that although it is classed on the basis of a combination of floristics and structure, its area does not overlap either of the other two maps; in fact, it is well separated from them.

Classification problems of the type just described are common and require the student of GIS to become more than a student of the terminology. Before you can master the technology, you should first master its concepts. We will begin this first step in the journey in the next chapter, where we will look closely at the nature of geographic data and the methods by which they can be represented on map documents. This first step will give you a better appreciation of the fundamental building blocks of GIS and will ensure a more cautious approach when you begin implementing geographic analysis and cartographic modeling.

Terms

attributes
Canada Geographic Information System
computer-aided drafting (CAD)
computer-assisted cartography (CAC)
digitizer
global positioning system (GPS)
mental map
Morton Matrix
retrieval monitor
spatial
tiling
topology

Review Questions

1. What was the initial impetus for the development of the first GIS?
2. Provide a description of the change in geographic thought from pure exploration through to prediction and planning. Describe the place of GIS in this context.
3. What is a GIS? Why is the software-only definition of GIS incomplete? What other components make up a GIS?
4. What are the six basic types of analytical techniques generally found in a geographic information system? Can you think of some examples of each?
5. What is the difference between GIS and CAC? Between GIS and CAD?
6. What are some of the technical difficulties encountered by Roger Tomlinson's team in the early development of GIS?
7. What is the Morton Matrix? Topology? Retrieval monitor? What role did each of these play in the early development of GIS?
8. Who would normally use a GIS? What accounts for its popularity?

References

Cowen, D.J., 1988. "GIS Versus CAD Versus DBMS: What Are the Differneces?" *Photogrammetric Engineering and Remote Sensing*, 54(11):1551–1554.

DeMers, M.N., 2005. "Geographic Information Systems." *Encyclopaedia Britannica*.

Tomlinson, R.F., 1984. "Geographic Information Systems: The New Frontier." *The Operational Geographer*, 5:31–35.

University Consortium for Geographic Information Science, 2007. *Geographic Information Science and Technology: Body of Knowledge 2006*. Washington, D.C.: Association of American Geographers.

U.S. News & World Report, 1995. "20 Hot Job Tracks: 1996 Career Guide," October 30, pp. 98–108.

UNIT 2

DIGITAL GEOGRAPHIC DATA AND MAPS

CHAPTER 2

Basic Geographic Concepts

Unlike the real world, our digital environments are composed of cartographic objects that represent the individual components of the earth. These objects differ in size, shape, color, pattern, scale of measurement, and importance. They can be measured directly by instruments on the ground, sensed by satellites hundreds of miles above the surface, collected by census takers, or extracted from the pages of documents and maps produced in ages past. Some will be very important, others merely useful, and some will need to be discarded as unimportant for selective tasks. Some data will need to be sampled because there is too much to collect, while other data will need to be aggregated or disaggregated to provide a sense of order. Thus, to effectively explore the modeled world, cartographic objects must be collected, organized, and synthesized.

The nature of the data often dictates not only how we will later represent the earth with a GIS but also how effectively we will analyze and interpret the results of that analysis. In turn, how we view and experience our environment affects what features we note and how we will eventually represent them. The points, lines, and areas we encounter are all different. Moreover, their representation and utility depends, in large part, on our ability to recognize what features are important and to identify those that may be modified by the temporal and spatial scales at which we observe them. This information, in turn, dictates how the data are stored, retrieved, modeled, and finally output as the results of analysis.

In addition to the temporal scale and the physical sizes of objects stored in a GIS, we must consider the measurement level we will use to represent them. At one geographic scale, for example, large cities could be considered to be points, while at larger scales objects as small as insects or even microbes will be the important point data under consideration. The cities may also include descriptive attributes, such as their names (**nominal** measurement scale); whether their viability for placing an industry would be considered major, moderate, or minor (**ordinal** measurement scale); their average annual temperature (**interval** measurement scale); or the average annual per capita income (**ratio** measurement scale). Some data, such as the relative difficulty of traversing a particular type of terrain, are scaled to conform to a given situation or circumstance and are called **scalar** values because the scale is what determines their inherent value. Each of these types of data represents a fundamentally different criterion, measured with a substantially different measuring device and with a different level of data precision. The same can be said for lines, areas, and surfaces.

The first step toward better GIS skills is to begin to think spatially. We operate in a spatial environment, but we are often oblivious to the space around us, paying little attention to how we and other objects occupy, traverse, interact with, and even modify our space. Become familiar with all the possible patterns, interconnections, distances, directions, and spatial interactions in your world. As you become more sensitive to the objects themselves, you will find it an easy next step to imagine how the objects and interactions can be measured and what measurement tool you will need to record them.

LEARNING OBJECTIVES

When you are finished with this chapter you should be able to:

1. Explain, with examples, how an improved spatial or geographic vocabulary improves your perception of the world.
2. Illustrate, using mapped examples, the impact of scale on our perceptions of the objects we encounter in the world.
3. Describe the difference between discrete and continuous geographic data using real-world examples.
4. Give an example of where discrete data might be displayed using symbols normally employed for continuous surfaces. Explain why someone might want to do this.
5. Explain the relationships among nominal, ordinal, interval, and ratio scales of data measurement and give examples for point, line, area, and surface features.
6. Describe, with at least one concrete example, the necessity of a structural framework or grid system for locating yourself and your relationships to other earth features in both absolute and relative terms.
7. Define the terms *orientation, arrangement, diffusion, pattern, dispersion, density*, and *spatial association* and be able to use them when discussing geographic phenomena.
8. Illustrate the use of ground sampling methods and the advantages and disadvantages of directed ground sampling, probability-based ground sampling, and remote sensing for gathering data about the earth.
9. Explain the concept of the modifiable area unit problem (MAUP) and demonstrate how different spatial sampling schemes will result in totally different results.

DEVELOPING SPATIAL AWARENESS

The focus on one underlying principle—the examination of spatial patterns—separates geography from all other fields of study. Because of this focus, geographers have developed a language that reflects the way they think about space. This spatial language, like any language, allows geographers to think more clearly and communicate more concisely about space, examining only the essential structures and arrangements that explain spatial phenomena.

The spatial language is an intellectual filter through which only the necessary information passes (Witthuhn et al., 1974). It modifies the way we think, what we acknowledge as important, and how we make decisions. A child's early speech patterns illustrate a limited ability to communicate about their world. As the language grows and becomes more sophisticated, so does the ability to communicate with others. In addition to the growth of language, children also grow in their experience of their environment through spatial exploration. These experiences include recognizing where objects are relative to each other (Piaget et al., 1960), and later, the concepts of movement and speed (Piaget, 1970). The more we traveled, the more we developed route-finding abilities that allowed us to recognize shortcuts. We began to comprehend that places and things were either near or far, straight ahead or at an angle, and to take note of obstructions that had to be avoided, hills that slowed us down or allowed us to coast on our bicycles. In short, we were beginning to think geographically.

We will begin exercising our geographic skills by examining the types of objects and features we encounter. Spatial objects in the real world can be thought of as occurring as four easily identifiable types: points, lines, areas, and surfaces (Figure 2.1). Collectively, they represent most of the tangible natural

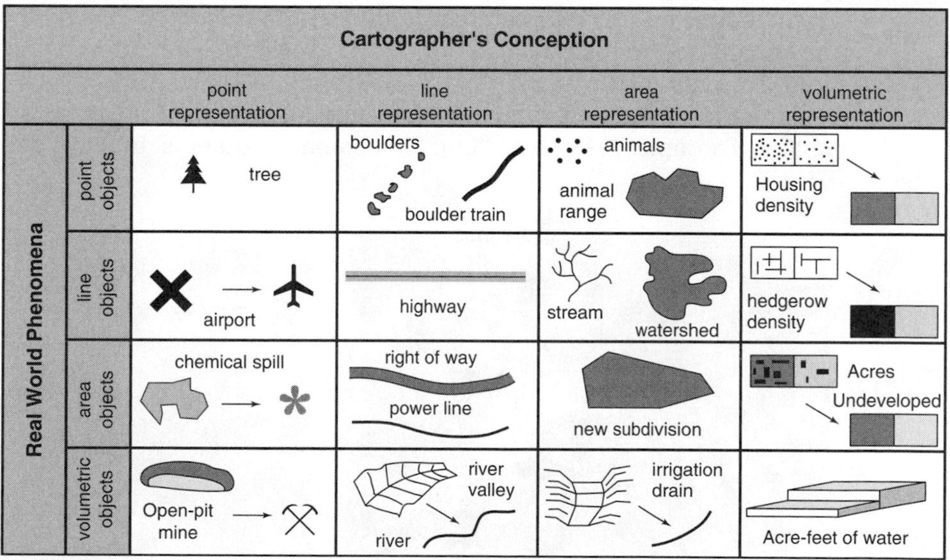

Figure 2.1 Comparison of real-world phenomena and the cartographer's conception. Point, line, area, and surface features with examples. *Source:* P.C. Muehrcke and J.O. Muehrcke, *Map Use: Reading, Analysis and Interpretation*, 3rd ed., J.P. Publications, Madison, WI, © 1992, Figure 3.18, page 84. Used with permission.

and human phenomena that we encounter. Inside the GIS software, real-world objects are represented explicitly by three of these object types. Points, lines, and areas can be represented by their respective symbols, which we will study in Chapter 3, whereas surfaces are most often represented either by point elevations or other computer structures, which we will cover in Chapter 4. What is most important now is that in the GIS software, data are explicitly spatial. Phenomena that are by their nature aspatial (ideas, beliefs, etc.) cannot directly be explored in a GIS unless a way can be found to assign them a representative spatial character. The process of finding **spatial surrogates** is difficult, and is a skill we will examine in later chapters. For now we restrict our discussion to more tangible forms of geographic data.

Point features are spatial phenomena, each of which occurs at a single location. From your own experiences you can easily recognize such features as trees, houses, road intersections, and many more. Each feature is said to be **discrete** in that it can occupy only a given point in space at any time. For the sake of conceptual modeling, these objects are assumed to have no spatial dimension—no length or width—although each can be referenced by its locational coordinates. Points are said to have "0" dimensionality. When we apply symbols to points on maps they take up space, but the point itself has no spatial dimensions. Thus measurements of air pressure, for example, are characterized by a potentially infinite number of points each of which occurs at a distinct location with absolutely no overlap with its neighbors. The spatial scale at which we observe these objects or features imposes a framework within which we determine whether to think of them as points or not. For example, if you observe a house from only a few meters away, the structure occupies length and width. The house appears less like an areal object and more like a point object the farther away you are (Figure 2.2). You choose your spatial scale on the basis of different criteria: whether you want to examine the **arrangement** of people and furniture in the house, for example, or whether you are interested in the house only in relation to other houses. In the latter case the house would be considered to be a point. Your observation has been filtered based on your frame of reference.

Linear objects are perceived as occupying a single dimension in coordinate space. Examples of these "one-dimensional" objects include roads, rivers,

Figure 2.2 Effect of scale on spatial dimensions. A house observed in a close-up aerial view appears to have length and width, but as we pull back, its length and width dimensions disappear, leaving us with the impression of a house as a point.

political boundaries, fences, and hedgerows. The observational scale prevents us from perceiving widths for these objects. When observed up close many of these objects have width—political boundaries would be an exception. The farther away they are, the skinnier they become. Eventually they look so thin that it is impossible to imagine that they are anything but one-dimensional objects. We cannot measure their width at this scale of observation (Figure 2.2). Other lines, such as many political boundaries, have no width dimension to be concerned about. In fact, these lines are often not physical entities at all but rather a construct of political convention and agreement. Despite their lack of tangibility, however, they can be thought of as explicitly spatial because their existence separates regions of geographic space.

Linear objects, unlike point objects, have a measurable length. In addition, they require at least two coordinate pairs to describe their spatial location. The more complex the line, the more points we need to describe them. If we take a stream as an example of a linear object, the description of its many twists and turns may require many points. Because we have described the stream's geometry, we can also measure the comparative shapes and **orientations** of linear objects.

Objects observed closely enough to be clearly seen to occupy both length and width are called areas. Examples of "two-dimensional" objects include ball fields, the areal extent of a city, and even an area as large as a continent. In GIS we conceive of areas as closed graphic forms composed of lines that begin and end at the same location. With areas we can now add the area occupied as a measurable feature not available for points or lines.

Adding the dimension of height to our area features allows us to observe and record surfaces. Although we could observe a house at close range and describe it in terms of its overall length and width, we often want to know whether it is a one-story or two-story structure. In this way we observe the house not as an area, but rather as a "three-dimensional" object, having length, width, and height. Surfaces occur all around us as natural features. Hills, valleys, ridges, cliffs, and a host of other features can be described by citing their locations, the amount of area they occupy, how they are oriented, and now, with the addition of the third dimension, by noting their heights. As it turns out, surface features are composed of an infinite number of possible height values. We say that they are **continuous** because the possible values are distributed without interruption continuously across the surface (Figure 2.3). In fact, because the height of a three-dimensional object varies from one place to another, we can also measure the rate of change in height with a change in distance from one edge to another (e.g., feet per mile). With this we can also calculate the volume of material contained in the feature itself. The ability to make such calculations is very useful if, for example, you wish to know how much water is contained in a reservoir or how much surface material (overburden) lies on top of a coal seam.

All of these objects, whether point, line, area, or surface features, occur in space. Their positions can be exactly located. But how can we communicate the relative importance of a rectangular areal feature located in a particular location, occupying 10 hectares of space and oriented north–south? We need a way of classifying these features based on other observable properties and using terminology that others can understand.

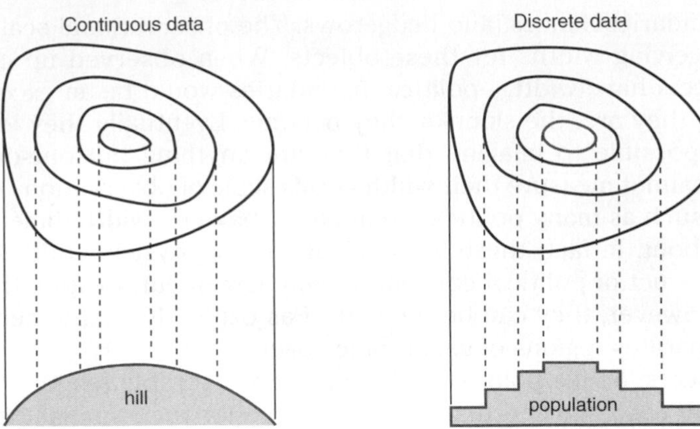

Figure 2.3 Continuous versus discrete surfaces. The difference between discrete and continuous data types.

SPATIAL MEASUREMENT LEVELS

So far we have examined the types of spatial features and their locations in space. The objects themselves are called **entities**, and as you have seen, they have associated with them a set of coordinates that allow us to describe where they are located. All entities contain information not only about how they occupy space, but also about what they are and how important they are to what we are studying. A tree, for example, viewed as a point feature, might be classified as a pine or an oak. We could also enumerate the age of the tree, its height, its condition, the presence of insect infestations, and so on. Such additional nonspatial information that helps us describe the objects we observe in space comprises the feature's attributes. With this information we can now indicate that a particular feature, with a specific name and with some measurable attributes, exists in a documented location.

But before we assign these properties and attributes, we must know how to measure them so we can compare them to those at other locations. What meterstick do we use? How accurately can we describe the objects we encounter? What effects will these different levels of measurement have on our ability to compare them?

Fortunately, there is already a well-established measurement framework for nearly all forms of data, including geographic data. These so-called levels of **geographic data measurement** range from simply naming objects to precise measurements that allow us to directly compare the qualities of different objects. The measurement level we use will be determined by what we are classifying, by what we are investigating, and by our ability to make measurements at our selected scale of observation. Figure 2.4 illustrates the levels of measurement in terms of three commonly used geographic features.

The first level of measurement is the nominal ("named") scale. Nominal geographic data cannot be compared to others because they are a different kind. For example, at one location we have a maple tree and at another we have an oak. Although this statement certainly separates the objects, we cannot say that one is better than the other because the two are inherently different.

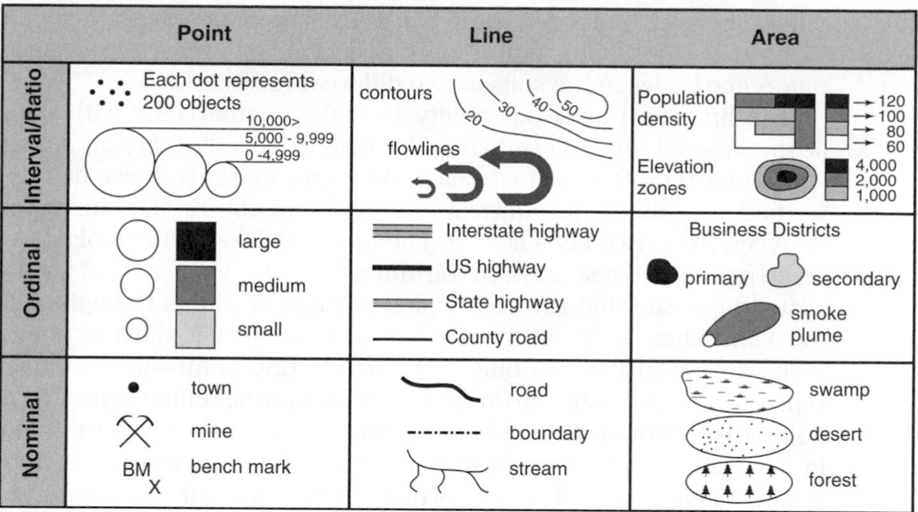

Figure 2.4 Measurement levels for cartographic objects. Point, line, area, and surface features at the nominal, ordinal, interval, and ratio levels. *Source:* Robinson et al., *Elements of Cartography*, 6th ed., John Wiley & Sons, Inc., New York, © 1995, modified from Figures 16.1, 16.2, 16.3, pages 272, 273, 274. Used with permission.

If we want to compare two objects, we must be more precise in our level of measurement. Using our tree example, if we were to compare how well a maple tree, an ash tree, or a pine tree might serve as a setting for a picnic, we could place each on an ordinal scale from best to worst. Because pine trees often have low branches and tend to drop sap on the ground, we might class pines as the worst of the three choices for our picnic area. Although ash trees produce less debris and don't have low branches requiring us to hunker down while we eat, their leaves are small and some sunlight is likely to get through. We might class ash trees as moderate for our picnic setting. The maple tree produces very dense shade, doesn't produce abundant sap, and doesn't have low branches. Therefore, we would assign a classification of excellent to the maple tree.

In our picnic example we produced a spectrum of values ranging from best to worst. Clearly, however, the spectrum is based on the intended use of the information. Because we classed maple as best, ash as moderate, and pine as worst, the classification does not extend to other uses of the trees. Neither the maple nor the ash would serve well as a Christmas tree, for example, so our classification is based on a single spectrum representing a single set of circumstances. Clearly, then, ordinal data can give us insights into logical comparisons of spatial objects, but comparisons are limited to the specific spectrum our question entails. Put another way, your history grade cannot be compared to your calculus grade. Each uses different criteria with its own spectrum of results.

The next more exact level of measurement is the interval scale. Data measured at this level can be compared at a more precise interval than ordinal data. An example of interval scale spatial data is soil temperatures across a study area containing widely different soil types. We may find that temperatures of some very dark, organic-rich soils are much higher than those of other parts of our study area that lack organic material and are a lighter color. We might find a difference of 8°F between these two soils. The dark soil, for example, might be 85°F, while the lighter soil is only 77°F. We now have a precisely calibrated difference between the soils at two different locations.

One limitation is in our ability to make comparisons with data measured at the interval level. Take two additional, extremely different soils, one nearly white and the other almost black. When we measure these at the same time, we find very different temperatures: 50°F and 100°F. The numerical difference between these two soils is 50°F. The numbers 50 and 100 look appealing as we try to read more meaning into the numerical difference. At first glance it appears that we can say that the dark soil is twice as warm as the light soil. However, the Fahrenheit temperature scale has an arbitrary starting point. For us to make a ratio of the two numbers, our starting point—in this case 0°F—must represent an absolute starting point for measuring temperature. To make a ratio comparison would require us to convert the temperatures to the Kelvin scale. In the Kelvin scale, the starting point, or zero, represents a total lack of the atomic movement that produces heat. Converting our two temperatures of 50°F and 100°F to Kelvin, we get 283°K and 311°K, respectively. When we form a ratio of these two numbers, we see that the dark soil is not twice as warm as the light soil, but rather only about 9 percent warmer.

In our conversion of Fahrenheit temperatures to Kelvin, we moved to the most exact level of data measurement—ratio. By converting our temperatures, we were able to calculate a meaningful ratio between the two numbers because they are at the ratio scale. As the name implies, this is the only measurement level that allows a ratio comparison between two spatial variables.

A final level of data measurement is often ignored because of its highly variable nature—the scalar level. As you might guess from the name, scalar values are based on a selective scale of measurement. The scale itself is set by the user, frequently because it is based on heuristics or seat-of-the-pants experience. Take, for example, the case of estimates of the level of roughness of terrain on a scale of zero to ten. A value of ten might indicate a level of terrain too rough for any off-road vehicle to traverse, while zero indicates a perfectly smooth surface. However, the interim values (e.g., 4, 5, and 6) are only measurable "relative" to the bounding values because no exact measurements have been taken. Scalar values are used frequently in GIS analysis where only vague notions of the absolute or relative values are known.

SPATIAL LOCATION AND REFERENCE

Until now we have indicated that we can **locate** objects in space. But to locate objects means that we need a structured mechanism to communicate the coordinates of each object. One type of location, **absolute location**, gives us a definitive, measurable, fixed point in space. But first we must have a reference system, and this reference system must have a fixed relationship to the earth we measure.

The earth is a roughly spherical object. Around that spherical shape we can use simple geometry to create a spherical **grid system** that corresponds to the rules of geometry. This grid system, known as the geographic grid, places two sets of imaginary lines around our earth (Figure 2.5). The first set of lines

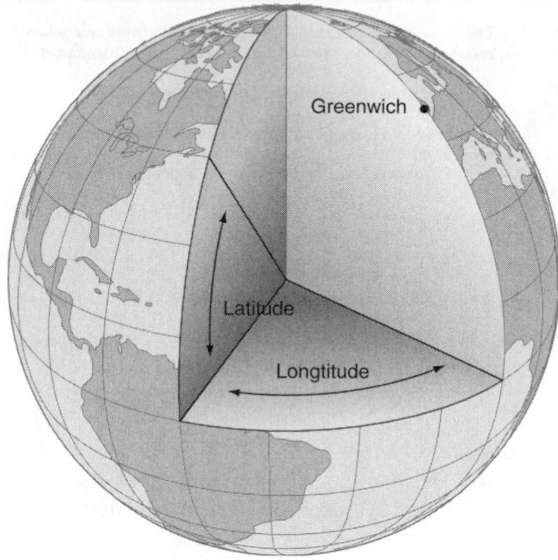

Figure 2.5 The geographic grid. Spherical grid system showing parallels and meridians. Parallels allow us to measure angular distance north and south (latitude) from the equator (0° latitude) up to a maximum of 90° north (North Pole) and 90° south (South Pole). Meridians start at the prime meridian and allow us to measure angular distance east and west (longitude) up to a maximum of 180° where they would meet at the international date line. *Source:* Robinson et al., *Elements of Cartography*, 6th ed., John Wiley & Sons, Inc., New York, © 1995, modified from Figure 4.4, page 47. Used with permission.

for our geographic grid starts at the middle of the earth, or equator. These lines are called **parallels** because they are parallel to each other. They circle the globe from east to west. At the equator, the first of these parallel lines is assigned a value of 0. As we move both north and south from the equator, we draw additional parallels until we reach the poles. Because each of these lines is a given angular distance, measured between the center of the earth and the intersection of the line, each is used to measure the angular distance from our starting point at the equator to where the last line would occur at the poles. The angular distance, called **latitude**, from the equator to either pole is equal to one fourth of a circle, or 90°. So we can measure the angular distance from the equator north or south up to a maximum value of 90°.

To complete the grid we need another set of lines running exactly perpendicular to the first. These lines, called **meridians**, are drawn from pole to pole. The starting point for these lines, called the **prime meridian**, runs through Greenwich, England, and then circles the globe, becoming the **international date line** on the opposite side of the earth. The prime meridian is the starting point—or zero point—for angular measurements east and west, called **longitude**. Longitude is measured east 180° of the prime meridian until it reaches the international date line. It is also measured west of the prime meridian up to 180°, again reaching the international date line.

This system of angular measurements allows us to state the absolute location of any point on the earth by calculating the degrees of latitude north or south of the equator and the degrees of longitude east and west of the prime meridian. With this system we can describe the location of any object we wish (Figure 2.6). In addition, these angular measurements can be converted to linear distance measurements. Although we must make adjustments to adapt this system to flat maps that closely approximate the locations and arrangements of objects on the globe, what we have works well for determining the absolute location on a globe.

It would also be useful to describe not only the absolute locations of objects but also their positions relative to other objects in geographic space. In fact, this **relative location** becomes quite prominent in our GIS analyses, especially when relative location impacts the objects in question. With our absolute grid

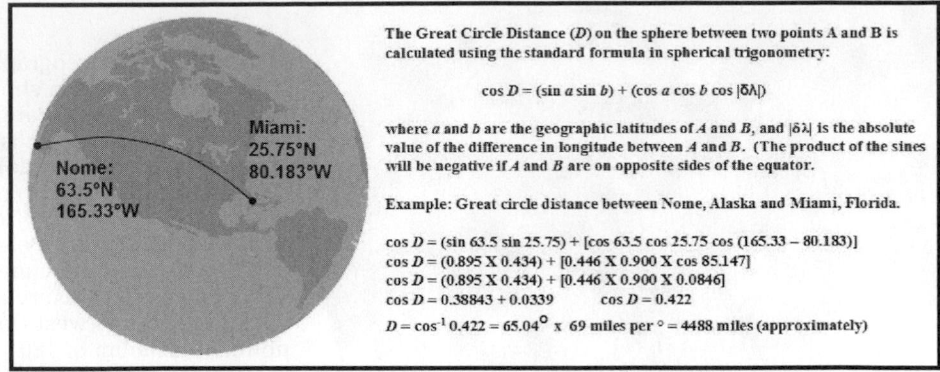

Nome:
63.5°N
165.33°W
Miami:
25.75°N
80.183°W
The Great Circle Distance (D) on the sphere between two points A and B is calculated using the standard formula in spherical trigonometry:
cos D = (sin a sin b) + (cos a cos b cos |δλ|)
where a and b are the geographic latitudes of A and B, and |δλ| is the absolute value of the difference in longitude between A and B. (The product of the sines will be negative if A and B are on opposite sides of the equator.
Example: Great circle distance between Nome, Alaska and Miami, Florida.
cos D = (sin 63.5 sin 25.75) + [cos 63.5 cos 25.75 cos (165.33 – 80.183)]
cos D = (0.895 X 0.434) + [0.446 X 0.900 X cos 85.147]
cos D = (0.895 X 0.434) + [0.446 X 0.900 X 0.0846]
cos D = 0.38843 + 0.0339 cos D = 0.422
D = cos-1 0.422 = 65.04° x 69 miles per ° = 4488 miles (approximately)

Figure 2.6 Calculating great circle distance. The illustration shows the calculation of spherical or great circle distance between two points on a sphere. *Source:* P.C. Muehrcke, and J.O. Muehrcke, Map Use: Reading, Analysis and Interpretation, 3rd ed., J.P. Publications, Madison, WI, © 1992, Figure 12.3, page 255. Used with permission.

system we can calculate relative locations by knowing the absolute distances between any two objects. We do this by subtracting the smaller coordinate values from the larger ones. It would also be useful to know the difference in direction between the objects as well as their distances. We could, for example, indicate that a landfill is located 1,500 meters southeast of the center of the city. This measure of distance and direction provides us with a framework for describing the precise location of the landfill relative to the city. From the standard Pythagorean theorem (Figure 2.7a) for relating the parts of a triangle that has one 90° angle, we can determine the distance of the hypotenuse by the distance theorem (Figure 2.7b), which is expressed as follows:

$$d = \sqrt{(X_2 - X_1)^2 + (Y_2 - Y_1)^2}$$

where

$(X_2 - X_1)$ = difference in the X direction or longitude
$(Y_2 - Y_1)$ = difference in the Y direction or latitude
d = distance between the two points

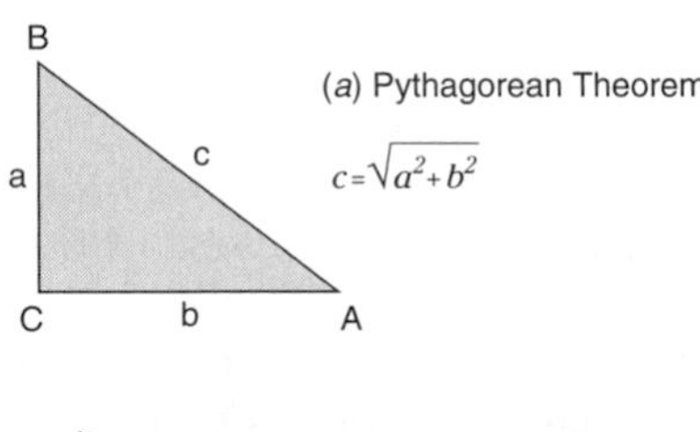

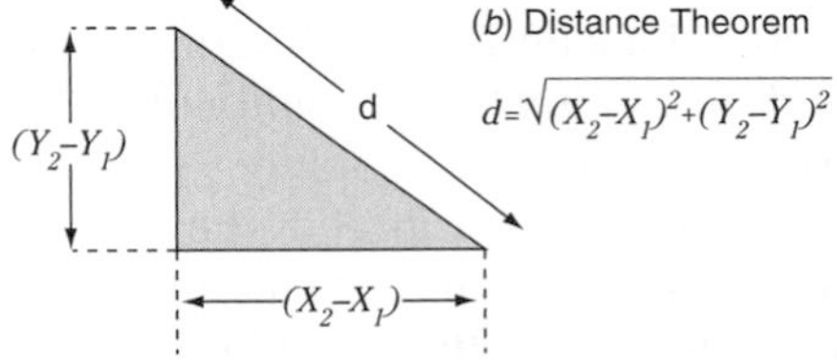

Figure 2.7 Calculating planar distance. The calculation of distance between two points using the Pythagorean theorem (*a*) and the distance theorem (*b*).

However, we could provide additional relative information that would be of use to people living in the city. If we know that the prevailing winds are from the northwest, we can say that the landfill is located 1,500 meters downwind of the city, thus illustrating that the odor of the dump will not normally be a problem for city dwellers. This latter approach, although less precise in terms of measurement, has more practical utility. This also provides us with a means of interpreting what we have observed—determining the important relationships between and among features.

SPATIAL PATTERNS

We have alluded to the importance of knowing the relationships among objects in space. A primary purpose of GIS is to analyze these relationships. We might begin by examining our last measure of location, in which we determined the relative location of a landfill to the nearby city. This easy calculation was a measure of **proximity**, the quality of being near something. Proximity can be measured as the absolute distance between features or objects. We can also set limits of acceptable proximity—a distance within which it is acceptable for two objects to be located. In our example, if the landfill is too close to the city, even if it is downwind, some people will have to look at this unsightly feature. Beyond 2,000 yards, however, we may assume that it is not visible to any city resident. We can then give a minimum acceptable proximity of 2,000 yards to ensure that any landfill will be placed at least 2,000 yards beyond the edge of the city.

Our landfill example shows the interaction of two spatial objects. However, many features and objects occur in far greater numbers: cities in a country, houses in cities, animals in natural areas, natural areas in states, trees in forests, roads across nations, tributaries along streams, and even plants in a garden all occur as multiples. But they do not all occur uniformly within these areas. Each set of objects exhibits a particular spacing or set of arrangements. We begin to notice that these arrangements and spacing seem to have underlying controls or processes that dictate their placement. The instructions on a package of tomato seedlings may tell us to place the plants three feet apart in the garden. By following the directions, we produce a **regular** or uniform pattern of evenly spaced plants (Figure 2.8). This uniform pattern differs considerably from the locations of trees in a forest, which seem to be scattered at **random**, with no apparent underlying design. Alternatively, when we look at the locations of cities, we often see that they are near lakes, oceans, or streams. Knowing that water bodies provide sources of drinking water and recreation and are useful in commerce, we can easily see that the tendency of cities to cluster near such features is driven by these needs. The clustered distribution of cities demonstrates a type of distribution with very high **density** of features, whereas the more sparse distribution of objects, such as farmhouses scattered about a nation's rural landscape, demonstrates a more **dispersed** pattern.

Certain features seem to be organized in yet another way. For example, the vegetation on the north side of a steep slope tends to be decidedly different from that on the south side. In other words, vegetation assemblages have a particular orientation (Figure 2.9). The property of orientation is also found in planted shelterbelts that are oriented at right angles to the direction of the wind.

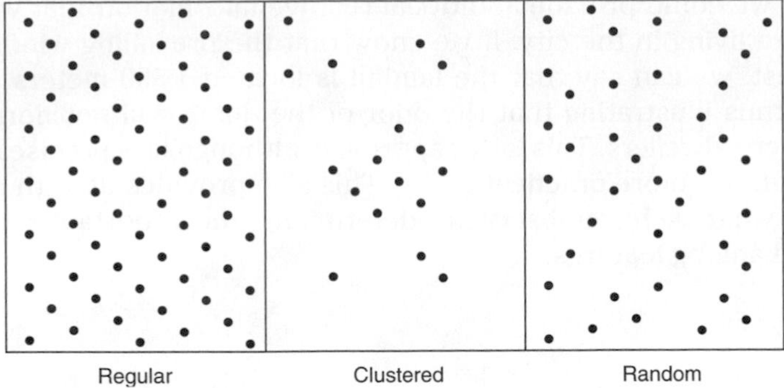

Figure 2.8 Point distribution patterns. Regular, clustered, and random distributions of spatial objects.

Some city streets are oriented along a relatively square grid, whereas others seem to be scattered about with no particular sense of direction. Some of you may have noticed that a great many American Civil War statues are oriented with the figures facing in the direction from which the foe would have arrived. In some regions of the world that experienced continental glaciation, we find piles of debris called moraines oriented at right angles to the flow of ice, or lines of giant rocks called boulder trains that give evidence of the direction of the ice's retreat.

Still other spatial processes give us clues about changing geography. Returning to your hometowns after a lengthy absence, you may see that it is larger than it was when you left. The town has experienced growth and **diffusion** into the neighboring farmlands. Or you may see that a former downtown area is less important as a center of commerce, and instead there has been a diffusion

Figure 2.9 Orientation illustrated by the north-south differences in vegetation on these slopes in Big Bend National Park.

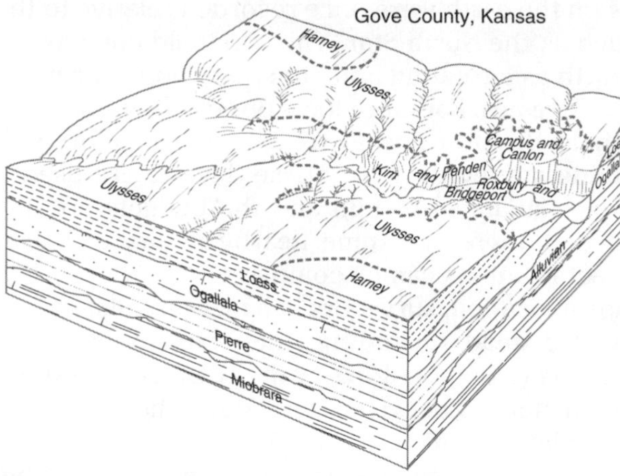

Figure 2.10 Spatial relationships between spatial phenomena. Spatial correlation between geological formations and the types of soils formed from them. *Source:* U.S. Department of Agriculture county soil survey for Gove County, Kansas.

of stores into malls on the outskirts of town. The process of diffusion can occur daily as people concentrated in the residential areas of large cities travel outward toward their workplaces elsewhere in the city.

But as we view all these spatial configurations, we also examine the relationships that might occur among sets of features. We saw this when we recognized the locations of many cities near water bodies. What we discovered is that one spatial pattern may be related to another. The spatial arrangement of slopes had a strong **association** with its vegetation. With such associations we can now ask questions about the causes not only of single distributions but also of **spatially correlated** distributions of phenomena (Figure 2.10). This is among the more powerful capabilities of modern GIS: the ability to illustrate, describe, and quantify spatial associations and to examine their causes. This has long been a major concern among geographers (Sauer, 1925).

You are beginning to acquire the spatial language of geography, and with GIS you have the ability to automate these geographic concepts. Because you know the language, you begin to think in spatial terms—to filter your thinking to observe the important spatial features, to identify spatial patterns, and to ask the questions that will elicit explanations for pattern interactions.

GEOGRAPHIC DATA COLLECTION

There are many ways of collecting the myriad of geographic data. Many are collected through ground survey methods not unlike those used by our predecessors. While tape measures have been replaced by laser range finders, other observations, such as qualitative or categorical data, are still acquired by direct visual observation or by collecting specimens for later identification. These observations should be evaluated carefully before they are accepted as factual, because their quality is often a function of the observer's abilities. Early surveyors used to base landownership maps on so-called **witness trees**, trees left behind after lands had been cleared for agriculture. It was often up to the surveyor to determine the adequacy of surveys based on these markers.

Absolute positions on the earth were once recorded relative to the locations of celestial bodies such as the North Star. The handheld compass depends on the tendency of the earth as a rotating body to create opposing magnetic poles at its north and south extremes. For more local observations, locations are also obtained using handheld Brunton pocket transits, or the more sophisticated **plane table and alidade** survey devices. These devices are used less today, mostly where both the amount of terrain and the budgets are limited. Each device assumes also that there are some nearby points of known location against which the object locations can be compared.

From a point of known location, surveyors can measure distances and angles through a process called **dead reckoning**, they can intersect a number of line-of-sight measurements (**triangulation**), or they can create a known baseline against which only the distances between objects and the ends of the baseline need to be measured (**trilateration**) (Figure 2.11).

Differences in elevation at different places on the landscape can be measured with the use of a device called a **dumpy level**, which has a telescopic sight much like the alidade but is placed on a taller stand and is capable of rotating up and down to examine elevation differences at a distance. A number of traditional survey devices have been updated to offer much greater accuracy, as well as digital readouts and superior ease of use. The theodolite is one example. Even the improved versions of the traditional survey instruments are of limited utility for surveying large areas.

Technological innovations have improved the methods by which we can obtain positional information, especially for large portions of the earth (Figure 2.12). Circling the globe today are satellites whose positions are known with great accuracy. These satellites receive radio transmissions from field units on earth and return a signal that is processed by a nearby ground station and then sent to the field unit as a set of coordinates. The coordinates give positional location as well as elevational location, providing a very useful set of data to the user. Perhaps the most promising of these devices, and now among the most widely used, was mentioned in Chapter 1—the NAVSTAR global positioning system (GPS) (Figure 2.13). GPS depends for its accuracy and precision on the number of satellites, the amount of detail or service provided, the locality of the

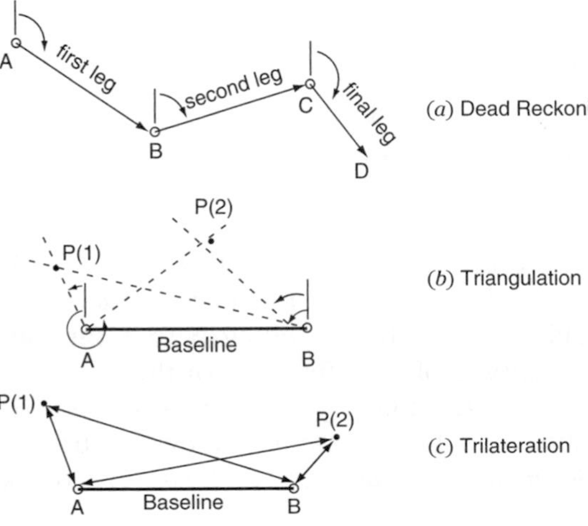

Figure 2.11 Methods of surveying the land. Measurement by dead reckoning (*a*), triangulation (*b*), and trilateration (*c*). *Source:* P.C. Muehrcke and J.O. Muehrcke, *Map Use: Reading, Analysis and Interpretation*, 3rd ed., J.P. Publications, Madison WI, © 1992, Figure 2.2, page 43. Used with permission.

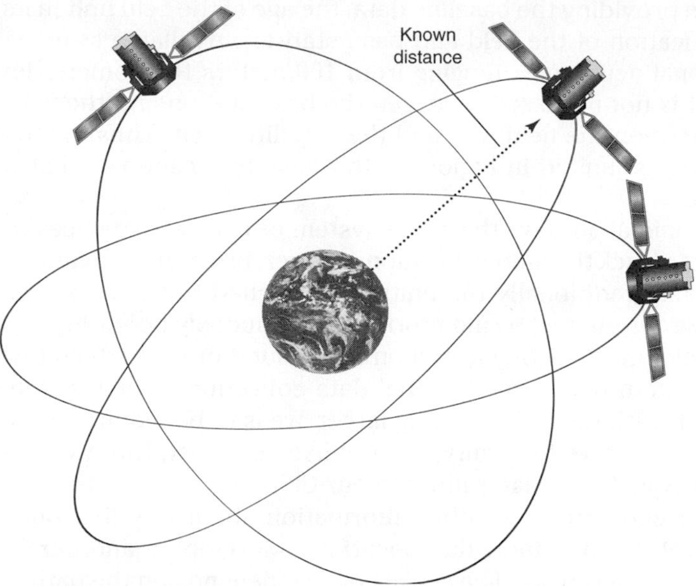

Figure 2.12 Schematic of global positioning system (GPS) showing multiple satellites, each of which is at a precisely known distance above the earth.

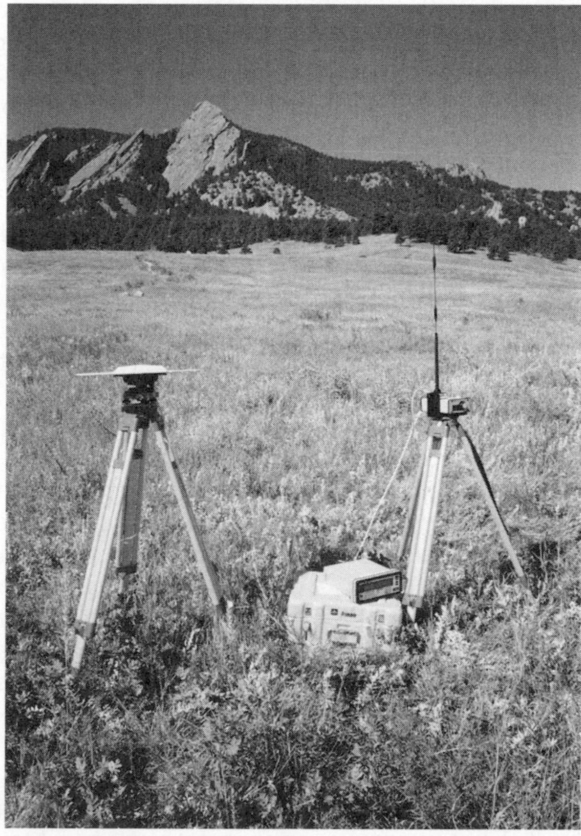

Figure 2.13 Global positioning system (GPS) field unit.

base station providing the baseline data, the age of the field unit in its place, and the sophistication of the field and base stations. Available systems today can give locational accuracies ranging from 100 meters to submeter levels. Direct line of sight is not necessary between the base station and the field unit but is required between the field unit and the satellite itself. Thus, the usefulness of these devices is limited in regions with dense tree canopies that obscure the line of sight.

For field animal surveys there is a system of telemetry devices using similar technology to track the mobile locations of deer, birds, small mammals, and even flying insects. Traditionally the units are attached to the creatures, and their locations are determined by monitoring a continuously operating signal at a field unit. The field unit may be located on the ground or carried aloft with aircraft. A recent trend in this form of spatial data collection, called radiotelemetry, is to combine it with the GPS, allowing larger areas to be monitored. As these two technologies improve in quality, ease of use, and cost, there will be more and better data available for later input to our GIS.

When our concern is to gather information about the distribution patterns of objects rather than individual locations, we employ another form of data collection called a census. The census device depends on the nature of the data to be collected. We might use direct contact to note the exact locations and characteristics of all individual shrubs in an area. A common example, of course, is the census of population conducted by governments. Such survey devices gather both locational and attribute data about people for later aggregation and inferencing about selected areas or districts. Data obtained often include marital status, income, housing, age, and gender. From these data aggregations it is possible to estimate which areas have the lowest per-capita annual income, whether the overall population is aging, whether there are more people living in apartments or in houses, and so on. These results, in turn, allow governmental bodies to suggest ways of planning for changing circumstances based on changes in spatial distribution.

The potential power of the census to describe spatial relationships among a nation's people has prompted a move toward computerization of the data themselves. The census has become computerized, both gaining from and contributing to digital geography. In Chapter 5 you will see how the U.S. Census Bureau has created and modified its methods of keeping track of data to make census information accessible to most commercial GIS products.

At times, however, neither ground survey nor census is an appropriate means of gathering data and providing contextual spatial information about a large portion of the earth. This statement applies particularly to natural phenomena, but it also holds true for anthropogenic features and cultural data under selected circumstances. Obtaining data about phenomena on the regional or even the continental scale may require the use of secondary or indirect methods of data collection. These indirect methods often rely on sensing devices far removed from the observer and are therefore collectively called **remote sensing**. Although the term most often implies some form of satellite sensing of radiometric data from the earth's surface, we will use the term more loosely to include the use of aerial photography.

Point-remote sensing devices are normally placed at strategic locations indicative of the general character of their surroundings. In addition, sensors are scattered throughout the study region so that data can be obtained for as much of the region as possible. Weather stations are perhaps the best-known point-remote sensing devices. However, sensor networks also collect continuous or periodic data on soil temperatures, moisture content, and other soil factors. These networks are connected through a telecommunications network to a base station that receives and stores the information, either as hard-copy displays in the form of paper polygraphs or as digital data inside a remote computer. Even more exotic devices have been developed to record the interception of flying insects, crawling animals, and moving sand grains or to monitor water or atmospheric pollution, seismic changes, and, of course, the prowling jewel thief. Collectively these point-specific sources of information can provide a large regional picture of the phenomena they sense.

The use of aerial photography permits the collection of data describing a synoptic view of the environment rather than discrete observations. This type of remote sensing often relies on the use of aircraft to carry a photographic device designed to sense and record portions of the electromagnetic spectrum. These devices come in various sizes and can use a range of films, from black and white to color to false-color infrared, depending on the data being collected. In many cases a particular film and filter combination is used to eliminate unwanted portions of the spectrum and enhance the visibility of regions more indicative of the features being sensed.

Aerial photography has a long tradition of use for forest and range evaluation and management because the photos allow analysts to see substantial portions of areas at a glance. Soil scientists have used aerial photographs to assist in perceiving the subtle changes in soil type over large areas and also as base data on which soil maps are constructed (Figure 2.14) (http://soils.usda.gov/technical/manual/images/fig4-15_large.jpg; last visited 8/13/07). Urban specialists have used aerial photographs to estimate populations by

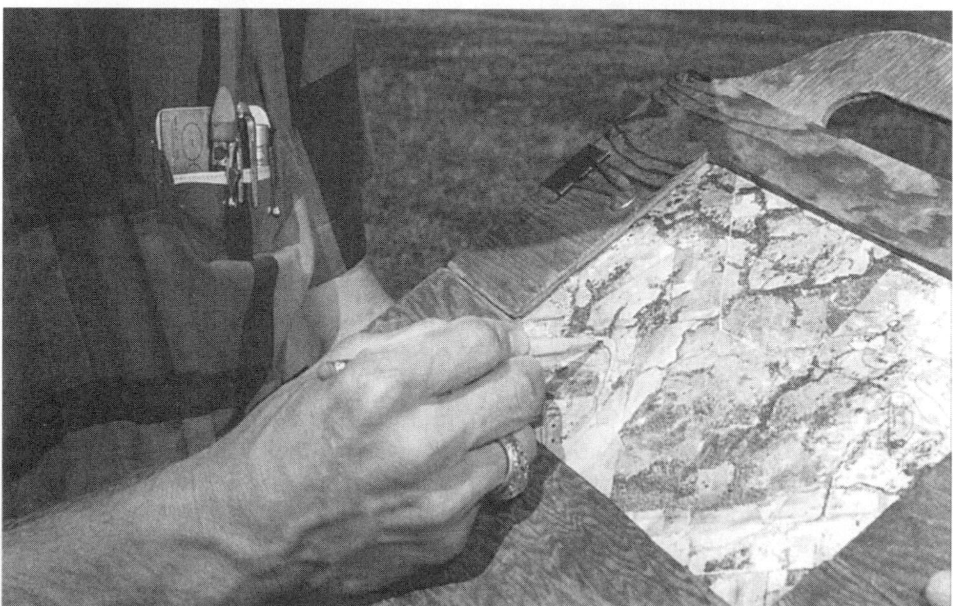

Figure 2.14 Hand-held picture board illustrating the use of aerial photography to support the development of soil maps. *Source:* U.S. Department of Agriculture.

counting dwellings and interpolating from known averages per household. Geologists use aerial photography to view the spatial distribution of surface features as well as subsurface phenomena such as salt domes and fault zones. The military use aerial photography as a means of reconnaissance. In fact, color infrared film, originally called camouflage detection film, was developed largely under military auspices. Thus the use of aerial photography as a means of examining geographic data is still common for areas that are not overly large. For extremely large areas, such as states, regions, or countries, the costs in money and time are prohibitive. Similarly, such airborne devices as **side-looking airborne radar (SLAR), scanning radiometers**, **color video**, and **digital photography** can be ruled out for large area applications for the same reason.

Large areas can be surveyed by deploying satellites, hundreds of miles from the surfaces they are viewing. The great distances between the sensing platform and its target allow the satellite to view large areas synoptically. In addition, because satellite remote-sensing devices are orbiting the earth, they are able to sense much of the planet in a fraction of the time that would be needed by a conventional airborne sensing device. As indicated, a wide range of sensing devices is available, each with its own spectral, temporal, and spatial characteristics. The following Web site provides a complete history of the commercial remote-sensing satellites: www.tbs-satellite.com/tse/online/mis_teledetection_res.html (last visited 8/13/07).

Remember that remote-sensing data are not gathered directly. Data from aerial photography, radar, and digital remotely sensed imagery are all more or less representative of what is on the surface. Rather than directly sensing types of vegetation or kinds of human activity, these are interpreted by professionals from the electromagnetic signals. Often the classified data are used as input to a GIS database, rather than as the raw data themselves.

POPULATIONS AND SAMPLING SCHEMES

While remote-sensing devices allow us to see large regions at a single glance, the resolution with which they view the earth often limits the sizes of objects that can be observed. Some features—for example, animal burrows in a prairie—are well below the ability of many remote-sensing devices to detect. And because such features occupy large areas, it is prohibitively expensive to obtain detailed aerial photography, so we must sample on the ground. Viewing them is easy because they are numerous enough to visually dominate the landscape, but there are far too many for us to count each burrow and assign it an absolute location. To explain the patterns of these features requires us to recognize a pattern in the first place, which ideally would be done by examining and recording their locations. Because we can't obtain a complete census of all the holes, we must estimate the population from a smaller, representative subset or sample.

Sampling can be performed in a number of ways; some are more difficult than others, and some give us a better ability to make inferences about the larger population. Although sampling is effective for either aspatial or spatial data, we will restrict ourselves to spatial sampling because the GIS deals with explicitly spatial data. Within the spatial domain we sample data in one of two primary ways, **directed** and **nondirected sampling**, determined both by the limitations of acquiring spatial data and by the inferences we wish to make about the population.

Directed sampling, as the name implies, involves making decisions about what objects are going to be viewed and later catalogued. In other words, we direct our sampling based on our ability to make samples from and about a **target population** of objects or features. This target population occupies a given **sampled area** within which our samples are taken (McGrew and Monroe, 1993). Together, the target population and the sampled area comprise the **sampling frame**, which includes the types of object of interest bounded by specific and identifiable spatial coordinates. With directed sampling, sometimes called purposive or judgmental samples (McGrew and Monroe, 1993), we use a combination of experience with the study area and its target population, accessibility to the objects we want to investigate, likelihood that we can obtain pertinent information about each individual (e.g., using only data from survey forms returned by people to whom they were sent), and focused (nonrandomly selected) study areas or case studies designed to demonstrate a particular phenomenon. Although this form of sampling is often discouraged as being "nonscientific," it is often necessary. Limited access due to natural or anthropogenic obstructions or lack of roads will often force you to direct your sampling where access is available. Another reason for directing a sample is to target a portion of the population—for example, those people in your community who have cable television—rather than asking cable TV-related questions to those who do not have cable.

Although a nondirected, probability-based sample is generally preferable to directed sampling because it eliminates bias, at times this approach is impossible. Before you begin your sampling procedure, determine whether the data can be collected by means of a probabilistic sampling procedure. If you are able to use a probabilistic spatial sampling methodology, each object or feature you select from your sampling frame is expected to have a known chance of being selected for study. You use the known probability to set up a sampling methodology that allows all objects the same probability of being selected.

The methodology for probabilistic sampling can easily be divided into four general categories: **random sampling**, **systematic sampling**, **stratified sampling**, and **homogeneous sampling** (Figure 2.15). These could be combined to create a hybrid design (e.g., stratified random). Our first approach, random spatial sampling, allows each point, line, area, or surface feature to be selected with identical probability. If the data you are sampling are discrete, such as trees, lakes, or people, your purpose is to randomly collect data from a sample of them to represent the population. If the data are continuous, such as with topography, barometric pressure, or soil temperature, we will randomly select locations at which these properties can be measured. In both cases it is possible to select random points, random areas called quadrats, or line transects for use in selecting our objects of study.

Systematic designs operate almost exactly the same as random designs, but now we decide on a repeatable pattern rather than random locations. For point data we could, for example, select every tenth tree, or trees located approximately 20 meters apart. To examine small plots or quadrats, we could select every *n*th object or an object every *n* meters. Likewise, if our sampling involves

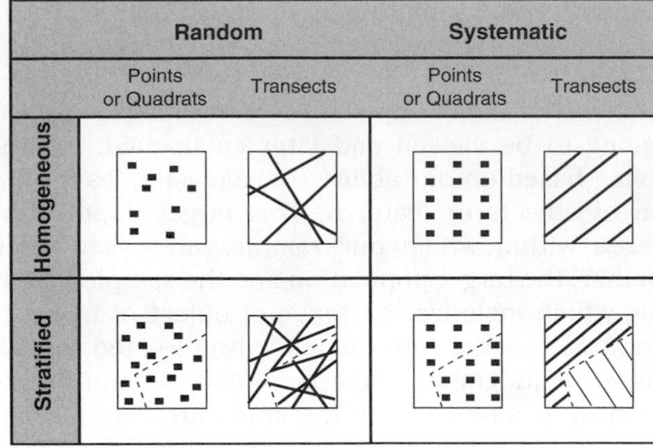

Figure 2.15 Spatial sampling methods. Random, systematic, stratified, and homogeneous sampling designs. *Source:* P.C. Muehrcke and J.O. Muehrcke, *Map Use: Reading, Analysis and Interpretation*, 3rd ed., J.P. Publications, Madison, WI, © 1992, Figure 2.6, page 51. Used with permission.

using line transects, a popular method for examining vegetation assemblages, we would systematically determine where each line transect would fall and take our census of the vegetation along each of these transect lines. Or, if we wanted to completely survey individual plots or quadrats, we would once again select these using a systematic, repeatable pattern for selecting each quadrat for study.

Stratified spatial sampling adds another dimension by selecting small areas within which individual spots, objects, or features are sampled. Stratifying simplifies the process of sampling by dividing the task into small regions that can, for example, be examined by one person, or in one day of sampling. Within each stratified area we can decide whether a random or a systematic design will be used to examine the objects or features. In a modification of this technique called **clustering**, we initially determine whether the objects are **clustered** or dispersed throughout the entire study area. Each of these groups can then be selected as a separate study subarea, much as we did when we stratified our overall study area. Again, we can use point, quadrat, or transect approaches, and we can choose a systematic or a random method of sampling within each subarea. The advantage of this approach is that homogeneous patterns often imply an underlying process. Selecting these areas out for individual study is likely to give us more accurate results because we acknowledge that the processes operating throughout the whole study area are generally uniform. This approach, however, has the disadvantage of requiring us to make decisions about which subareas are more clustered than others that may turn out to be inaccurate.

MAKING INFERENCES FROM SAMPLES

Spatially sampled data yield three possible major types of subsequent manipulation: (1) data located at unsampled locations can be predicted from those that were sampled; (2) data within regional boundaries can be aggregated and assigned a single class or category; and (3) data from one set of spatial units can be converted to others with different spatial configurations (Muehrcke and Muehrcke, 1992). All three of these techniques are forms of predictive modeling.

Although sampling shortens the time necessary to collect data about a region, it also leaves gaps in our knowledge about the unsampled locations. This commonly arises when we gather information about surfaces using point elevation samples. When we view topographic maps, either as contour lines or as three-dimensional representations, the missing data are not measured, but are instead spatially predicted (Muehrcke and Muehrcke, 1992). Under these circumstances there are two general types of predictive models. **Interpolation** is used to predict missing values when we have values bounding, or on both sides of, the gap. If there are values on one side of the missing data but none on the other side, we call the methodology **extrapolation**. Interpolation can be as simple as assuming a linear relationship between the known values and filling in the sequence. More sophisticated methods are based either on assumptions of a nonlinear relationship between these known values or on a weighted distance or local operator model, where nearer values are more likely to be useful in predicting missing values than points farther away. **Surface fitting models** involve fitting the observed values to an equation and then solving the equation for each missing value (Muehrcke and Muehrcke, 1992). Surface fitting models are more likely to be useful for extrapolation than the others, because the equation can easily be extended beyond the known data values. All these methods allow us to predict missing values, but remember—predictions are not measurements. Each prediction has its own set of problems and errors.

Extrapolation is more appropriate when we are sampling point samples as estimates of areas rather than surfaces. From these areas we want to be able to make further predictions about point locations in nonsampled areas. Let's say that we are sampling the density of trees in a number of small areas, and we want to be able to predict the densities in other surrounding areas. Such a task usually requires three tasks. First, we calculate the average densities on a region-by-region basis to prevent region size from affecting the values. Next we assign each density to a single location inside each region—usually some central point. Finally, we perform some form of numeric prediction (e.g., density of parts) of density values for each of our missing areas.

A final predictive problem based on sampling requires the prediction of area quantities on the basis of a set of quantities sampled from a different set of areas with different sizes and shapes. In the simplest approach, which entails a form of spatial overlay, we assume that the data inside each area are uniform and homogeneous. When these two sets of areas are overlaid, the amount of overlap should be directly proportional to the amount of data in each polygon pair.

One final conversion needs to be considered here that relates to sampling. Suppose you are sampling discrete objects, such as the locations of animals. Once you have located each (and for the sake of simplicity we will say that they don't move around much), you want to know what area of the earth they occupy. In other words, you need to know their range. This requires us to convert from point samples to area maps. Some relatively easy computer techniques can be applied, as well as some biological approaches, and we will consider these in more detail later. For the time being, it is enough that you be aware of this particular problem.

Terms

absolute location
arrangement
association
clustered
clustering
color video
continuous
dead reckoning
density
diffusion
digital photography
directed sampling
discrete
dispersed
dumpy level
entity
extrapolation
geographic data measurement
grid system
homogeneous sampling
international date line
interpolation
interval
latitude
locate
longitude
meridians
nominal
nondirected sampling
ordinal
orientation
parallels
pattern
plane table and alidade
prime meridian
random proximity
random sampling
ratio
regular
relative location
remote sensing
sampled area
sampling frame
scalar
scanning radiometers
side-looking airborne radar (SLAR)
spacial association
spatial surrogates
spatially correlated
stratified sampling
surface fitting models
systematic sampling
target population
triangulation
trilateration
witness trees

Review Questions

1. Why is it important that you understand the language of geography before you can be effective in GIS analysis? What impact does an increased spatial vocabulary have on your ability to work with spatial phenomena?
2. What impact does scale have on how we experience our world and on how we model it? Give an example, other than the one in the text, of how scale changes might permit you to view an object with length and width as if it were a point.
3. Why is it important to develop a spatial vocabulary for GIS? What impact does an increased spatial vocabulary have on how you explore your environment? How does a geographic language help you in filtering the data you view in your explorations?
4. What are discrete data? Can you give some examples of them for point, line, area, and surface features?
5. What are continuous data? Give some examples of these, especially with regard to surface data.
6. Why is it important to understand the levels of data measurement when you are observing and recording attribute data for the objects you encounter?
7. Give some concrete examples of nominal, ordinal, interval, and ratio data for each of the following features: point, line, area, and surface.
8. Why do we need a structural framework such as the latitude–longitude grid system? What does it add to the way we view our world?

9. What is the difference between absolute location and relative location? Give some examples of each.
10. Some of the terms added to your vocabulary in this chapter are *orientation, arrangement, diffusion, dispersion, and density*. How do these terms improve the way you observe earth?
11. What does *spatially correlated* mean? What does it have to do with GIS?
12. What is the impact of modern technology on ground sampling? Give some concrete examples.
13. What has remote sensing added to our geographic tool kit that ground sampling lacks? How will improvements in remote sensing impact GIS?
14. Why would you choose to use directed sampling, as opposed to probability-based sampling, if the former is considered to be less "scientific"? What are some of the conditions that force you to use directed sampling?

References

McGrew, J.C., Jr., and C.B. Monroe, 1993. *An Introduction to Statistical Problem Solving in Geography*. Dubuque, IA: Wm. C. Brown.

Muehrcke, P., and J. Muehrcke, 1992. *Map Use: Reading, Analysis and Interpretation*, 3rd ed. Madison, WI: J.P. Publications.

Piaget, J., 1970. *The Child's Perception of Movement and Speed*, translated by G.E.T. Holloway and M.J. Mackenzie. New York: Basic Books.

Piaget, J., B. Inhelder, and A. Szeminska, 1960. *The Child's Perception of Geometry*, translated by E.A. Lunzer. New York: Basic Books.

Sauer, C.O., 1925. *Morphology of Landscapes*. University of California Press, Berkeley.

Witthuhn, B.O., D.P. Brandt, and G.J. Demko, 1974. *Discovery in Geography*. Dubuque, IA: Kendall/Hunt Publishing Company.

CHAPTER 3

Map Basics

The map is the primary language of geography. It is, therefore, the fundamental language of digital geography. This graphic form of spatial data abstraction is composed of different grid systems, projections, symbol libraries, methods of simplification and generalization, and scales. If you are comfortable with the map as a method of modeling your environment, you may be able to skip this chapter and move on to Chapter 4, especially if your background includes coursework in map use and cartography. If you have not had such courses or if your experience in mapping or map reading is minimal, you should spend some time focusing on this material. You might also find it useful to select a couple of good books on cartography or map reading to supplement the short descriptions given here. As you evaluate your map familiarity, remember that inside a GIS you are likely to encounter an abundance of maps far greater than you expect. The thematic content of the maps there include geology, topography, ownership, and soils. It will also contain vegetation, transportation, animal distributions, utilities, urban plans, zoning, land use, land cover, cancer mortality, remotely sensed imagery, and a host of others. These maps may be prism maps, choropleth maps, point distribution maps, dasymetric maps, surface maps, graduated circle maps, and many others. If some of these terms are new to you, then read and study this chapter.

Our examination of the earth through GIS is predicated on our ability to think spatially. Spatial thinking enables us to select, observe, measure, catalog, and characterize what we encounter. But the depiction of objects in cartographic form depends on what questions are being asked, whether we are trying to display the cartographic form or analyze it in a GIS, whether we are observers or users, whether our data are collected in the field or through remote sensing, whether old maps are going to be analyzed, and many other factors. The more we know about the possible combinations and manipulations of cartographic elements, the stronger our geographic language. Knowledge about cartographic methods will increase that part of our spatial vocabulary called **graphicacy**—that ability to communicate with and about maps.

When we put existing maps into the GIS, we will be aware of the impacts of different levels of generalization, scales, projections, symbolization, and the like on what is input and how it is done. Once inside the GIS, we will be able to identify potential problems that will require editing. As we begin to analyze our data, we will be aware of the potential for error of some map layers that were created from very small-scale maps. Finally, as we produce output from our GIS, we will know how best to display the results of analysis, because we will be familiar with the cartographic method and its design criteria. We will also learn how spatial features and their relationships can be best displayed to allow us to readily view these relationships, thus helping us formulate analyses.

LEARNING OBJECTIVES

When you are finished with this chapter you should be able to:

1. Describe, with map examples, the role of graphicacy and cartographic communication in improving our understanding of the world around us.
2. Explain the relationship between the analytical and holistic cartographic paradigms as they apply to GIS.
3. Describe, with examples, how scale is illustrated on a map. Explain the pros and cons of each method.
4. Demonstrate, using your GIS software, the possible impact on the results of analysis of putting differently scaled maps into a single GIS database.
5. Show how the map legend acts to link entities and attributes on a map.
6. Describe the different families of map projections and suggest how they modify the properties of the maps.
7. Explain the relationships among different map projections and the map uses.
8. Describe at least three basic grid systems, their operation, and their advantages and disadvantages for GIS work.
9. Illustrate the impact of scale, class interval selection, symbolism, and simplification on the development of cartographic databases.
10. Explain, using existing map layers in your GIS databases, the difference between cartographic and geographic databases, and describe the potential problems of intermingling these layers.

ABSTRACT NATURE OF MAPS

The map is a model of spatial phenomena—an abstraction. It is not a miniature version of reality that is meant to show every detail of a study area. There are limits to what we can do with cartographic skills. Our misinterpretation of the limits of maps to display reality is because they are elegantly designed graphical instruments created to communicate spatial data. Maps have been around so long we have become very familiar with them. Such familiarity, combined with the compactness of information and their powerful visual appearance, lend a map a sort of authoritative infallibility. Because of this we must know the basic character, level of abstraction, symbology, and production methods that comprise the map document.

Maps come in many forms and subject matter. The two primary types are the general reference map and the thematic map. In GIS we deal primarily with thematic maps, although reference maps are used as GIS input. Although we will largely limit our discussion to thematic maps, much of what is covered in this chapter can be readily applied to the reference map.

A Paradigm Shift in Cartography

The adoption of the computer for mapping has had a profound impact on how we think about maps. The traditional approach to mapping, called the **communication paradigm**, assumed that the map was the final product, designed to communicate spatial patterns through symbols, class limits, and so on. This method of viewing cartography is limited. Raw, preclassified data are not accessible to the map user. The map can only display the set of criteria for which it was made.

An alternative approach to cartography that maintains the raw data for later reclassification developed concurrently with mapmakers' attempts to apply major advances in computer technology. This method, known as either the **analytical paradigm** (Töbler, 1959) or sometimes the **holistic paradigm**, maintains the unclassified attribute data inside a computer storage device and displays data based on user needs and user classifications (DeMers, 1991). Its advantage over the communication paradigm is the maintenance of the raw data for later manipulation. With these data readily accessible within the computer, the user can perform both communication and analysis.

The analytical paradigm was originally designed for use with value-by-area mapping, where each area has its own unique color or shading pattern relative to the raw data it represented. These maps, called classless choropleth maps (Töbler, 1973), were difficult to interpret because there were so many categories. In this respect, they were not unlike unclassed satellite images. With the computer as a storage and classification device, the method expanded in concept to allow multiple classifications of the data, and each set could be viewed readily by the viewer. This enhancement prompted some to change the name from analytical to holistic because it signaled the return of the map's role as a communication or visualization device.

An example using a single study area but for different purposes illustrates the different cartographic paradigms. Suppose you are creating a map for a public park. Initially you create a map that allows visitors to identify activities and sights offered by the park. The map highlights the lake, boat docks, camping areas, cabins, park ranger station, geysers or other natural features, walking trails, fishing sites, restricted areas, food vendors, and so on (Figure 3.1a). The fundamental reason for providing this map is to show the spatial distribution of the phenomena of interest in a way that indicates how each one may be reached. This is illustrative of the communication paradigm.

The park rangers will need more quantitative information to manage the park facilities and resources. They may need the average tree size within the forested regions, the amount of fire fuel in the forests, the numbers of rare or

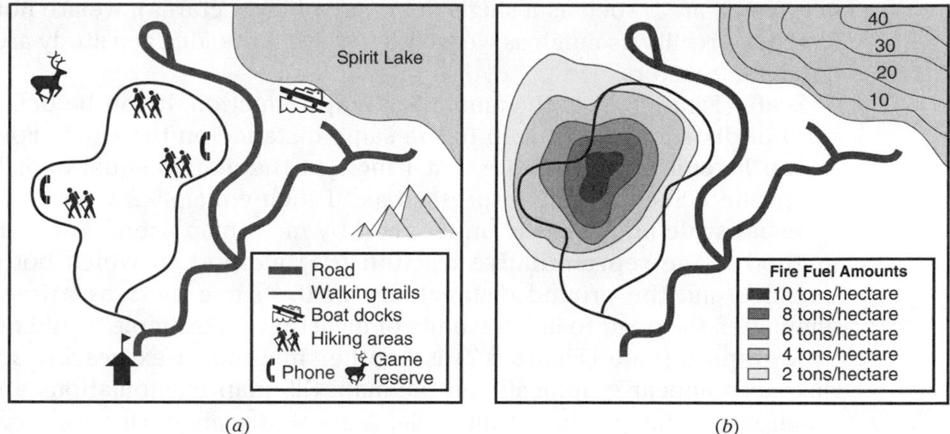

Figure 3.1 Cartographic paradigm shift. Two maps of the same study area—a state park—showing (*a*) the use of the communication paradigm to create a visual display of spatial relationships and (*b*) the added utility of the analytical paradigm for the same region.

endangered species in restricted areas, or the sizes and areal distributions of gaps in the forest that provide needed environments for browsing species. With these data, the park rangers can begin to predict whether deer populations will rise or fall, whether some trees may need to be cut to reduce the threat of fire, or where old and dying trees should be thinned to allow smaller trees to grow. These predictions require additional attribute data about tagged species, diseased trees, or annual differences in animal or plant species composition.

Eventually the amount of attribute information becomes so dense that a single map is incapable of illustrating the necessary conditions for the park rangers to develop management strategies. The data must first of all be detailed for each location. If, for example, there are 200 different types of vegetation assemblage, data must be recorded for each, but displaying all these areas on a single map will visually obscure the information. Multiple maps may have to be produced for each variable at each area. The primary function of these cartographic representations is analysis rather than simply display of spatial distributions (Figure 3.1b). The communication paradigm is inadequate for this application. It must be augmented to include more analytical capabilities such as are now available in the holistic paradigm with a GIS.

Map Scale

No matter which paradigm we choose, when we consider converting our spatial concept of space into a map, we must always remember that maps are reductions of reality. Although it might be intellectually appealing to envision a map that physically covers our entire study area, such a map would require us to explore the planet on foot. A primary purpose of any thematic map is to allow us to view important detail for a large region at a single glance, without the distractions of inconsequential or extraneous detail. The amount of reduction is primarily a function of the level of detail we need to examine our area. If we are looking at a very small area, such as a single field (say of 20 hectares), we are not required to reduce reality as much as we would if we were looking at a study area of 1,000 square kilometers.

Scale (Figure 3.2) is the amount of map reduction. It can be defined as the ratio of distance on the map to the same distance on the earth. For example, a **map legend** might indicate that 1 inch on the map is equal to 5,280 feet on the ground. Such a scale, expressed as "1 inch equals 5,280 inches," is called a **verbal scale** and is easily understood by most map users. Another common method is the **representative fraction (RF)** method, in which both the map distance and the ground distance are in the same units as a fraction. This eliminates the need to include units of measure. An example would be 1:24,000.

A **graphic scale** (Figure 3.2) is another method of expressing scale; here, distances appear graphically on the map. GIS map manipulations are likely to entail many changes in output scale. A graphic scale device can be placed on the map during input and, as the map scale changes on output, so will the scale bar itself.

As you begin working with GIS, you will find that most software can easily accomplish a change of scale. The input scale may differ from the display scale. The ability of the software to convert to nearly any scale imparts a feeling of confidence that can translate into subsequent problems. Reliable analysis is greatly influenced by the quality of the data that are input to the system. This reliability is, in turn, affected to a large degree by the scale of the cartographic documents you input. Keep in mind that on a very small-scale map, say of 1:100,000, a line of 1 millimeter covers over 100,000 millimeters, or 100 meters, on the ground. That's approximately the length of a football field. When the scale is changed in the GIS for later output at, say a scale of 1:1,000, the line that is drawn will be 1 millimeter in width, giving the reader the impression that the line was very accurately located when it was input. In fact, its location is far from accurate.

Another example might be quite useful in illustrating the problem of scale as it affects analysis. Someone not familiar with maps presents you with a map of the 48 contiguous states of the United States depicted on a piece of paper about the size of a CD. The person needs to have the map blown up to a much larger size, say 1 meter on a side, so that the areas of each state can be measured and the results used to calculate population density by state. Whether this map is to be enlarged through some xerographic approach or input to a GIS to have the areas measured, the end result is the same. The values obtained through dividing the number of people for each state by the highly error-filled measures of area are certain to be useless (Figure 3.3). Thus the following simple, long-held

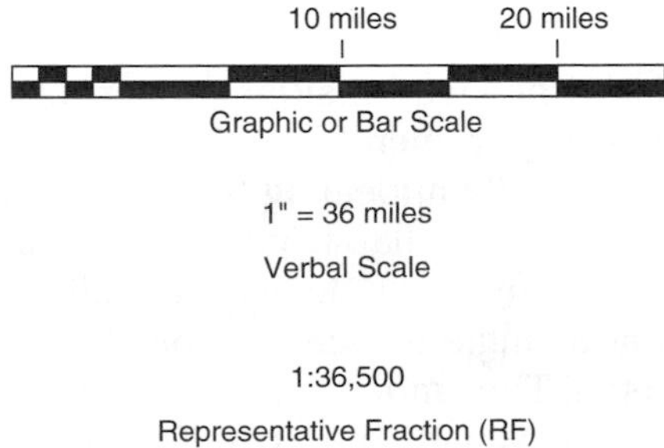

Figure 3.2 Methods of illustrating map scale. Examples of the three most common methods of illustrating scale on a map. The verbal scale, the representative fraction (RF), and the graphic or bar scale all have advantages and disadvantages for analog or digital mapping projects.

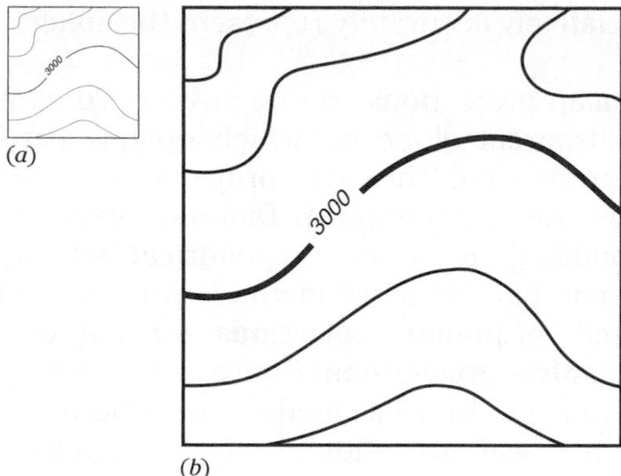

Figure 3.3 Effect of scale on accuracy. A portion of an extremely small-scale map blown up to a larger scale. The lines have become thicker and show a high level of generalization, areas are less precise, and measurement and analysis are practically impossible.

rule of thumb: it is always better to reduce a map after analysis than to enlarge it for analysis. This rule applies as much to the automated environment as to the manual one.

MORE MAP CHARACTERISTICS

Maps graphically represent the locations of objects in geographic space. They also describe their characteristics or magnitudes. These are called **entities** and **attributes**, respectively, and both are as necessary in the cartographic document as they were when their data were collected. Regardless of whether entities and objects represent points, lines, areas, or surfaces in the real world, they cannot be displayed as detailed miniaturizations because of the limitations of scale. Instead, we store them in the computer and then, upon output, assign a set of representative symbols. These symbols must have an interpretive key called a map legend. The legend unites the entities with their respective attributes. Each displayed entity represents a real feature with measurable attributes. This way the map reader can envision what was actually seen when the original data were collected.

MAP PROJECTIONS

When viewed up close the earth appears to be relatively flat except for the terrain. We all know that earth is more or less spherical. Maps are reductions of reality, designed to represent not only earth's features, but its shape and spatial configurations as well. Traditionally globes have been used to represent the earth's shape. Globes largely preserve the earth's shape and illustrate the spatial configuration of continent-sized features. They are, however, very difficult to carry in one's pocket, even at extremely small scales (often as little as 1:100,000,000). Most thematic maps are of considerably larger scales, say on the order of from 1:100,000 to as large as 1:1,000. A globe of this size would be very impractical. As a result, cartographers have developed a set of techniques called **map projections** to relatively accurately represent the spherical earth in two-dimensional media.

The process of creating map projections can be envisioned as positioning a light source inside a transparent globe on which opaque earth features (e.g., the geographic grid) are placed. Then one projects the outlines onto a two-dimensional surface surrounding the globe. Different ways of projecting could be produced by surrounding the globe in a cylindrical fashion, as a cone, or even as a flat piece of paper. Each of these methods produces a **projection family**. Thus, there is a family of **planar projections**, a family of **cylindrical projections**, and a family of **conical projections** (Figure 3.4). The actual process of projecting a spherical surface onto a flat medium is done using spherical geometry that recreates the physical projection of light through the globe.

Projections are not absolutely accurate representations of geographic space. Each imposes its own map distortions. The necessary analytical operations dictate which geometric map characteristics must be retained during the projection process. These characteristics or properties include (1) angles (or shapes), (2) distances, (3) directions, and (4) areal sizes. It is impossible to preserve all these properties at the same time when performing a map projection.

The process of map projection is twofold: first a scale change converts the actual globe to a **reference globe** based on the desired scale, and then the reference globe is mathematically projected onto the flat surface, where the globe's three-dimensions are transformed to a flat surface (Robinson et al., 1995). When we reduce the scale from the actual globe to its reference globe, we have changed the representative fraction to reflect this scale change. The representative fraction for the reference globe, called the **principal scale**, is calculated by dividing the earth's radius by the radius of the globe. We now have a representative fraction that is uniform throughout the reference globe because the globe, like the earth, is roughly spherical. This means that the actual scale will be the same as the principal scale everywhere on the globe.

Before we perform the second step of projecting the reference globe to its flat map counterpart, we should note that the **scale factor (SF)**, defined as the

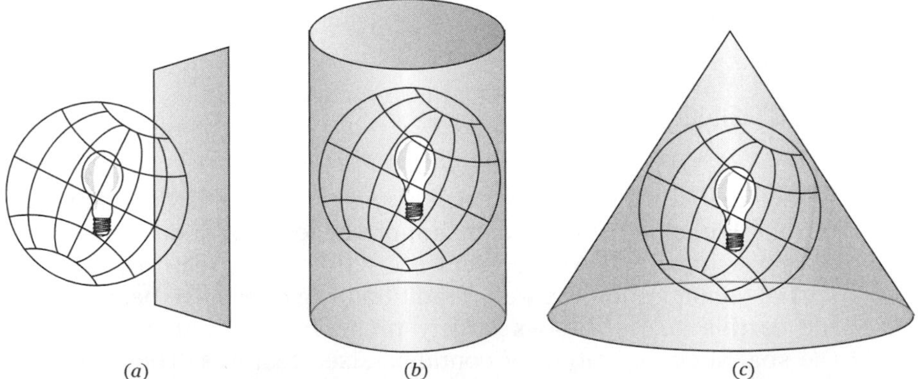

Figure 3.4 The three families of map projections. They can be represented by (*a*) flat surfaces, (*b*) cylinders, or (*c*) cones. *Source:* P.C. Muehrcke, and J.O. Muehrcke, *Map Use: Reading, Analysis and Interpretation*, 3rd ed. © 1992, Figure C.9, page 573. Used with permission.

actual scale divided by the principal scale, is by definition 1.0 at every location on the reference globe. When we move from a spherical reference globe to the two-dimensional map, the scale factor will change because the flat surface and the spherical surface are not completely compatible. Therefore, the scale factor will differ in different places on the map (Robinson et al., 1995). The distortion or change in scale factor is a function of the projection used.

On a globe the cardinal directions always occur at 90 degrees of one another. East will always occur at a 90-degree angle to north. This property, called **angular conformity**, can be preserved on some map projections. Such projections are called **conformal** or **orthomorphic** map projections. Remember, on the reference globe, the SF is always 1.0 in every direction for every point. Conformal projections mathematically arrange the stretch and compression so that, within the projected map, the SF is kept the same in every direction. Because the SF will no longer be equal to 1.0 at every point, the areas will be distorted, but, as with the reference globe, the parallels and meridians will still be at 90° of one another (Robinson et al., 1995). Maintaining true angles is difficult for large areas and should be attempted only for small portions of the earth.

The conformal type of projection distorts areas. We can preserve areas through projections called **equal area** or **equivalent projections**, in which the product of scale factors in cardinal directions equals 1.0 (Robinson et al., 1995). This ensures that if you find the areas of, for example, square surface features, their two dimensions multiplied together will result in an area identical to what would be calculated on a reference globe. This is because the product of these two dimensions results in the identical SF. However, once we have achieved this identity, we find that although the amount of area is correctly measurable, the scale factor will vary in every direction about a point *except* the cardinal directions. So by preserving area we distort angles. Thus areas and angles (shape) cannot be preserved at the same time.

If we wish to accurately measure distances, we must select a projection that preserves distances. Such projections, called **equidistant projections**, maintain a constant scale; they must also be the same as the principal scale on the reference globe (Robinson et al., 1995). There are two ways to do this. The first maintains a scale factor of 1.0 along one or more parallel lines called **standard parallels**. Then distances measured along these lines will be accurate. The second approach maintains a scale factor of 1.0 in all directions from either one or two points. Distance measured from those starting points will then be accurate. A distance measured from any other point on the map will be inaccurate. The choice of starting point is vital here. Usually you choose the point from which most of your distance measurements are to be made.

For navigational maps the primary interest is to preserve direction. The preservation of true direction on a map is limited to preservation of great circle arcs that define the shortest distance from point to point on the earth. Normally we depict these great circle routes as straight lines. There are two primary methods of doing this. The first uses small areas to show great circles as straight lines among all points on the map. The result is that angular intersections of meridians with these great circles will be incorrect. Both the limited area and the erroneous angular intersections between meridians and great circles severely limit the use of this projection. An alternative, called the **azimuthal** projection, is more commonly used for preserving directions. Here we begin by selecting one or two reference points from which our directions will be preserved. Straight lines drawn from these reference points will have true direction. Direction measured from any other points will be incorrect.

We need some rules for determining which of the many map projections we might want, depending on the types of analysis to be performed. If an analysis requires us to determine motion or changing directions of objects—for example, in tracking hurricanes—the conformal projection is preferred. This projection is also preferred for navigational charts, meteorological maps, and orienteering maps. This group of projections includes the **Mercator**, **transverse**, **Lambert's conformal conic**, and **conformal stereographic projections**.

General reference and educational maps use equal area projections, but our interest is in thematic maps. For GIS equal area projections are best when calculations of area dominate our analysis. If, for example, your interest is in the calculation of the changing percentage of a land cover type with time, or if you are trying to analyze a particular land area to determine whether it is large enough to be considered for a shopping mall, equal area projections are employed. When considering the use of an equal area projection, you will need to take into account the size of area involved and the distribution and amount of angular deformation. Small areas will display less angular distortion when equal area projections are used, and this might be relevant if both shape and size are important characteristics for your analysis. Alternatively, the larger the area you are analyzing, the more precise your area measures will be if you use an equal area projection rather than another type. Types of equal area projection for use with medium-scale maps encountered in GIS work include **Alber's equal area** and **Lambert's equal area projections**.

GIS projects that require determination of shortest routes, especially for long distances, call for the use of azimuthal projections because of their ability to show great circle routes as straight lines. These projections are most often used for creating maps for airline traffic, determining ranges of radio signals, keeping track of satellites and satellite data, and mapping other celestial bodies (Robinson et al., 1995). The most common azimuthal projections include Lambert's equal area, conformel stereographic, **azimuthal equidistant**, **orthographic**, and **gnomonic projections**. Some azimuthal projections maintain both direction and area. This property may prove useful for analyzing aerial distributions such as volcanic plumes that are likely to travel great circle routes as they disperse into the atmosphere and follow the general circulation patterns of the earth.

There are many map projections from which to choose—far more than are listed here. Some are special-purpose projections especially useful for depicting the entire earth or very large portions of it. Other projections allow for better coordination of large-area mapping programs such as the topographic mapping of whole continents done piece by piece. The selection of projection is a fundamental process of designing a GIS and should be considered carefully (Nyerges and Jankowski, 1989; Snyder, 1988.

GRID SYSTEMS FOR MAPPING

A coordinate system is necessary to reckon distance and direction. The geographical coordinate system based on latitude and longitude is useful for locating objects or features when they are confined to the spherical earth or its reference globe. Because we will usually use two-dimensional maps projected from a reference globe, we need coordinate systems that correspond to the distortions introduced through map projection. These **rectangular coordinates** or **plane coordinates** allow us to locate objects correctly on these projected maps.

The basic **Cartesian coordinate system** consists of two lines, an abscissa and an ordinate. The abscissa is a horizontal line that contains equally spaced numbers starting from 0, called the origin, and extending as far as we wish to measure distance in either of two directions (Figure 3.5). The values are called X coordinates; they are positive if we move to the right of 0 and negative if we move to the left. The second line, the ordinate, allows us to move vertically from the same point of origin in a positive or negative Y direction. Together the X and Y coordinates allow us to locate any point or feature by combining the values of X and Y. Graphical input devices called **digitizers** are based on this simple system. It is possible to obtain reasonably accurate results with such a system as long as the map exhibits the property of **conformality**. Conformal maps are most likely to cover only small portions of the earth's surface. Each of these large-scale maps must maintain its own coordinate system to ensure measurement accuracy (Robinson et al., 1995).

When reading maps using rectangular coordinates, we give the X value first and the Y value second. When a map is oriented with north at the top, the X value is called an **easting** because it measures distances east of the origin or starting point. In like manner, the Y value is called a **northing** because it measures distances north of the origin. There are no westings or southings because the origin is placed so that all references are positive. This allows us to read first right, then up from the origin, a process called reading right-up. In some cases, the size of the area requires us to construct **false origins** to ensure that each portion of the earth has limited distortion. Measurements from false

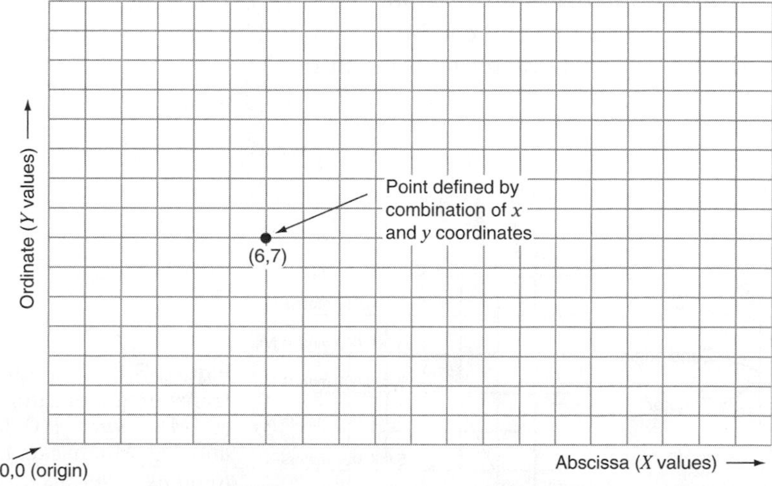

Figure 3.5 A Cartesian coordinate system. Classic rectangular grid system used to represent the number system. The abscissa (X values) is horizontal and the ordinate (Y values) is vertical. Any point can be defined by combinations of these two.

origins are called **false eastings** on the abscissa and **false northings** on the ordinate line.

Plane coordinates are not normally used on small-scale maps because of the potential for distortion. When they are used, adjustments must be made to compensate for the distortions introduced during projection. Despite the large number of map projections available, the vast majority of plane coordinate systems attempt to adjust for conformality by using only conformal map projections, typically the transverse Mercator, polar stereographic, and Lambert's conformal conic. This will not always be the case, however. If, for example, your study area is located in equatorial locations, a **Mercator projection** may prove more useful (Robinson et al., 1995).

In the United States, five primary coordinate systems are used—some based on properties of the map projections and others on historical land subdivision methods. When you encounter maps of other nations in your GIS work, you will need to ascertain the projections, coordinate structure, datums, and other properties of the systems of those nations. Many countries use one or another of the types discussed in this section but will require familiarity with their points of origin and the zones of the earth they occupy before beginning your GIS input procedures.

Perhaps the most prevalent plane grid system used in GIS operations is the **universal transverse Mercator (UTM) grid** (Figure 3.6). UTM divides the earth from latitude 84° north and 80° south latitude into 60 numbered vertical **zones** that are 6 degrees of longitude wide. To allow for all coordinate locations to be positive, the UTM has two false origins, one at the equator and the other 100,000,000 meters south of the equator (80° south latitude). For reference, these zones are numbered starting at the 180th meridian in an eastward direction. Each zone in turn is divided into rows or sections of 8 degrees of latitude each, with the exception of the northernmost section, which is 12 degrees, allowing all the land of the northern hemisphere to be covered with the system.

Relatively small portions of the earth can be isolated because each section can be located by its number and letter combination (read right-up as before). With the exception of the northernmost group, each of these sections occupies 100,000 meters on a side and can be designated by eastings and northings of up to five-digit accuracy (1 meter level resolution).

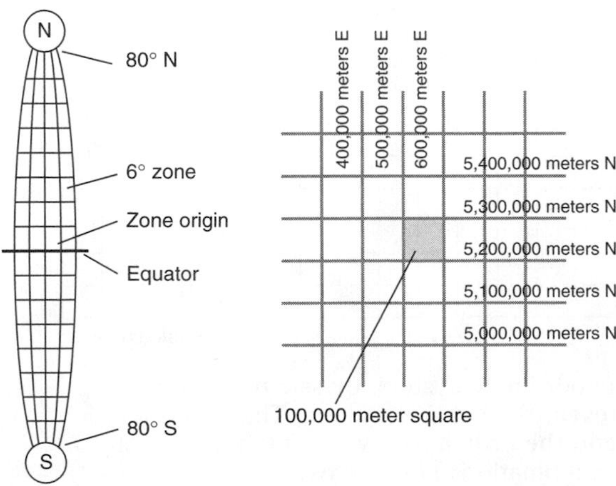

Figure 3.6 The universal transverse Mercator system (UTM). *Source:* P.C. Muehrcke, and J.O. Muehrcke, *Map Use: Reading, Analysis and Interpretation*, 3rd ed. J.P. Publications, Madison, WI © 1992, Figure 10.8, page 215. Used with permission.

The UTM grid uses a transverse Mercator projection. For each of the 60 longitudinal zones, a separate Mercator projection is applied to reduce distortion. The Y-coordinate origin is placed at the exact center of each zone, and a false origin is offset 3 degrees west of that. The scale factor does not vary from its 0.99960 in the north–south direction, but it does vary in the east–west direction. Still, at its farthest point from the Y-coordinate origin, the scale factor is very nearly the same at 1.00158. This near equivalence illustrates the minimization of distortion possible with UTM that results in accuracies approximating only 1 meter variation in every 2,500 meters distance (Robinson et al., 1995).

For polar regions that extend beyond the area covered by UTM we use a **universal polar stereographic (UPS) grid**. This system divides the polar regions into concentric zones—one each for the northern and southern poles. It then splits each into two equal halves east to west at longitude 0 and 180 degrees. These zone halves are assigned different designations for the North and South Poles. In the northern hemisphere the west half is designated grid zone Y and the east half grid zone Z. In the southern hemisphere the west half is designated zone A, and the east half zone B. As in the case of UTM, measurements are made as eastings and northings of up to 2,000,000 meters, corresponding to 180 degrees in longitude. Also like UTM, the zones can be divided into 100,000-meter squares, each with its own projection, resulting again in an accuracy of approximately 1 meter in 2,500. Collectively, the UTM and the UPS systems provide a global coverage with only minor distortion and reasonably accurate measures.

The United States Geological Survey (USGS) has found it convenient to simplify the UTM system for its topographic mapping. Because the United States falls entirely inside the northern hemisphere, you only need a zone number (1–60) and a single pair of easting and northing values to locate a point on the map. The equator is designated as the origin for northing values, and the meridian in the center of each meridian zone is given a value of 500,000 meters as a false easting. These simple modifications make the designations simpler, but the locations are identical to what would be found on the UTM. Most USGS topographic maps have this coordinate system printed in blue lettering on the margins, together with another grid system we will discuss a little later.

A system devised in the 1930s by the U.S. Coast and Geodetic Survey (now the U.S. Chart and Geodetic Survey), called the state plane coordinate (SPC) system, uses a unique set of coordinates for each of the 50 states (Figures 3.7, 3.8) (Claire, 1968; Mitchell and Simmons, 1974). This grid system uses either a transverse Mercator or Lambert's conformal conic projection tied to a national geodetic framework. Its original design was to provide a permanent record of land survey monument locations. Like the UTM system, the SPC uses two or more zones, in this case overlapping, to maintain an accuracy of one part in 10,000. Each zone has its own projection and coordinate grid that is measured in feet rather than meters, although metric equivalents have recently been included. To define a location using this system, you give the state name, the zone name, and the easting and northing values (in feet). The advantage of SPC is in its accuracy, estimated at four times that of the UTM system.

A common rectangular grid system in the United States is the **U.S. Public Land Survey System (PLSS)**, established in 1785 as a method of land subdivision (Figure 3.9) (Pattison, 1957). The PLSS is not formally tied to any particular map projection or reference globe. It is a tool for recording land ownership, and its basic unit of measure is an area of 1/640 of a square mile, or 1 acre. Each

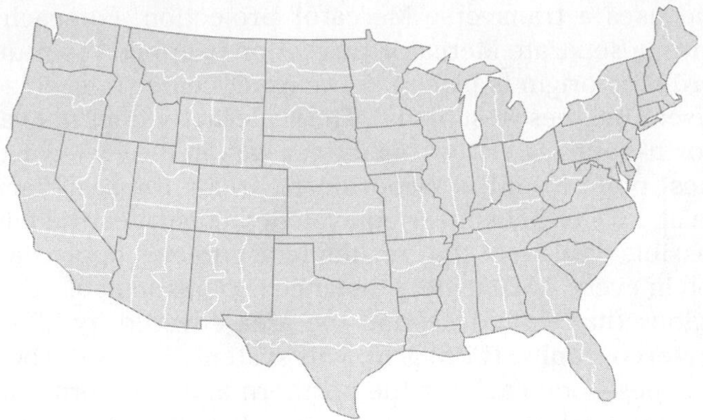

Figure 3.7 State plane coordinate system zones for the United States.

square mile area is a **section** (640 acres), and these are grouped in larger, 36-square-mile groups that collectively are called **townships**. The sections are numbered in a serpentine fashion starting in the upper right with section 1 and ending in the lower right with section 36. Each of the 36 sections can be divided and subdivided into halves and quarters, permitting designation of ownership of smaller portions of land. For example, a person can own a quarter of a quarter of a quarter section—meaning that he or she owns 1/4 of 1/4 of 1/4 of 640 acres, or 10 acres. One could also own a quarter of half a section, or 1/4 of 1/2 of 640 acres, or 80 acres. By defining the location of the subdivisions as to north, south, east, or west for halves, or northeast, northwest, southeast, or southwest for quarters, individual landowners can determine which lands belong to them.

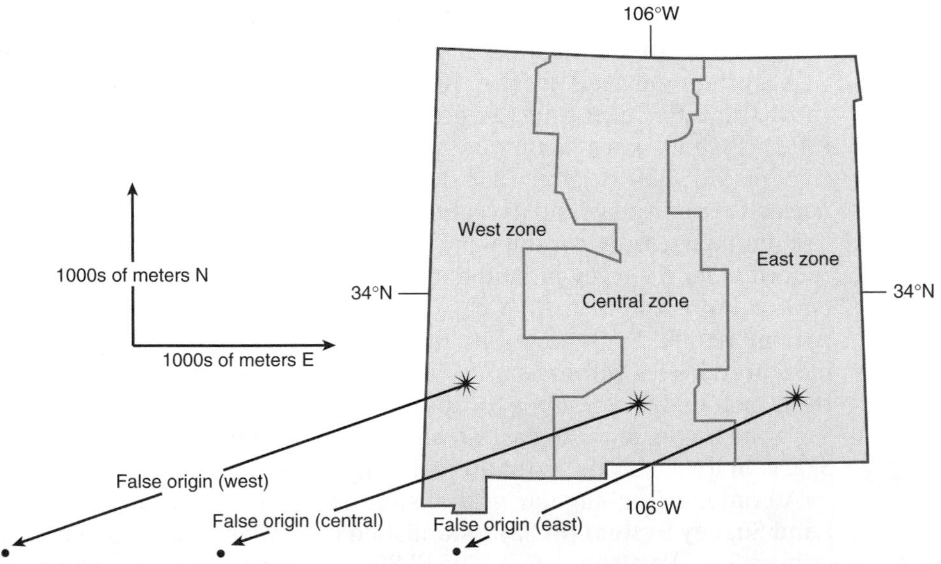

Figure 3.8 The United States state plane coordinate (SPC) system. Applied to New Mexico, this illustrates the potential for problems at cross-state boundaries when using SPC for geographic information systems (GIS) work in large regions.

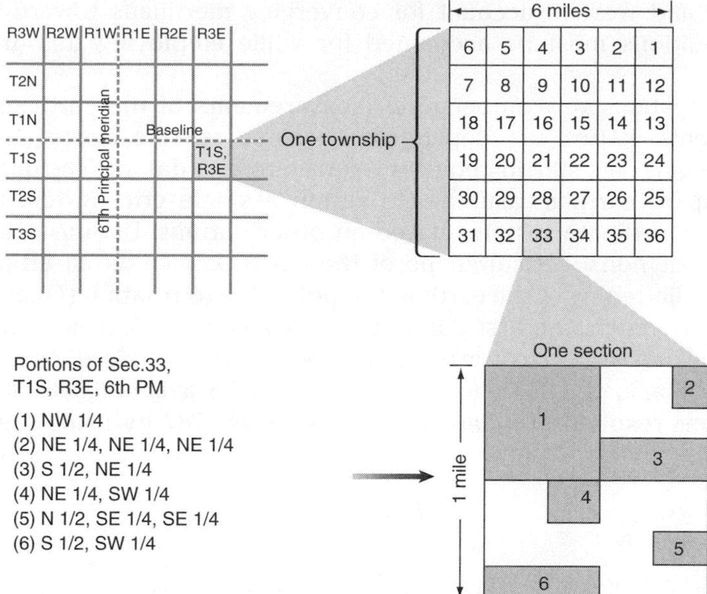

Figure 3.9 The Public Land Survey System (PLSS) of the United States. Township (T) and range (R) lines intersect at the center of a township. The township occupies 36 square-mile sections (S), each of which is 640 acres in area. When more than one fractional portion of a lot is given, the position in the section map is located by reading the subportions from right to left. *Source:* P.C. Muehrcke, and J.O. Muehrcke, *Map Use: Reading, Analysis and Interpretation*, 3rd ed. J.P. Publications, Madison, WI, © 1992, Figure 10.20, page 226. Used with permission.

Townships and their smaller, square-mile sections are located within a larger grid of established horizontal and vertical lines. The horizontal lines, called **baselines**, allow measurements of **township lines** (not the same as townships themselves, although they bound townships), whereas the vertical lines, called **principal meridians** (not the same as prime meridian), measure the longitudinal bounds of land through measurements called ranges. Township and **range lines** run through the center of each township of land. Together, these lines, chosen as conventions by the U.S. Continental Congress, allow the designation of landownership by stating first the subportion of the numbered section (e.g., NE 1/4 N 1/2 of section 18), and then the township and range numbers that identify the township they intersect.

Although the PLSS as just described is not particularly difficult to understand, a number of problems have been encountered in its implementation in the GIS environment. First, boundaries with Canada and Mexico interrupt and often distort the system, preventing a designation of land without substantial minute subdivisions, especially along meandering stream borders. Second, past surveyors' errors are compounded when markers are lost, creating many situations in which landownership disputes occur. Finally, because the earth is round and maps are flat, and the PLSS is often portrayed on projected maps together with other rectangular grid systems, the square sections are often offset east and west to account for converging meridians toward the poles. These conditions must be accounted for while employing this grid system in GIS.

All grid systems rely on accurate measurements of the earth's size. Such measurements in turn are dependent on some reliable starting point from which to begin. This starting point is referred to as a **datum**. The datum allows us to compare earth measurements within a single grid system as well as with other coordinate systems based on other datums. Datums are based on the somewhat nonspherical shape of the earth, known as an **ellipsoid**, that resembles a flattening of the earth at the poles due to rotation (Figure 3.10).

In North America, the first datum was devised in 1866, based on reference points along the earth's surface that characterized the earth's ellipsoid as having an equatorial axis of 6,378,206.4 meters and a polar axis of 6,356,538.8 meters. These values resulted in a flatness ratio of 1/294.9787 by using the following formula:

$$f = 1 - \mathrm{b}/\mathrm{a}$$

where

a = the equatorial semimajor axis
b = the polar semimajor axis

Although a simpler flatness ratio of 1/297 was already adopted as the international standard, the original was used for the 1927 **North American datum (NAD27)** because a substantial amount of mapping in the United States had already taken place using the older standard.

Newer datums have been established that are based on estimates of the distance to the center of the earth. In 1983, for example, a new datum was established for the United States (**NAD83**) that became the standard **geodetic reference system (GRS80)**. A modified version of GRS80, developed by the U.S. military in 1984, resulted in the **world geodetic system (WGS84)**. There are many more of these datums being used throughout the world; refer to the National Geospatial Intelligence Agency (NGA) Web site for information (http://earth-info.nga.mil/GandG/coordsys/onlinedatum/index.html; last visited 8/13/07).

The principal reason for knowing the datum with which you are working is to allow for the development of a proper geodetic framework for your GIS maps. To ensure that multiple maps of the same study area are located precisely at the same locations on the earth, we must use a common datum. All GPS receivers

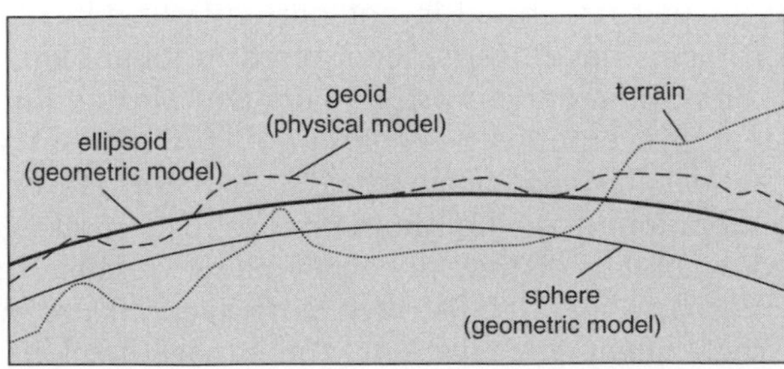

Figure 3.10 Relationships among terrain and geoid, ellipsoid, and spherical representations of the earth.

refer to a particular datum and ellipsoid to obtain correct locational values. You should be aware, however, that geodetic scientists also employ another model of the earth based on compilations of all the variations from the ellipsoid. This is called a **geoid**, meaning "earth shaped." All topographic measurements on the earth are based variously on a reference sphere, an ellipsoid, a geoid, or the actual terrain (Figure 3.10).

MAP SYMBOLISM

Maps are not exact miniaturizations of reality, but rather abstractions we need to employ graphic symbols to represent geographic points, lines, areas, and surfaces. In Chapter 2 you learned that they can be described by four different levels of measurement. When we move to cartographic abstractions, we need to represent all these objects, no matter what their individual measurement level, by careful selection, categorization, and symbolization so that the results will physically fit on the space provided and in a way the reader can see and understand.

Figure 3.11 expands the relationships between spatial data types and their levels of measurement (Muehrcke and Muehrcke, 1992). We now have some sample sets of symbols that correspond to points, lines, areas, and surfaces at all levels of measurement. Figure 2.4 grouped the last two levels of measurement as interval/ratio, rather than keeping them separate, because the symbol set is generally the same for these measurement levels. It is important to note this because it adds another level of abstraction. If we are not aware of how the data were originally gathered, we may not know whether they are interval or ordinal. The symbols do not tell us. We may find ourselves performing a ratio-type analysis on a set of data that do not have an absolute starting point. Results from such an analysis would be meaningless.

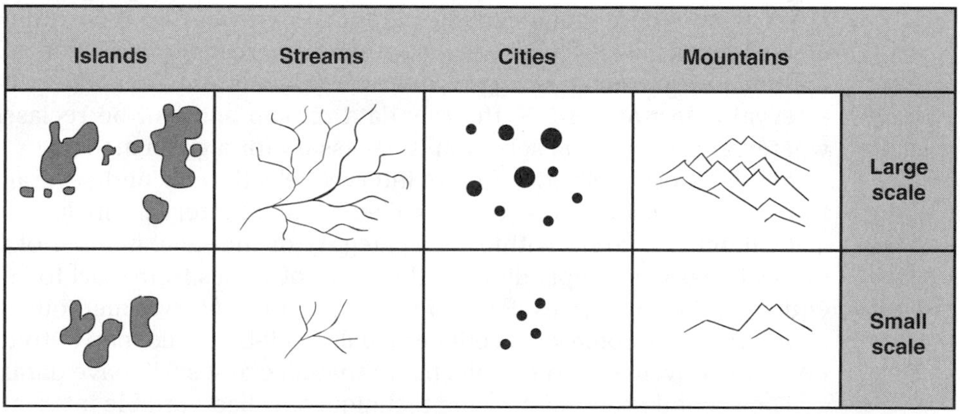

Figure 3.11 Scale change and feature elimination. Comparison points, lines, areas, and surfaces as these features are detailed in large-scale maps and eliminated for small-scale documents. *Source:* P.C. Muehrcke, J.O. Muehrcke, *Map Use: Reading, Analysis and Interpretation*, 3rd ed. J.P. Publications, Madison, WI © 1992, Figure 3.13, page 81. Used with permission.

Note that lines can be used to illustrate surfaces. This may add some confusion as well, for two reasons. First, there is the tendency of the novice to misinterpret lines as one-dimensional features rather than as symbols of surfaces. This form of confusion is easily overcome by continued familiarity with map features and symbols as you encounter more and more maps. The second problem, due to the use of line symbols to represent surfaces, is that of interpreting the lines themselves as accurate representations of point elevation values (see Chapter 2). As you have seen, the lines are estimates or predictions of specific elevations made through interpolation. The only nonpredictive elevation values are those that were actually measured, as exemplified by benchmarks on topographic maps.

The use of lines to represent surfaces is one of a large number of changes of dimensionality that can occur when the data are abstracted with the use of symbols. Figure 2.1 clearly indicates the wide range of dimension changes that can be symbolized on a map. Good judgment will always be necessary in deciding whether the symbol geometry and dimensionality given are truly representative of an object. For example, if an area symbol is used to depict a point feature, you must be aware that despite the two-dimensionality of the symbol, only a single point needs to be recorded or input for the feature.

The same can be said of levels of measurement. Symbols such as graduated circles, indicative of attributes at point locations, are often manipulated to achieve a particular visual response (Robinson et al., 1995). This change in symbol size is not directly proportional to the changes in the data; rather, it is designed to allow the viewer to visualize proportionality. Acknowledging these symbol manipulations, the GIS analyst should examine the map carefully before proceeding with data encoding. As always, if the raw data are available, the attributes should be input from them rather than from their graphic depictions.

A major difference between the communication and holistic paradigms is the classification-oriented manipulation of data prior to map production. Because a single thematic map is meant to serve a single purpose, this classification process normally takes place only once and the raw data are no longer available. Consider the case of choropleth or **value-by-area mapping**. When mapping areas under the communication paradigm, we want to group the areas into meaningful as well as visually appealing aggregations. The many common methods of grouping or aggregating these areas are collectively called **class interval selection**. In GIS, the raw data remain and can be reclassified using whatever class interval selection is necessary for a given map use.

Among these methods of class interval selection we find several categories that merit examination. The first, called constant intervals, includes having the same number of areas within each category or the same number of data points for each class, or simply dividing the range of values from start to finish into an equal number of classes. Each of these constant interval methods has its own characteristics. Some will produce visually well-balanced map output; some will be more convenient; some will ensure that all classes will have data.

The second group of classing techniques, called variable intervals, produce maps that are less intuitive but may be highly useful for isolating certain high or low values, or for highlighting variations in thematic value. Variable interval methods can be systematic, including arithmetic, logarithmic, or other mathematical series; or they can be unsystematic, with the clustering of the data into natural groupings used to determine where the class intervals will be designated.

Both general types of class interval selection can be used for point, line, and area symbols to depict all spatial data types. Take, for example, the use of contour intervals on topographic maps. The selection of contour intervals is as much a method of class interval selection as is grouping areas for choropleth maps. Likewise, creating a discrete set of point symbols to show variation in attribute variable is also a method of class interval selection, because the cartographer knows that the human eye is incapable of discerning extremely subtle changes in sizes.

All these interval selection techniques create documents that, to varying degrees, disguise the original data and, if poorly selected, can obscure the original distributional patterns. Keep these principles in mind when you prepare to produce GIS data output. If, on the other hand, you intend to use these classified maps as input to a GIS database, you must be very careful about the analyses you perform with the aggregated data.

Of course, symbolism and classification are not the only cartographic compilation methods of which we need to be aware. Among the most important compilation processes the GIS analyst will encounter is the graphical simplification that takes place on the map as these classified data and their symbols are transferred to paper. This process presents a problem during GIS input, but it also affects the results of subsequent measurements and other analyses. Simplification goes further than classifying the map data or determining the types and levels of abstraction for the symbols used. Simplification eliminates some features that are not wanted, or smooths, aggregates, or further modifies the features on the map. Ultimately the purpose of map simplification is to provide for readability of the cartographic document once it is produced. Two basic methods are used: feature elimination and feature smoothing. We will discuss each separately.

In feature elimination we perform a function that is very much like the process of spatial data gathering itself (Figure 3.11) (Muehrcke and Muehrcke, 1992). When we observe a portion of the earth, we use our geographic filter to make decisions about which features we will note and which we will ignore. The importance of the features during the data gathering process is determined before we begin, and it is largely controlled by our reasons for gathering the data in the first place. In fact, the selection of objects for investigation will, by default, act as a passive process of feature elimination on our maps, because only the features selected will be placed in the database or map document. In some cases passive feature elimination occurs because we are unable to view objects in the field with the instrumentation at hand. Sensitivity to features is also a function of the current state of scientific knowledge. For example, changes in plant and animal species affect what is recorded during biological census activities; mammal locations that previously could not be pinpointed specifically now can be registered by **radio telemetry**; prior to initiation of a national population census, we were unable to collect large-scale information on population changes and socioeconomic variables. Sensitivity to environmental factors and their interactions affects how we conceptualize them, which in turn affects what we select for investigation and later mapping.

Unlike passive feature elimination, active feature elimination can be used in data collection and in mapping and cartographic database development as well.

When we select certain electromagnetic radiation bands for remote sensing, we are actively eliminating certain portions from our data set. The sampling schemes we discussed earlier also actively promote selectivity by eliminating large portions of the objects that could be collected. We perform active feature elimination on the map or digital cartographic database itself, prior to final map production. Here the map, rather than the person collecting the data, acts as the spatial filter. Very small towns often do not show up on maps of highly populated states or nations, whereas towns of the same size may appear on maps of areas low in population density. Similarly, we may eliminate some smaller or less important river tributaries, lakes, or islands during the mapping process because of lack of display space. In all these cases, a set of rules is formulated to determine which are selected and which are not. The rules may be as simple as eliminating a certain proportion of the objects or perhaps every other one; or they may include a set of decision rules (e.g., eliminate towns below a certain population, eliminate the smallest tributaries of stream networks). Whichever set of techniques is applied, the result is a less detailed output. If the product of feature elimination is to be input to a GIS, you create a cartographic database with some data missing.

Another useful method of simplification is called smoothing (Figure 3.12). This process abstracts detailed geometric objects into objects reduced in detail. Much as in caricatures of famous people, important geometric features are retained by representing a given detail as a simplified geometric shape. On maps showing coastal regions, boundaries, sinuous streams, or islands, we may generalize the lines that represent these irregular features so that their presence is recorded, but their spatial detail is minimized to fit on the document. GIS input from these maps will result in less than satisfactory measures of length, shape, area, or other geometric property we may wish to calculate. Because GIS analysis frequently results in cartographic output, however, we may find these two forms of simplification useful when we produce the final results of an analysis.

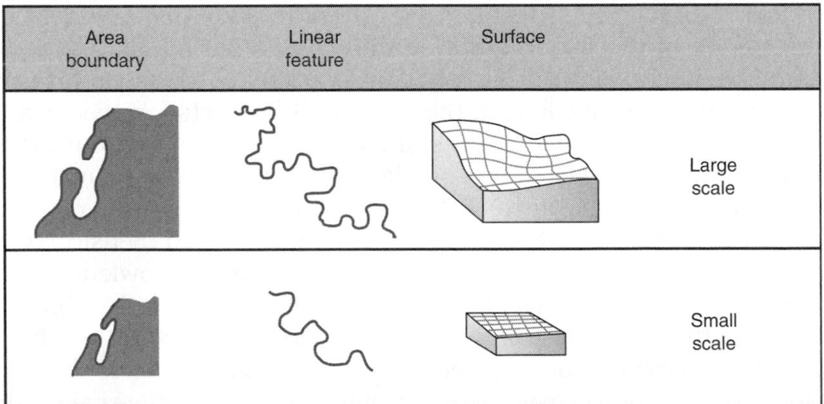

Figure 3.12 Scale change and smoothing. Use of smoothing, including the use of change in dimensionality. Note how features are simplified to only the most representative caricatures. *Source:* P.C. Muehrcke, J.O. Muehrcke, *Map Use: Reading, Analysis and Interpretation*, 3rd ed. J.P. Publications, Madison, WI, © 1992, Figure 3.14, page 81. Used with permission.

MAP ABSTRACTION AND CARTOGRAPHIC DATABASES

As you have already seen, map scale restricts the amount of data that can be contained in a single map document. Base maps used for GIS data input have had to be adjusted during compilation to account for these limitations. It is largely on the basis of scale that features must be generalized, spatially displaced, or even abstracted further to allow for map readability (Robinson et al., 1995). When these graphic entities and their associated attributes, sometimes collectively called a **cartographic database**, are collected from existing cartographic documents rather than from field or remotely sensed data, they must somehow be digitally captured to a computerized database.

Cartographic databases move from a higher level of abstraction (low accuracy) to an *apparent* lower level of abstraction, giving us an incorrect view of improved accuracy. Their counterpart **geographic database** data move from an actual lower level of abstraction (high accuracy) to a higher level of abstraction (lower accuracy).

When data are digitally encoded from map documents, the symbols representing points, lines, areas, and sometimes surfaces are highly abstract representations of reality. Their size and placement are themselves graphic abstractions that are less accurate than the ability of computerized input devices to define them. Therefore, when we move from map document to digital database, we must decide exactly which part of a point, line, or area symbol should be recorded. If the placement of, for example, a point symbol is physically offset from its actual location to make room for another feature, our GIS input will indicate a locational accuracy that does not exist. In addition, the point symbol itself occupies space. When we attempt to locate an exact point for GIS input for this zero-dimensional object, we must decide whether the center of the symbol is its most accurate location. Hence, the "apparent" lower level of abstraction for digital cartographic databases.

Alternatively, if we are creating a geographic database from, say, a survey instrument, we are now faced with the opposite problem. The survey instrument may be accurate to within centimeters or even millimeters of some absolute location. However, the digitizing input device is often not capable of reproducing this information with the same level of accuracy.

Although each of these issues needs to be addressed individually as we create our GIS database, some of the more difficult problems arise when these two sets of compilation rules interact. For example, a wildlife scientist attempting to create a database indicating the location of roads relative to a species of bird that inhabits regions in or around the roads. The expedience of creating a road network from existing 1:24,000 USGS topographic maps, rather than collecting road data in the field, made the decision to create a cartographic database for this GIS theme a straightforward one. The birds, in this case burrowing owls, create individual burrows on the ground that do not appear on maps, so a geographic database was created by using a GPS unit with 1 meter accuracy and then entering the data into the GIS. But when the two digital maps were displayed, the locations of burrows that were known to lie well within 10 meters of the roads appeared to be offset by at least 100 meters from the roads. This incompatibility of different compilation rules resulted in failure of the GIS to analyze the interactions between roads and burrowing owls.

The same problem can arise when creating a cartographic database from maps of different scale. Maps at very small scales will require symbols that occupy much larger proportions of space on the ground than will very large-scale maps with the same symbol sizes. Thus, the scales of input for a cartographic database should be as nearly identical as possible.

Terms

Alber's equal area projections
analytical paradigm
angular conformity
attributes
azimuthal equidistant projections
azimuthal projections
baselines
Cartesian coordinate system
cartographic database
class interval selection
communication paradigm
conformal
conformal stereographic projections
conformality
conical projections
cylindrical projections
datum
digitizers
easting
ellipsoid
entities
equal area (equivalent) projections
equidistant projections
false eastings
false northings
false origins
geodetic reference system (GRS80)
geographic database
geoid
gnomonic projections
graphic scale
graphicacy
holistic paradigm
Lambert's conformal conic projections
Lambert's equal area projections
map legend
map projections
Mercator projections
North American datum (NAD27)
North American datum (NAD83)
northing
orthographic projections
orthomorphic
planer projections
plane coordinates
principal meridians
principal scale
projection family
Public Land Survey System (PLSS)
radio telemetry
range lines
rectangular coordinates
reference globe
representative fraction (RF)
scale
scale factor (SF)
section
standard parallels
township
township lines
transverse projections
universal polar stereographic (UPS) grid
universal transverse Mercator (UTM) grid
value-by-area mapping
verbal scale
world geodetic system (WGS84)
zones

Review Questions

1. What is graphicacy? What impact does improved graphicacy have on our ability to function as GIS specialists?
2. What is the communication paradigm? What is its primary purpose? What impact does it have on GIS? Give an example of the use of the communication paradigm for traditional maps.
3. What is the analytical or holistic paradigm? How does it differ from the communication paradigm? Give an example of its use in cartography. What impact does this paradigm have on GIS?

4. What are the basic methods of illustrating scale on a map? Describe them. What are the relative merits of each type of scale as used in GIS?
5. What are some potential problems of putting multiple scales of maps into a GIS database? How could they affect analysis and measurements?
6. What is the purpose of the legend on a map? What impact does a map legend have on the relationships between entities and attributes?
7. What are map projections? What is their purpose? What are the three basic families of map projections?
8. What basic properties of the spherical earth are affected by using map projection? Which types of projection are best for preserving each of these properties? Provide one example of the application of a map for each preserved geometric property.
9. On the basis of your answer to Question 8, suggest some basic decisions you will need to make in deciding which projections to use for a variety of different GIS analyses.
10. Describe the UTM grid system. What are its advantages and disadvantages for GIS use?
11. What is the state plane coordinate system? Describe how it works. Explain its possible limitations (hint: how would it work in Vietnam or Uzbekistan?).
12. Describe the U.S. Public Land Survey System. What is a section? A township? What are the problems associated with using the PLSS for GIS work?
13. Using a real map, provide a legal description of a single property using the PLSS system.
14. What is a datum? Why do we need to know which datum we are using? Describe NAD27. Describe NAD83. What is GRS80?
15. What impact does class interval selection have on GIS input and later analysis?
16. What impact does the physical size of cartographic symbols have on map accuracy? What about the interaction of two or more map symbols? Explain the impact of scale on GIS input.
17. What is the difference between a cartographic and a geographic database? What potential problem exists in attempting to produce a GIS database from both of these sources?

References

Claire, C.N., 1968. "State Plane Coordinates by Automatic Data Processing." Coast and Geodetic Survey, Publication 62–4, U.S. Department of Commerce. Washington, DC: U.S. Government Printing Office.

DeMers, M.N., 1991. "Classification and Purpose in Automated Vegetation Maps." *Geographical Review*, 81(3):267–280.

Mitchell, H.C., and L.G. Simmons, 1974. *The State Coordinate Systems*. U.S. Coast and Geodetic Survey, Special Publication No. 235. Washington, DC: U.S. Government Printing Office (originally published in 1945).

Muehrcke, P.C., and J.O. Muehrcke, 1992. *Map Use: Reading, Analysis and Interpretation*. Madison, WI: J.P. Publications.

Nyerges, T.L., and P. Jankowski, 1989. "A Knowledge Base for Map Projection Selection." *American Cartographer*, 16:29–38.

Pattison, W.D., 1957. "Beginnings of the American Rectangular Land Survey System, 1784–1800." Chicago: University of Chicago, Department of Geography, Research Paper 50.

Robinson, A.H., J.L. Morrison, P.C. Muehrcke, A.J. Kimerling, and S.C. Guptill, 1995. *Elements of Cartography*, 6th ed. New York: John Wiley & Sons.

Snyder, J.P., 1988. *Map Projections—A Working Manual*. U.S. Geological Survey, Professional Paper 1935. Washington, DC: U.S. Government Printing Office.

Töbler, W., 1959. "Automation and Cartography." *Geographical Review*, 49:526–534.

Töbler, W., 1973. "Choropleth Maps Without Class Intervals." *Geographical Analysis*, 5:262–265.

Vrana, R., 1989. "Historical Data as an Explicit Component of Land Information Systems." *International Journal of Geographical Information Systems*, 3(1):33–49.

CHAPTER 4

GIS Computer Structure Basics

A basic understanding of the computer data structures and models that compose a GIS is essential to enable us to select the most appropriate software, perform the correct analyses, and conduct our everyday GIS work efficiently. Each system has its own unique computer data structures, methods of representation, and ways of analyzing spatial data. Fortunately, these can be grouped into a relatively small number of basic types. Because you will likely move from system to system throughout your career, and because data models change as new computer methodologies change, you should become familiar with all the basic types. You will have plenty of opportunity to focus on the system you use most often as you work with it.

Today's sophisticated commercial GIS software is so complex that it extends well beyond a package that can be used "out of the box." Instead, these systems require selecting subsets of the tools that are available for a particular job and merging them with other internal or external programs to address particular problems. In short, the GIS is no longer a tool, but a toolbox from which the appropriate tools for the task must be carefully chosen. When the tools themselves are inadequate we may need to add to the toolbox or change the way the tools work. This requires us to return to the roots of GIS—computer science. GIS is, after all, linked to the computer software we employ and the limitations of the computer code we develop for our work.

The computer representation of spatial data is a type of formalism not unlike the ones we have already examined when moving from actual earth features to a limited set of definable objects called points, lines, areas, and surfaces. Nor is it unlike the cartographic abstractions of these objects and their measurement levels into mappable objects. The difference is merely in how we represent the data inside the computer in such a way that they can be edited, measured, analyzed, and output in some useful form. In this chapter we will first examine some basic **computer file structures**, and then we will discuss **database management systems** that enable large amounts of geospatial data to be organized, searched, and analyzed.

LEARNING OBJECTIVES

When you are finished with this chapter you should be able to:

1. Define the following computer-related terms and be able to recognize them when you observe them: *records, fields, parent-child relationship, pointers, stack, array, simple list, direct files, inverted files, indexed files*, and *ordered sequential files*.
2. Illustrate, with examples, the different types of file structures and indicate the advantages and disadvantages of each for computer search operations. State how this will impact the GIS software you are employing by giving an example of a long process you might use in your own software.
3. Explain the operation of the stack in the movement of pieces of computer information, being sure to include the Last In/First Out (LIFO) concept.
4. Describe how the terms *records, fields*, and *arrays* are related to one another and how they differ.
5. Identify and explain the differences among network, hierarchical, and relational database structures and enumerate the advantages and disadvantages of each for manipulating geospatial data.
6. Create a diagram that illustrates how a hierarchical database management system works.
7. Create a diagram that illustrates how a network database management system works.
8. Create a diagram that illustrates how a relational database management system works.
9. Define and illustrate (create a diagram) the following terms related to relational database management systems: *primary key, tuple, relation, foreign key, relational join*, and *normal forms*.
10. Using data within your own GIS software, describe the method of database management used.
11. Differentiate among data models, data structures, and file structures. Explain how these terms are related.
12. Describe the methods of storage of each of the basic data structures. Explain and contrast the relative efficiency of each.
13. Describe the retrieval methods required for each of the basic data structures. Explain and contrast the relative efficiency of each.

A QUICK REVIEW OF THE MAP AS AN ABSTRACTION OF SPACE

Before we move into another level of abstraction, one that allows the computer to operate on spatial data, it is best to review how we have moved from the real earth to the abstract. We begin the process of abstraction by conceptualizing what we encounter as a group of points, lines, areas, and surfaces. This process, as you remember, is filtered by the questions we want to ask and how we intend to answer them. We make decisions about which objects to take note of and which to ignore. Then we decide on a method of data collection, whether it is a complete census or a sampling procedure. Some objects are assigned names; others are measured at higher levels of measurement (i.e., ordinal, interval, and ratio).

Once we have obtained our data, we proceed to make decisions about representing them in graphic form. We collate and group our data, decide on which projection we will need (if any), what grid system would be best, and so on. In some cases, especially in the absence of GIS, we produce a map directly from the data, only later entering these data into a cartographic database for subsequent analysis. More often today we do enter the data into the GIS directly, creating a geographic database based on direct observations.

The processes just summarized involve examining the environment, deciding what a necessary conception of that reality should be, and abstracting it further, either as a map or in computer-compatible form for direct GIS entry (Figure 4.1). However, the computer forces us to modify how we envision our data. Computers do not operate directly on visual or graphical objects as we would draw on a piece of drafting paper; instead, the computer must be addressed by means of a programming language.

When we develop our concept of space and spatial relationships, we are able to organize our data with a view to making sense of it. The map has long been the primary graphic language we have used to visualize space and its objects. But our graphic language has a distinctly different structure from what is available inside the computer. We must create an explicit, rule-driven language that allows the computer to use its digital (0 and 1) view of the world to identify the

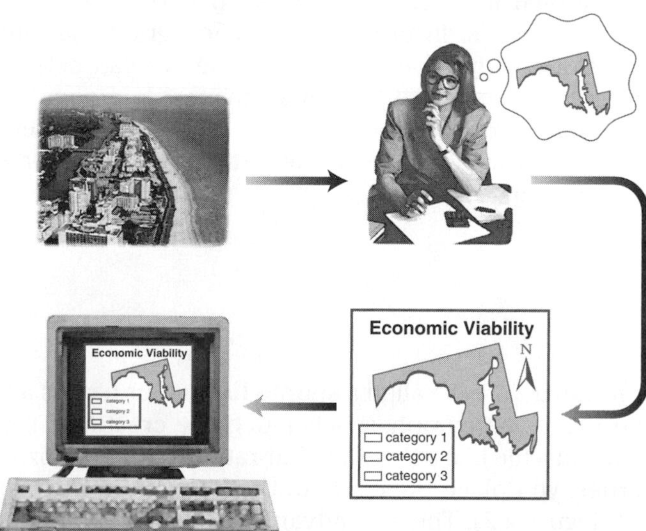

Figure 4.1 Transforming geographic space into a geographic information system (GIS) database. Flow diagram showing change in abstraction level from real world to cartographer's conception to cartographic abstraction, and finally to computer abstraction.

spatial extent of each object, to locate it in a system of coordinates, to separate adjacent objects from one another, and to be able to identify and sort objects by orientation, size, location, and so on. Such matters that require analog thought for their accomplishment are simple to us but difficult for the computer.

Fortunately, we as students of GIS are not going to have to begin from scratch to explain all the details of representing and operating on geographic objects in a spatial context. But to be good at GIS, we should be aware of some of the basic ways others have found to allow the computer to do so. We will begin at an elementary level of computer structures—but not at the machine level—and examine traditional computer file structures that allow for the storing, ordering, and searching of pieces of data. Then we will move to a higher level of organization in the computer called database structures—combinations of file structures that allow more complex methods of managing large volumes of related data. Next we will examine the way in which geographic space can be represented explicitly in a graphic data structure. Finally, we will extend this to include multiple graphic data layers and their databases into what we call GIS.

SOME BASIC COMPUTER FILE STRUCTURES

If you have had a basic computer course or a computer language course, among the first things you learned was simple computer file structures. Files are nothing more than simple accounting systems that allow the machine to keep track of the records of data you give it and retrieve them in any order you wish. For GIS this is no different. Much of what we do in GIS consists of storing entity and attribute data in a way that permits us to retrieve (for display, for example) any combination of these objects. This requires the computer, using a representational file structure, to be able to store, locate, retrieve, and cross-reference records. In other words, each graphical entity must be stored explicitly, along with its attributes, so that we can select the correct combinations of entities and attributes in a reasonable amount of time. This is identical to having a cross-referenced list of names, addresses, and phone numbers. The list items will, of course, need to be sorted—for example, by alphabetically organizing the names.

Simple Lists

The most basic file structure is called a **simple list**. In our names and addresses example in the previous paragraph, this is much like creating a separate index card for each name in a file like a Rolodex. But rather than organizing the names in any formal order, you place the cards in the Rolodex in the order in which they are entered (Figure 4.2). The only advantage to such a file structure is that to add a new record, you simply place it behind all the rest. Clearly, all the cards are there, and an individual name can be located by examining the cards, but the lack of structure makes searching inefficient. Suppose your database contains 200,000 records. If your basic file structure is a simple, unstructured, unordered list structure, you may have to search 200,000 cards to find what you are looking

Figure 4.2 Simple list. File structure illustrated as an unordered Rolodex.

for. If it takes, for example, 1 second to perform each search, searching will require you to perform as many as (n + 1)/2 search operations (Burrough, 1983). This translates into a maximum of (200,000 + 1)/2 seconds, or nearly 28 hours to search for one point. Obviously we need to get our cards organized.

Ordered Sequential Files

As you know, most Rolodexes, like telephone directories and PDA contact lists, are ordered according to the sequence of alphabetic characters (Figure 4.3). This method of ordering allows each record to be compared to each record that follows or precedes it to determine whether its alphabetic character is higher or lower in sequence. **Ordered sequential files**, as these are called, can use alphabetic characters, as in our Rolodex example, or numbers, which also occur in recognizable sequences against which individuals can be compared. The normal search strategy here is a sort of divide-and-conquer approach. A search is begun by dividing the file in half and looking first at the item in the middle. If it exactly matches the target combination of numbers or letters, the search is done. If not, the item of interest is compared to each of its neighbors to determine whether the alphanumeric combination is lower or higher. If it is lower, the half containing higher numbers or letters is searched and the other half is discarded. If it is higher, the half containing lower numbers or letters is searched and the other half is ignored. The half that is searched uses this same divide-and-conquer method until the individual card is located. In this way, the program avoids having to search large portions of the file. The number of search operations using this strategy is defined as $\log_2 (n + 1)$ operations. In the case of the simple list of names and addresses, we reduce the maximum amount of time to just over 2 hours rather than the nearly 28 hours that would have been needed to search the entire list if each operation took a second to perform. Of course, any computer that takes a full second to perform this operation is not

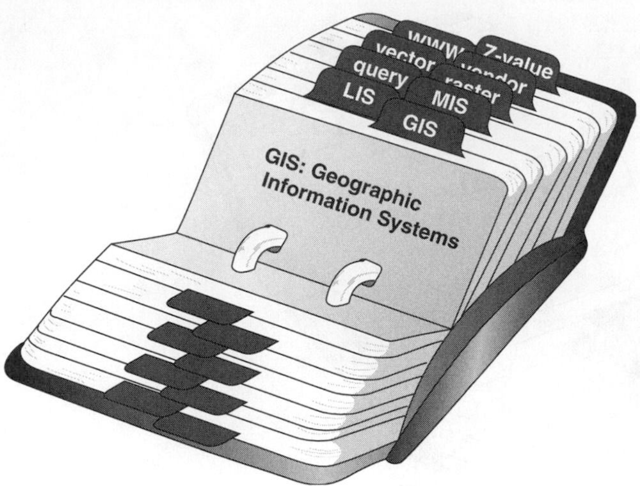

Figure 4.3 Ordered sequential lists. File structure illustrated as an ordered Rolodex. In this case the ordering is produced using alphabetical listings of names.

likely to be very useful for GIS. But the proportional savings in time is made quite obvious by this example. The search speed does come at a cost, however, because now every new record must be entered in the correct sequence or your file will quickly resemble the simple list structure, defeating the binary or divide-and-conquer search strategy.

Indexed Files

In both of the preceding examples, the records were retrieved by examining and comparing a key attribute, say a number sequence or alphabetic sequence. The search strategy was based on the key attributes themselves. In GIS, as in many other situations, the items you want to search are primarily points, lines, and areas. However, you will seldom search for a particular point, line, or area based on its code numbers. Instead, each entity will often have assigned to it a number of descriptive attributes, just as we have seen before. Typically, a search will consist of finding the entities that match a selected set of attribute criteria. Thus, you might ask the GIS to find all study plots in excellent condition for subsequent display or analysis. Or you might want to examine or analyze all study areas in poor condition that have slopes less than 25 percent. Because of the potentially large numbers of attributes linked to each entity, a more efficient method of search will be necessary if you are to find specific entities with associated, cross-referenced attributes. Your search method otherwise will rapidly deteriorate into an exhaustive search of all attributes associated with all entities—the same tedious process employed with the simple list file structure (Burrough, 1983). In short, you need an index to your directory (Figure 4.4), much like the index located at the end of your textbook. Imagine how difficult it would be to find every occurrence of the word "orthophoto" in your text without using an index.

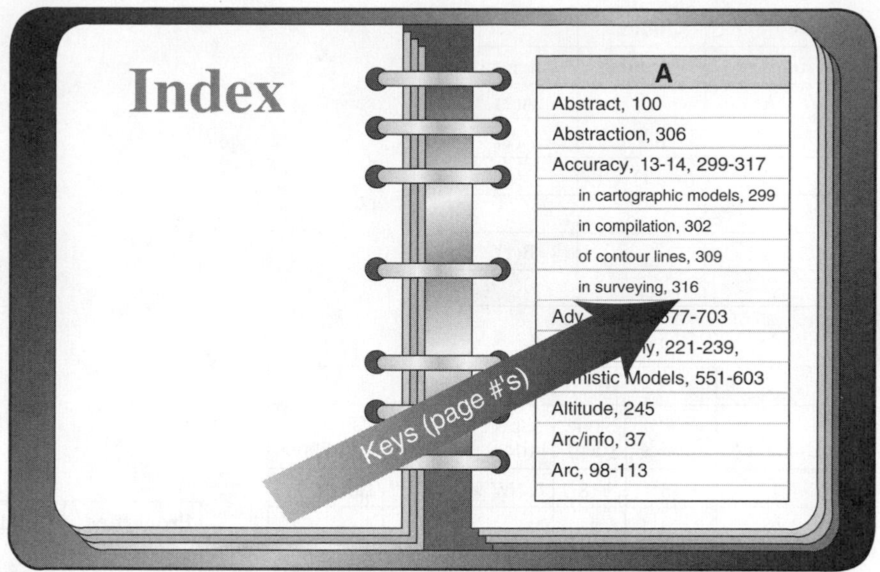

Figure 4.4 Indexed data structure. Files illustrated as an index to a book. The index shows how to find information in the larger file by selecting key features that can be searched for.

Indexed files begin with a set of records and item keys (Figure 4.5a) and can be indexed either as **direct files** (Figure 4.5b) or **inverted files** (Figure 4.5c). In direct indexed files the records (properties) themselves are used to provide access to other pertinent information. For example, if you search for "excellent conditions" (properties) of a set of quadrats, the computer will compare each quadrat with its recorded condition and isolate each record that contains the excellent condition. The program then compiles a list or index of all quadrats that contain excellent conditions. In this way it is now able to relate this to the other properties such as number of species, slope, aspect, or % bare. By creating this index the program is able to ignore all other records not satisfying these search criteria.

Further improvements in search speed can be obtained if a formal permanent index is created for a selected attribute to be searched. From our previous example we could create an index of condition properties and have it explicitly associated with specific entities (e.g., quadrats) and their computer locations (Figure 4.5c). Because the index is based on anticipated search criteria rather than the entities themselves, the information is literally inverted—the attributes are the primary search criteria and the entities rely on them for selection. For this reason we call this an inverted index file structure.

To create an inverted file structure requires an initial sequential search of the data to order the entities on the basis of the attributes of interest. This search, in turn, has three requirements. First, it requires you to know beforehand the criteria you are likely to be searching. Second, any additions of data will require you to recalculate the index as well as the original data. Finally, if you fail to indicate a particular criterion needed for your search—for example, if you forget to include accident rates as one of your search criteria—you will be required to use sequential search methods to obtain the information you want.

	Index	
Item key	Record #	Items
A	1	A(1)
B	$n+1$	A(2)
C	$(n+1)+1$	.
.	.	.
.	.	B(1)
.	.	.

(*a*)

	Properties				
Quadrat #	# species	Slope (deg)	Aspect (deg)	% Bare	Condition
1	15	8	N-NW	14	Good
2	3	27	N	35	Poor
3	21	5	NE	<5	Excellent
4	11	10	S	20	Fair
5	6	18	SW	15	Poor
6	18	7	NW	10	Good

(*b*)

Properties	Quadrats
Poor	2, 5
Fair	4
Good	1, 6
Excellent	3

(*c*)

Figure 4.5 Direct and inverted files. Comparison of (*a*) direct indexed files and (*b*) inverted index files. Note the improvement made by ordering the file to select the index items in the form of an inverted file for the single category of property "condition" (*c*).

DATABASE MANAGEMENT STRUCTURES

We have seen three basic types of file structure for the storage, retrieval, and organization of data. However, we seldom store a single file; rather, we typically compile and work with multiple files. A collection of multiple files is called a **database**. The complexity of working with multiple files in a database requires a more elaborate structure for management, called a **database structure** or a **database management system**. Although new forms of database structure are being created all the time, there are three basic types with which you should be familiar in the early stage of your GIS education. These are **hierarchical data structures**, **network systems**, and **relational database management systems**. Each will be examined separately below.

Hierarchical Data Structures

In many cases there is a relationship among data called a one-to-many or parent–child relationship (Burrough, 1983). The parent-to-child relationship implies that each data element has a direct relationship to a number of symbolic children, and, of course, each "child" is also capable of having an association

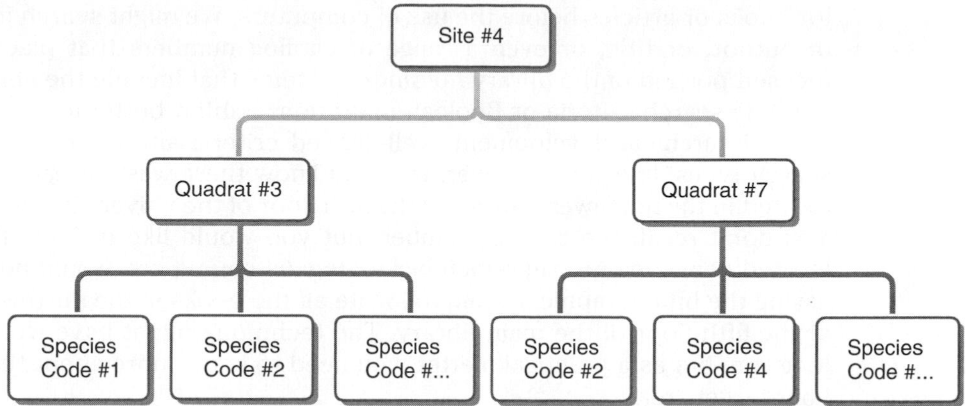

Figure 4.6 Hierarchical database structure. This illustration shows parent-to-child branching based on key attributes. *Source:* Modified from R.G. Healey, "Database Management Systems," Chapter 18 in *Geographical Information Systems, Principles and Applications*, D.J. Maguire, M.F. Goodchild, and D.W. Rhind, Eds., Longman Scientific and Technical, Essex, England, © 1991, Figure 18.3, page 256. Used with permission.

with his or her "offspring," and so on. As the name implies, the parents and children are directly linked, making access to data both simple and straightforward (Figure 4.6). This type of system is perhaps best exemplified by the hierarchical system of classifying plants and animals found in taxonomic literature. Animals, for example, are separated into vertebrate and invertebrate forms—those with a backbone and those without. In turn, the vertebrates have a group called mammals, creatures that nurse their young. Mammals, of course, can be further broken down into other groups. The structure then begins to resemble a family tree, and in fact taxonomists use a nearly identical graphical structure to illustrate relationships among organisms. A major feature of hierarchical systems exemplified by the taxonomic hierarchy is a direct correlation between one branch and another. The branches are based on formal and quite specific criteria or key descriptors that act as decision rules for moving from one branch to another through the structure.

In a hierarchical database structure, if your criterion or key descriptor information is incomplete, your ability to navigate through the net is severely hampered. In fact, the nature of the hierarchical system requires that each relationship be explicitly defined before the structure and its decision rules are developed. A major advantage of such a system is that it is easy to search because the structure is well defined, and it is relatively easy to expand by adding new branches and formulating new decision rules. If your initial description of the structure is incomplete, however, or if you want to examine the structure on the basis of valid criteria that are not included in the structure, a search becomes impossible. Explicit knowledge of all possible questions that might be asked is absolutely necessary to develop a hierarchical system, because questions are used as the basis for developing the decision rules or keys. If the types of search are identified, it is probably safe to assume that the system has a well-established set of linkages based on long-term use under nonautomated conditions, or that the number of questions is relatively limited.

You may have used a modern computerized bibliographic search system. This type of system was developed to simulate the ways in which people looked for books or articles before the use of computers. We might search for subject, or author, or title, or even a range of catalog numbers that places us in a focused portion of the library holdings. Systems that include the ability to add multiple search criteria or Boolean operations exhibit both the necessities for easy hierarchical development: well-defined criteria and a limited number of search types. Imagine, however, that you know there was a great book on GIS located in the northwest corner of the fifth floor of the university's main library. You don't recall the catalog number, but you would like to find other books located near the one you remember seeing. Of course you would not consider asking the bibliographic system to locate all the books in the northwest corner of the fifth floor of the main library. The technique might have worked in the long run, but as a practical matter, you need to know more about the books in your target area.

Situations like the extremely vague search just described occur more often than not when you are working with information inside a GIS. Among the most difficult things to do is to anticipate all the possible searches you might perform. After all, a GIS database normally contains widely varying types of data stored as points, lines, or areas, and there are many different thematic maps in the database. One of the most enjoyable features of a GIS is that you can try out searches or test out relationships you had not envisioned before you began. Unfortunately, the hierarchical structure is not very good at this because of its rigid key structure.

Burrough (1983) cited a classic example of how a system based on a hierarchical structure fails to comply with a user's request because the information needed is not even included in the system (for additional details, see Beyer, 1984). Not unlike our fifth-floor library search, the director of the Royal Botanical Gardens in Kew wanted to query the institution's botanical database system for all the available plants native to Mexico so that he could examine them before an upcoming field visit. This particular search criterion had not been anticipated, however, and the geographic locations of the plants in the system were not recorded; thus, the program was incapable of selecting plants based on location. Clearly the rigid structure of the hierarchical system also would make it difficult to restructure to allow further searches of this nature. Beyond this severe limitation, the hierarchical structure creates large, often cumbersome, index files, frequently requiring repeated entries of important keys to perform its searches. This adds substantially to the amount of memory needed to store the data and sometimes contributes to slow access times as well.

Network Systems

Searches performed in hierarchical data structures are restricted to the branching network itself. In graphic databases, users frequently need to jump to different portions of the database to acquire entity information based on queries of attributes. Since attribute and entity data may very well be stored in different locations, the creation of a hierarchical structure requires that direct links be made between search criteria and the graphic devices used to illustrate the locations in space. Even though this can be done, the potential numbers of hierarchical branchings and associated keys can get unwieldy. Such awkwardness is experienced principally because the hierarchical data structure is most useful when there is a one-to-one or a many-to-one relationship among the variables.

Many GIS databases have, in addition to one-to-one and many-to-one relationships, many-to-many relationships, in which a single entity may have many attributes, and each attribute is linked explicitly to many entities. (An area representing a research quadrat for a study site, for example, will have numerous point locations, with multiple plant or animal species associated with each. In addition, each species might be found in more than one quadrat.) To accommodate these relationships, each piece of data can have associated with it an explicit computer structure called a pointer that directs it to all of the other pieces of data to which it relates (Figure 4.7). Each individual piece of data can be linked directly anywhere in the database, without the existence of a parent–child relationship. Pointers are common features in computer languages such as C and C++, and a basic knowledge of these languages will aid in your understanding of exactly how the devices are used. For our purposes, a graphic visualization should suffice. Figure 4.7 illustrates 2 quadrats (#3 and #7) for study site #4. Notice how the pointers are used to relate individual point locations to their representative species identified for each. Pointers also look back from the species to the locations and finally back to the quadrats in which they are located.

Network systems are less rigid than hierarchical structures for GIS work and, unlike hierarchical structures, can handle many-to-many relationships. As such, they allow much more flexible search strategies than do hierarchical structures. They also have less data redundancy (e.g., coordinate pairs), thus saving computer space. Their major drawback is that in very complex GIS databases, the number of pointers can get quite large, often comprising a substantial portion of storage space. In addition, although linkages between data elements are more flexible, they must still be explicitly defined with the use of pointers. The numerous possible linkages may become an extremely

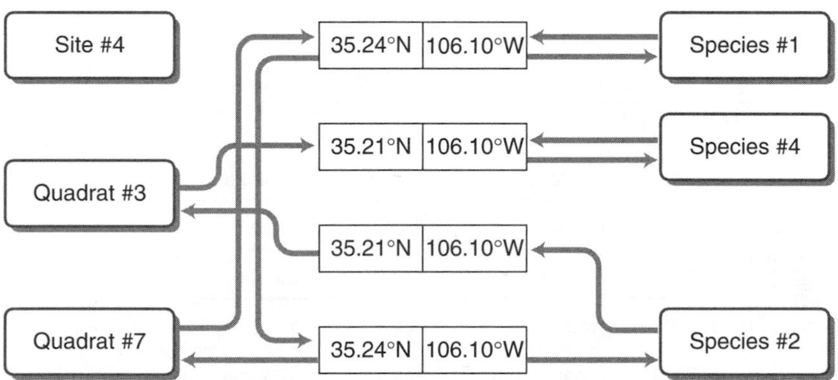

Figure 4.7 Network database structure. This structure allows users to move from data item to data item through a series of pointers. The pointers indicate relationships among data items. *Source:* Modified from R.G. Healey, "Database Management Systems," Chapter 18 in *Geographical Information Systems, Principles and Applications*, D.J. Maguire, M.F. Goodchild, and D.W. Rhind, Eds., Longman Scientific and Technical, Essex, England, © 1991, Figure 18.4, page 257. Used with permission.

tangled web, often resulting in confusion and missed and incorrect linkages. Novice database and GIS users often become overwhelmed by these conditions, although experienced users can become quite efficient with such systems and often prefer them over other types.

Relational Database Management Systems

The disadvantages of large numbers of pointers can be avoided by using another database structure. In relational database structures, the data are stored as ordered records or rows of attribute values called **tuples** (pronounced "tooples," to rhyme with quadruples) (Figure 4.8). In turn, tuples are grouped with corresponding data rows in a form collectively called **relations** because they retain their respective row positions in each column and are related to one another (Healey, 1991) (Figure 4.9). Each column then represents the data for a single attribute for the entire dataset. For example, you could have a column of quadrat numbers (a single attribute) organized numerically. In a separate column you would have additional information pertaining to the collector, yet another column showing collection date, and finally a column showing the site number that corresponds to each of the other columns. In this way, each item in each column corresponds and can then be related to additional tables as well.

Relational systems are based on a set of mathematical principles called relational algebra (Ullman, 1982) that provides a specific set of rules for the design and function of these systems. Because relational algebra relies on set theory, each table of relations operates as a set, and the first rule is that a table can't have any row (tuple) that completely duplicates any other row of data. Because each of these rows must be unique, a single column or even multiple columns can be used to define the search strategy. Thus, as an example of using a single column to decide your search strategy, you might search for a social security number, telephone number, home address, and so on available in other

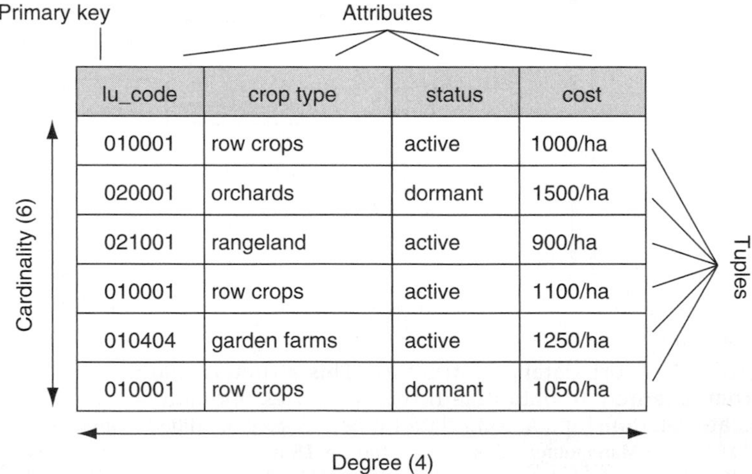

lu_code	crop type	status	cost
010001	row crops	active	1000/ha
020001	orchards	dormant	1500/ha
021001	rangeland	active	900/ha
010001	row crops	active	1100/ha
010404	garden farms	active	1250/ha
010001	row crops	dormant	1050/ha

Figure 4.8 Elements of relational tables. All relational database systems share these elements.

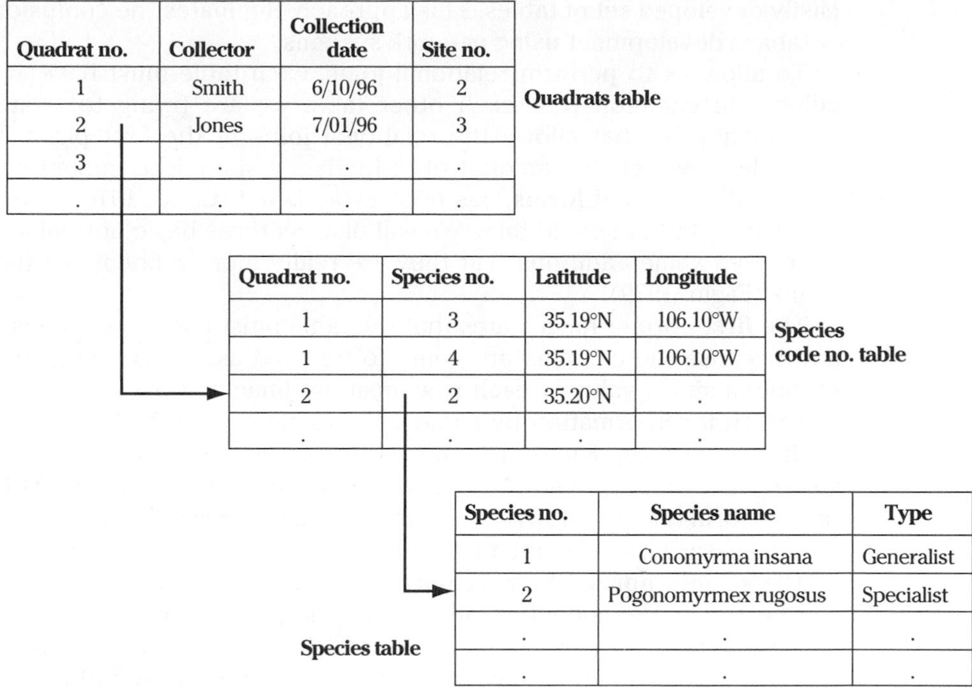

Quadrat no.
Collector
Collection date
Site no.
1
Smith
6/10/96
2
2
Jones
7/01/96
3
3
Quadrats table
Quadrat no.
Species no.
Latitude
Longitude
1
3
35.19°N
106.10°W
1
4
35.19°N
106.10°W
2
2
35.20°N
Species code no. table
Species no.
Species name
Type
1
Conomyrma insana
Generalist
2
Pogonomyrmex rugosus
Specialist
Species table

Figure 4.9 Relational database structure. Note the tuples, relations, and primary keys. *Source:* Modified from R.G. Healey, "Database Management Systems," Chapter 18 in *Geographical Information Systems, Principles and Applications*, D.J. Maguire, M.F. Goodchild, and D.W. Rhind, Eds., Longman Scientific and Technical, Essex, England, © 1991, Figure 18.5, page 258. Used with permission.

columns of the same table by selecting a particular name from the first column. This search criterion is called a **primary key** for searching the other columns in the database (Date, 1986). No primary key row can have missing values because missing row values could result in permitting duplicate rows to be stored, thus violating our first rule.

Relational systems are useful because they allow us to collect data in reasonably simple tables, keeping our organizational tasks equally simple. When we need to, we can match data from one table to corresponding (same row) data in another table by the use of a linking mechanism called a **relational join**. Because of the predominance of the use of relational systems in GIS, and because of the rather large databases produced for GIS, this process is a common one and you should pay close attention to it. Any number of tables can be "related." The process involves matching the primary key (column) in one table to another column in a second table. The column in the second table to which the primary key is linked is called a **foreign key**. Again, the related row values are assumed to be in the same positions, to ensure that they correspond. This link means that all the columns in the second table are now related to the columns in the first table. In this way, each table can be kept simple, making accounting easier. You can link a third table by finding a key column in the second table that acts as the primary key to a corresponding key column (now called the foreign key) in the third table. The process can continue, connecting each of the simple tables to allow for quite complex searches while maintaining a simple, well-defined, and easily developed set of tables. This approach eliminates the confusion found in database development using network systems.

To allow us to perform relational joins, each table must have at least one column in common with each other table we are trying to relate to. This redundancy is what allows the relational joins in the first place. Whenever possible, however, the amount of redundancy should be minimized. A set of rules, called **normal forms**, has been established (Codd, 1970) to indicate the forms your tables should take. We will discuss three basic normal forms; there have been some additions, but they are really more refinements than normal forms (Fagin, 1979).

The **first normal form** states that the table must contain columns and rows, and because the columns are going to be used as search keys, there should be only a single value in each row location. Imagine how difficult it would be to search for information by name if the name column had multiple values in each row location. Figure 4.10 shows two examples of databases: one which violates the first normal form (Figure 4.10a) and one that conforms to the first normal form (Figure 4.10b). Consider how difficult a search would be if your GIS database was more like *a* than *b*.

The **second normal form** requires that every column that is not a primary key be totally dependent on the primary key. This simplifies the tables and reduces redundancy by imposing the restriction that each column of data be findable only through its own primary key. If you wish to find a given column using other relationships, you can use the relational join rather than placing the column over and over again in separate tables to make sure it can be

In a database, this means that each column should only be designed to hold one and only one piece of information. Consider the following examples:

Bad

Name	Address
Ben Goren	Post Office Box 964, Tempe, Arizona 85280-0964
Jane Doe	1234 Main Street, Mesa, Arizona 85345

Good

First_Name	Last_Name	Street_Address	City	State	ZIP
Ben	Goren	Post Office Box 964	Tempe	Arizona	85280-0964
Jane	Doe	1234 Main Street	Mesa	Arizona	85345

Figure 4.10 The principle of first normal form is violated in (*a*), where the data are not atomic in each cell but conform to the first normal form in (*b*), where each cell has a unique value.

Bad

ID	First_Name	Last_Name	Spouses_Birthday
3	John	Smith	02/07/65
4	Sue	Jones	07/26/79

(*a*)

The problem is that the last column relates not to John Smith, but to another person entirely. John has only one spouse, who only has one birthday . . . but that's too tenuous a relationship to embed into a single table. Rather, you would want to structure your tables as follows:

Good

Table: People			
ID*	**First_Name**	**Last_Name**	**Birthday**
3	John	Smith	08/01/62
4	Sue	Jones	10/12/75
5	Mary	Smith	02/07/65

Table: Spouses	
Husband*- >People:ID	**Wife*-** >People:ID
3	5

(*b*)

*Primary key

Figure 4.11 The principle of second normal form is violated in (*a*), where the data have no relationship to other columnar data but conform to the second normal form in (*b*), where each column is directly related to data in the other columns.

found. Figure 4.11a illustrates a clear violation of the second normal form and Figure 4.11b conforms to the normal form. As before, consider the difficulties involved in searches where the second normal form is violated.

The **third normal form**, which is related to the second normal form, states that columns that are not primary keys must "depend" on the primary key, whereas the primary key does not depend on any nonprimary key. In other words, you must use the primary key to find the other columns, but you don't need the other columns to search for values in the primary key column. Again, the idea is to reduce redundancy—to ensure that the smallest number of columns is produced.

The rules of the normal forms were summed up by Kent (1983), who indicated that each value of a table represents something important "about the primary key, the whole primary key, and nothing but the primary key." For the most part, these rules are highly useful and should be rigorously enforced. There are always situations, however, when rigid enforcement will be impossible or will hamper system performance (Healey, 1991).

SOME BASIC COMPUTER TERMINOLOGY

There are some basic terms you need to understand because of the frequency of their occurrence in everyday GIS usage. One common temporary abstract

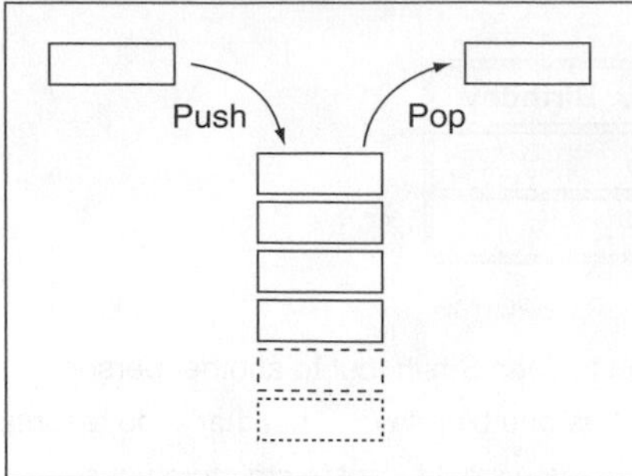

Figure 4.12 The stack. This is a data structure designed for temporary storage of data while processing is going on. It is based on the idea that the last item placed in the stack will be the first item "pushed" off the stack.

data type that will be shown in Chapter 6 to have an impact on the data input processes as well as some data transformations is the **stack**. The stack stores and moves computer data in "stack-based" computers (the most common type) as opposed to CPU registers in "register-based" computers. Employing a process known as the "Last In/First Out (LIFO)," or Push-Pop (Figure 4.12) approach, the stack provides space for the movement of data through the computer. Each new piece of data is pushed onto the stack and displaces (pops) an existing piece of information. The order in which this occurs may have unexpected results on the order and placement of spatial data that would otherwise be predictable. We will discuss this briefly again in Chapter 6.

When the pieces of data form complex groups, their parts can be separated into components called **fields**. The classic example of this is the common notation mm/dd/yyyy, which is code for recording a calendar data by month (2 numbers), followed by day (2 numbers), followed by year (4 numbers). In this example *mm* is one of three fields, together with the fields *dd* and *yyyy*. Most computer languages have a record data type that represents such complex data types as a series of fields. For example, a series (an **array**) of Boolean (0 and 1) data might be represented in the computer as a bit field. This is often used in GIS representations of directional and/or surface data. When data are collated in a relational database management system, they are arranged as sets of database **records** (rows). Each record consists of several fields, and the fields of all records form the columns. We will use these terms consistently throughout the remainder of the text.

Terms

array
computer file structures
database
database management system
database structures
direct files
field
first normal form
foreign key
hierarchical data structures
indexed files
inverted files
network systems
normal forms
ordered sequential files
primary key
record
relational database structures
relational join
relations
second normal form
simple list
stack
third normal form
tuples

Review Questions

1. Since we're mostly going to use GIS software to answer questions, why do we need to know about basic computer file structures, database structures, and graphic data structures?
2. What is the difference between simple list file structures and ordered sequential file structures? Which is more efficient at adding records? Which is more useful for sorting and retrieving records? Give an example of how an ordered sequential file structure works.
3. What are indexed files? How do they differ from ordered sequential files? What are the advantages of indexed files? What is the difference between direct and inverted indexed file structures? Which is more efficient at handling data retrieval?
4. What is a hierarchical database structure? How does it work? Give an example. What are its limitations, especially for GIS?
5. What are network database structures? How do they keep track of records without a hierarchical structure? What are their advantages or disadvantages over hierarchical systems?
6. What is a relational database management system? How does it work? What advantages or disadvantages might it have over database management systems of other types?
7. Provide a simple diagram illustrating the LIFO movement of data on and off of a computer stack.
8. Within the context of a database management system, create a sentence that employs the following terms used correctly: *records, fields,* and *arrays*.
9. What is a primary key? A tuple? A relation? A foreign key? A relational join?
10. What are normal forms? List the first three normal forms and describe the restrictions they place on the database management system.
11. Create a table based on geographic data that violates the first normal form. Create a table based on the same data that conforms to the first normal form. Explain the differences in terms of ease of searching.
12. Using data similar to those in Question 11, create a table that violates the second normal form, and another that conforms to the second normal form. Explain the differences in terms of ease of searching.

References

Beyer, R.I., 1984. "A Database for a Botanical Plant Collection." In *Computer-aided Landscape Design: Principles and Practice*, B.M. Evans, Ed., Scotland: Landscape Institute, pp. 134–141.

Burrough, P.A., 1983. *Geographical Information Systems for Natural Resources Assessment*. New York: Oxford University Press.

Codd, E.F., 1970. "A Relational Model of Data for Large Shared Data Banks." *Communications of the Association for Computing Machinery*, 13(6):377–387.

Date, C.J., 1986. *An Introduction to Database Systems*, Vol. II. Reading, MA: Addison-Wesley.

Fagin, R., 1979. "Normal Forms and Relational Database Systems." In *Proceedings of the ACM SIG-MOD International Conference on Management of Data*, pp. 153–160.

Healey, R.G., 1991. "Database Management Systems." In *Geographical Information Systems: Principles and Applications*, D.J. Maguire, M.F. Goodchild, and D.W. Rhind, Eds. Essex: Longman Scientific & Technical.

Kent, W. (1983), "A Simple Guide to Five Normal Forms in Relational Database Theory." *Communications of the Association for Computing Machinery*, 26 (2):120–125.

Ullman, J.D., 1982. *Principles of Database Systems*. Rockville, MD: Computer Science Press.

CHAPTER 5

GIS Data Models

The data structures we examined in Chapter 4 are not specific to geographic information systems and can be applied to a wide array of computer programs. There are, however, more complex structures and data models that are necessary for the computational complexity of the GIS programs and the spatial nature of the data. These data models will have a profound impact on the efficiency, method of operation, output, analytical power, and even the nature of the data used within the GIS software. The choice of GIS software can often depend on these data models and necessitates a basic familiarity with them.

This chapter examines two types of computer data models: those related to the graphical representation of the map data (entities) and those that combine the graphics with the analytical power to manipulate the descriptive (attribute) variables of the maps. The graphic representation in the first instance involves how one conceptualizes the representation of space either as a series of uniform grid cells, or as a point, line, and area representation. In addition, the point, line, and area representation can explicitly code spatial relationships between and among objects, or those relationships can be calculated through subsequent analysis. All of these have advantages and disadvantages for cartographic representation and analysis. We will also examine some graphic data structures that are specifically designed to conserve computer space by a variety of compression techniques. Some of these allow direct analysis and some must be converted to another form first.

The analytical capabilities of the software are also related to how the geographic descriptors themselves are envisioned and subsequently represented in the computer. In some cases they are linked to the graphics through numerical codes assigned to their grids; in others they are linked through tables. The tables themselves can be based on traditional relational database management systems or they can be based on the idea of property inheritance through object-oriented data structures that allow a more exact description of the actual geographic phenomena.

This chapter examines some of the established methods of representing geographic space without an effort to be exhaustive. Additionally, we will explore the more common types of data models related to system construction and analysis. You should be able to identify which computer data models are available in your own GIS software as you become more familiar with it.

LEARNING OBJECTIVES

When you are finished with this chapter you should be able to:

1. Describe the basic methods of representing graphic entities on a map using raster, vector, quadtree, and MrSID, and understand their advantages and disadvantages.
2. Diagram and explain the map analysis package data structure for raster representation, particularly with regard to how relational database management systems can be added to enhance this structure.
3. Explain the limitations of the typical relational database management system for representing geographic data.
4. Explain, with examples, the difference between topological and nontopological data models.
5. Explain the differences between topological data models based purely on graph theory (e.g., Census DIME files) and those that are more geographically accurate (e.g., TIGER files).
6. Explain where and why the spaghetti model is still used even though it is a rather unsophisticated, nontopological data model.
7. List and describe the methods of compact input and storage of raster data (e.g., run-length codes, block codes, quadtrees, and chain codes) and know how they work.
8. Provide an actual sample analysis that takes advantage of the topological data structure and indicate why it is superior to the nontopological counterpart.
9. Diagram and describe how the shapefile model works.
10. Argue for the geodatabase model over the old ArcInfo Coverage model in terms of how it represents geographic objects. Provide a concrete example in doing so.
11. Describe how the vector data model can be compacted to save computer space.
12. Describe how the triangulated irregular network (TIN) model works to store surface data.
13. Understand the basic differences between hybrid and integrated GIS systems, especially with regard to how the data are stored and accessed.
14. Have a basic understanding of how object-oriented GIS systems organize data and analytical techniques.

GRAPHIC REPRESENTATION OF ENTITIES AND ATTRIBUTES

Thus far we have concentrated on data structures that have little to do with the graphic representation of cartographic or geographic objects as we have envisioned and abstracted them. Although all three of the systems mentioned in Chapter 4 could be used to manage graphics, they tell us little about how the graphics themselves are represented in the GIS. We know that the human mind is capable of producing a graphic abstraction of space and objects in space. Although commonplace, this representation is actually quite sophisticated, as you will see when we attempt to make the jump to computer representation from hard copy. A primary difficulty is that our paper graphic devices contain an implied set of relationships about the elements displayed on the paper rather than an explicit set. Lines are connected to other lines and together are linked to create areas or **polygons**. The lines are related to each other in space through angles and distances. Some are connected, whereas others are not. Some polygons have neighbors, whereas others are isolated. The list of possible relationships that can be contained on a graphic diagram is quite large. We need to represent each object and each geographic relationship as an explicit set of rules. This allows the computer to "recognize" that all these points, lines, and areas represent something on the earth, that they exist in an explicit place in space, and that those explicit locations are also related to other objects within space in an absolute as well as a relative sense. We may even want to explain to the computer that a polygon has an immediate neighbor to its left, and that neighbor may share points and lines. In other words, we need to create a language of spatial relationships.

There are two fundamental methods of representing geographic space. The first method **quantizes** or divides space into a series of packets or units, each of which represents a limited but defined amount of the earth's surface. This **raster** method or field-based model can define these units in any reasonable geometric shape, as long as the shapes can be interconnected to create a planar surface representing all the space in a selected study area. Although a wide variety of raster shapes are possible—for example, triangles, parallelograms, or hexagons—it is generally simpler to use a series of rectangles or, more commonly, squares called **grid cells**. Grid cells or other raster forms generally are uniform in size, but this is not absolutely necessary, as you will see when we briefly review a less often used system called **quadtrees**. For the sake of simplicity, we will assume that each grid cell is square, and the same size as all others, and therefore occupies the same amount of geographic space.

Raster data structures do not provide precise locational information because geographic space is now divided into discrete grids, much as we divide a checkerboard into uniform squares. Instead of representing geographic points with their absolute coordinate locations, each is represented as a single grid cell (Figure 5.1). The assumption is that somewhere inside that grid cell we will find the represented point object. This requires a certain amount of suspension of disbelief because it employs a change-of-dimensionality abstraction that illustrates a zero-dimensional object with a two-dimensional data structure. Likewise, lines, or one-dimensional objects, are represented as a series of connected grid cells. Again, we are changing our dimensionality from one-dimensional objects to two-dimensional data structures. Each point along the line is represented by a grid cell, meaning that any point along the line must occur somewhere within one of the displayed grid cells. This form of data structure produces a stepped appearance when it is used to represent very irregular lines (Figure 5.1). This stepped appearance is also obvious when we represent areas with grid cells (Figure 5.1). All points inside the area that is bounded by a close

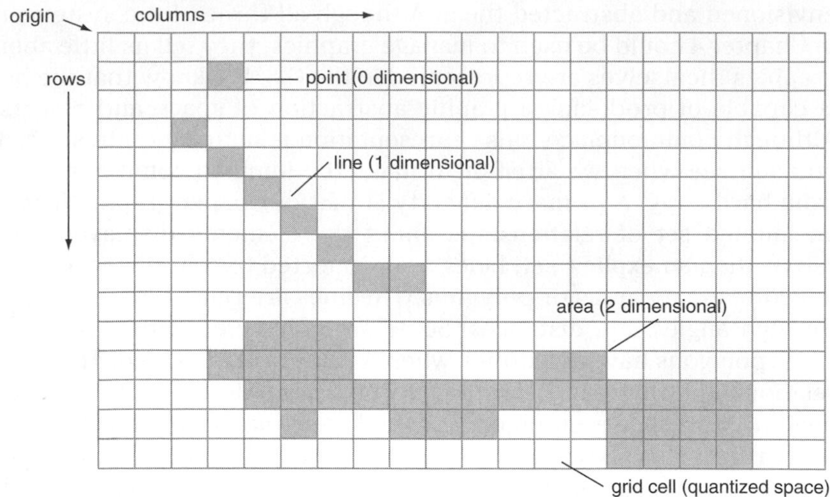

Figure 5.1 Basic raster graphic data representation. The structure illustrates points, lines, and areas as quantized units of geographic area. Grid structures do not allow precise locational information.

set of lines must occur within one of the grid cells to be represented as part of the same area. The more irregular the area, the more stepped the appearance.

In grid-based or raster systems there are two general ways of including attribute data for each entity (object). The simplest is to assign a single number, representing an attribute (e.g., a class of land cover), for each grid cell location. By positioning these numbers, we ultimately are allowing the position of the attribute value to act as the default location for the entity. For example, if we assign a code number "10" to represent water, then list this as the first number in the X or column direction and the first number in the Y or row direction, by default the upper left grid cell is the location of a portion of the earth representing water. In this way, each grid cell can hold only a single attribute value for a given map. An alternative approach, actually an extension of the one just discussed, is to link the grid cells to a database management system, with the result that more than one attribute can be represented by a single grid cell. This approach is becoming more prevalent because it reduces the amount of data that must be stored and because it can easily be linked to other data structures that also rely on **database management systems** to store, search, and manipulate data.

Although absolute location is not explicitly part of the raster data structure, it is implied by the relative locations of the grid cells. Thus, a line is represented by grid cells at particular locations relative to one another, and areas are represented by the grid cells that form as a result of being adjacent to one another or connecting at points of one another. The larger the grid cell, the more land area is contained within it—a concept called **resolution**. The more land area contained—that is, the coarser the grid cell resolution—the less we know about the absolute position of points, lines, and areas represented by this structure.

Grid cells are pieced together to represent areas. To do this we assume that the area is more or less flat. In other words, the normal coordinate system we are using is a Cartesian coordinate system, and the grid cells themselves approximate the locations in space. As you have seen, Cartesian coordinate systems employ a map projection to permit an approximation of the three-dimensional shape of a portion of the earth. More sophisticated grid system representations have a coordinate system based on a map projection embedded in them, and this more accurately represents absolute location. For example, the pixels from satellite imagery have an associated projection, and a more exact grid can be placed on them for reference. In general, however, exact measurements based on any raster structure are problematic. When accurate locational measures are needed, raster structures are less often used than are other types.

Raster data structures may seem undesirable because of their spatial inaccuracy, but what they lack in spatial accuracy they make up for in ease of conceptualization, compatibility with other grid cell-based systems, and some strong analytical capabilities. Envisioning our environment as a series of small grid cells can be likened to the ability of television monitors to portray images as a series of illuminated phosphor dots (grid cells). In remote sensing the relationship between the pixel and grid cell used in GIS allows satellite data to be readily incorporated into raster-based GIS with few changes. This is another advantage of raster data structures over the alternatives. Analytical operations, especially those involving surface representation and overlay operations, are simple to perform with this type of data structure.

Among the disadvantages of raster data structures are the aforementioned reduced spatial accuracy, which decreases the reliability of area and distance measures. Raster data also need large amounts of storage because they must record every grid cell as a numerical value. Although storage is no longer a major limitation, even the fastest computers can be slowed to a crawl if highly complex calculations are performed on very large raster databases.

The second method of indicating geographic space, called the **vector** or polygon-based model, assigns accurate spatial locations explicitly. In the vector model it is assumed that geographic space is continuous rather than quantized as discrete grid cells. With the vector model, data points are represented as a single set of coordinates (*X* and *Y*) in coordinate space, lines as sequences of coordinate pairs, and areas as sequences of lines whose first and last coordinate points are the same (Figure 5.2).

The vector data structure is much more representative of the actual size, shape, and location of geographic features as they appear on maps. To give it the utility of a map, however, we link the entity data with associated attribute data kept in a separate file, perhaps in a database management system. This extends the structure beyond a simple graphic caricature of the objects, making it more map-like and more representative of the geography we model. Remember that in raster we explicitly stored the attributes and implied location (based on position

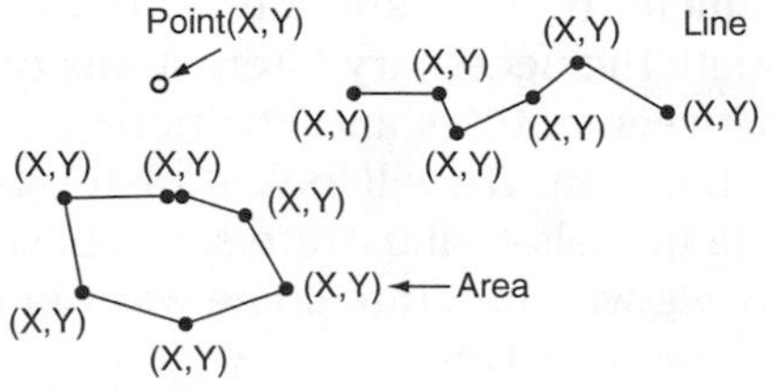

Figure 5.2 Basic vector graphic data representation. The structure shows points as individual coordinate pairs, lines as groups of two coordinate pairs, and areas as connected lines having identical beginning and ending coordinates.

in the grid); in vector we explicitly store the entities without their attributes, relying on computational pointers to link to the separate attribute database.

In vector data structures, a point is represented by a single pair of *X* and *Y* coordinates with its descriptive attributes stored in a separate file. A line consists of two or more coordinate pairs, again storing the attributes in a separate file. For straight lines, two coordinate pairs are enough to show location and orientation in space. More complex lines will require a number of line segments, each beginning and ending with a coordinate pair. For complex lines, the number of line segments must be increased to accommodate the many changes in angles. The shorter the line segments, the more exactly they will represent the complex line. Thus, we see that although vector data structures are more representative of the locations of objects in space, they are not exact. They are still an abstraction of geographic space.

Although some lines act singly and contain specific attribute information that describes their character, other, more complex collections of lines, called **networks**, add a level of attribute complexity. For example, not only does a road network contain information about the type of road as well as other characteristics, it will also indicate, for instance, that travel is permitted only in a particular direction and at a particular speed. This information must be extended to each connecting line segment to advise the user that movement can continue along each segment until the attributes change—perhaps when a one-way street becomes a two-way street. Other codes linking these line segments might include information about the **nodes** that connect them. For example, one node might indicate the existence of a stop sign, a traffic signal, or a sign prohibiting U-turns. All these attributes must be connected throughout the network so that the computer knows the inherent real-world relationships that are being modeled within the network. Such explicit information about connectivity and relative spatial relationships is called **topology**, a topic we will return to when we look at the vector data models we can produce from the basic vector data structure.

Vector areas (polygons) are composed of lines connected into a looping form, where the first coordinate pair on the first line segment is the same as the last coordinate pair on the last line segment. As with point and line entities, the polygon also has associated with it a separate polygon attribute file. Again, this convention improves the simple graphic illustration of area entities, making it possible for them to better represent the abstraction of area patterns we observe on the earth's surface.

GIS SYSTEM DATA MODELS

Although simple raster and vector data structures allow us to depict spatial phenomena in a single map environment, there is still a need to develop more complex data models, to accommodate the necessary interactions of objects in the database, to provide the link between entities and attributes, and to allow multiple maps to be analyzed in combination. We will look at raster data models first and then proceed to vector data models to illustrate some basic methods that allow multiple analytics. Finally we will illustrate some ways of combining these data models into systems—in our case, GIS.

Raster Models

The simplest raster systems are those where each grid cell is represented by a single attribute value. To create a thematic map of grid cells, we collect data about a particular theme arranged as a two-dimensional array of grid cells. This two-dimensional array is called a **grid** and represents a thematic layer (e.g., land use, vegetative cover, soil type, surface geology, hydrology, etc.) In this simple approach, the objects, patterns, and interrelationships of each theme are recognized without unnecessary confusion. Because we will most often be concerned about the relationships of one theme, such as soil type, with one or more additional themes, say vegetation and topography, we stack these maps into a three-dimensional structure where the combinations of themes represent the interactions of the factors we selected based on our geographic filter or study framework. So if we are interested only in physical phenomena, each of the important components of the physical geography will be represented separately, but together they will give us a more complete, three-dimensional view of the study area.

There are several ways for a computer to store and reference individual grid cell values and their attributes, map names, and legends. Most of these systems are older and may be of historical significance, but we will focus on more current approaches. Among the most copied grid cell system data model for GIS is the **map analysis package (MAP)** model, which is named after the system developed by C. Dana Tomlin (Burrough, 1983). In this data model (Figure 5.3), each thematic map grid is recorded and accessed separately by map name or title. This is accomplished by recording each variable, or mapping unit, of the map's theme as a separate number code or label, which can be accessed individually when the map is retrieved. The label corresponds to a portion of the legend and has its own symbol assigned to it. In this way, it is easy to perform operations on individual grid cells or groups of similar grid cells, and the resulting value changes require rewriting only a single number per mapping unit, thus simplifying the computations. The MAP method allows ready manipulation of the attribute values and grid cells in a many-to-one relationship.

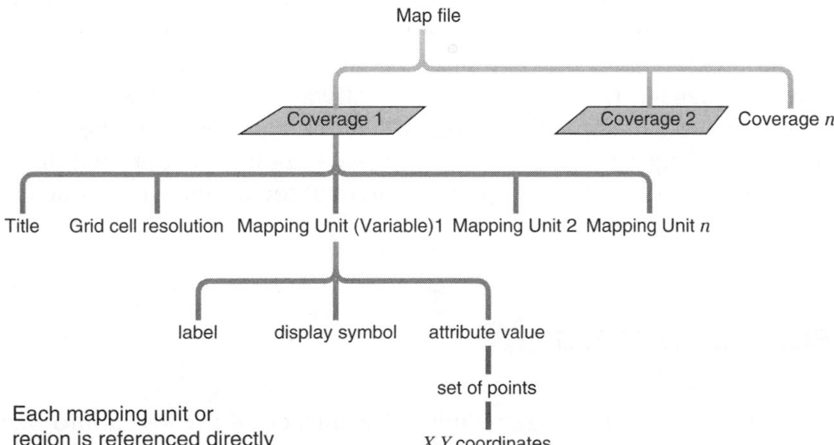

Figure 5.3 The map analysis package (MAP) GIS data model.

3	3	3	2	2	1
3	3	2	4	4	4
4	0	0	0	4	4
0	0	0	4	4	1
0	0	4	4	1	1
3	3	3	4	4	1

Value	Count	Type	Condition	Farmer
0	8	Fallow	Fair	McMillan
1	5	Barley	Fair	Denmeyer
2	3	Wheat	Poor	Bushee
3	8	Corn	Excellent	Denmeyer
4	11	Oats	Good	Denmeyer

Figure 5.4 Extended raster data model. Note how each grid cell value has additional information stored in the database management system tables.

Although raster GIS have traditionally been developed to allow single attributes to be stored individually for each grid cell, most today have evolved to include direct links to database management systems (Figure 5.4). This extends the utility of the raster GIS by minimizing the number of maps by substituting multiple variables for each grid cell in each grid. Such extensions to the raster data model also allow direct linkage to existing GIS programs that use a vector graphic data model. Because such integrated raster/vector systems include modules that convert back and forth from raster to vector, the user is able to operate with all the advantages of both data models. The conversion process is often quite transparent, allowing the user to perform the analyses needed without concern for the original data structure. This feature is particularly important because it strengthens the relationship between the digital image processing and image segmentation software used to manipulate remotely sensed raster data and GIS software. Many software systems already have both sets of capabilities, and still more are likely to in the future. Together with the linkage to statistical software, we are rapidly approaching systems that operate with a superset of spatial analytical techniques, resulting in a maturing of digital geography.

RASTER SURFACE MODELS

Because raster data models divide the surface of the earth into distinct units, each unit or grid cell can contain only a single absolute elevational value. This in effect takes a continuous data variable where values could occur

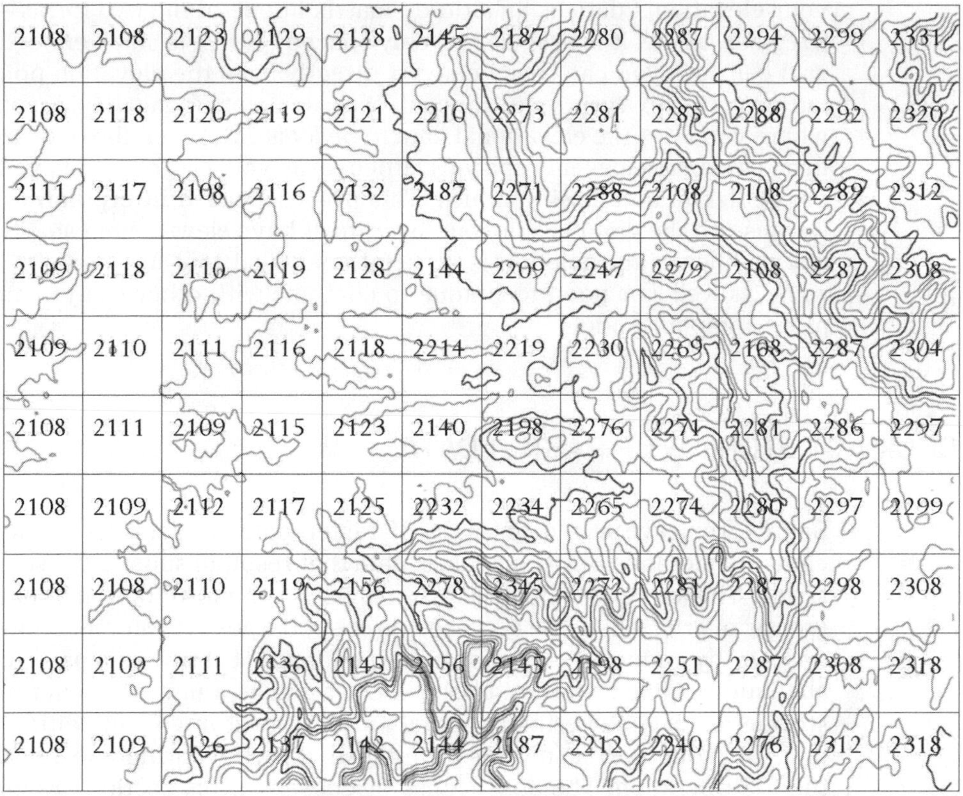

2108	2108	2123	2129	2128	2145	2187	2280	2287	2294	2299	2331
2108	2118	2120	2119	2121	2210	2273	2281	2285	2288	2292	2320
2111	2117	2108	2116	2132	2187	2271	2288	2108	2108	2289	2312
2109	2118	2110	2119	2128	2144	2209	2247	2279	2108	2287	2308
2109	2110	2111	2116	2118	2214	2219	2230	2269	2108	2287	2304
2108	2111	2109	2115	2123	2140	2198	2276	2271	2281	2286	2297
2108	2109	2112	2117	2125	2232	2234	2265	2274	2280	2297	2299
2108	2108	2110	2119	2156	2278	2345	2272	2281	2287	2298	2308
2108	2109	2111	2136	2145	2156	2145	2198	2251	2287	2308	2318
2108	2109	2126	2137	2142	2144	2187	2212	2240	2276	2312	2318

Figure 5.5 Raster representation of surface data.

everywhere, and converts it to a discrete representation where values only exist in selected places (Figure 5.5). Although each grid cell contains a single absolute elevational value, it also occupies space. To more accurately represent elevation in raster, you may have to select a relatively small grid cell size to increase the accuracy with which the point elevational values for each grid cell represent an appropriate elevation throughout the entire space occupied by the grid cell.

Although grid cell size is important in coding continuously varying data into a raster GIS, it is equally important to decide where within the grid cell area you want the actual elevation point to be located. You have options for indicating the exact location of the elevation point: at the center of the cell or in one of the four corners (Figure 5.5). For analyses that require the use of topography, however, the location will have an impact on the results of the calculation. For example, when we measure least-cost paths in Chapter 9, we begin in the center of each grid cell because we assume that the elevational value is stored in that location. If your grid cells are coded so that the elevational value is at one of the four corners, your calculations of distance will be off by at least half a grid cell distance. Thus, in deciding where to place the elevational value within each grid cell, you should first examine how the raster GIS you are using actually calculates such measures as functional distance, slope, and aspect.

In many instances data will be available for only a sample of grid locations, whether you are using raster or vector. In raster, you are most likely to obtain your elevational data as an altitude matrix in one of the two forms—regular or irregular lattice. If the regular (evenly spaced) lattice is small enough to fit your grid cell size, you can easily convert directly from the elevation points at each lattice vertex to a grid value for each of these points (again deciding beforehand where you want the elevational data to be located). When the data are provided in an irregular (unevenly spaced) lattice form, you are going to have to estimate or predict all the missing values. This process, called **interpolation**, is needed because all the grid cells in your layer must have elevational values. But as you will see next, interpolation is a useful analytical tool for modeling in its own right, as well as a good technique to combine with other analysis methods for more complex models.

COMPACT STORING OF RASTER DATA

Before we leave our discussion of raster data models, let's examine some common methods of storing raster data that result in substantial savings in disk space. These require a data model similar to the MAP data model that allows groups of grid cells to be accessed directly, because the compactions generally operate by reducing the information content of these groups of cells to the absolute minimum needed to represent them as a unit. Compact methods for storing raster data certainly operate under the storage and editing subsystem of a GIS, but they can also be applied directly during the input phase of the GIS operation. The approaches discussed in this section are illustrated in Figure 5.6.

The first method of compacting raster data is a process called **run-length codes** and is a common method of grid cell data input (Figure 5.6a). Historically, raster data were input into a GIS by preparing a clear acetate grid, referencing it, and overlaying it on the map to be encoded. Each grid cell had a numerical value corresponding to a category of data on the map that was input (generally typed) into the computer. For a map of perhaps 200 by 200 grid cells, someone would have to type 40,000 numbers into the computer. As one begins typing, you quickly see patterns emerging from the data that present opportunities for reducing your typing. Specifically, there are long strings of the same number in each row. Think how much time you could save if, for a given row, you could just tell the computer that starting at column 8 all the numbers are 1s, representing some map variable, until you get to column 56, then at column 57 the numbers are 2s until the end of the row. Indeed, you could also save a great deal of disk space by giving starting and ending points for each string and the value that should be stored for that string. That's run-length codes.

Of course, this method limits you to operation on a row-by-row basis. If you could tell the computer to begin at a single grid cell with value 1, then go in a particular direction, say, vertically 27 grid cells and then change to a different grid cell value, this would allow you to code strings in any direction. But the principle can be extended even further. Suppose you see large groups of grid cells that represent an area. If you started at one corner, giving its starting position and grid cell value, then moved in cardinal directions along the area, storing a number representing the direction and another indicating the number of grid cells moved, you could then store whole areas using very few numbers.

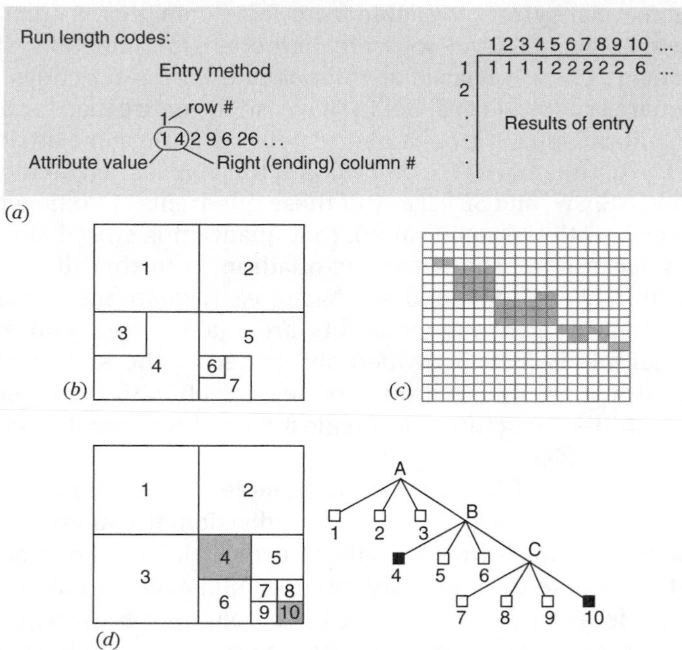

Figure 5.6 Compact data models. Methods of compacting raster data to preserve storage: (*a*) run-length codes, (*b*) raster chain codes, (*c*) block codes, and (*d*) the unique structure called quadtrees.

In this way you would save even more space and, of course, typing time. This method relies on what are called **raster chain codes**, which literally run a chain of grid cells around the border of each area (Figure 5.6b). In other words, you assign origins based on *X* and *Y* positions, a grid cell value for the entire area, and then directional vectors showing where to move next, where to turn, and how far to go. Usually the vectors include nothing more than the number of grid cells and the vector direction based on a simple coding scheme, where 0, 1, 2, and 3 could indicate north, south, east, and west, respectively.

There are two remaining approaches to reducing the storage necessary for grid-based systems, both relying on a square collection of grid cells as the primary unit of storage. The first, called **block codes**, is a modification of run-length codes (Figure 5.6c). Instead of giving starting and ending points plus a grid cell code, we select square groups of cells and assign a starting point, say the center or a corner, pick a grid cell value, and tell the computer how wide the square of grid cells is based on the number of cells. As you can see, this is really a two-dimensional run-length code. Each square group of grid cells, including individual grid cells, can be stored in this way with a minimum amount of numbers. Of course, if your grid has very few large square groups of cells, this method is not a major improvement for storage. Run-length codes also become somewhat cumbersome if there are few long runs or strings of the same value. Most thematic maps, however, have fairly large numbers of these groups, and block code methods are effective methods of reducing storage.

Quadtrees, our final method of compact storage, is a somewhat more difficult approach, and your instructor may elect not to cover it (Figure 5.6d). Still, at least one commercial system available from PCI Geomatics, a spatial analysis add-on to their Geomatica Core software, and one experimental system called Quilt (Shaffer et al., 1987) are based on this scheme. Like block codes, quadtrees operate on square groups of cells, but in this case the entire map is successively divided into uniform square groups of grid cells with the same attribute value. Starting with the entire map as the entry point, the map is then divided into four quadrants (NW, NE, SW, and SE). If any of these quadrants is homogeneous (i.e., contains grid cells with the same value), that quadrant is stored and no further subdivision is necessary. Each remaining quadrant is further divided into four quadrants, again NW, NE, SW, and SE. Again, each quadrant is examined for homogeneity. All homogeneous quadrants are again stored, and each of the remaining quadrants is further divided and tested in the same way until the entire map is stored as square groups of cells, each with the same attribute value. In the quadtree structure, the smallest unit of representation is a single grid cell (Burrough, 1983).

Systems based on quadtrees are called variable resolution systems because they can operate at any level of quadtree subdivision. If you do not need high resolution for your computations—in other words, if detail is not essential—you can use a rather coarse level of resolution (or quadtree subdivision) in your analysis. Thus, users can decide how fine the resolution needs to be for various manipulations. In addition, because of the compactness of storage from this method, very large databases, perhaps at continental or even global scales, can be stored in a single system.

The major difficulty with the quadtree structure is in the method by which it separates the grid cells into regions. In block codes, the decision is based entirely on the existence of square groups of homogeneous grid cells, regardless of where they were located on the map. With quadtrees, the subdivision is preset to the four quadrants (NW, NE, SW, and SE), resulting in some otherwise homogeneous regions appearing in two or more different quadrants. This results in computational difficulties for analysis of shape and pattern that must be overcome through rather complex computational methods that are beyond the scope of this book. GIS software using the quadtree data model operates under workstation and PC platforms and uses multiple operating systems. Such programs are in use worldwide and offer some interesting opportunities, especially for those needing to analyze very large databases.

COMMERCIAL RASTER COMPACTION PRODUCTS

With the advent of the Internet and the rapid increase in downloading of raster imagery, a vast array of new methods of data compression techniques and file types has arisen. We see adaptive binary optimization (ABO) files, graphics interchange format (GIF) files, portable network graphics (PNG) files, Joint Photographic Expert Group (JPEG) files, Tagged Image File Format (TIFF) files, and many more. Each of these employs some form of graphic compression and each has its own features, strengths, and limitations. GIF files, for example, are lossless (i.e., there is no information loss) but provide a limited number of display colors. TIFF employs run-length encoding but is known to produce large file sizes. JPEG files produce smaller file formats but have a tendency to produce blocky output. A newer version of the existing JPEG file format, called JPEG 2000, is based on a spatial resampling method using biorthogonal wavelet technology and provides a readily available method of image compression for image tasks including GIS and remote sensing. Unlike the original JPEG compression technology based on the discrete cosine transformation (DCT), it avoids the blocky appearance.

One lossless data compression technique that seems to be having a profound impact on the GIS and remote sensing communities is called **MrSID**. This image compressor, which is very common in the industry, is extremely powerful in its ability to compress files so they can be passed easily from user to user over the Internet. It has been commonly employed in GIS because of its supposed lossless characteristics.

MrSID, which stands for Multiresolution Seamless Image Database, was developed at Los Alamos National Laboratory and is now a commercial package exclusively developed and marketed by LizardTech Corporation. While the specific techniques are proprietary, it is important to be aware of MrSID, especially if your GIS operations use very large datasets or if your GIS delivery will be online.

Among the benefits of MrSID are its ability to handle large images and to display them in seconds. It allows instantaneous browsing of these large images and reduces the storage space needed to archive them. Typical compression ratios are 15–20:1 for grayscale and 30–50:1 for color images without loss of resolution. Additionally, the file format is compatible with any platform or operating system.

As GIS and remote-sensing data become both more voluminous and more commonplace, the need for such technology becomes increasingly important. Most GIS software already has the ability to work with JPEG, TIFF, PNG, GIF, and many other common graphics formats. It will not be long before virtually all commercial and most noncommercial GIS software will operate fully with MrSID, and other lossless data compression file formats will become commonplace.

VECTOR MODELS

Vector data structures allow the representation of geographic space in a way that is more visually intuitive and more reminiscent of the familiar analog map. Remember that they represent the spatial location of entities explicitly and most often store the attributes in another file. Because the relationships between individual entities are implicit rather than explicit, the intervening space between graphical entities does not have to be stored. There are several ways that vector data structures can be created to enable us to examine the relationships between variables in a single map or among variables in different maps. We will look at a range of these using three basic types as examples: (1) **spaghetti models**, (2) **topological models**, and (3) **vector chain codes**. Although there are others, and many variations on each type, these should suffice to give you an overview of what is available for vector GIS.

The simplest vector data structure, called the spaghetti model (Dangermond, 1982) (Figure 5.7), is a one-for-one translation of a graphical map image. It is natural and logical because the map is maintained as the conceptual model. The

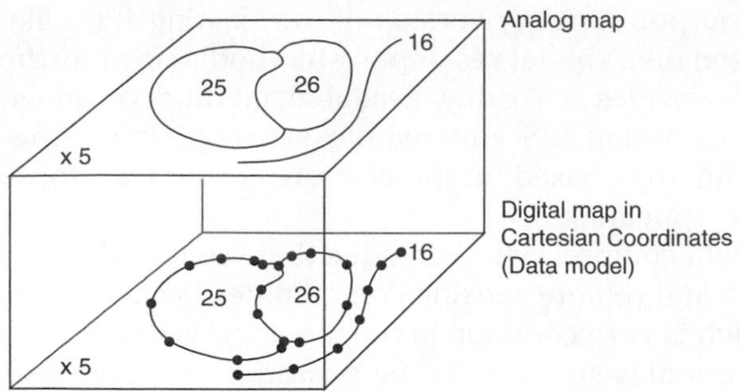

Data Structure

Feature	Number	Location
Point	5	x,y (single pair)
Line	16	(string of x,y coordinate pairs)
Polygon	25	(closed loop of x,y coordinate pairs where first and last pair are the same)
	26	(closed loop sharing coordinates with adjacent polygons to form a data structure)

Figure 5.7 Spaghetti vector data model. There is no explicit topological information, but the model is a direct translation of the graphic image. *Source:* Figure derived from Environmental Systems Research Institute, Inc. (ESRI) drawings and tabular data.

name is quite descriptive of how the spaghetti model works. I imagine covering each graphic object on a map with a piece of spaghetti (al dente). Each piece of spaghetti acts as a single entity—very short ones for points, longer ones for straight-line segments, and collections of line segments that come together at the beginnings and endings of surrounding areas. Each entity is a single, logical record in the computer, coded as variable length strings of (X, Y) coordinate pairs.

Imagine a large collection of pieces of spaghetti, each straight piece with a beginning and an ending set of coordinates. Any polygons that lie adjacent to each other must have separate pieces of spaghetti for adjacent sides. That is, no two adjacent polygons share the same string of spaghetti. Instead, each side of each polygon is uniquely defined by its own set of lines and coordinate pairs. Of course, adjacent sides of polygons, even though they are recorded separately in the computer, should have the same coordinates.

Because the data model is conceptualized as a one-for-one translation of the analog map, the geometric relationships (**topology**) among objects—for example, the locations of adjacent polygons—are implied, rather than explicitly coded in the computer. In addition, all relationships among all objects must be calculated independently. A result of this lack of explicit topology is enormous computational overhead, making measurements and analysis difficult. Because it so closely resembles the analog map, however, the spaghetti model is relatively efficient as a method of cartographic display and is still used quite often in CAC when analysis is not the primary objective, and in plotter languages for output because it is faster than many other methods.

Because the absence of topology in the spaghetti structure impedes its analytical capabilities, we need a vector data model that incorporates it. Such models are called topological models (Dangermond, 1982) (Figure 5.8). To allow advanced analytical techniques to be performed easily, we want to provide the

Topologically Coded Network and Polygon File

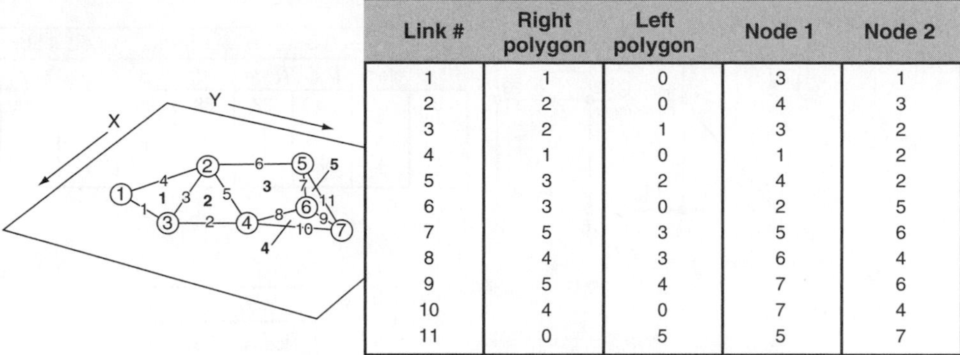

Link #	Right polygon	Left polygon	Node 1	Node 2
1	1	0	3	1
2	2	0	4	3
3	2	1	3	2
4	1	0	1	2
5	3	2	4	2
6	3	0	2	5
7	5	3	5	6
8	4	3	6	4
9	5	4	7	6
10	4	0	7	4
11	0	5	5	7

X,Y Coordinate Node File

Node #	*X* Coordinate	*Y* Coordinate
1	19	6
2	15	15
3	27	13
4	24	19
5	6	24
6	20	28
7	22	36

Figure 5.8 Topological vector data model. Note the inclusion of explicit information concerning connected points, lines, and polygons. *Source:* Figure derived from Environmental Systems Research Institute, Inc. (ESRI) drawings and tabular data.

computer with as much explicit spatial information as possible. The topological data model incorporates solutions to some of the more often used advanced GIS analytics. This is done by explicitly recording adjacency information into the data structure while the data are input. Each line segment—the basic logical entity in topological data structures—begins and ends when it either contacts or intersects another line, or when there is a change in direction of the line. Each line has two sets of numbers: a pair of coordinates and an associated node number. The node is more than just a point; it is the intersection of two or more lines, and its number is used to refer to any line to which it is connected. In addition, each line segment, called a link, has its own identification number that is used as a pointer to indicate which set of nodes represents its beginning and ending. Polygons, composed of these links, also have identification codes that relate back to the link numbers. Each link in the polygon can now look left and right at the polygon numbers to identify its adjacent polygons. In fact, the "left and right polygons" are also stored explicitly, so this step is eliminated. This allows the computer to know the actual relationships among all its graphical parts, thus approximating the map reader's ability to identify spatial relationships.

A number of topological data models are in common use. Perhaps the best-known topological data model is the **GBF/DIME** (geographic base file/dual independent map encoding) model created by the U.S. Census Bureau to automate the storage of street map data for the decennial census (U.S. Department of Commerce, Bureau of the Census, 1969) (Figure 5.9a). In this case

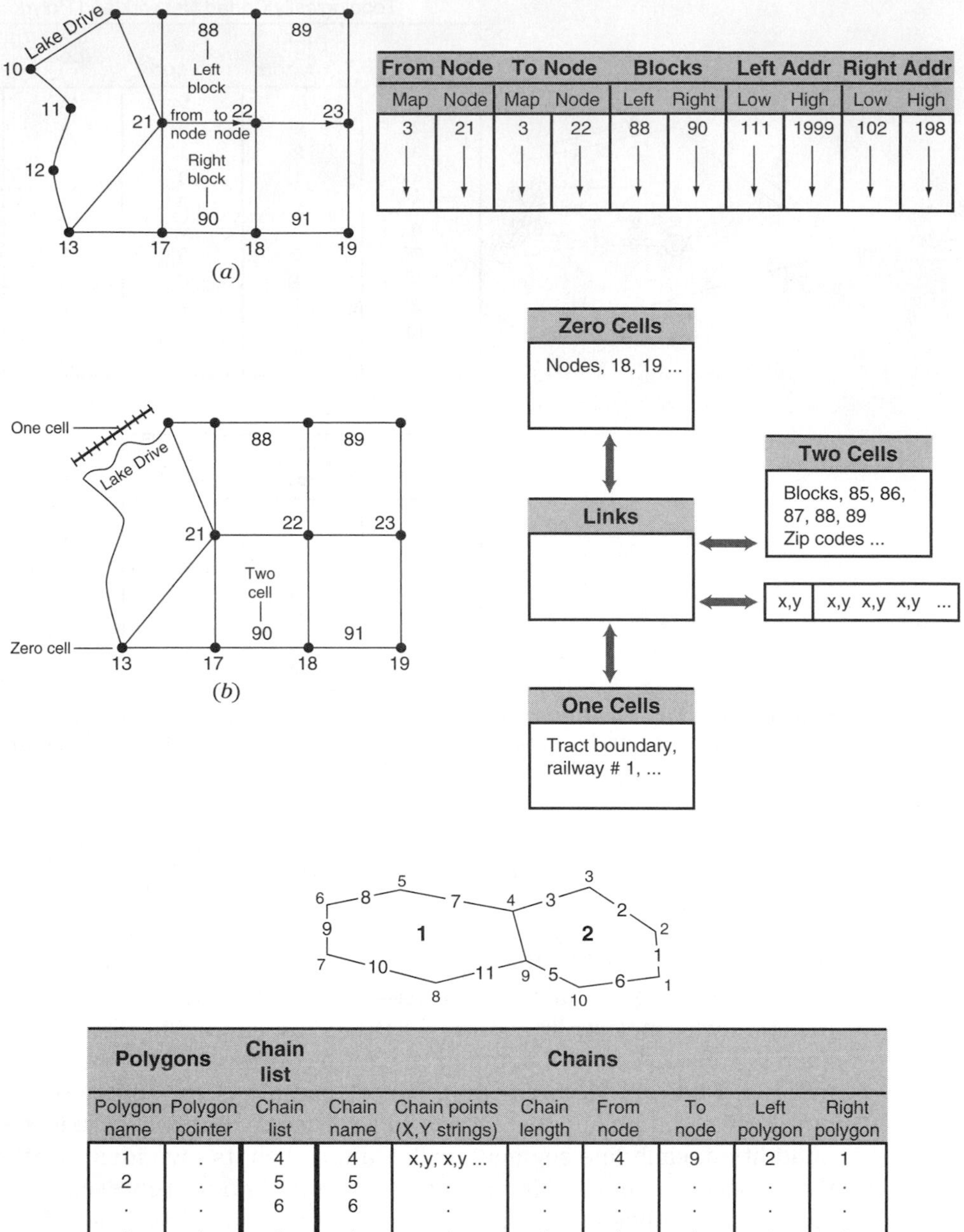

From Node		To Node		Blocks		Left Addr		Right Addr	
Map	Node	Map	Node	Left	Right	Low	High	Low	High
3	21	3	22	88	90	111	1999	102	198

Polygons		Chain list	Chains						
Polygon name	Polygon pointer	Chain list	Chain name	Chain points (X,Y strings)	Chain length	From node	To node	Left polygon	Right polygon
1	.	4	4	x,y, x,y ...	.	4	9	2	1
2	.	5	5	.	.	.	.	.	.
.	.	6	6	.	.	.	.	.	.
.	.	.	.	.	.	.	.	.	.
.	.	.	.	.	.	.	.	.	.

Figure 5.9 DIME files and TIGER files. Example of topological vector data models: (*a*) the GBF/DIME model, (*b*) the TIGER model, and (*c*) the POLYVRT model. *Source:* Parts (*a*) and (*b*) modified from K.C. Clark, *Analytical and Computer Cartography*, © 1990, Prentice-Hall, Inc., Englewood Cliffs, NJ. Used with permission.

the straight-line segments represent streets, rivers, railroad lines, and so on (Peuquet, 1984). In this model each segment ends when it either changes direction or intersects another line, and the nodes are identified by codes. In addition to the basic topological model, the GBF/DIME model assigns a directional code in the form of a From-node and a To-node (i.e., low-value node to higher-value node in a sequence). This approach makes it possible to identify missing nodes during the editing process. If polygonal nodes don't completely surround an area, a node is missing (Peuquet, 1984).

As an additional feature of GBF/DIME, both the street addresses and UTM coordinates for each link are explicitly defined, permitting street addresses to be accessed by geographic coordinates. However, this data model suffers the same basic problem that dogs the generic topological model and, of course, the spaghetti model as well. Because there is no predefined line segment order, to search for a particular line segment, the program must perform a sequential search of the entire database. Remember this is the slowest computer search strategy. The GBF/DIME system, moreover, is based on **graph theory** in that it doesn't matter whether the line connecting any two points is curved or straight. Thus, a side of a polygon serving to indicate a curved river boundary would be stored not as a curved line but rather as a straight line between two points, and the resulting model would lack the geographic specificity we expect of an analog map.

Some problems of the GBF/DIME system have been eliminated with the development of the topologically integrated geographic encoding and referencing system (**TIGER**) (Marx, 1986), designed for use with the 1990 U.S. census (Figure 5.9b). In this system, points, lines, and areas are explicitly addressed, and therefore census blocks can be retrieved directly by block number rather than by relying on the adjacency information contained in the links. In addition, since the model does not rely on graph theory, real-world features such as meandering streams and irregular coastlines are given a graphic portrayal more representative of their true geographic shape (Clarke, 1990). Thus, TIGER files are more generally useful for noncensus-related research than graph-theory based systems.

Another data model, developed by Peucker and Chrisman (1975) and later implemented at the Harvard Laboratory for Computer Graphics (Peuquet, 1984), was called the **POLYVRT** (POLYgon conVeRTer) model (Figure 5.9c). As with TIGER, it eliminates storage and search inefficiencies of the basic topological model by separately storing each type of entity (points, lines, and polygons). These separate objects are then linked in a hierarchical data structure with points relating to lines, which in turn are related to polygons, all through the use of pointers. Each collection of line segments, collectively called **chains** in this model, begins and ends with specific nodes (intersections between two chains). And as in the GBF/DIME system, each chain contains explicit directional information in the form of To–From nodes as well as left–right polygons (Figure 5.9c).

Like TIGER, POLYVRT has the advantage of allowing selective retrieval of specific entity types: you can select points or lines or polygons at will by identifying them based on their codes (which of course are connected to records of their attributes). An additional advantage of POLYVRT is that because lines bounding polygons are explicitly recorded as chains of individual line segments, the individual line segments do not have to be accessed to find the beginning and ending of a particular polygon. Instead, the chains can be accessed directly, saving time for searches (Peuquet, 1984). Because in POLYVRT chain lists bounding polygons are explicitly stored and linked through pointers to each polygon code, the size of the database is largely controlled by the number of polygons rather than by the complexity of the polygon shapes. This makes storage and retrieval operations more efficient, especially when highly complex polygonal shapes found in many natural features are encountered (Peuquet, 1984). The major drawback of POLYVRT is that it is difficult to detect an incorrect pointer for a given polygon until the polygon has actually been retrieved, and even then you must know what the polygon is meant to represent.

Despite the efficiencies of topological data structures, advances in computer processor speeds and massive increases in the capacity of storage devices may have reduced the need for explicit topology. The shapefile was a reaction to these computer advances. It is a nontopological data model that stores the geometry and attribute information for geographic features in a data set. The geometry for each feature is stored as a shape comprising a set of vector coordinates and is linked to their attributes. There are 14 shape types, each describing particular entities or entity combinations (Figure 5.10).

The shapefile generally has lower processing overhead than its topological counterpart. This results in faster drawing speed and editing, allows for the handling of overlap and noncontiguous features, reduces disk space requirements, and makes the files themselves easier to read and write. Shapefiles are actually not single files, but three separate and distinct types of files: main files, index files, and database tables. The main file (e.g., counties.shp) is a direct-access, variable record length file that contains the shape as a list of vertices. The index file (e.g., counties.shx) contains character length and offset (spaces) information for locating the values, and a dBASE© table (e.g., counties.dbf) that contains the attributes that describe the shapes.

Value	Shape type
0	Null shape
1	Point
3	PolyLine
5	Polygon
8	MultiPoint
11	PointZ
13	PolyLineZ
15	PolygonZ
18	MultiPointZ
21	PointM
23	PolyLineM
25	PolygonM
28	MultiPointM
31	MultiPatch

Figure 5.10 Examples of the shape types of geographic features in a data set for a shapefile.

AN OBJECT-ORIENTED DATA MODEL

A new variety of vector data model takes advantage of the properties of object-oriented programming techniques—specifically, those of property inheritance. Adaptable to both topological and nontopological settings, the geodatabase data model permits the handling and specification of property-rich data types (e.g., a set of pipes might include sizes, materials, capacities, etc.). These properties themselves suggest a set of rules and relationships that can also be included in the data model (e.g., there might be limitations on the capacity pipes must have for particular tasks). Additionally, the geodatabase model allows access to large volumes of geographic data stored in both the files themselves and in associated databases.

Because geographic objects are both numerous and industry specific, geodatabases can be created for selective groups of tasks. For example, there might be a geodatabase model for transportation systems in which specifics of construction materials, road conditions, weight restrictions, and other factors are linked to the road objects as properties that are inherited. Another geodatabase might be employed for environmental impact statements, where soil properties limiting its usefulness for human use might be included as properties. Or you might see a geodatabase for crime mapping in which neighborhood economic factors, existence of gang activity, and other data might be inherited properties for neighborhood polygons. Electrical infrastructure might have its own set of inherited rules regarding above-ground or underground power line restrictions. Each of the objects for each of the models possesses specific properties that provide structure and place limitations on how they can be used for analysis. These properties define the relationships that are permitted among geographic entities as they are represented by the software. The following is a sample of geodatabases suggested by Arctur and Zeiler (2004): (1) streams and river networks, (2) census units and boundaries, (3) addresses and locations, (4) parcels and the cadastre, and (5) federal lands survey. Many more are available, for example, including wildlife management, geological mapping, and urban infrastructure.

COMPACTING VECTOR DATA MODELS

Although vector data models are generally more efficient at storing large amounts of geographic data, it is still necessary to consider reductions. In fact, a simple codification process developed more than a century ago by Sir Francis Galton (1884) is relatively similar to the compaction technique you are about to see. It might be useful to travel back in time to accompany the English scientist as he tried to develop a shorthand scheme for recording directions during geographic excursions. The form Galton devised is simplicity itself. He simply applied eight numerical values: one for each of the cardinal compass directions and one for each of the intermediate directions, northeast, southeast, southwest, and northwest (Figure 5.11).

A surprisingly similar coding scheme, developed in our time, is known as **Freeman—Hoffman chain codes** (Freeman, 1974) (Figure 5.11). Eight unique directional vectors are assigned the numbers 0 through 7. As Galton had done

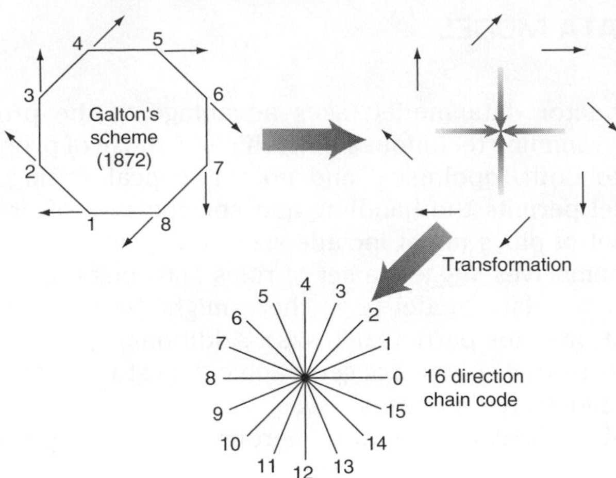

Figure 5.11 Chain codes. Comparison of compact models for direction and path finding developed by Sir Francis Galton and refined in the Freeman–Hoffman chain codes. Note the strong similarity between the older and more recent models.

for ground navigation on his journeys, the Freeman–Hoffman method assigned these vectors in the same four cardinal directions and their diagonals. By assigning a length value to each vector, individual line entities can be given a shorthand to show where they begin, how long they are, in which direction they are drawn, and where the vector changes direction. There are many variations on this theme, including increasing the codes to 16 (Figure 5.11) or even 32 values, rather than 8, to enhance accuracy. But the result is the same—reduced storage for vector databases.

Although the chain-code models produce significant improvements in storage, they are essentially compact spaghetti models and contain no explicit topological information. This limits their usefulness to storage, retrieval, and output functions because of the analytical limitations of nontopological data structures. In addition, the way the lines and polygons are encoded as vectors, performing coordinate transformations—especially rotation—leads to heavy computational overhead. Chain-code models are good for distance and shape calculations because much of this information is already part of the directional vectors themselves. In addition, because the approach is so similar to the way vector plotters operate, the models are efficient for producing rapid plotter output.

A VECTOR MODEL TO REPRESENT SURFACES

We have largely ignored surfaces thus far, even though they are a fundamental feature we will want to model with a GIS. They differ significantly in their manner of representation, especially in vector. In raster, the geographic space is assumed to be discrete in that each grid cell occupies a specific area. Within that discretized or quantized space, a grid cell can have encoded as an attribute the absolute elevational value that is most representative of the grid cell. This might be the highest or lowest value, or even an average elevational value for the grid cell. As such, the existing raster data structures can handle surface data.

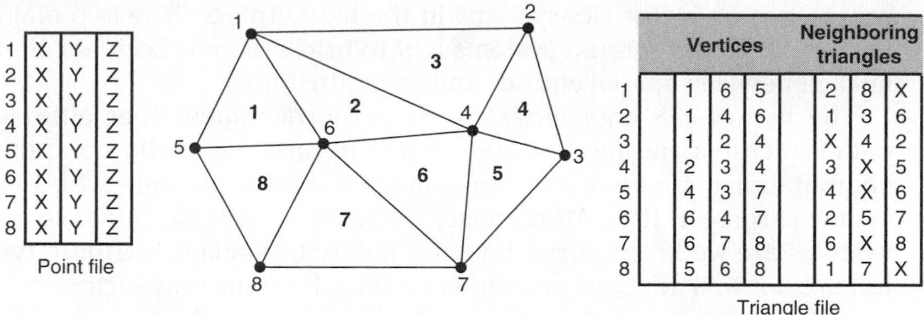

	X	Y	Z
1	X	Y	Z
2	X	Y	Z
3	X	Y	Z
4	X	Y	Z
5	X	Y	Z
6	X	Y	Z
7	X	Y	Z
8	X	Y	Z

	Vertices			Neighboring triangles		
1	1	6	5	2	8	X
2	1	4	6	1	3	6
3	1	2	4	X	4	2
4	2	3	4	3	X	5
5	4	3	7	4	X	6
6	6	4	7	2	5	7
7	6	7	8	6	X	8
8	5	6	8	1	7	X

Figure 5.12 TIN model. In a vector system surfaces are represented by connecting points of known elevation into triangulated flat surfaces. The model is called a triangulated irregular network (TIN) model, a specific form of tessellation.

In vector, much of the space between the graphical entities is implied rather than explicitly defined. To define this space explicitly as a surface, we must quantize the surface in a way that retains major changes in surface information and implies areas of identical elevational data. For a simple way to envision this, consider how mineralogists or crystallographers describe minerals. Each mineral is said to have a series of smooth faces connected by points and lines that show major changes in its structure. In a similar way, we can envision a topographic surface to be a crude mineral, with flat surfaces, edges, and vertices (Figure 5.12). Thus we can model a surface by creating a series of either regularly or irregularly placed points that act as vertices. Each vertex has an explicit elevational value. We can connect any three vertices to represent an area of uniform slope, thereby creating what essentially amounts to a crystallographic model of our surface.

This **tessellation** model, called a **triangulated irregular network (TIN)**, allows us to record topographic data as points in a regular or irregular grid. Then when a three-dimensional view is necessary, the grid can be converted to the TIN or crystalline shape. In addition, the point data can be used for performing the typical processes of representing the surface as a series of contour lines, with all the interpolation procedures needed for surface analysis. This particularly elegant means of representing surfaces was, in fact, used as the primary data structure for earlier systems that relied heavily on surface data (DeMers and Fisher, 1991).

SYSTEMS MODELS

We have discussed file structures, database management systems, and spatial models. We now need to examine how these are combined to produce complete systems. Most raster systems are so simple that the data model itself provides a relatively complete description. You have seen that the MAP data model is the most often employed raster system. In vector systems, however, there are two primary approaches to integrating the graphics elements of the data model to a database management system. It is useful to examine these two models not only because of the basic differences in approach, but also because of the overall prevalence of vector GIS systems in the marketplace. The two major forms of vector GIS are **integrated systems** and **hybrid systems**; both accommodate the linkage between spatial entities and their attributes.

The hybrid GIS data model is an acknowledgment that although graphic data structures and models are efficient at handling entity data, they lack the computational sophistication to manage the attribute data with the same efficiency (Aronson, 1985; Morehouse, 1985). Alternatively, database management systems are well recognized for their ability to manage attribute types of data but are not well adapted to graphic entities. It seems only logical that these two technologies, if linked through software, would provide the best of both modes. To implement this approach, the coordinate and topological data required for graphics are stored as a separate set of files (Figure 5.13). The attribute tables, carrying all the necessary attribute data for each graphic entity, are also stored separately within existing commercial database management system software. Linkage is performed by storing identification codes (i.e., polygon identification codes) as a column of data in the attributes database. In this way, the column is directly associated with the attribute codes contained in the table. Because the hybrid model allows raster and vector data types to be operated on in the same system, this approach is sometimes used to link grid-based modules to the database management system within the overall GIS structure. Moreover, since multiple attributes can be stored in the database management system, analytical capabilities are increased and storage space is saved.

The second major type of GIS model is the integrated data model, another spatial database management system approach. In this case the GIS serves as the query processor, but it is more closely integrated with the database management system than in the hybrid system (Guptill, 1987; Morehouse, 1989). Usually based on vector/topological data models, the integrated system stores map coordinate data (entities) as relational tables, together with separate tables containing topological data within the same database (Figure 5.14). Attributes are stored in the same database and can be placed in the same tables as the graphic entities. Or, as noted earlier, they can be stored as separate tables and accessed through the use of relational joins (Healey, 1991).

There are two ways of storing coordinate information as relational tables. The first records individual (*X*, *Y*) coordinate pairs, representing points as well as line and polygon terminators and vertices as individual atomic elements or rows in the database. This approach conforms to Codd's normal forms, but it makes searching quite difficult because each element must be recomposed from its atomic format to create whole polygons or groups of polygons. Most GIS applications access large groups of entity elements for display purposes,

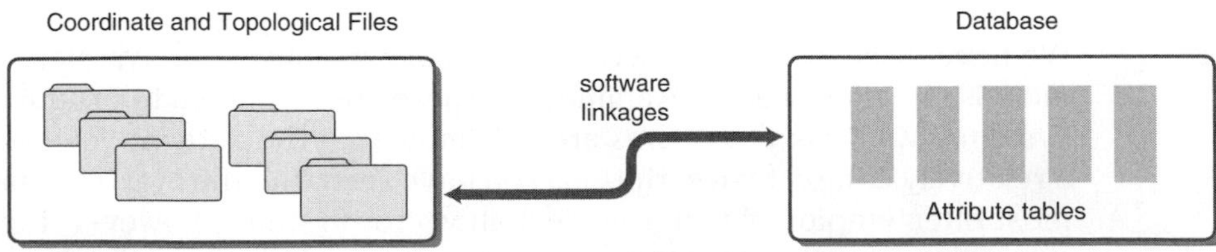

Figure 5.13 Hybrid vector geographic information system (GIS). Pointers connect the entity files to the database management system that contains the attribute information.

Point ID	x	y
1	25.4	27.5
2	24.8	45.8
3	27.8	50.4
4	28.9	43.7
.	.	.
.	.	.
.	.	.

Point table

Polygon ID	Line ID
34	12
34	14
34	17
34	21
36	78
36	84
.	.

Polygon table

Polygon ID	Attribute
34	104
36	624
37	108
38	107
30	53
41	91
.	.

Attribute table

Relational join

Figure 5.14 Integrated geographic information system (GIS). Note how a single database can be configured to contain separate files for entities and attributes.

a function used more often than users might think as they review the results of intermediate analytical steps. To avoid this approach, the integrated model could code whole strings or collections of coordinate information in the tables. As such, a single polygon could be described with its ID code in one column and a list of lines, addressed by code, in another. Then the lines, which would be identified by code in a separate column of a line table, would describe the polygon's locations with a number of coordinate pairs. This approach reduces overhead for retrieval and display purposes, but it violates the first normal form. From the user's perspective this is generally not a serious problem, and grouping these nonatomic strings of data as one-dimensional arrays provides the advantage of enhanced system performance (Dimmick, 1985) while more closely following the rules of first normal form (Sinha and Waugh, 1988).

The choice of hybrid versus integrated system for the vast majority of users is less technical than pragmatic. Each has some advantages over the other, but especially as we move toward higher-powered workstations, networking, and distributed computing, both can supply a wide spectrum of analytical power. For most of us, at least as novices, the choice of system will be made for us. Those in the enviable position of choosing their system want to decide on which one best fits present equipment and future networking needs. Both system types will continue to improve, and it will be necessary to ask the vendors for detailed specifications and even performance evaluation tests on given hardware configurations.

In addition to the typical high-level models already discussed, a third, called **object-oriented database management systems**, is emerging as an alternative approach. This model is an extension of the integrated GIS model that incorporates a spatial query language (Healey, 1991) and reflects recognition of the importance of being able to access both the cartographic database and the operations to be performed. Conceptually, such a GIS is nearly identical to the object-oriented approach to computer programming (Aronson, 1987).

There is little agreement about the exact meaning of object oriented, but we know that an object is an entity that has a condition or state that is represented by both local variables and a set of operations that will act on the object. Because it belongs to a set of objects and operations, each individual object can be thought of as a member of that set (i.e., a set defined by local variables and operations at the same time). Each of these classes of objects belongs to a larger group called a superclass and inherits properties from that superclass—in the same manner that humans (as a class of objects) inherit characteristics of a larger set called mammals. For GIS the concept might be exemplified by an object class called polygon that gives to each polygon in the database its many properties (e.g., local variables like number of polygons; lists of nodes, arcs, and areas; operational procedures for calculating centroids, drawing polygons, and overlaying polygons; etc.) (Figure 5.15).

In addition, in the GIS context the object class called polygon acts as a superclass for another set of objects called land parcel. Thus, land parcel objects also inherit the variables and operational instructions for the superclass polygon, as well as having many of their own characteristics (e.g., categories of land parcels, values, owners and operational procedures allowing transfer of ownership, and rezoning operations). This explicit linking of variables and operations, together with the property of inheritance, is meant to more closely resemble the way actual geographic queries are performed. It also provides a method by which changes in one set of objects can be immediately reflected by changes in related objects.

Current systems that rely to one degree or another on the object model include ESRI's Map Objects and GE Smallworld's GIS. ESRI's new geodatabase (short for geographic database) represents a radical change in data models for that company. This system is a shift from the hybrid model that has for years been used by its ArcInfo software, to an integrated software system which it now calls ArcGIS to indicate a clear shift to an alternative data model. One major change in this software system is that it has added object orientation, with all the flexibility that entails. This new model allows raster, vector, address, measures, CAD, and a wide variety of file types to be stored together (integrated system) in a commercial off-the-shelf DBMS. It also supports intelligent features, rules, and relationships. The geodatabase model supports, as standard, a wide selection of objects (rows in the database tables) and features (objects with geometry).

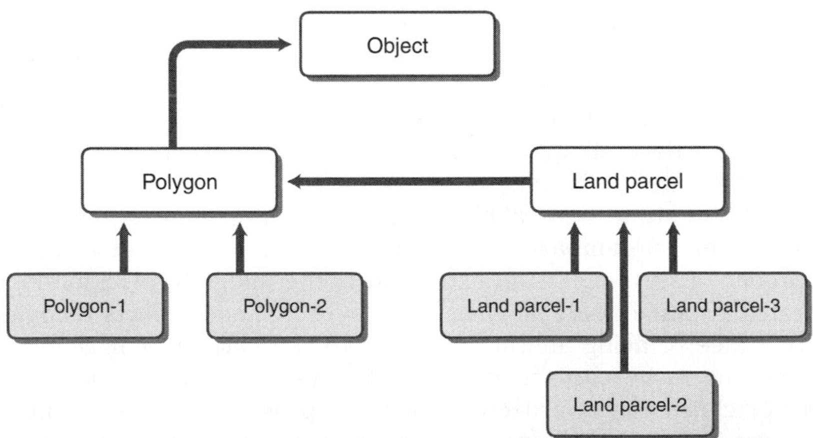

Figure 5.15 Object-oriented geographic information system (GIS). Example of a hierarchy of object classes as they might be configured by an object-oriented GIS. *Source:* Modified from R.G. Healey, "Database Management Systems," Chapter 18 in *Geographic Information Systems, Principles and Applications*, D.J. Maguire, M.F. Goodchild, and D.W. Rhind, Eds. Longman Scientific and Technical, Essex, England, © 1991. Used with permission.

All of these suggest that object-oriented GIS models are likely to increase in number. This will mean that the ability to model with the GIS software will become easier and model complexity will increase. It also suggests that the idea of a GIS as out-of-the-box software is being replaced with the concept of GIS as a toolkit. Under this concept users will select the objects they need for particular tasks and will design the software and its user interface for particular applications. It also suggests that students of GIS will, of necessity, need to learn to program, particularly in object-oriented languages, so their effectiveness with the new generation of GIS software will be the best possible.

Terms

block codes
chains
database management system (DBMS)
Freeman-Hoffman chain codes
GBF/DIME
geodatabase
graph theory
grid
grid cells
hybrid systems
integrated systems
interpolation
map analysis package (MAP)
MrSID
networks
node
object-oriented database management systems
polygon
POLYVRT
quadtrees
quantize
raster
raster chain codes
resolution
run-length codes
spaghetti model
tessellation
TIGER
topological models
topology
triangulated irregular network (TIN)
vector
vector chain codes

Review Questions

1. Describe the process of quantizing space into equal-sized rasters called grid cells. What impact does grid cell size have on the locational accuracy? How would you store points, lines, and polygons using a raster system?
2. What are some possible advantages and disadvantages of using a raster GIS as opposed to vector?
3. Describe the basic vector data structure. How does it differ from raster in its ability to locate objects in space? How does it deal with the space between objects and other spatial relationships compared to raster?
4. Diagram and describe how surface data are coded in a raster data model.
5. Describe the map analysis package raster GIS model. Explain why this model is so important to raster GIS modeling as well as compact raster models.
6. Describe the methods of compact storage for raster models. Why are they needed, anyway? What is the major problem with quadtree representation of earth features as squares compared to the block codes method?
7. Describe the spaghetti vector data model. What are its advantages and disadvantages?

8. Diagram, label, and describe the general topological vector data model. How does it differ from the spaghetti model? How is this difference achieved? What are some examples of topological models? Describe their differences. What advantages or disadvantages does each have?

9. Describe the geodatabase model and explain its advances over other data models. Why is there a need for a different geodatabase for different industries? Describe one case that supports your answer.

10. What is a common method of compacting vector data models? How does it work? What nineteenth-century geographer developed a similar scheme?

11. Diagram and explain the DIME model and the TIGER model. Suggest which you might prefer for tax assessment or flood zone mapping and explain why.

12. Describe the TIN model. How does it quantize space differently from raster models? Why is it necessary to develop such a model for vector GIS?

13. What is the major difference between hybrid and integrated GIS systems? Create a diagram of data storage and access in each system.

14. Describe the shapefile model and explain why it did not include topology.

15. Explain, in general terms, what an object-oriented GIS is and indicate its potential advantages over other systems.

References

Arctur, D.K, and M. Zeiler, 2004. *Designing Geodatabases: Case Studies in GIS Data Modeling*, Redlands, CA: ESRI Press.

Aronson, P., 1985. "Applying Software Engineering to a General Purpose Geographic Information System." In *Proceedings of AUTOCARTO 7*. Falls Church, VA: ASPRS, pp. 23–31.

Aronson, P., 1987. "Attribute Handling for Geographic Information Systems." In *Proceedings of AUTOCARTO 8*. Falls Church, VA: ASPRS, pp. 346–355.

Burrough, P.A., 1983. *Geographical Information Systems for Natural Resources Assessment*. New York: Oxford University Press.

Clarke, K.C., 1990. *Analytical and Computer Cartography*. Englewood Cliffs, NJ: Prentice-Hall.

Dangermond, J., 1982. "A Classification of Software Components Commonly Used in Geographic Information Systems." In *Proceedings of the U.S.–Australia Workshop on the Design and Implementation of Computer-Based Geographic Information Systems*. Honolulu, HI, pp. 70–91.

DeMers, M.N., and P.F. Fisher, 1991. "Comparative Evolution of Statewide Geographic Information Systems in Ohio." *International Journal of Geographical Information Systems*, 5(4):469–485.

Dimmick, S., 1985. *Pro-FORTRAN User Guide*. Menlo Park, CA: Oracle Corporation.

Freeman, H., 1974. "Computer Processing of Line-Drawing Images." *Computing Surveys*, 6:57–97.

Galton, F., 1884. *Art of Travel; or, Shifts and Contrivances Available in Wild Countries*. New York: David & Charles Reprints.

Guptill, C., 1987. "Desirable Characteristics of a Spatial Database Management System." In *Proceedings of AUTOCARTO 8*. Falls Church, VA: ASPRS, pp. 278–281.

Healey, R.G., 1991. "Database Management Systems." In *Geographical Information Systems: Principles and Applications*, D.J. Maguire, M.F. Goodchild, and D.W. Rhind, Eds. Essex: Longman Scientific & Technical.

Herring, J.R., 1987. "TIGRIS: Topologically Integrated Geographic Information System." In *Proceedings of AUTOCARTO 8*. Falls Church, VA: ASPRS, pp. 282–291.

Marx, R.W., 1986. "The TIGER System: Automating the Geographic Structure of the United States Census." *Government Publications Review*, 13:181–201.

Morehouse, S., 1985. "ARC/INFO: A Geo-relational Model for Spatial Information." In *Proceedings of AUTOCARTO 8*. Falls Church, VA: ASPRS, pp. 388–397.

Morehouse, S., 1989. "The Architecture of ARC/INFO." In *Proceedings of AUTOCARTO 9*. Falls Church, VA: ASPRS, pp. 266–277.

Peucker, T., and N. Chrisman, 1975. "Cartographic Data Structures." *American Cartographer*, 2:55–69.

Peuquet, D.J., 1984. "A Conceptual Framework and Comparison of Spatial Data Models." *Cartographica,* 21:66–113.

Shaffer, C.A., H. Samet, and R.C. Nelson, 1987. "QUILT: A Geographic Information System Based on Quadtrees." College Park: University of Maryland, Center for Automation Research, CAR-TR-307, CS-TR-1885, DAAL02-87-K-0019.

Sinha, A.K., and T.C. Waugh, 1988. "Aspects of the Implementation of the GEOVIEW Design." *International Journal of Geographical Information Systems*, 2:91–100.

U.S. Department of Commerce, Bureau of the Census, 1969. "*The DIME Geocoding System*." In Report No. 4, Census Use Study. Washington, DC: Government Printing Office.

UNIT 3

INPUT, STORAGE, AND EDITING

CHAPTER 6

GIS Input

The most important tasks associated with GIS are selecting, acquiring, and converting data into a compatible digital format the software can use. The many decisions that are made prior to GIS analysis will have a profound long-term impact on the quality and utility of the GIS. These decisions include determining what data need to be included in the database, where they can be obtained, their quality, costs, methods of input, documentation for future users, and many more. While most of us view GIS as a tool for data analysis and are often impatient to get started analyzing them, the most time-intensive and costly operations deal with the many aspects of data entry.

GIS data come in two general forms — primary and secondary. Often referred to as **primary data acquisition** and **secondary data acquisition**, this terminology is somewhat misleading because we actually do more than just acquire the data. Primary data refer to data that are created in-house, while secondary data are obtained from outside data providers. Most operational GIS will use some combination of primary and secondary sources and will acquire both analog and digital data, at different scales and projections, in many different formats. We must, therefore, convert them to data models and data structures that are compatible with our own GIS software package. This allows us to store, edit, retrieve, analyze, and output all of them regardless of their original form or source.

In the following pages we will examine primary data input first and then secondary data input. Next we will look closely at the role of data documentation to describe the many aspects of our data. This particular topic is becoming increasingly important, especially because much GIS data are now being used for long-term projects and are often shared among different users both within single organizations and among different organizations.

LEARNING OBJECTIVES

When you are finished with this chapter you should be able to:

1. Define the difference between primary and secondary data acquisition and describe the unique problems of each.

2. Know the four primary functions of GIS data input.
3. Know how to decide on the input devices you will need.
4. Be aware of the potential problems of converting to and from different data structures.
5. Understand the transformations that take place during data input.
6. Know the procedures for translation, rotation, and scale change necessary for modifications in GIS data layers.
7. Understand the procedures of map preparation and the importance of those map preparation procedures in map input.
8. Have a feel for what data and how much data need to be input and why they need to be input to a GIS.
9. Understand the relationship between input scale and projection on GIS error.
10. Be familiar with the basic procedures for digitizing and the importance of attribute data to the overall quality of the GIS database.
11. Be familiar with the four basic methods of raster data input, with their advantages and disadvantages, and with the data types that should be input with the respective method.
12. Describe what metadata are, in detail, and discuss their role in maintaining data integrity and sharing.
13. Understand some of the technical problems of using aerial photography for GIS input.
14. Be aware of the potential for and the problems of external databases for GIS input.

PRIMARY DATA

Primary data are those that are collected or developed by the intended user. Because the collectors are going to be the primary users, they are the most familiar with the necessary quality, scales and resolutions, classifications and aggregation levels, and all other aspects of the data. Primary data generally experience a higher level of quality control than secondary data. This higher level of quality control often requires more intensive collection (e.g., field collection) and conversion (e.g., digitizing) than secondary data. As a result, they are often better quality for the specific applications and more costly to produce.

There are many forms of primary data, including field observations, biological specimen sampling, human interviews and surveys, aerial photographs and satellite remote sensing, field mapping, GPS surveys, police calls and accident reports, and literally thousands more. The specific collection methods and techniques are as varied as the types of data themselves. Some primary data sources will result in GIS-ready digital files; others will need to be converted.

We will limit our discussion here to analog to digital conversion of mapped data through digitizing or scanning.

Input Devices

Many different types of devices are available for general input of any data into a computer. For inputting spatial data manually, the use of a digitizer is standard. The digitizer is essentially a sophisticated type of mouse, but the movement is recorded through the use of an electronically active grid within the digitizing tablet. A mouse-like device (puck) is connected to the tablet and is moved to different locations on a map that is attached to the tablet. Digitizing pucks contain a crosshair device, encased in glass or clear plastic, that allows the operator to place the puck exactly over individual map elements. In addition, the puck has buttons that indicate the locations of points or nodes, the beginnings and endings of lines or polygons, explicitly define left and right polygons, and so on.

Digitizing tablets can range from small page-sized formats to very large formats capable of accepting good-sized maps with room to spare (Figure 6.1). Some of the larger format digitizers also have adjustable stands that can vary the elevation of the tablet from the ground and change the angle of the tablet to make digitizing easier for the operator. Sizes and formats are determined in part by both the general size of documents to be input to the GIS and budgetary constraints. Modern digitizers can provide **resolutions** of 0.001 inch, with an accuracy that approaches 0.003 inches for an area of 42 inches by 60 inches (Cameron, 1982).

Factors that will prove useful in selecting a digitizer include **stability**, **repeatability**, **linearity**, resolution, and **skew**. *Stability* deals with the tendency of the exact reading of the digitizer to change as the machine warms up. For the first-time digitizer operator, it can be most disconcerting to watch the values change while the puck sits in one place. The simplest solution is to allow the digitizer table to come up to operating temperature before using it. If the drift continues when the tablet is warm, the tablet may need to be repaired or replaced.

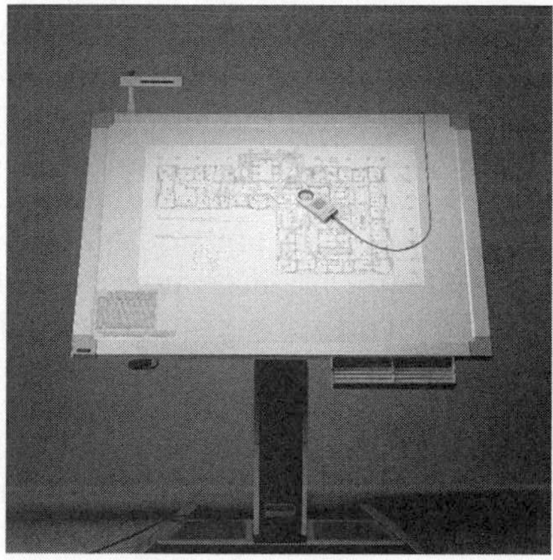

Figure 6.1 Large-format digitizing table. Most large tablets allow the user to raise and lower the table and adjust the angle.

Repeatability, a synonym for precision, describes how close successive puck placement readings of a single location will be to each other. Good digitizers should be repeatable to about 0.001 inch (Cameron, 1982). *Linearity* is a measure of the ability of the digitizer to be within a specified distance (tolerance) of the correct value as the puck is moved over large distances. A linearity of 0.003 inch measured over 60 inches is common with today's equipment. *Resolution* is the ability of the digitizer to record increments of space. In other words, the smaller the units of measure it can handle, the better its resolution. This is not unlike the resolution of a camera or a remote-sensing system. Resolutions of 0.001 inch are quite good but may prove to be unnecessary for much GIS work. Finally, *skew* is a measure of the squareness of the results on a tablet: Do coordinates located at the four corners of your digitizer produce a true rectangle, as intended? Some portions of the digitizer may begin to wear out, especially toward the edges, reducing the ability to digitize on the entire tablet, perhaps even compromising the quality of the input.

Scanners are more advanced input devices that perform much of the work of the entity input process with little help or intervention from the user. Note that the scanners we are examining here are not the types of scanners we employ for scanning snapshots for our personal use, but do employ a similar although more sophisticated technology. Professional scanners come in three general types: line-following scanners, flatbed scanners, and drum scanners. Line-following scanners, as the name implies, are placed on a line and move on small wheels, tracking by a laser or other guiding mechanism. At different intervals (either time intervals or distance intervals), these devices send a signal to the computer that records the digitizer coordinates at each of the sample locations. Line-following scanners require more technician intervention because they must be manually placed on each new line to keep the scanning process going. They also suffer from two shortcomings. First, complete automation requires the device to use a simple time or distance measure within which to record or sample the lines followed. Remember that this is not always the best way to abstract a line because complex lines require more sampling than straight lines. Additional operator intervention is required to ensure that sufficient points are sampled. The second problem is a more difficult one for the line scanner to deal with. Say you are using a line scanner on a topographic map to digitize contour lines that converge along a cliff (a conventional method of depicting cliff faces is with contour lines). When the lines diverge on the opposite side of this feature, however, it is quite common for the scanner to get confused about where it should go next. This problem can get even worse if the topography under an overhang is represented by dashed lines, which the scanner may not be able to find because of the gaps in the lines or because the color is lighter and has less contrast than the original contour line.

Flatbed scanners are more commonly available today because of major reductions in cost. These scanners range from small units designed for scanning business cards, photographs, or documents, to large commercial scanners (Figure 6.2). Using a charged coupled device (CCD) scanning technology not unlike that commonly available in digital cameras and remote-sensing devices, the commercial flatbed scanner most often scans the map or image one line at a time and then compiles all the lines as a composite raster image. Depending on the amount of memory and the quality of the original image or map, these devices can produce very good resolutions, certainly far better than the smaller

Figure 6.2 Large-format flatbed scanner.

office variety. Of course, the increased resolution results in very large images that can range up to many gigabytes.

The third basic form of scanner, the drum scanner, uses a raster-like approach similar to the flatbed scanner, but the mechanics are somewhat different. The map document is placed on a drum that rotates at the same time that a sensor device moves along the frame at right angles to the direction of rotation (Figure 6.3). In this way, the entire document is scanned one line at a time. As with flatbed scanners, each location on the map or image is recorded, even if there are not graphical objects present. The result is a detailed raster image of the entire document.

Both drum scanners and flatbed scanners can give monochromatic or color output. For color, each of the three primary colors is scanned either individually and then recombined or all at once, depending on the technology used. Whether the output is monochromatic or color, they can be converted to vector format (usually for cartographic scanning) or can be retained as raster (usually for imagery).

Scanning has taken on a new importance, primarily for the use of aerial photographic imagery as GIS thematic maps from which onscreen or heads-up digitizing can take place. In many situations, scanned aerial photography can

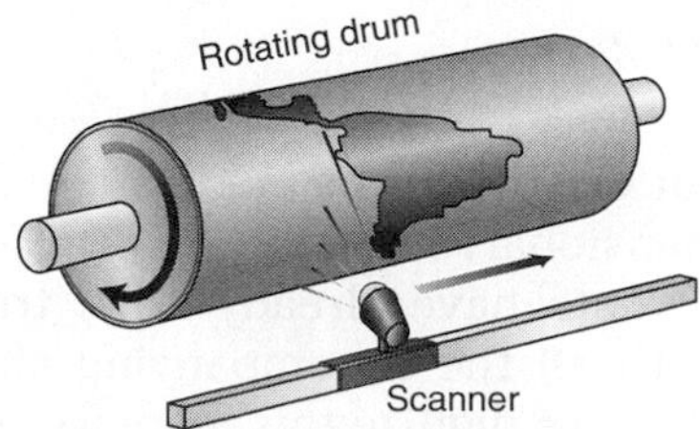

Figure 6.3 Principles of map scanning. Diagram of a drum scanner showing the rotation of the drum and the movement of the scanning device along the drum.

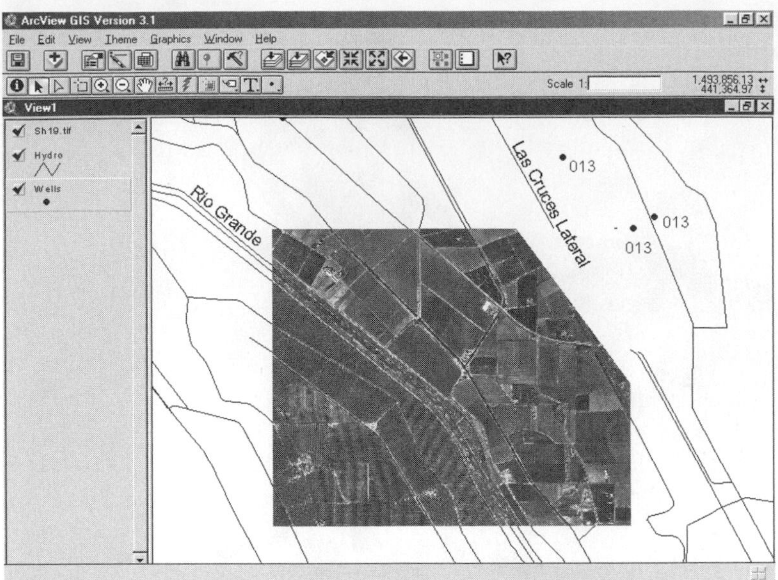

Figure 6.4 Scanned aerial photograph. The relationship between the point and line features and the natural environment is much clearer when they appear in context.

provide a much improved method of producing additional layers by simply tracing the data directly as observed on the aerial photography. Additionally, for the GIS analyst, the digital photography also provides context for the digital cartographic data, thus enhancing the visualization and improving understanding (Figure 6.4).

Scanning, especially of complex maps, is not a panacea for digitizer woes. Even with large-format commercial scanners, the editing process requires human intervention. And these scanners are still more expensive than most small GIS operations can afford, so scanning the occasional piece of large cartographic and imagery data will typically be performed by professional scanning firms for the near future. Trade shows are showing more and more large-format scanning devices at prices that are far below those of only a few years ago. Beyond the technical specifications of these devices, you need to examine the volume of scanning that you or your company is likely to perform before deciding on the purchase of one of these devices. Still, you are likely to be doing manual digitizing for many maps for which the large-format scanning devices are either unsuitable or for which the cost is prohibitive.

Reference Frameworks and Transformations

Digitizers are typically designed to input existing maps, and as you have seen, maps are representations of a three-dimensional reference globe projected onto a flat surface. That is, our geographic data have already been transformed from their original 3-D form to 2-D, with all the accompanying shape, area, distance, and angular deformations. When we digitize this map, we reduce this sophisticated projection to a set of Cartesian (in this case digitizer) coordinates. Before we do this, we normally provide the GIS software with information about the input projection and specific information about the datum, the grid system, and the zone or zones of origin within which it was produced. This preparation allows us to transform from digitizer coordinates to the projected map coordinates after input. In fact, the GIS will produce a number of transformations of this kind as we project from the Cartesian coordinates on the digitizer to the two-dimensional map projection coordinates, and from there through a process called the **inverse map projection** to three-dimensional latitude and longitude coordinates. From there we will eventually need to reverse the process to produce Cartesian coordinates for the output device (Figure 6.5).

To perform these projections and manipulations, the GIS software will need to perform a number of graphical manipulations, all of which can be accomplished by combinations of three basic transformations: **translation**, **scale change**, and **rotation**. *Translation* is the movement of a graphic object to a different location on the Cartesian surface. This is done by adding or subtracting the coordinate values necessary for the X and Y coordinates of the object (Figure 6.6a). In other words, the new X coordinate X' for each graphic object will be equal

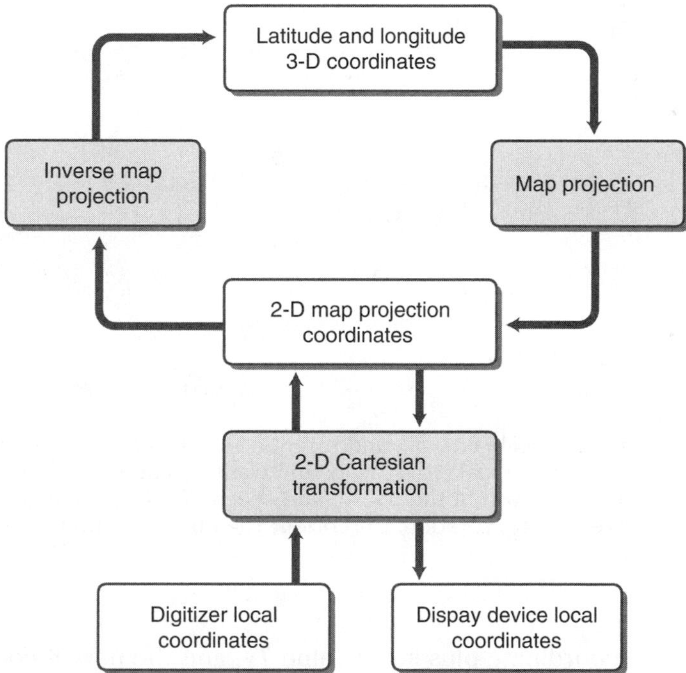

Figure 6.5 Geographic information system (GIS) coordinate transformations. Transformation steps from Cartesian (digitizer) coordinates to two-dimensional projected map coordinates, through an inverse map projection to longitude and latitude coordinates. To produce map output, the process is reversed through the projection process and finally to the necessary Cartesian (display device) coordinates. *Source:* Duane F. Marble, Department of Geography, Ohio State University, Columbus, OH.

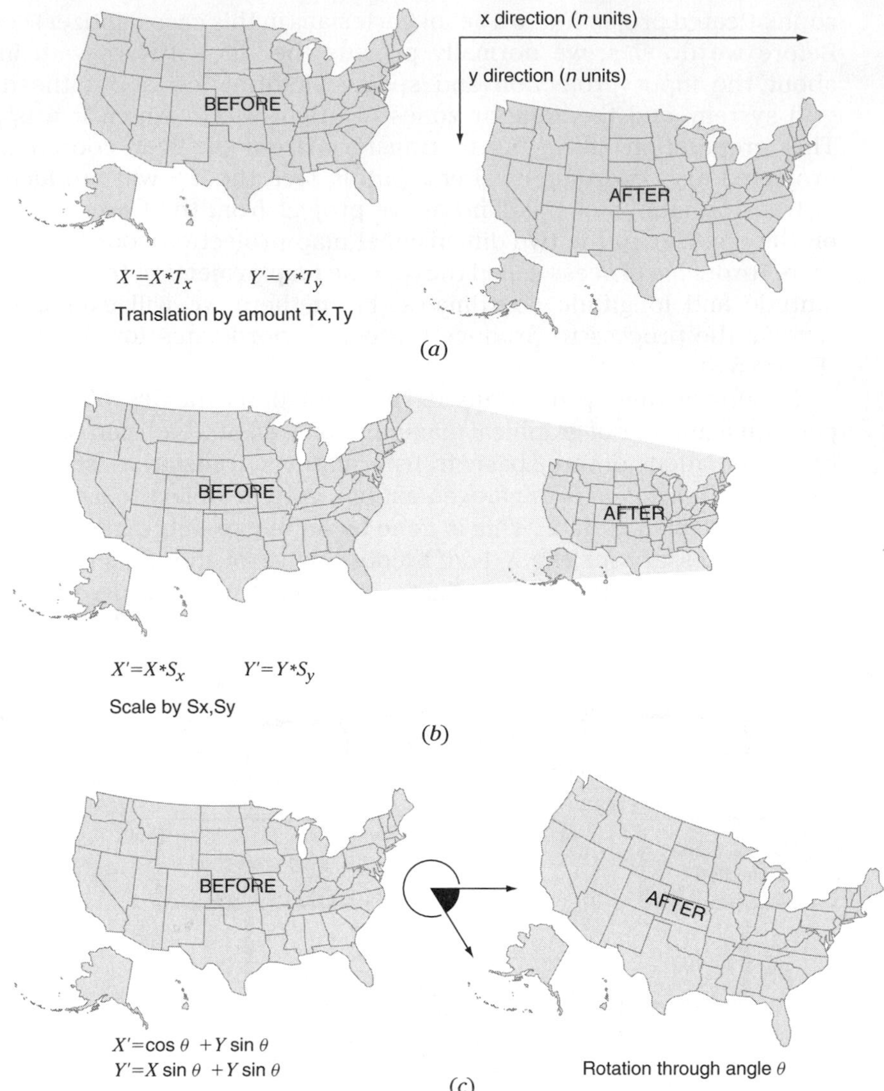

Figure 6.6 Translation, scale change, and rotation. The three basic graphic transformations necessary to make the projection transformations: (*a*) translation, or object movement within the coordinate space; (*b*) scale change, to adjust the object size; and (*c*) rotation, to reorient the object within the coordinate space.

to the original X coordinate plus some value T_X, and the new Y coordinate Y' for each graphic object will be equal to the original Y coordinate plus some value T_Y (see Box 6.1). *Scale change* is also relatively useful because of the need to compare differently scaled maps and to output in different scales as well (Figure 6.6b). This is done by multiplying the overall X coordinate extent (e.g., the two sets of X coordinates for a line segment) by a scale factor s_X, and each set of Y coordinates by a Y scale factor s_Y. *Rotation* is used frequently during the process of projection and inverse projection. It is accomplished by using basic trigonometry (Figure 6.6c). For X coordinate locations, the new X location

BOX 6.1 Translation, Scale Change, and Rotation

Translation Formula

$$X' = X + T_X \quad \text{and} \quad Y' = Y + T_Y$$

where
T_X and T_Y can be either positive or negative

Scale Change Formula

$$X' = Xs_X \quad \text{and} \quad Y' = Ys_Y$$

where
S_X and S_Y represent the amount or percentage of scale change

Rotation Formula

$$X' = X\cos\theta + Y\sin\theta \quad \text{and} \quad Y' = -X\sin\theta + Y\sin\theta$$

where
θ is the desired angular displacement

X' is found by multiplying it by the cosine of the new angle (θ), then adding that value to the original Y coordinates multiplied by the sine of theta (sin θ). The new Y coordinate locations Y' are found by multiplying the negative of the original X value by the sine of the angle and again adding that to the product of the Y coordinate and sin θ.

These three equations illustrate the three basic transformational operations, but in practice we don't use the cosine and sine formulas because of the difficulties with angular computation. For ease of computation we normally use what is called an **affine transformation** formula, which will take care of translation, rotation, and scale change just by solving the equations with control points.

Map Preparation and the Digitizing Process

We cannot begin digitizing until we have provided the GIS software with pertinent information regarding the projection, the grid, and so on. This part of the map preparation process is often neglected, yet it is so important to developing a useful database. Many software packages require you to provide this information before you can begin, although some will allow you to input the information later. In either case, you should prepare the information beforehand and keep it handy so that you will always know what it is and where to find it.

It is also a good idea, before you place your map on the digitizing tablet to begin the input process, to prepare the map by making appropriate marks directly on the map document or on a firmly attached clear plastic covering, to identify the exact locations you will be digitizing (Figure 6.7). Remember, there will be many curved lines, and you will have to reduce the curves to short,

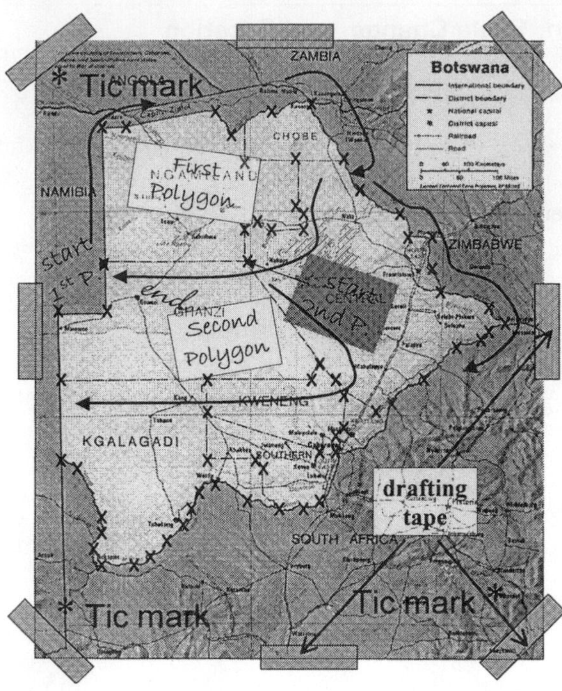

Figure 6.7 Map preparation. Notice the markings used to illustrate where the tic marks are, locations of digitizing points, the order of digitizing, and the start and ending for each polygon.

straight-line segments. Although many prefer to digitize freehand, eliminating this useful step, if you know all the points that will be digitized and their sequencing, you will not need to make these decisions while you are digitizing.

Because digitizing is tedious work, you will probably digitize the map in portions rather than all at once. This is all the more reason to prepare the map ahead of time by outlining directly on the map portions you expect to be digitizing at each session. Because you will probably be digitizing in multiple sessions, sometimes requiring you to remove the map and later replace the map, you will need to provide an adaptable coordinate set. The first necessary points for this coordinate set, called **registration points** or **tic marks**, will be entered in digitizer inches as well as in map coordinates. They should also be marked on the map as part of your map preparation, so you, as well as the computer, know your starting points. The registration points provide an outside frame for the document and should be outside any graphic object you will digitize, including a neat line if you will include that in your database. Usually three points located on the corners of a rectangle are needed to define the map area for the software. Some software can get by with only two points if they are located on a diagonal. In this case, the software assumes that the outside boundary is a rectangle and infers the other two corners. No matter which technique you are required to use for your software, the accurate locations of these registration points or tic marks is absolutely essential to ensure good quality. Special care should be taken to locate your tic marks precisely. It is a good idea to double-check these, because if registration marks are misplaced, virtually all the remaining digitizing will be erroneous.

Most digitizing software will assist you in assuring the accuracy of your tic marks during map registration prior to digitizing. As you select the tic marks, the software will provide you with an error measurement using a method called

root mean square (RMS) error. Mathematically this is the spatial equivalent of standard deviation in that it provides a measure of deviation or discrepancy between the known point locations and their digitized locations. The lower the RMS error, the more accurate the digitizing or transformation (Walker, 1993). Because RMS error is expressed as a single number, it does not provide information about each tic and its real coordinates. Rather, it indicates that the actual error might vary across the map depending on the number, placement, and accuracy of the tics used to register the map.

Other map preparations include a clear definition of the order in which you intend to digitize objects; a systematic approach to identify the portions of the map to be digitized at each session has been mentioned. It is also a good idea to develop a method of identifying which areas (sections, lines, points, etc.) have already been digitized. Taking periodic breaks to mark the map helps you to keep track of your progress. Your software may require you to identify nodes, left and right polygons, and so on, depending on its sophistication and the graphic data model it uses. These designations should be placed directly on the map document as well; this spares you from having to stop frequently to retrieve the information.

Most digitizing software provides editing capabilities to help identify mistakes, and some allow you to edit as you digitize. Many packages include a feature that allows for the occasional shaky hand, namely, a sort of fudge factor, sometimes called a **fuzzy tolerance**. The fuzzy tolerance may have to be set before digitizing, or it may be set during editing. Small fuzzy tolerances are less forgiving of digitizing errors and may result in gaps between points that were meant to be connected. Alternatively, a fuzzy tolerance that is too large will result in the merger of points and lines that are real because the software assumes the gaps were simply mistakes. Marble et al. (1990) discusses the entire digitizing system, especially with regard to organizations and commercial operations.

Another map preparation function deals with the tendency of the source material to shrink and swell with changes in temperature and humidity. A stable material, such as Mylar, is preferred to paper for digitizing. Although Mylar also shrinks and swells, it reacts far less than paper. Although this matter of digitizing media may seem trivial, consider that a change in only one millimeter on a map at a scale of 1:50,000 means that you have a potential error of 50,000 mm on the ground (50 meters).

There are several ways to limit digitizing errors based on media fragility. First, the room should be air-conditioned to maintain a standard cool temperature and low humidity level. The material you are about to digitize should be allowed to stay in the room for several hours, unrolled (it is a good idea to avoid using folded maps because creases reduce the accuracy of such documents severely). Once acclimated to the room conditions, the map can be taped to the digitizer tablet using a removable tape material, such as drafting tape. Do not use masking, clear plastic, or other very sticky tapes that may tear the document or even remove information when the document is moved. In addition, very sticky tape can make it difficult to remove the document, stressing it unnecessarily and possibly resulting in stretched media. When you tape the map to the digitizer, place it several inches from the edges to avoid any possible skew (horizontal distortion), which is most likely to occur there. The map should also be placed so that it does not cause you to stretch to locate the puck on the map objects, since this can put stress on the document and can also limit your freedom of movement, adding more error to your database. When you digitize a map in multiple sessions, be sure to keep the document in the climate-controlled room to avoid expansion and contraction.

WHAT TO INPUT

Now that you have some basic guidelines on how to digitize, you can begin to select the appropriate data for input. A major factor suggesting what to input is the target audience—the user. Historically, a common practice in many GIS, including those designed for whole states, has been to input everything (DeMers and Fisher, 1991; Fisher and DeMers, 1990) regardless of whether it will actually be useful. This often destines a system to failure. Thus, *rule one* is to determine why you are building the GIS database in the first place. Although a really nice map of quaternary geology may seem natural as an input theme, it is unlikely to be useful for a study of atmospheric pollution caused by factory smokestacks. In short, the themes should be directly related to the modeling and analysis you plan to perform and to the results you desire (sometimes called a **spatial information product (SIP)**. This suggests *rule two*: define your goals as specifically as possible before selecting the layers. It is easy to see how these two rules relate to each other.

Even with very specific goals and known spatial information products, there will be multiple ways of obtaining the available data in some cases. For example, coordinate locations and elevation variables can now be obtained with the use of GPS units. But they may also already be available on maps in a reasonably accurate format. Or data on current land use may be obtainable from ground surveys, aerial photographs, satellite output, airborne scanners, or any number of other sources. There is no easy answer to which should be used. While there is no absolute recipe for success, there is a recipe for failure, which leads us to *rule three*: avoid the use of exotic sources of data when conventional sources are available, especially if the latter provide a similar level of accuracy. You will need to define exotic for your own particular project. In general, I like to use a practical definition, applying the term to any data source I am not familiar with. If you or another team member is familiar with a given set of data and can comfortably use it correctly, and if it increases the utility or accuracy of your database, it should be applied. If all your multiple data sources for a particular theme or coverage are in traditional form, then invoke *rule four*: use the best, most accurate data necessary for your task.

Accuracy here refers to the necessary accuracy, not an absolute accuracy. If you don't need 1-centimeter contours for your topography, for example, use topographic data that most closely matches your level of observation. Although having an extremely detailed map of any coverage may seem advantageous, it is costly to input; it also slows analysis and may even make analysis more difficult. An example of the use of 30-meter TM data from the LANDSAT satellite as compared to 80-meter MSS data from the same source might prove instructive. Suppose your purpose is to identify large fields of grain. Since the enhanced spatial resolution of 30-meter data has been known to produce many difficult-to-separate categories over an area that is essentially all grain fields, the better resolution would confuse your situation rather than simplify it. And, of course, the computing and human resources needed to provide clarification would increase the overall cost of the system. Thus we have *rule five*: remember the law of diminishing returns when deciding on data accuracy levels.

Most thematic maps contain ancillary data on roads and other anthropogenic features that may very well be useful as input to the GIS. Whenever possible, and when the quality of the data dictates it, you should input these data as separate layers from the same map sheet—that's *rule six*. This rule does not negate the use of other sources when they are of superior quality or accuracy, but it has two advantages. First, because the data are on a single map, you don't need to go to multiple map sheets and then repeat all the preliminary steps in map preparation. Second, because the data are on the same map sheet they are already co-registered, reducing the need to perform this sometimes difficult task later.

The final general rule, *rule seven*, is that each layer should be as thematically specific as possible, without relying on a binary (0 and 1) system. The more specific a theme, the easier it is to search if you need to know something pertaining to data contained in a single layer. In addition, when you perform complex operations it is easier to keep track of the process if you are completely familiar with the original categories and can identify their transformations.

We can summarize the seven rules in a very few simple statements. First, define your purpose. Then be sure your maps address the purpose. Use the most accurate maps needed for the purpose—not too accurate for your needs and not too inaccurate to do the job. Keep your themes simple, and use the same map to obtain these simple coverages whenever reasonable and prudent, to avoid the need to co-register them. Above all, think about your project before you begin inputting data. Data input takes time and costs money.

HOW MUCH TO INPUT

Related to the question of what to input is the question of how much. Too much data input and your GIS must bear the added weight of the data throughout the project life cycle; too little data and you may find yourself unable to answer questions about matters you had planned to cover. The input of data is a sampling process. In vector GIS, each line you input will likely have some curves. To produce a reasonable facsimile of the line using straight-line segments, you will decide thousands of times where to place the digitizer puck and where to record data. This process is very similar to that of line generalization, encountered in our earlier discussion of cartography. A simple rule of thumb is to take more samples (i.e., record more points) for very complex objects than for simple lines (Figure 6.8). The locations of a straight line can be recorded accurately with only two points. Additional data points are unnecessary and will result in slower computation.

Line and polygon complexity can be compared to information. The more the line changes direction, the more information it conveys. The more densely packed the points, lines, and areas, the more information content the map has. Higher information content necessitates a higher sampling rate. This construct, called **information theory** (Shannon, 1948), provides a useful measure of sampling rate that can easily be applied to the digitizing process. In the original theory the smallest object to be included in a system should be recorded at least

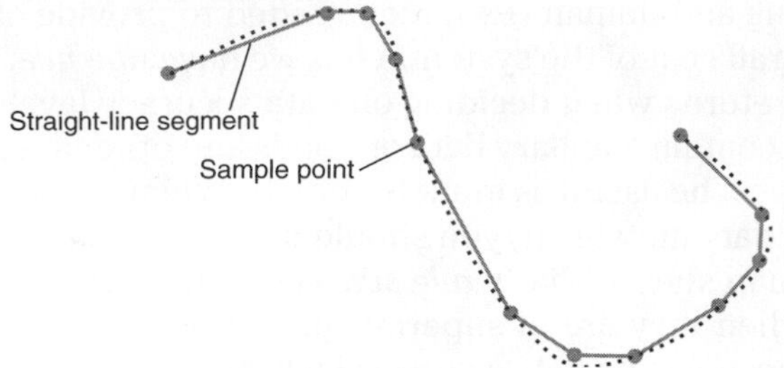

Figure 6.8 Sampling a complex line. An example of sampling for digitizing a line. Points to be digitized are selected based on change of direction of the line. Each direction change indicates an additional piece of information the map contains.

twice. So, for each change of line direction (i.e., for each piece of information), you should record at least two data points. This is all the more reason for careful map preparation. You should also keep in mind that there is a direct relationship between the complexity of the map, or the volume of data in the map, and the spatial data handling problems (Calkins, 1975).

The idea of information content can also be applied to raster data. Once again, the general rule is this: the smaller the object to be identified in your database, based on your modeling needs, the smaller the grid cells need to be (DeMers, 1992). This principle often determines the selection of grid cell size (resolution) for the entire database. While employing information theory in raster you need to remember that grid cells are two-dimensional, requiring you to think of information theory in two dimensions. For example, let's say that you want to use grid cells to represent farms displayed on a map. If the smallest farm is, say, 40 hectares, you would need to sample this area at least four times (twice in each dimension) to ensure that it has been captured in your GIS. Stated differently, it means that the grid cells will need to be 10 hectares or smaller to allow the capture of the 40-hectare object in your GIS database. This also assumes, by the way, that four grid cells will surround the field itself. Suppose the field lies along a long, skinny riparian area. Although it is 40 hectares in total area, it is also spread out as a linear object, reducing the chances that the entire field will be input into your GIS. This aspect of the process is somewhat dictated by the method you input the grid cells (covered in a later section). The general rule of thumb, however, remains the same: sample more for more information content.

Whether in raster or vector, sampling is dependent on the amount of area covered by the map and the use for which the data are input. Small-scale maps, those covering large amounts of space, contain a much more abstract view of the land surface being mapped. In addition, the lines and symbols on the map take up space. The amount of error contained in a symbol is dependent on the scale of the map on which it is placed. Lines on small-scale maps take up more land space than same-sized lines on large-scale maps. This physical condition, called **scale-dependent error**, is an indication that the amount of error is directly related to the scale of the map and needs to be considered during the map preparation phase prior to digitizing.

METHODS OF VECTOR INPUT

You have seen that there are numerous instruments available to input vector data into a GIS. We will restrict our discussion to manual digitizing because of the frequency with which it occurs. After preparing your map and placing it on the digitizing tablet, you will need to use the digitizer puck to locate and record the registration marks. Some software requires these marks to be recorded in a specific sequence, others do not. Your particular software package will indicate which numbered keys on the digitizer puck you need to enter for specific object types. Some numbered keys will be used to indicate the location of a point entity, others for beginning and ending of line segments, still others for the closure of a polygon. Many digitizing errors, especially those made by novices, are due to pushing the wrong-numbered button.

The exact method of digitizing is also related to the data structure on which the software operates (Chrisman, 1987). Some (e.g., the POLYVRT structure) require you to indicate the locations of nodes, whereas others do not. Some require you to encode explicit topological data while you digitize; others will use software to build in the topology after the database has been produced. The rules are different for each package, and you will have to look over the appropriate documentation to determine these strategies beforehand. This should be considered part of the map preparation for digitizing rather than the digitizing itself. Paying attention to the appropriate button codes will ensure that your lines don't become nodes and your nodes don't become just points. Some find it useful to tape a crib sheet to the corner of the digitizing tablet until they become more familiar with the process.

In vector GIS the attribute data are most often typed in using the computer keyboard. Although simply typing data in is extremely simple, the task requires the same care used for inputting the entities. This is true for two reasons. First, it is very easy to make typing errors, and second, and more importantly, the attributes must be correctly attached to their entities. These latter situations have the potential to produce some of the more difficult errors to detect because they cannot always be visually identified and often do not appear until an analysis is under way. A good practice is to check attributes while they are being input, perhaps by taking frequent short breaks to look them over. Time spent doing this on input will save a great deal of time during editing.

METHODS OF RASTER INPUT

Raster data input employs some different strategies than its vector counterpart. In the first place we must decide how much area should be occupied by each grid cell. This decision must be made prior to digitizing and is also based on the overall size of study area and scale of input. You will also need to decide how each grid cell will represent the different themes that will occur. Beyond the grid cell resolution, this may be the most important decision you will have to make.

Normally in raster data input the digitizer records vectors that are then converted to raster format. The entities and attributes are generally entered simultaneously. This approach, requiring vector-to-raster conversion, can sometimes be iffy depending on the sophistication of the digitizing software used. The problem arises primarily in the conversion process, which is often not well documented by the software vendor. Most often, difficulties extend from digitizing adjacent areas using vector lines that are then converted to two separable polygons. In such cases the software must decide which polygon will contain the grid cell through which the line runs. (Remember that in many grid-based systems a grid cell can contain only a single value.) The decision is sometimes based on the "last come, last coded" rule. That is, when the same line is digitized first for one polygon, and then again for the second polygon, the second polygon will be assigned to that grid cell. In a more computer-oriented approach, the assignment will depend on which value occurs first inside the computer. This is based on the push-pop strategy of the computer stack. Although these computationally convenient methods are fast, they often produce unacceptable errors along edges, especially if the grid cell resolution is coarse. For very large raster databases this level of inaccuracy may not present a substantial percentage of error compared to the entire database, but measurements of shape and area will be compromised. A better method, at least from the standpoint of the utility of the database, would be to allow the system to select from one of four systematic input methodologies outlined by Berry and Tomlin 1984.

In the first of these methods, the **presence/absence method**, for each grid cell on each coverage a decision is made on the basis of whether the selected entity exists within the given grid cell, hence the name "presence/absence" (Figure 6.9a). A major advantage of this method is that decisions are easy. No measurements are necessary. A simple Boolean operator either is there or isn't.

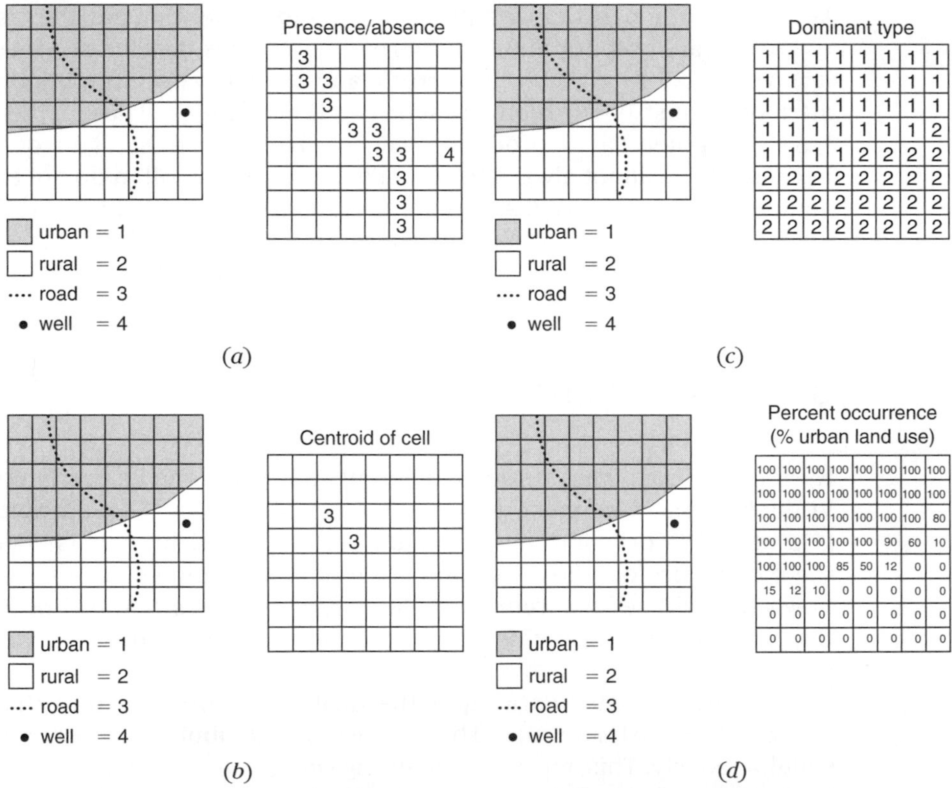

Figure 6.9 Methods of raster data input. The four basic methods of raster data input as defined by Berry and Tomlin (1984): (*a*) presence/absence, (*b*) centroid-of-cell method, (*c*) dominant type method, and (*d*) percent occurrence method.

Presence/absence is the best method, in fact the only useful method, for coding points and lines for grid systems, because these entities do not normally take up a large portion of a cell's area. Thus, if a road crosses through a grid cell, its presence is recorded with an attribute code (a number); if it does not, it is ignored.

The second method of raster data input is called the **centroid-of-cell method** (Figure 6.9b). Here the presence of an entity is recorded only if a portion of it occurs directly at the central point of each grid cell. Clearly this requires substantial calculations, since each central point will need to be calculated for each grid cell, and then the object will have to be compared with the location of that point. Of course, if you are using a clear Mylar grid overlay this becomes a visual rather than a computational approach, but still you must evaluate each grid cell centroid against each entity. As you might imagine, the chances of point entities and line entities passing directly through the center of a given grid cell are minute compared to the chances they will occur anywhere else. Therefore, the use of the centroid-of-cell method should be restricted to polygonal entities.

A more common method of coding area data is also generally considered to be the best method. This so-called **dominant type method** (Figure 6.9c) encodes the presence of an entity if it occupies more than 50 percent of the grid cell. Under most circumstances the decision is straightforward and the coding is reasonably representative of what is there. It seems logical that if you are restricted to a single category for each grid cell, the one that occupies the most space should be coded. Computationally, this requires the computer to determine the maximum amount of each polygon for each grid cell—again a computationally expensive strategy. Performed manually with a Mylar grid, the method is reasonably easy, since each grid cell can be given a quick visual inspection. In two cases, however, this approach may not be satisfactory. First, there may be highly irregular or elongated polygon shapes, and features that start out long and sinuous, like a stream, covering little of the grid cell, then become quite bulbous within the same cell. It is difficult to decide such cases by visual inspection, and computational approaches making use of a digitizer are likely to be preferable. In the second problem case, three or more polygon types converge in an irregular pattern within a single grid cell. With three or more polygons in a grid cell, the chances are slim that any one will occupy more than half of the area under the cell, and so you cannot use the 50 percent value as your cutoff point for input. Instead, the amount of area for each must be visually or computationally determined and the largest one encoded. This situation is likely to arise when you are encoding a map that is at a small scale but retains a high amount of areal detail—for example, a detailed land use or land cover map produced as a hard-copy map from satellite data at a scale of perhaps 1:1,000,000. Difficulties may arise, as well, when you are trying to encode a map of physical features such as soils or vegetation. Natural features tend to be quite irregular and their polygons much more sinuous and intermingling. Although these examples do not mean that all detailed or complex maps are to be avoided as raster data input, they may suggest the use of the centroid-of-cell method as an alternative to simplifying the input process. This is especially true if the complexity occurs uniformly throughout the map.

A final method of raster data input, called **percent occurrence**, is also used exclusively for polygonal data (Figure 6.9d). The idea is to give more detail, not by coding just the existence of each attribute but rather by separating each attribute out as a separate coverage, then recording the percentage of the area of each grid cell it occupies. For example, a map of land use divided into urban and rural categories would be separated into two more specific themes, one urban and one rural. The percentages urban and rural would be recorded for each grid cell, with 0s entered to indicate nonappearance of a category. If the percentages are accurately calculated, the urban and rural maps should be perfect complements.

The percent occurrence method provides detailed data about each attribute. However, the disadvantages are numerous. First, of course, both for automated and visual approaches, the amount of decision making is greatly increased. Second, we need to cross-check both complementary themes to ensure that the percentages for each grid cell add up to 100 percent. Finally, as in the case of the dominant type method, we have the problem of multiple categories in a single grid cell. In this case, however, the problem may be even more severe because the more categories we have, the more themes we produce: a map of land use with 15 attributes explodes into 15 themes, each one representing a single attribute. Before the percent occurrences method is selected, the advantages of the additional data detail should be weighed carefully against the input problems.

Some raster GIS software allows a fifth, nonsystematic method of raster data input beyond the four systematic types. Known as the **most important type** method, this approach employs the user's knowledge of the data being input. With this approach the user may make deliberate decisions as to which category or class of data should be assigned to a given cell. If, for example, you are digitizing a set of land cover categories and water is not going to be included in the analysis, any other type of category may take precedence over water when coding. This way all possible dry land uses will be included in the database and may even be overcounted, while the less important "water" category will be underrepresented because it is of limited, or in this scenario, no use.

No matter which of the five methods of raster data input is used, the choice usually falls to the software to determine. The GIS will either have its own predetermined input rules, or more commonly it will prompt the user to select which method is to be used. Additionally, when data are converted from vector to raster, the same decisions apply. For quality control it is best if the software allows the user to define the rules rather than rely on random or unspecified methods. The larger the grid cell size used, the more impact the method of raster data input or raster data conversion has on the final product and on any modeling that takes place subsequent to input or conversion.

REMOTE SENSING DATA INPUT

Among the primary data becoming increasingly available are those received from aircraft and satellite remote sensing. The utility of digital and analog forms of remotely sensed data sources for GIS is unquestioned, especially for tasks such as rapid data updating and building temporal databases for large areas. Most digital remote sensing data are based on a raster format, with each grid cell (called pixels) recording electromagnetic radiation as a number of radiometric values. The number is dependent on the type of system used. For example, LANDSAT data have a radiometric resolution of 256 grayscale levels, whereas AVHRR data derived from the NOAA satellite have 1,024 radiometric levels. In either case, input to a raster GIS is easily accomplished because of the similarity of data structure. Even so, the raster data structure of satellite remotely sensed data should not drive the choice of a raster GIS data model over a vector model. Rather, the choice should be based on the use for which the database is being built. In addition, when any kind of remotely sensed data are being considered as input to the GIS, the data must be evaluated for cost, utility, and accuracy compared to data from alternate sources. Remember input rule three: avoid exotic data whenever possible. Of course, a familiarity with remotely sensed data and their spatial, spectral, and radiometric characteristics may make the choice of remotely sensed data advantageous.

Aerial photography has long been a primary source of base map data for many common products. The USGS topographic maps are largely compiled and revised from viewing stereo pairs of aerial photographs, and soil maps depend on aerial photography as well. Because many maps come from aerial photography, it is wise to find out whether maps derived from such photographs already exist before choosing to use this mode as direct input. Of course, the choice may also reflect the data that were mapped from the aerial photography as well as the classification system used. The two primary stumbling blocks to the use of aerial photography as direct input to the GIS are the relationship between the classification needed and an ability to obtain those classes from the photography and, of course, the problems of rectification and lack of reference grid. Prior to input you need to determine how much distortion your project can tolerate and what specific categories of data you will need to perform your analysis. When these categories are known, they can easily be matched to the scale and spectral sensitivity of your photography.

A special type of aerial photograph deserves mention because the images do not contain the scale, relief, and tilt distortions normally characteristic of aerial photographs. These products, called orthophotographs, or orthophotoquads if they are based on the areas occupied by topographic quadrangle maps, are photographic images of the earth that resemble maps in that they have a single scale. Orthophotographs are subjected to a process called **differential rectification**, which involves point-by-point correction of the scale and relief displacements normally caused by differences in elevation between the aircraft and the topography over which it flies. Although a detailed description is beyond the scope of this book, it is important to note the availability of these products (both in analog and now in digital forms) as sources of input. Refer to Lillesand and Kiefer (1995) for more information on rectification. If digital orthophotographs are not available, the analog versions can serve as an excellent source of manual data input to the GIS. In our discussion of digital remote sensing in Chapter 2, we noted that there are generally two major products derived for input to the GIS: digitally enhanced imagery (designed to highlight certain features for analysis, such as edges) and classified images (obtained through complex computer manipulations designed to replace the visual analyst as a classifier of features). As an input to GIS, these classified images will most likely be used to update and/or compare their classification to classified data already inside the GIS. Even when two sets of classified data are obtained through digital remotely sensed image sources, the comparison is difficult. In fact, remote sensing scientists often prefer that these comparisons be made on the raw, unclassified images because of the potential for confusion in classification (Haddad, 1992). Indeed, direct comparisons of classified digital data to maps classified from aerial photography or historical data are very difficult. Let's look at some of the technical difficulties associated with remotely sensed data input to the GIS as defined by Marble (1981) and Marble and Peuquet (1983).

Satellite data require preprocessing to remove geometric and radiometric flaws resulting from the interaction of two moving bodies (the satellite and the earth), sensor drift as the satellite systems age, and differences in atmospheric conditions. Techniques for correcting radiometric difficulties are readily available with most digital image processing software, and the necessary equations are quite easily obtained. For GIS input, the major problem related to preprocessing is a need to obtain geometrically correct ground positions for the imagery. This geometric correction requires a number of **ground control points (GCPs)** within the image to place it in a correct spherical coordinate space on the surface of the earth. There should be a reasonable number (the more the better) of GCPs, and they should be evenly spaced. In some areas it is reasonably easy to use devices such as GPS units to obtain very precise locations, but the ground points must also be observable on the imagery.

Obtaining adequate GCPs may be quite difficult, especially in areas like the tropics where the forest canopy is so dense that the GPS has no direct line of sight to the satellite. In addition, even if line of sight exists, it is often difficult in such areas to find features that are distinguishable from the imagery. Care should be taken when using imagery lacking in GCPs because their absence degrades the coordinate accuracy severely, especially at the margins of classified areas. The issue of GCP accuracy should indicate the importance of coordinate accuracy to the overall functioning of a GIS. A proper geodetic framework will always improve the utility of a GIS to perform its measurement and other analytical functions.

The second major technical problem with using digital remotely sensed data is classification. If your project does not employ direct biophysical inputs to the GIS, you most likely will have to convert from interval or ratio data to a nominal classification scheme. Remember that the classification process used in image-processing software is often based on the ease with which it can be obtained. That is, the classification methods are designed to improve the ability to obtain an adequate classification on the basis of the image content, while the utility of the image in an integrated GIS database is not always addressed. Supervised classification, requiring human intervention, is generally an improvement over unsupervised classification, because the process can be more easily controlled to conform to user needs. Nevertheless, the question of correspondence to classifications for comparable themes remains a major consideration.

The classification of satellite imagery also implies that the results are accurate, not just compatible with existing map layers. Lillesand and Kiefer (1995) have shown that the ability of image-processing software to produce classifications far exceeds our ability to assess the accuracy of the classification. This is especially true of comparisons of multiple-date imagery, where the accuracy of each data set must also be compared to the accuracy of the amount of change between the two dates. It is often best to use ancillary data to produce the classes from digital imagery for input to a GIS. The insertion of topographic data, preclassified data, rule sets, and other techniques will generally result in a much improved classification, and one that more easily conforms to the corresponding GIS map layers.

GPS DATA INPUT

An increasingly important source of primary data for GIS databases is GPS data. Not only does the modern GPS allow for the development of a highly accurate geodetic framework, it is often linked explicitly to the input of other primary data sources such as field data. With the advent of new handheld computer technology in the form of PDAs and iPAQs with full color displays and software packages like ArcPad, we are able to display digital map layers in full color on handheld units to link those same layers to GPS coordinates, and to include a complete set of field notes. The field notes can contain anything from descriptions of conditions at any given location, to digital voice recordings, digital pictures, and even geometry. By taking multiple readings, for example, one could easily describe the accurate area and perimeter of area objects (such as forest patches or subdivision boundaries for urban subdivisions). This is especially useful as a total mobile solution in areas that are not well surveyed or for field checking satellite imagery for environmental surveys. The computer and GPS technologies have already been integrated to allow out-of-the-box single-unit solutions (Figure 6.10). The advantages of mobile GIS/GPS solutions include reductions in time and costs in development of geographic databases, the simultaneous coregistration of datasets, rapid GIS and remote sensing data verification, incorporation of ancillary data and descriptions, and the easy transfer of these data to full GIS databases composed of both primary and secondary data.

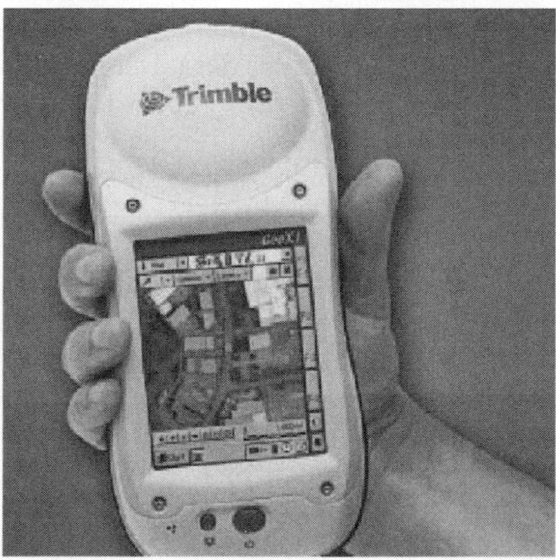

Figure 6.10 Integrated iPAQ and GPS technology illustrated by this Trimble® GeoXT GPS unit with built-in Hewlett Packard iPAQ® and ESRI ArcPad software.

SECONDARY DATA

An efficient method of building a GIS database is to limit the amount of time and costs necessary to develop databases. Fortunately, an increasing supply of digital databases is becoming available. Digital elevation models, digital orthophotoquads, and digital line graphs are available from the USGS as well as third-party vendors (Appendix A). The U.S. Census Bureau has TIGER and dual independent map encoding (DIME) files and associated attributes via their summary tape files (STF), and similar data are available from the Canadian Census Bureau (Appendix A). The U.S. Department of Agriculture makes soil maps available in digital form. There are many more examples, and there are increasing numbers of sources for shared data via the Internet.

The availability of databases also introduces other problems, including some you will also encounter as you input data to your GIS. The first is the physical format of the media. Countless hours can be spent trying to obtain digital data in the proper format. Computer data formats are changing rapidly, and this poses problems for the acquisition of data, especially archived data. Older data storage devices involved, for example, 9-inch tape, 8-millimeter tape formats, floppy disks (8″, 5 1/2″, 3 1/4″), and ZipDisks. Today we have various sizes and formats of CD-ROM and DVD technologies, memory sticks, and flash drive technologies, all of which are changing very rapidly. Our task is to acquire the data in a format compatible with our retrieval equipment. Even when the technologies are the same the data may be stored in incompatible formats. This suggests that you be aware of the available formats and data exchange standards (Moellering, 1992).

A more insidious problem with external databases deals with quality. Although some third-party data vendors may provide easier access for data than could be obtained from government bodies, you need to be aware that the service may not supply data in the original format. Some data, no matter the source, will be filled with easily viewable errors, some systematic and correctable, some not. You need to be aware of the quality control procedures used by each vendor. In addition, you need to know what your options are concerning the return of poor-quality data. Ask where the data were obtained. Were they created in-house by qualified professionals, or were they obtained from digital sweatshops, frequently operated by severely underpaid and untrained individuals? All these questions are vital to the utility of the data. Details should include the specific format provided, the quality control procedures under which it was developed, the quality you can expect, the return policies of the vendor, and any other pertinent information that will help guarantee successful integration of the data into your GIS. Vendors that fail to comply with these requests should be considered suspect.

A major problem that is often encountered with the use of an external database is one you should also take to heart as you prepare your own databases. Databases require information about their own content; such **data dictionary** material amounts to a catalog of your data content. There are two general forms: active and passive. Passive data dictionaries might include scale, resolution, names of the database fields, codes used, and what they mean. Imagine obtaining a database from a vendor that includes a category called "wetlands." The definition may seem self-evident to you, but you need to know more about the vendor's criteria for establishing this category. Remember, one person's wetland may be another person's watered lawn. The data dictionary should provide enough detail to ensure that any analysis based on it will be valid. This, of course, should remind you to keep a clear and concise record of your operations in a form that will enable someone unfamiliar with your original database input procedures to recreate them.

Active data dictionaries operate on the GIS database by performing checks for correctly coded inquiries. For example, if your vector GIS database management system is set up to allow only a four-digit code for a particular entity, the active data dictionary could check each inquiry to determine whether this four-digit limitation is uniformly met. Such checks are quite useful for allowing proper functioning of the system and for preventing erroneous results from incorrect input requests.

Beyond their technical problems, external databases are accompanied by certain fundamental legal and institutional problems. A major institutional problem is that external databases are often hard to find, especially if they were produced by government agencies that may be tasked with dissemination but not promotion and advertising. There are an increasing number of efforts to consolidate GIS database catalogs at a variety of private, government, and commercial levels to facilitate searching. This is currently being done on a piecemeal basis, however, and ignorance of available databases often results in costly data redundancy. The cost of the data is also an institutional problem that may limit access. It is not the overall cost per unit that is prohibitive, but the frequent practice of providing data in large units that cover far more than a user needs.

Among the thornier issues facing the GIS user today is the fairness of having to pay for data produced through public funds. Tied to this, of course, are the twofold problems of data access and data security (Dando, 1991; Davies, 1982; Rhind, 1992). Although many believe that public data should be readily accessible by the general public, the problem of sensitive data, such as the specific locations of endangered species or military storage facilities, makes the issue much less simple. Even when data can be obtained through the U.S. Freedom of Information Act, the time required to complete the legal documentation may exceed the life cycle of the GIS project for which the data are needed. These problems will not be solved easily, but you will encounter them as you continue on your journey in automated exploration.

METADATA AND METADATA STANDARDS

While data dictionaries are useful, they are often not enough for the growing set of GIS users, particularly where data are to be shared by many users. Many federal, regional, and local government organizations are now adopting the need for a rigorous set of standards that provide a detailed record of the datasets that they use and share with others. Such **metadata**, defined as data about data, are a major extension of simple data dictionaries and are becoming a major component of the GIS data input process. In the United States, for example, the Federal Geographic Data Committee (1998) (FGDC) has adopted a set of metadata standards that extend to a wide array of possible data types and require those who intend to share or provide data to the federal government adhere to them.

Metadata describe in some detail the content, quality, condition, and other important characteristics of the data, thus helping others to both locate and understand them. Major uses of metadata are as follows:

1. Organize and maintain an organization's investment in the data.

   In this case, as personnel change and time passes, the integrity of the data is maintained because the content and appropriate uses of the data are cataloged.

2. Provide information to data catalogs and clearinghouses.

   Among the most difficult tasks of the GIS professional is to identify what data are available for their particular needs. This need has prompted the development of both private sector and government organizations that keep catalogs of metadata, essentially acting as an index of available datasets. The FGDC, for example, is developing a network of geographic data sharing, called the National Geospatial Data Clearinghouse, through which the metadata can be shared. This allows potential users to become aware of what data are available and how to obtain them.

3. Provide information to aid in data transfer.

   Any time the actual datasets are to be shared with others it is vital that the metadata be shared as well. This allows the users to process and interpret data correctly and to match the data with its own data holdings.

Metadata describe many different aspects of the geospatial data and may require the use of software for their creation. Most of these software packages are either free or relatively inexpensive and can easily be found by searching the web under the keyword "GIS metadata software." A GIS neophyte might easily ask, "What can be so difficult about metadata that I would need software to help me compile it?" To answer this let's first take a look at seven basic categories of metadata that should be included, and then parse out the more general individual data that should be included in each (Federal Geographic Data Committee, 2001).

1. Identification
   a. What is the database called?
   b. Who developed it?
   c. What geographic area is covered?
   d. What categories or themes of data are included?
   e. When were the data collected and how current are they?
   f. What, if any, restrictions are there for data access?
2. Data Quality
   a. How good are the data? (This can be a very complex question.)
   b. How does the user know if the data are applicable for their use?
   c. What is the positional and attribute accuracy level?
   d. How complete are the data?

   e. Was data consistency verified?

   f. What were the source data from which the existing data were created?

3. Spatial Data Organization

   a. What data model was used to encode the spatial data (raster? vector?)?

   b. How many spatial objects are there?

   c. Are methods other than coordinates (e.g., street addresses) used to encode locations?

4. Spatial Reference

   a. Are coordinate locations encoded using latitude/longitude?

   b. What, if any, projection was used?

   c. What geographic datums (both vertical and horizontal) were employed?

   d. What parameters should be used to convert the data from one coordinate system to another to preserve integrity?

5. Entity and Attribute Information

   a. What specific attribute information is included?

   b. How is the information encoded?

   c. Were codes used?

   d. What do the codes mean?

6. Distribution

   a. Where and from whom do I obtain the data?

   b. What formats are available

   c. What media types are available?

   d. Can I obtain the data online?

   e. How much does it cost?

7. Metadata Reference

   a. When were the data compiled?

   b. Who compiled the data?

Each of these questions may comprise multiple individual questions themselves, thus making the task of developing metadata a much larger one than you might have originally anticipated. As it turns out, three supporting categories of metadata will also prove useful: (1) citation information with details of the originator (source) of the data and publication dates, (2) time period information indicating whether the data represent a single date or multiple dates, and what the range of dates is, and (3) more detailed contact information including primary contact person (or organization), addresses, and phone numbers. This latter information is particularly important for the clearinghouses that are attempting to make GIS data accessible to others.

Terms

affine transformation
centroid-of-cell method
data dictionary
differential rectification
dominant type method
fuzzy tolerance
ground control points (GCPs)
information theory
inverse map projection
linearity
metadata
most important type method
percent occurrence method
presence/absence method
primary data acquisition
registration points
repeatability
resolution
root mean square (RMS) error
rotation
scale change
scale-dependent error
secondary data acquisition
skew
spatial information product
stability
tic marks
translation

Review Questions

1. What four characteristics are shared by every method of GIS data input?
2. What are the five factors, beyond cost, that should be examined in deciding on a digitizer? Define each.
3. What are the fundamental differences between line-following scanners and drum scanners? What potential problems relating to the map document itself act as sources of error for line-following scanners?
4. Converting between raster and vector data structures can produce results that degrade the quality of the initial input. What primary problem might occur during conversion from vector to raster? Where is this most likely to occur? What problem often arises during conversion from raster to vector?
5. Describe the transformation processes involved in moving from the digitized map to three-dimensional coordinates and finally to an output map. What is an inverse map projection? How does it relate to the map projection process?
6. Illustrate the processes of translation, rotation, and scale change? Why are these important to the input subsystem of a GIS?
7. Why is map preparation important to GIS data input? What are registration points or tic marks used for? Why are they needed?
8. Why should you mark your map before digitizing? What kinds of information should you include on your prepared map document? Why do you need to provide information about projection and the grid system when the digitizer is set up for planar or Cartesian coordinates?
9. What is fuzzy tolerance? Why is it important? What difficulties can arise if it is set too low? Too high?
10. What is the potential impact of the cartographic medium on the digitizing process? How does this relate to temperature and humidity? What can be done to reduce errors due to media distortion?

11. How do you decide what to input to the GIS? What is a spatial information product and how does it relate to GIS input?
12. List and explain the seven rules determining what should be input to the GIS.
13. What is a good rule of thumb to help you determine how much to input? What do we mean when we say that digitizing is a sampling procedure?
14. What does data input have to do with information theory? What general rule does information theory suggest for data input? Suggest situations in which this general rule is less useful than one might suspect.
15. What is scale-dependent error? How does it relate to the input subsystem of a GIS?
16. Why do you need to read the software manuals in regard to vector data input? Isn't this just a simple matter of pointing to a point and pushing a button? What are the numbered buttons on the digitizer puck used for? What can you do while you are digitizing to reduce editing time later on?
17. What is so critical about attribute data input in the vector domain? What is the primary problem that can arise if it is not done carefully? Why is this situation so hard to detect later?
18. What are the four basic methods of raster data input? How do they differ? What are the advantages and disadvantages of each? Which is (are) best for point and line data? Which results in data explosion?
19. What are the technical problems involved in using aerial photography for GIS input? How about digital satellite data? What are ground control points, and why are they important in using digital satellite data? What institutional problems are involved in using remotely sensed data as GIS input?
20. What positive impact are external databases likely to have on the growth of the GIS industry? What are some of the major technical and institutional problems with using external databases? Why is the data dictionary or metadata so important? What is the difference between active and passive data dictionaries?

References

Berry, J.K., and C. Dana Tomlin, 1984. *Geographic Information Analysis Workshop*. New Haven, CT: Yale School of Forestry.

Calkins, H., 1975. "Creating Large Digital Files from Mapped Data." In *Proceedings of the UNESCO Conference on Computer Mapping of Natural Resources*, Mexico City.

Cameron, E.A., 1982. "Manual Digitizing Systems." Paper presented to the ACSM/ASP National Meeting.

Chrisman, N.R., 1987. "Efficient Digitizing Through the Combination of Appropriate Hardware and Software for Error Detection and Editing." *International Journal of Geographical Information Systems*, 1:265–277.

Dando, L.P., 1991. "Open Records Law, GIS, and Copyright Protection: Life After Feist." In *URISA Proceedings*, pp. 1–17.

Davies, J. 1982. "Copyright and the Electronic Map." *Cartographic Journal*, 19:135–136.

DeMers, M.N., 1992. "Resolution Tolerance in an Automated Forest Land Evaluation Model." *Computers, Environment and Urban Systems*, 16:389–401.

DeMers, M.N., and P.F. Fisher, 1991. "Comparative Evolution of Statewide Geographic Information Systems in Ohio." *International Journal of Geographical Information Systems*, 5(4):469–485.

Federal Geographic Data Committee, 1998. Content Standard for Digital Geospatial Metadata (revised June 1998). Federal Geographic Data Committee, Washington, DC. http://fgdc.gov/metadata/constant.html, page 85.

Federal Geographic Data Committee, 2001. "Content Standard for Digital Geospatial Metadata Workbook Version 2.0." Federal Geographic Data Committee, Washington, DC.

Fisher, P.F., and M.N. DeMers, 1990. "The Institutional Context of GIS: A Model for Development." In *Proceedings of AUTOCARTO 9*. Falls Church, VA: ASPRS, pp. 775–780.

Haddad, K.D., 1992. "CoastWatch Change Analysis Program (C-CAP) Remote Sensing and GIS Protocols." In *Global Change and Education*, Vol. 1, ASPRS/ACSM 92 Technical Papers, Bethesda, MD, pp. 58–69.

Lillesand, T.M., and R.W. Kiefer, 1995. *Remote Sensing and Image Interpretation*. New York: John Wiley & Sons.

Marble, D., 1981. "Some Problems in the Integration of Remote Sensing and Geographic Information Systems." In *Proceedings of the LANDSAT '81 Conference*, Canberra, Australia.

Marble, D., and D. Peuquet, 1983. "Geographic Information Systems and Remote Sensing." In *The Manual of Remote Sensing*, 2nd ed. Falls Church, VA: American Society of Photogrammetry, Chapter 22.

Marble, D., J.P. Lauzon, and M. McGranaghan, 1990. "Development of a Conceptual Model of the Manual Digitizing Process." In *Introductory Readings in Geographic Information Systems*, D.J. Peuquet and D.F. Marble, Eds. London: Taylor & Francis, pp. 341–352.

Moellering, H., Ed., 1992. *Spatial Data Transfer Standards: Current International Status*. London: Elsevier Applied Science.

Rhind, D., 1992. "Data Access, Charging and Copyright and Their Implications for GIS." *International Journal of Geographical Information Systems*, 6(1):13–30.

Shannon, C.E., 1948. "The Mathematical Theory of Communication." *Bell System Technical Journal*, 27:379–423, 623–656.

Walker, R., Ed. 1993. *AGI Standards Committee GIS Dictionary*. Association for Geographic Information.

CHAPTER 7

Data Storage and Editing

The GIS storage and editing subsystem provides a variety of tools for storing and maintaining the digital representation of a study area. The large size of databases may require us to create methods of segmenting or tiling them so that portions can be examined rather than the entire database. As a result, these will also require a coding scheme that enables us to piece these tiles back together when needed.

The storage subsystem also provides tools for examining each theme for mistakes that may have crept into its preparations. Before we can successfully use these tools, we need to know what these possible mistakes are and how they can be discovered and corrected. If we have been careful in our input, we should encounter relatively few errors. However, even the selection of an improper fuzzy tolerance level can produce errors. Many of these will not appear until the GIS has completed organizing complete themes or grids. In raster, for example, we may need to display each grid to isolate illogical or out-of-place grid cells as we compare them to the input documents. In vector systems, we may have to build in topology after the initial data input to help us pinpoint any polygons that don't completely connect, lines that end in the wrong place, or points that occur where they should not. In the case of entity–attribute agreement, we may need to output sample portions of our map for comparison against the original input material.

There are many aspects to error detection and correction. Experience allows us to increase our efficiency in both error identification and correction. The detection and correction of errors depends on the software capabilities, the type of errors present, and even the data models employed—especially raster versus vector models. Most errors are a function of input, but some will appear as a result of data interactions during analytical operations. Some errors are purely entity errors, while others are attribute errors. Whatever the source or type of error, the more errors that are present, the greater the opportunity for poor analytical results.

LEARNING OBJECTIVES

When you are finished with this chapter you should be able to:

1. Know what tiling is and what its purpose is.
2. Explain how Morton Sequencing works, especially with regard to the numerical coding scheme.
3. Understand the basic types of errors that can occur in GIS, how these errors are edited, and the importance of editing in GIS databases.
4. List and describe six areas of entity errors with specific examples and suggest how they might occur, how they are edited, and how they can be avoided.
5. Illustrate and describe the types of attribute errors in both vector and raster, suggesting how they might occur and how they are detected and corrected.
6. Describe the process of converting projections using a vector GIS.
7. Understand the idea of edge matching and why it is needed.
8. Describe the process of conflation, explain why it is needed, and discuss conceptually how it is executed.
9. Describe the process and the purpose of templating. Provide real examples of templating.

GIS DATABASE STORAGE

An analysis of the precise computer methods for storing GIS databases is well beyond the scope of this text, as are the ever increasing types of hardware technology used to record the data. The methods themselves are also highly dependent on the data model used. In raster systems, the attribute values for the grid cells are the primary data stored in the computer, usually on a hard drive. The locations of each grid cell are cataloged by the order in which they are placed in columns and rows. For this reason, editing is primarily concerned with the correct relative positions of each grid cell. Some raster systems employ a variety of compact storage methods such as run-length codes, block codes, raster chain codes, and quadtrees. To effectively examine the relative positions of individual grid cells, usually you must be able to retrieve the data from storage for display in a manner that allows each individual grid cell to be identified separately by column and row position as well as by attribute code.

If your raster system allows a linkage to a database management system, the matter becomes somewhat more complicated in that each grid cell has attached to it a number of different attribute codes. Depending on your GIS, you may have to display and analyze each set of attributes as a separate map. Others may provide you with the ability to list the attribute codes for each grid cell as you examine it.

In vector, the entities and attributes are either stored as individual tables within a single database or as separate databases linked by a series of pointers. The separation of entities and attributes requires you to look at the editing procedures applied to entities, attributes, and databases. You can retrieve the graphic entities separately and display them to identify missing objects, incomplete links, and polygons. By retrieving the attribute tables, you will be able to examine them apart from their linked entities to determine whether you have typos, incorrect code sequences, or even the wrong attributes in the wrong columns of your table. Finally, you will be able to retrieve all or part of your database (i.e., parts of the graphics and/or parts of the database) to examine both the entities and the attributes for agreement. You will most often be able to isolate individual entities and display, on the same screen, the attribute values you desire.

Many vector GIS systems also allow you to separately store portions of your database as large, predefined subsections for archival purposes. This process, called **tiling**, is most often used to reduce the volume of data needed for the analysis of extremely large databases. Say, for example, that you are creating a detailed database for an entire county. You may wish to divide the database into smaller tiles based on the coordinates of the individual maps (such as topographic sheets) that you used for input. Tiling means that you retrieve just the portion of the overall database of interest, thus reducing your computational overhead and increasing system response. Another important purpose of tiling is to allow a system administrator to have final control over the editing and updating process by permitting only certain sections of the database to be operated on when needed. Even when small portions are released for editing and updating, the system maintains an original copy of the pre-edited database until the system administrator is satisfied that the updates and edits are correct. Thus, by limiting access to those who are qualified to make changes, corruption of an entire database can be prevented.

A classic method of tiling that provides a formal sequential coding scheme is called **Morton sequencing**. Morton sequencing has been copied and modified, but the original is still quite common and very useful. The sequencing is based on quarters and provides a highly flexible way to segment databases, moving in a four-fold left to right zig-zag fashion (Figure 7.1). Note the first four segments numbered 0 through 3 from left to right, then down, and left to right again. By repeating this sequence we move to the right to the next four numbered 4 through 7, then move to the lower left, and finally lower right. At the next level the sequencing continues again moving to the right. There are many types of sequences that could be employed. The power of the Morton sequence is that if your study area should expand, it is relatively easy to add new numbers to the sequence. This avoids the problem of having to completely redo the number sequences.

Whatever method is employed, tiling both provides needed data security and reduces the amount of data, as well as acting as a screening mechanism for the entire editing process. Most often the database is completely cleaned and edited prior to tiling and archiving, and as a rule access is gained primarily for updating and analysis. This is not always the case, however, and you will need to select the appropriate tiles to do your editing. In some cases you may need to perform the process called **edge matching**—operating on more than one tile at a time to ensure that there is a correct match between the two tiles for entities that extend across the tile boundaries.

In general, today's GIS software provides a visual display mechanism that will enhance your ability to visualize the errors. The exact methods depend on

y \ x	0	1	2	3	4	5	6	7
0	0	1	4	5	16	17	20	21
1	2	3	6	7	18	19	22	23
2	8	9	12	13	24	25	28	29
3	10	11	14	15	26	27	30	31
4	32	33	36	37	48	49	52	53
5	34	35	38	39	50	51	54	55
6	40	41	44	45	56	57	60	61
7	42	43	46	47	58	59	62	63

Figure 7.1 Morton sequencing. This is a common form of database tiling that employs a quadrant strategy and clockwise numbering.

the data model you use and the sophistication of your system. Because most systems allow interactive editing through visualization, it is possible to correct errors as they are encountered. Still, despite the sophistication of modern GIS and its ability to find some obvious errors, the process is not completely automatic. You will need to interact closely with the software to both detect and edit the maps. This is all the more reason for ensuring that your map preparations prior to input are complete. The prepared documents will often be used as a form of truth set against which to evaluate the digital database.

BASIC ERROR TYPES

Although some errors occur as a result of computational miscalculations and rounding errors in the GIS software, most database errors result from improper input. Even with meticulous map preparation procedures, the finest equipment, and the best-trained input technicians, mistakes will happen. Causes include simply pushing the wrong button on the digitizer puck, a shaky hand, typing errors during attributes input, and registration difficulties. The potential sources for errors are numerous and include problems with the input documents themselves (Laurini, 1994). But the most costly aspect of input errors is not their source but their correction. Because mistakes are generally very small and extremely difficult to find even with the best GIS software, correcting them is time-consuming and costly. It is not unlikely that more time will be spent correcting even a small number of errors than was used to prepare your maps and input them to the system.

There are two primary types of errors with which we need to be concerned. The first type of error deals primarily with vector systems and is called **entity error** (positional error). Entity errors can come in three different forms:

(1) missing entities, (2) incorrectly placed entities, and (3) disordered entities. We will discuss these in more detail later. The second type of error is **attribute error**. Attribute errors occur in both vector and raster systems. When attributes are typed in, the sheer volume of typing required for large databases often constitutes a major source of errors. In vector systems attribute errors include using the wrong code for an attribute, as well as misspellings, which make an attribute impossible to retrieve if a query uses the correct spelling. In raster, the input most often consists of attributes, so the result of typing a wrong code number or placing it in the wrong grid location is a map that displays these incorrectly coded grid cells in the wrong place. Such incorrectly placed attribute data result in problems with entity–attribute agreement (logical consistency), which also occurs in vector systems when correctly typed codes are attached to the wrong entities.

Attribute errors are the most difficult to find. Mistyped attributes placed in correct locations (e.g., in a particular place in an external database) might be found if an active data dictionary is part of the system. This active data dictionary feature is helpful if you have violated a rule already established for the data dictionary, such as putting in numbers in locations requiring letters, or putting 80 percent of a five-digit code in a four-digit slot. However, misspelled attributes may not be found until you actually perform an analysis. Entity–attribute agreement error is often even more difficult to find than misspellings or incorrect codes. In raster, the only way to observe problems of this type is to display the map to identify displaced grid cells. With vector, you will most often be able to point to entities and display their attributes on your monitor. However, the software is unlikely to be able to identify that you have incorrectly labeled an entity. Instead, you will need to have a copy of your input map available as you display or highlight each entity.

If you created a very complex database, you may have to spend months evaluating each of the thousands of entities and making comparisons to your input document. It is far better to check these errors in small groups as you input the data. You will be more familiar with the data as you enter it than you will be if you go back later. In addition, the input document is already there in front of you. For this reason, some software vendors allow you to use the editing portion of the GIS as a method of input rather than using the input subsystem. Some vendors take an alternate approach and build the editing capabilities into the input subsystem. In either case, you can examine your map for entity as well as attribute errors and determine agreement problems as they occur. Although these steps slow the input process, it is much cheaper and faster to do it right the first time.

CONSEQUENCES OF ERRORS

Error-prone data contribute to error-prone analysis, and although simple errors may seem quite innocuous, even the simplest can produce results that are grossly incorrect. As a simple example, imagine a database containing over 8,000 polygons, some of which illustrate the locations of highly toxic materials; a single polygon (e.g., polygon number 2,003) bears an incorrect attribute code indicating that the location contains no toxic material. Another theme shows cancer mortality. Map overlay should show that high cancer mortality is spatially correlated to the presence of toxic material (i.e., polygon number 2,003). Because polygon number 2,003 does not contain the correct code for the presence of toxic materials, results do not allow you to come to the correct conclusion that cancer mortality is related to high levels of toxic material. In the short term it seems to be a relatively harmless result. If, however, the polygon numbered 2,003 is a highly densely populated area, your results have potentially fatal consequences for the residents in the area. Such circumstances are among the topics covered in an ongoing discussion about the legal liability of GIS databases as tools for decision making (Epstein, 1989; Seipel, 1989). And although this hypothetical example may seem extreme, it should point out the distinct possibility that minor errors in data can produce major errors through analysis.

ERROR DETECTION AND EDITING

You have seen that GIS databases are subject to both entity and attribute errors. We will look at entity errors first, then consider attribute errors and the associated entity–attribute agreement errors next. Most often attribute errors are detected and identified because of a failure of the entities and attributes to agree. Although this is not always the case, the detection of pure attribute errors is most commonly performed by producing tabular listings of the attributes or their tables. Although this is certainly part of the process of error correction, a complete description is unnecessary. Examples for each type of GIS would take up a large portion of this book, so we will see how errors are detected for a common system or two, and you can modify your procedures on the basis of the GIS you use.

Entity Errors: Vector

Upon completion of data input, many GIS systems require you to perform an operation that builds topology, unless this was part of the input procedure itself. In either case, the topology, by providing explicit information about the relationships of the entities in your database, should permit you to identify the types of entity errors in your digital maps. Some of these errors will be pointed out through text-based error flags. Others must be interpreted by looking at database statistics concerning the numbers and types of entities or by inspecting the graphics displayed on the screen for errors the GIS is not designed to detect. You will be looking for six general types of entity errors, represented by the negative case of the following statements (Environmental Systems Research Institute, 1992):

1. All entities that should have been entered are present.
2. No extra entities have been digitized.
3. The entities are in the right place and are of the correct shape and size.

4. All entities that are supposed to be connected to each other are.
5. All polygons have only a single label point to identify them.
6. All entities are within the outside boundary identified with registration marks.

A commercial GIS should provide these general spatial relationships, and you can use them to identify errors. A useful procedure for comparing the entities you digitized with the original map is to produce a graphic output that can be physically overlaid with the original map. Alternatively, many GIS packages provide symbols to indicate the presences of some error types. We will now examine specific types of errors we can find related to the six general types just listed.

Remember that in vector data models, nodes are special points that indicate a link between lines composed of individual line segments. In some topological vector data models the nodes are often described as "To-nodes" and "From-nodes," indicating the overall extent of a line feature. Nodes are not just points between line segments; they carry specific topological and/or attribute meaning. Nodes may identify an intersection between two streets or a connection between a stream and a lake, but they should not occur at every line segment along a line or a polygon. Thus, the first type of error that can be detected entails false nodes, called **pseudo nodes**, which occur where a line connects with itself (an **island pseudo node**: Figure 7.2a) or where two lines intersect along a parallel path rather than crossing. A GIS should be able to flag the existence of pseudo nodes with an easily identifiable graphic symbol. When you build your first GIS database, you may be perplexed by an abundance of pseudo nodes showing up on your coverage. Some pseudo nodes are not errors but are merely flags indicating the presence of potential problems. A pseudo node connecting a line with itself may simply be the beginning and ending of an island polygon (sometimes called a **spatial pseudo node**), in which case its flag can be ignored. Or, for two line segments with an intervening node (also known as an **attribute pseudo node**), the node may indicate something as simple as a speed limit change (Figure 7.2b).

Nonpurposeful spatial pseudo nodes are most often due to a misplaced data point or to pushing the wrong button on the digitizing puck. Either you were

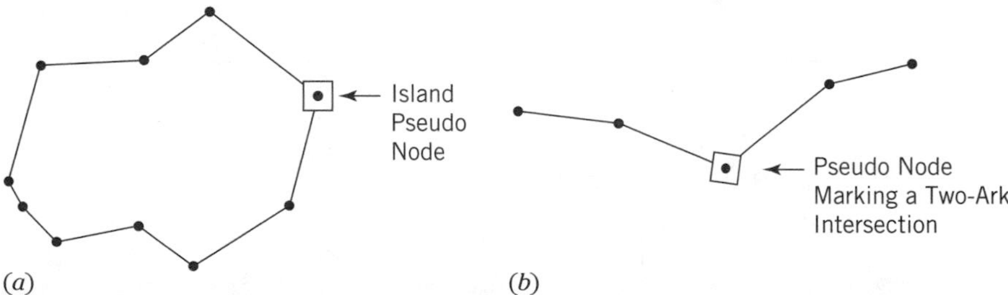

Figure 7.2 Pseudo nodes. Two types of pseudo nodes that are permitted are those that indicate the presence of an island, called **island pseudo nodes** (*a*), and those that indicate attribute changes along a line, such as a speed limit change (*b*). If, in *a* there isn't an island, this indicates an error and must be corrected. In *b*, if there is no change in attributes along this line, then this pseudo node condition also needs to be corrected.

trying to create a nonclosing structure but placed the puck in the wrong place, or you were trying to create a polygon that connected to other polygons (i.e., had neighbors that share line segments) but pushed the wrong button. To avoid improper spatial nodes you should number your points when preparing your map or use a special code or symbol to indicate where your nodes are going to be. A good practice is to use a numerical code that is identical to the numbers on your digitizing puck that correspond to nodes. This procedure will also reduce the likelihood of producing an erroneous pseudo node for line entities.

If your software indicates that your map has pseudo nodes, the prepared map document can be used to correct them. After first determining that the pseudo nodes are errors and not purposefully constructed, they can be corrected by either selecting each individually and deleting when necessary or by adding nodes where needed.

Another common node error, called the **dangling node**, can be defined as a single node connected to a single line entity (Figure 7.3). Some GIS packages require you to have a From-node and a To-node, rather than just a single node. Dangling nodes, sometimes just called dangles, result from three possible mistakes: (1) failure to close a polygon, (2) failure to connect the node to the target object (called an **undershoot**), or (3) going beyond the entity you were supposed to connect to (called an **overshoot**). In some cases the problem is a result of incorrect placement of the digitizing puck; in others the fault lies in an unnecessarily tight fuzzy tolerance setting. Setting the fuzzy tolerance properly is one way of avoiding this problem, and map preparation is another. It is generally easier to find overshoots than undershoots. If you tend to produce dangling nodes, a good practice is to overshoot rather than undershoot the line

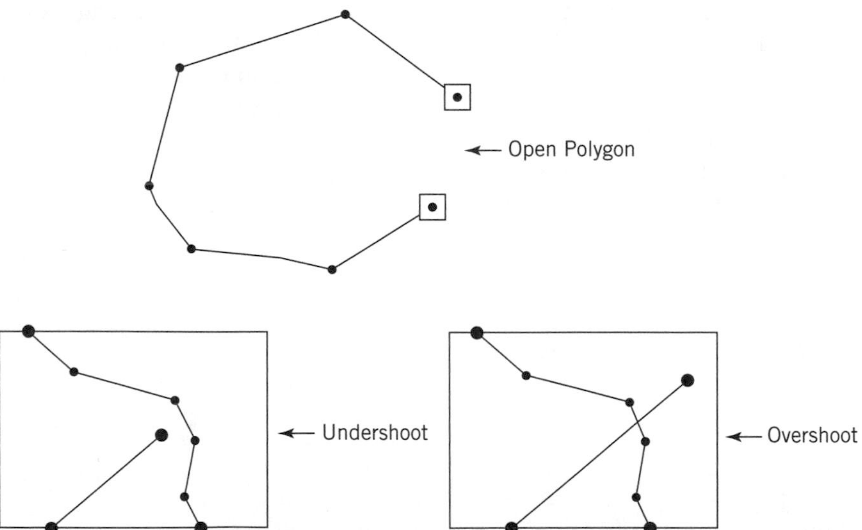

Figure 7.3 Node errors. Illegal dangling nodes come in three basic types: those that are a result of a failed polygon closure, those classified as undershoots because a node falls short of an object to which it is meant to be connected, and overshoots, where a node lies beyond an object to which it is supposed to be connected. *Source:* Figure derived from Environmental Systems Research Institute, Inc. (ESRI) drawings.

you are trying to connect. Although accurate digitizing would be better, this method has proven useful for some who produce this error frequently.

Dangling nodes are most often identified by a graphic symbol different than that used for pseudo nodes. In addition, if your dangles indicate an open polygon, the GIS will alert you by telling you the number of complete polygons in the database: if it differs from the count you had prepared prior to digitizing, you know you need to look for these dangles as incomplete polygons. Corrections are again quite simple. For undershoots, the node is identified and is moved or "snapped" to the object to which it should have been connected. Overshoot errors are corrected by identifying the intended line intersection point and "clipping" the line so that it connects where it is supposed to. In the case of an open polygon, you merely move one of the nodes to connect with the other.

As with pseudo nodes, some dangles are intentionally input to the GIS. Most often these nodes serve as indicators of some important attribute change at the end of a line or arc. For example, you might use nodes to indicate the locations of residential cul-de-sacs (Figure 7.4). In even more unusual circumstances, a line used to indicate the location of a multistory building may contain numerous nodes, each one indicating the location of a separate floor. Although this abstraction deviates from the normal cartographic form, it illustrates the potential for dangling nodes to be legitimate objects within the database, rather than existing as errors.

As you digitize polygons you need to indicate a single point inside each polygon that will act as a locator for a descriptive label. You need one, but only one label point per polygon. Two types of errors can occur relating to label points in polygons: **missing labels** and **too many labels** (Figure 7.5). Both are typically caused by the failure to keep track of the digitizing process. Although good map preparation will reduce the occurrence of label errors, most often the problem is caused by confusion, disruption in the digitizing process, or fatigue. Fortunately such errors are very easy to find via graphic indicators, and editing is simply a matter of adding label points where necessary and deleting them where superfluous.

Another type of digitizing error most commonly occurs when the software uses a vector data model that treats each polygon as a separate entity. In such cases you are required to digitize the adjacent lines between polygons more than once. Failure to place the digitizing puck at exactly the correct location for each point along that line will often result in a series of tiny graphic polygons called **sliver polygons** (Figure 7.6). Sliver polygons can also occur as a result of overlay operations or when each of two adjacent maps is input from a separate projection. We will limit our current discussion to the sliver polygons produced through the input process.

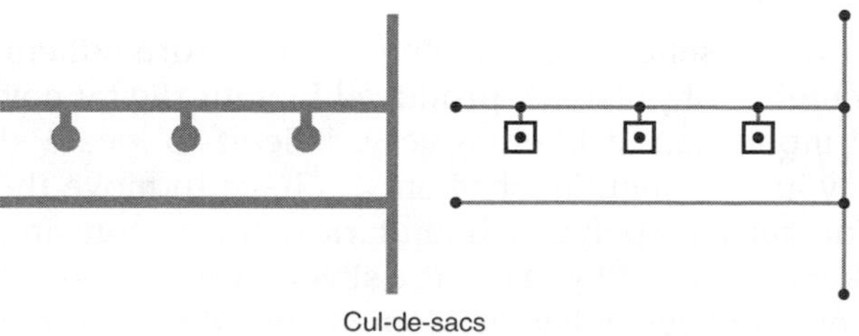

Figure 7.4 Acceptable dangling nodes. Legal dangling nodes created to indicate the existence of cul-de-sacs along a residential street. *Source:* Figure derived from Environmental Systems Research Institute, Inc. (ESRI) drawings.

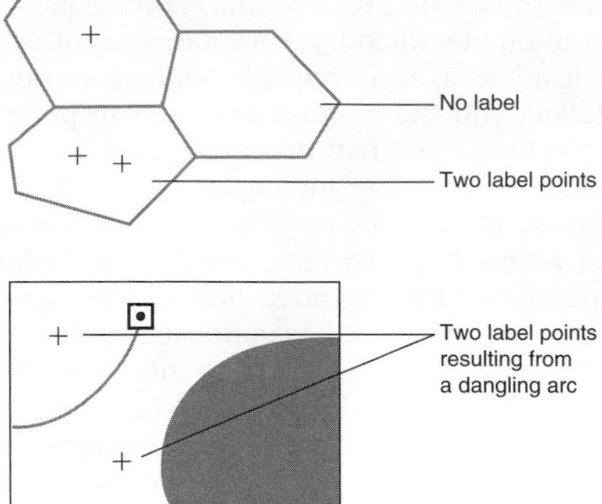

Figure 7.5 Label errors. Inside each polygon there must be a single point, somewhere, to which attributes can be attached. Errors result when polygons have no labels or more than one label.

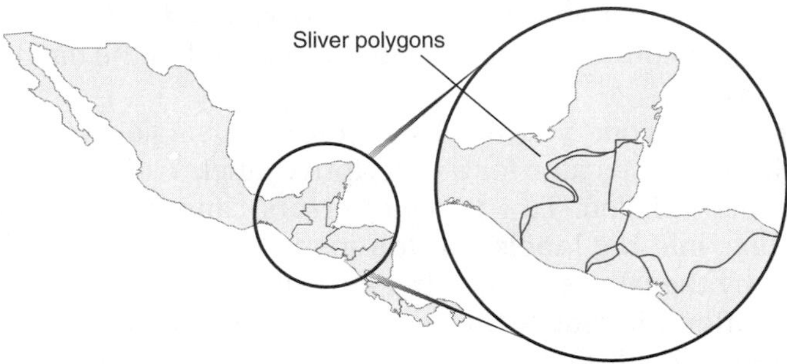

Figure 7.6 Sliver polygons. These result from poor digitizing along common boundaries where the line must be digitized more than once. Highly irregular national boundaries, as in Central America, are particularly vulnerable to such digitizing fuzziness.

The easiest way to avoid sliver polygons on input is to use a GIS input module that does not require digitizing the same line twice; in fact, this requirement is nearly obsolete. At times, however, you will accidentally digitize the same line twice. The result is the same: sliver polygons. The method of finding sliver polygons depends somewhat on whether you actually completed the adjacent line with nodes that are effectively placed on top of each other. If you digitized the same line twice by accident, you may also have a dangling node because an unneeded line has been created. In this case the line can be removed, correcting the problem.

Finding slivers in the absence of a dangling node is more difficult. One way is to compare the number of polygons produced in your digital coverage with that of the original input map. It is often very difficult to locate slivers even when you know they are present. Most often you have to move through your image, searching for suspect polygon boundaries, then zoom in to see the slivers. You often don't know until you see the slivers whether you are zooming in on a single line or on sliver polygons. Sometimes when you have a series

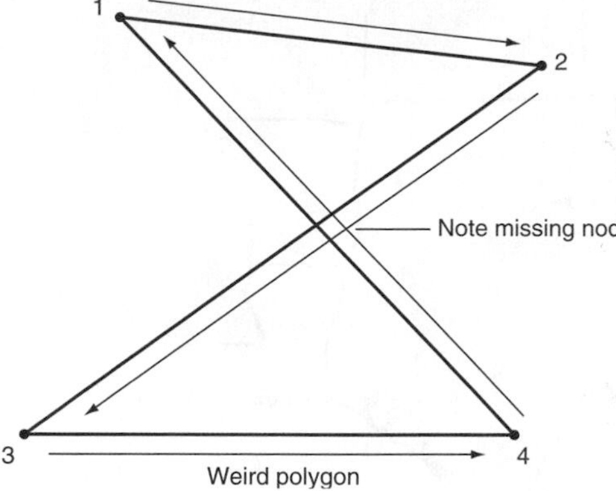

Figure 7.7 Weird polygons. Example of the creation of a weird polygon by digitizing points out of sequence. Although graphically this seems to be two polygons, the point at which the lines cross in the middle does not have a node.

of very tiny polygons but no dangling node, it is possible to adjust (loosen) the fuzzy tolerance you used during data input, which will remove the slivers automatically.

A separate problem related to polygons is the production of **weird polygons**, which are defined as polygons with missing nodes (Figure 7.7). In this case the polygon is a graphic artifact that appears to be a true polygon but is missing one or more nodes. Generally this occurs when two or more lines cross over, producing the semblance of a polygon. The most frequent cause of this error is a point digitized in the wrong place or in the wrong order. Although you digitized all the points, your polygon is twisted and the central point has no node. The software will indicate that you have one fewer polygon than what you see on the map.

A simple way of avoiding this problem is to number the input points before digitizing. Even if you don't do this, you can avoid the problem by establishing a set pattern for digitizing polygons. For example, you may decide to consistently digitize polygons in a clockwise direction. Such a systematic procedure for digitizing will most often keep you from omitting essential nodes.

Detecting weird polygons is difficult but not impossible. One straightforward method is to highlight the nodes and display them as part of the polygon map. The areas that should have nodes but don't will stand out from those that are properly digitized. Editing the error involves moving the lines to the correct locations, thereby placing the nodes in the correct sequence. Sometimes it is easier to simply remove the offending lines and use the editing subsystem to redigitize the points in the correct sequence.

The errors discussed thus far are among the easiest vector mistakes to find; as a rule, you can make the necessary adjustments without plotting out your map to identify the errors. A more annoying group of entity problems include missing objects, extra objects, and displaced or misshapen objects. These are easiest to detect by plotting the digitized map at the same scale used for input (Figure 7.8). If you overlay the original document and the plot on a light table, you can identify problem areas. Nearly all of these errors are due to lack of preparation, although the problems of disruptions and fatigue will always play a part in their occurrence. Correcting them requires you to mark the problem areas on the map document. Extra objects should be marked for removal. For objects that extend beyond the map extent borders defined by your registration marks, the points need to be removed and redigitized. Objects

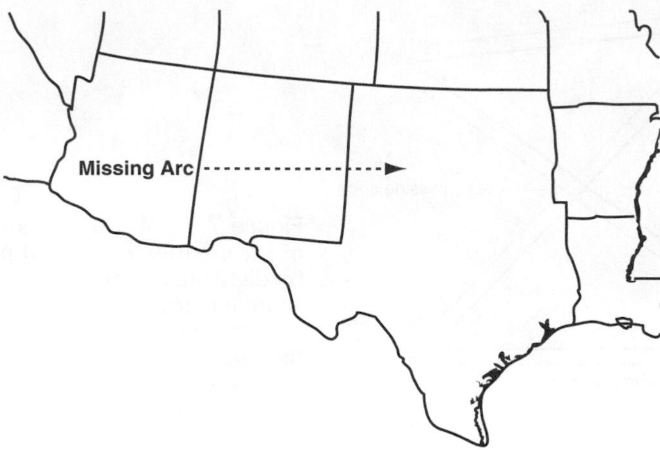

Figure 7.8 Missing Arc. Illustration of the use of source map comparison to identify common graphical errors. Notice the missing arc in the digitized version of this map of the south western U.S.

that are misshapen or incorrectly placed but still within the map extent can most often be individually selected and moved, without being redigitized.

Attribute Errors: Raster and Vector

As you have seen, attribute errors, including entity–attribute agreement, are among the most difficult to detect. This is primarily because the GIS does not know which attributes are correct and which are not. Because vector entity and raster grid attributes (and their assumed entity locations) differ significantly from application to application, and because there is no attribute equivalent of topology, there are no rules against which the GIS can check your accuracy. This means that with most systems there are no explicit rules to indicate that a particular attribute occurs with a particular, stable pattern with respect to its neighbors. One advantage of the new geodatabase data model is that it does have rules that can effectively prevent some attribute errors from occurring. In the absence of this newer data model we are forced to compare the attributes in our digital database to the original map document to determine most of the attribute errors that might occur. The following discussion assumes that we do not have an operational geodatabase available for our GIS layers.

Missing attributes are perhaps the only attribute errors that are detectable without direct comparison to the input document. In raster these can occur as missing rows or columns, or portions of rows or columns. These are detectable because familiarity with the original map shape alerts us to the absence of rows or columns of grid cells, which result in substantially altered shapes (Figure 7.9). Missing rows and columns in raster are most often caused by satellite sensor errors such as start and stop problems and dropped lines. Solutions to these problems most often involve examining neighboring grid cells and averaging or duplicating their numeric values to estimate the missing values. This is akin to the graphics software algorithms that eliminate scratches in a digital picture.

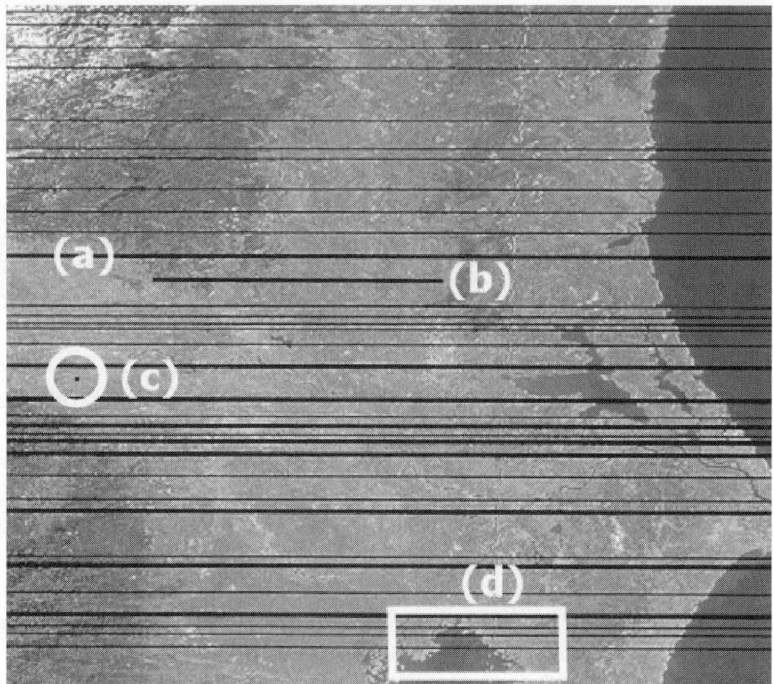

Figure 7.9 Raster attribute errors. Common raster attribute errors that are identified on the basis of how they distort the coverage: A, missing row; B, incorrect or misplaced attributes (appearing as one or more rows of vastly different values); C, incorrect attributes occurring singly; D, attributes errors occurring along area margins (most often caused by digitizing problems).

Incorrect attribute values are more difficult to detect, both in raster and in vector. In raster GIS, they occur as individual grid cell errors or horizontally displaced short row segments. They are, again, most common in digital remote sensing imagery and are often the result of start and stop problems in digital scanners, or temporary fluctuations in signal strength resulting in unusually high or low values (spikes). In the less common case where wholly incorrect attributes occur for large areas, they are typically a result of inputting the wrong attribute value, either through digitizing or a block encoding method. In raster, an incorrectly coded grid cell is most likely to be identifiable as one individual that seems "out of place" among the surrounding grid cells. Unusually high or low attribute values show up easily. They normally appear as out-of-place grid cells or groups of grid cells that disrupt the natural organization and integrity of portions of the map. If they occur as continuous strips of incorrect grid cell attributes, most software will allow you to use a run-length encoding strategy to edit these interactively. Individual grid cell errors can be selected individually and edited.

In raster maps that have few contiguous areas (e.g., maps of raw topographic data), incorrect values will not easily be seen on a two-dimensional display of the data because there are no large area patterns to disrupt. In such cases, a three-dimensional display of the raw data may show unusually high spikes or very low dips in the surface. Although these abnormal features may be errors, be sure to check them out—they may be real attribute anomalies. Generally such incorrect values occur as individual grid cells and can be selected and edited interactively as before.

Incorrect attributes in raster systems also may be found along the margins of the areal patterns. In such cases, the typical culprit is either unreliable digitizing algorithms that employ the push-pop strategy of the internal computer stack rather than relying on one of the four systematic grid cell input methods (Berry and Tomlin, 1984) or carelessness in the determination of the correct attribute codes along these margins. The pitfall here is that the incorrect values are most often identical to the neighboring area, thereby giving the impression that they are correct. You will need to compare the shapes of area patterns on the raster map to the original shape of the input map. Correcting such a problem usually means reevaluating each grid cell as to its correct attribute. In other words, which of the two adjacent areas does this grid cell really belong to? Once this has been determined, each can be selectively edited as before.

Incorrect attributes are often more difficult to identify in vector than in raster because they generally require intimate familiarity with the source map and its attributes and attribute patterns before these can be compared to the digital version. If you are using a coding strategy that replaces the actual names of the items by a code number, there are many chances to enter an incorrect number. In such cases the codes will sometimes not correspond to linked attribute tables. The software should be able to flag such inconsistencies. Take, for example, a numerical coding strategy devised to represent the names of individual plant species. Although this releases the user from having to spell out the species names exactly during query operations, it increases the likelihood of input errors. If you have established an appropriate set of rules in your data dictionary—especially if it is an active data dictionary—it should be possible to flag any code you enter that does not represent a real species name stored in the tables. Often, however, the code sequences become ingrained in your subconscious, resulting in plant codes that are real. They just aren't the correct ones for a given species. The only way to prevent such errors is to double-check each code as you enter it. Post-input identification of this type of error requires you to display the output and compare it to the original. Although tedious, by selecting the offending entities you can easily edit their attribute values as before, interactively.

A common source of the foregoing type of attribute error is failure to keep track of the attributes as you are recording them. In most cases the attributes are not incorrectly typed, but the attributes are mismatched with their entities. This mismatch can also be a result of unconsciously using the wrong code. This is distinctly different from placing a meaningless (i.e., a code that doesn't correspond to anything) code in your tables. In many cases, these misplaced codes occur predictably. For example, you may tend to lose track of your place as you are typing codes. This may provide a clue to finding these errors, particularly when the tables are printed and compared with the input data. Finding them using the graphical output requires you to compare it to the input map. If the attributes were input using the label points on the entities themselves, this would be the best way to identify the errors. Alternatively, if your attribute tables were created separately and later linked to the attribute codes, systematic errors are more easily detected by comparing tabular data from the database with the original data from which the tables were produced.

The systematic errors that are most likely to occur involve short periods during which incorrect codes are entered, then suddenly they begin to assume the correct order of entry. As before, a good way to avoid this is to get assistance while typing in the data. Of course, if your software permits you to input your attribute data interactively, you are likely to produce only occasional systematic attribute misplacements.

DEALING WITH PROJECTION CHANGES

Although a major function of the storage and editing subsystem is the correction of entity and attribute errors, it is also used to convert between Cartesian (digitizer) coordinates and real-world coordinates based on a reference globe. Quite often the software will require you to identify the projection of the map upon input, but some programs do not. In any event, the conversion to a set of projected map coordinates is a necessity for any analysis requiring real-world measurements. Moreover, because not all of your input maps will have the same projection, it will be necessary to standardize them to permit comparison of the different maps.

If you input the reference points using Cartesian as well as projected coordinates, you will use these as reference points for the coordinate transformations necessary to project your map. If your system operates on real-world coordinate systems but requires you to input the data strictly in Cartesian coordinates, you must define the reference points as latitude/longitude coordinates to make the projection possible. Another important factor is the method by which the GIS stores and manipulates these coordinates. Some use latitude/longitude coordinates recorded in **degrees, minutes, and seconds (DMS)**, whereas others require you to convert these to **decimal degrees (DD)**. The formula for converting from DMS to DD, which gives the numbers in degrees or fractions of degrees (Environmental Systems Research Institute, 1992), is

$$DD = \text{degrees} + \text{minutes}/60 + \text{seconds}/3600$$

In many cases you will operate first on the reference coordinates by creating a separate map layer with these values only, recorded in digitizer inches. The values can be read directly from the associated tables, then edited by typing in their longitude and latitude equivalents. The reference points are now in geographic coordinates. Upon completion of these edits and after double-checking for accuracy, you will save the data and link them to the geographic coordinate system (latitude/longitude). This tells the computer that the numbers you edited are defined by a real-world (geographic) projection. That is, you are telling the computer that you have changed the Cartesian coordinates to the geographic projection.

Remember, the map you are digitizing will be a projected map. You must get your coordinates, which are now in a geographic projection (3-D), into the same type of projection as the original input document. It is a simple matter to employ the GIS transformation procedures of scale change, rotation, and translation embedded in your software to produce the projection required for your reference points. These transformations require mathematical manipulation of your original geographic coordinates, and errors will always be part of the procedure. No projection procedure is without error. Many software packages provide some measure of this error, and you should look at these numbers. High amounts of errors resulting from this transformation are often a result of inaccurate digitizing, map sheet distortion, or incorrect recording of original reference coordinates (Environmental Systems Research Institute, 1992).

You now have a map of reference points, or tics, that represent the projection of the original map input. Upon completion, the reference points can easily be linked to the input coverage from which you extracted the reference points to create the reference coverage. Whenever you use the original coverage, it will be connected to the reference map, which also will have correctly transformed coordinates. Thus measurements can be produced in real-world units rather than in digitizer inches.

JOINING ADJACENT MAPS: EDGE MATCHING

Now that you have created a layer that contains all the pertinent coordinate transformation information, you must consider a closely related topic in the storage and editing subsystem—edge matching. In edge matching, two adjacent maps, usually of the same theme, are physically linked to permit the analysis of a larger study area. There are two sources of difficulty when two adjacent maps are input. First, two maps were input with the same projection, but because they were put in separately, they are likely to contain different levels of entity errors for the entities to be connected. Remember that the maps are registered to the digitizing tablet separately, the tic marks or reference points were input separately, and all the entities were input during a separate digitizing session. Therefore, although each map may be reasonably accurate in isolation, the differences in input errors between the two will most likely cause mismatches—sometimes subtle, sometimes obvious. You will need to link all the line and polygon entities that are supposed to be connected. For example, if a road digitized on one map sheet is supposed to be a straight line that runs across the two sheets, make sure that when the sheets are connected you don't have a jagged edge or a slightly offset road (Figure 7.10). Both portions of the road will have to be connected so that they exist as a straight line.

The second situation that is likely to cause problems requiring edge matching arises either when two adjacent maps are input from different projections (or with the same projection but based on different baselines or other starting

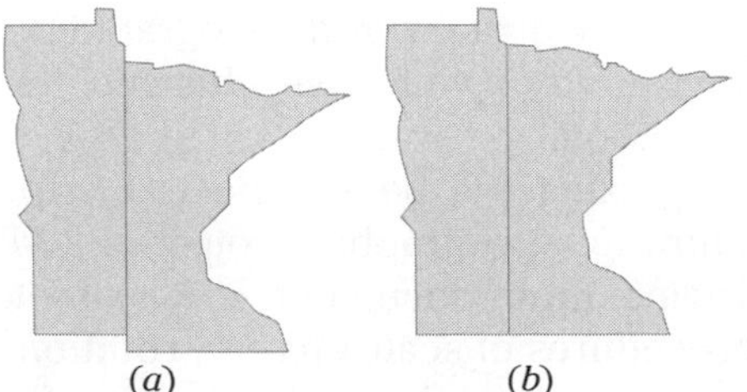

Figure 7.10 Edge matching. An example of edge matching to connect a polygonal object across maps. Notice how the road is offset before edge matching (*a*) and connects after (*b*).

points) or when the projection is applied specifically to that map layer without regard to its possible effects on the neighboring map layer.

Edge-matching problems can occur in raster systems as well as vector. A common example of edge matching in raster systems involves the use of remotely sensed data products, such as LANDSAT TM (30-meter resolution) images. Because horizontally adjacent scenes are sensed days apart (Lillesand and Kiefer, 1995), there is the possibility that the satellite will not be located at exactly the same latitudinal coordinates. This often results in a skew between the two images of one to several pixels (grid cells). Most often this is corrected by moving one of the data sets until the pixels match. Finding these errors is not unlike our example for finding missing lines from raster data input. Even very complex satellite images usually contain areas that produce relatively similar responses that, when classified, would often correspond to a single area class. When this group of pixels is offset, even before classification, the condition is relatively easy to detect because the regular patterns themselves show the incongruity along this line. In addition, geographic coordinates are most often provided with such data, allowing the user to match the edges by matching the coordinates along both adjacent images.

CONFLATION

There are frequent circumstances, particularly in vector, when some content of two digital source maps needs to be combined. This may be required to make a composite map that is superior to the originals. You may, for example, have a map showing hydrology that is extremely accurate, but you also have a map of vegetation where you wish to compare the vegetation along the river to the river. This requires you to combine both maps to create a map of riparian vegetation. Another condition when you may wish to compare multiple maps is where you have multiple dates for the same region, but each map was digitized from a different set of aerial photographs. Other circumstances include the consolidation of multiple thematic datasets within a single organization, or where you wish to add features to an existing map. All of these require a complex process called **conflation**, a special case of image registration (sometimes called **rubber sheeting**, although this terminology is most often applied to the edge mapping problem) to combine these datasets with an acceptable level of accuracy.

The process of conflation involves identifying features within one reference map that most reasonably correspond to accurate locations of real earth objects that are to be combined with one or more target maps. Next, one selects control points within the maps and their locations are reconciled with the selected objects. Some portions of the map will move a great deal while others may not move much or at all (Figure 7.11). In many cases the process may have to be repeated to guarantee that the "best" possible outcome is obtained. Each successive iteration may require the selection of additional control points, removal of ineffective ones, and an overall evaluation by the user. In short, the process is not fully automated, nor is it likely to become so in the near future.

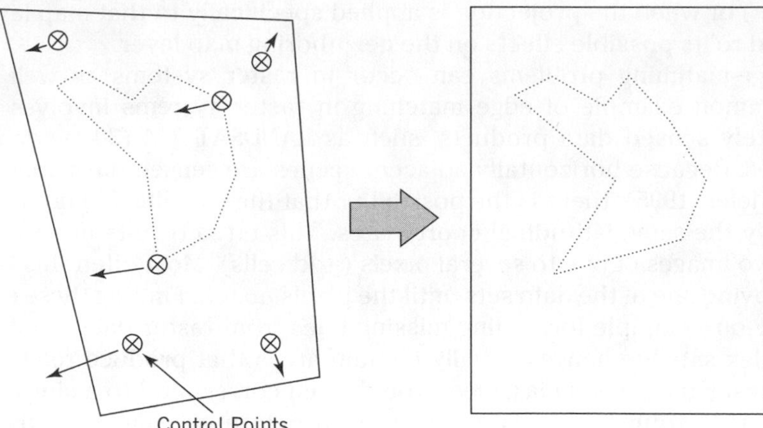

Figure 7.11 Conflation. The use of conflation to move objects in a map: some entities are tacked down to prevent them from moving, whereas the others are shifted to adjust to the same coordinates as the base data layer.

TEMPLATING

In the preceding section we mentioned the use of a single map layer as the most accurate among several for the same area. When you look at multiple maps of the same theme for different time frames, you notice a number of graphical discrepancies, including one we have thus far neglected. Viewing the same maps simultaneously, you also notice that, despite all your efforts to prevent it, the outside dimensions of all four study areas seem to differ slightly in shape. When you input these maps you chose certain data points as reference points and assigned them to true geographic coordinates, yet the four maps are not identical. Probably, then, the differences in location of these reference points from coverage to coverage, combined with the nuances of the projection algorithms and computer rounding errors, have produced slightly different results for each coverage.

If you must later perform an overlay of these four maps, there will be numerous areas along the margins of some maps that will not have associated areas for other maps. Once again, you must select the coverage you trust most to be representative and use it as a **template** (or cookie cutter). If the boundary of the template is within the boundaries of all the other maps, you simply use this pattern to cut out the study area of the rest. However, if any of its boundaries extend beyond those of another map, you will need to select coordinates somewhat inside the margins of the template to ensure that all subsequent maps will occupy all areas within the template. Once applied, the layers should all have the same shape, coordinates, and size.

A note here about the statistics reported from multiple themes is important. It is entirely possible that once you have applied a template to all of your themes and then produced output statistics, there will be slight differences in the total area for each of the four. This was the result in a study of land use change in which every attempt to remove the discrepancy failed (Simpson et al., 1994). Much of the incongruence can be attributed to a combination of computational rounding and the algorithmic methods by which the GIS calculates area. For the most part, if the error is a minor percentage of the overall database, it should simply be accepted. No amount of editing is going to remove it. Your task is complete, and you are now ready to begin analysis.

Terms

attribute error
attribute pseudo node
conflation
dangling node
decimal degrees (DD)
degrees, minutes, and seconds (DMS)
edge matching
entity error
incorrect attribute values
island pseudo node
missing attribute
missing labels
Morton sequencing
overshoot
pseudo nodes
rubber sheeting
sliver polygons
spatial pseudo node
template
tiling
too many labels
undershoot
weird polygons

Review Questions

1. What is the purpose of tiling in the storage and editing subsystem?
2. Draw a tiling system utilizing Morton sequencing. What impact does this method have if one is required to expand the study area?
3. What is so important about editing a database? What problems can occur as a result of even a simple error in the database? Give an example other than one from the text.
4. What are the two basic types of errors that need to be edited? Describe each. Which of these is the most difficult to find? Why?
5. What are the six major areas of entity errors that need to be addressed? Give an example of each.
6. What are pseudo nodes? How are they produced? How can they be avoided? Give an example or two of a pseudo node that is not an error.
7. What are dangling nodes? Overshoots? Undershoots? How are these caused? What can you do to avoid them?
8. Why is a missing polygon label point a problem? Why do we sometimes end up with missing or multiple label points? How can this result be avoided?
9. What are sliver polygons? Weird polygons? Describe methods of avoiding and correcting these conditions.
10. How do you find missing attributes in vector? Raster? How can you avoid creating a map that is incomplete in this respect?
11. How do you find incorrect or displaced attributes in vector? Raster? What are the principal causes of each?
12. Describe the process of converting projections using a vector GIS.
13. What is edge matching? Why is it needed?

14. What is rubber sheeting (conflation)? Why is it needed? How is it done?

15. What is templating? What is it used for? How do you decide which coverage to use as the template?

References

Berry, J.K., and C.D. Tomlin, 1984. *Geographic Information Analysis Workshop Workbook*. New Haven, CT: Yale School of Forestry.

Environmental Systems Research Institute, 1992. *Understanding GIS: The ARC/INFO Method*. ESRI: Redlands, CA.

Epstein, E.E., 1989. "Development of Spatial Information Systems in Public Agencies." *Computers, Environment and Urban Systems*, 13(3):141–154.

Laurini, R., 1994. "Multi-Source Updating and Fusion of Geographic Databases." *Computers, Environment and Urban Systems*, 18(4):243–256.

Lillesand, T.M., and R.W. Kiefer, 1995. *Remote Sensing and Image Interpretation*. New York: John Wiley & Sons.

Seipel, P., 1989. "Legal Aspects of Information Technology." *Computers, Environment and Urban Systems*, 13(3):201–205.

Simpson, J.W., R.E.J. Boerner, M.N. DeMers, L.A. Berns, F.J. Artigas, and A. Silva, 1994. "Forty-eight Years of Landscape Change on Two Contiguous Ohio Landscapes." *Landscape Ecology*, 9(4):261–270.

UNIT 4

SPATIAL ANALYSIS

CHAPTER 8

Query and Description

The analysis subsystem is the heart of the GIS: it is what the GIS is designed for. It is also the most abused subsystem of the GIS. The abuses range from attempts to compare nominal spatial data with ratio spatial data to statements about the causative nature of spatially corresponding phenomena made without testing for alternative causes. Much of the abuse of the analytical subsystem is a result of lack of understanding of the nature of the spatial data contained in the system. For example, some data are purposely ranked at the ordinal level on the basis of their perceived importance for a particular problem, and then these same ordinal data are applied to an unrelated analysis for which the ranking no longer applies. This is not unlike correlating history grades to calculus grades to compare the intellectual capabilities of a student. Another common abuse occurs when nominal grid cell categories coded with integers are multiplied or divided by grid cells numbers representing ordinal, interval, or ratio data, yielding numerical values that essentially have no meaning. Yet the results of such analysis are frequently used for decision making. The list of possible offenses is quite large, partly because the power of the GIS is too great to be effectively applied without a firm understanding of fundamental geographic, numerical, and statistical concepts and partly because of the inherent believability of map output.

Beyond the problems of GIS abuse, there is a common belief that GIS is a panacea for all geographic problem solving. Although the tool is quite powerful, its algorithmic content is more often driven by market pressures than by computational comprehensiveness. In short, GIS is an incomplete set of spatial analytical tools. In many cases the user is obliged to combine GIS tools with other computational tools to accomplish modeling tasks. Fortunately, there are many types of GIS software, each with its own strengths and weaknesses. The limitations of the GIS you choose should be weighed against the kinds of analysis you most often perform.

This chapter provides a general description of the analysis subsystem, and describes the potential abuses that are possible within it. Because of their importance to analysis and modeling it begins with a description of model flowcharting and explains how they are constructed and how they support GIS modeling. It explains the importance of the analytical capabilities of the software to both modelers and nonmodelers. Focusing on the simplest and often most used techniques found in the GIS, this chapter describes and provides examples of methods for retrieving points, lines, and polygons. It further provides a basic background in descriptions of the sizes and shapes of the objects as well as their attributes. Moreover, this chapter also shows how objects can take on a higher meaning by the nature of their positions, juxtapositions, or specific attribute characteristics. Finally, the chapter defines and explains how polygons can form different types of region, and how collections of objects can be described as communities, thus taking on a special meaning for search operations and description.

LEARNING OBJECTIVES

When you are finished with this chapter you should be able to:

1. Create, describe, and explain both formulation and implementation GIS flowcharts.
2. Discuss, with examples, why the analysis subsystem of the GIS is often the one most abused, and why it is considered an incomplete set of geographic analysis tools.
3. Understand the importance of the analysis subsystem even for those who will primarily be building databases for others.
4. Explain the process and the importance of isolating, identifying, counting, and separately tabulating and displaying individual items.
5. Describe the differences in the process necessary for isolating, counting, and identifying the different object types (points, lines, and areas) in a spatial database.
6. Explain the use of measurable attributes in searching for lines and areas.
7. Describe how shape can be used to isolate both linear and polygonal entities.
8. Explain why shape might be an important attribute for which to search.
9. Define *higher-level objects*, giving examples for points, lines, and areas, and describing their utility.
10. Describe the approach to identifying higher-level objects.
11. Calculate *mean center* and *weighted mean center*.
12. Define and describe the different types of formal region.
13. Suggest how collections of points, lines, and areas could be considered to be communities.

MODEL FLOWCHARTING

Before we begin to describe the GIS analytical capabilities, it is important to provide you with a preliminary introduction to flowcharting. We will develop

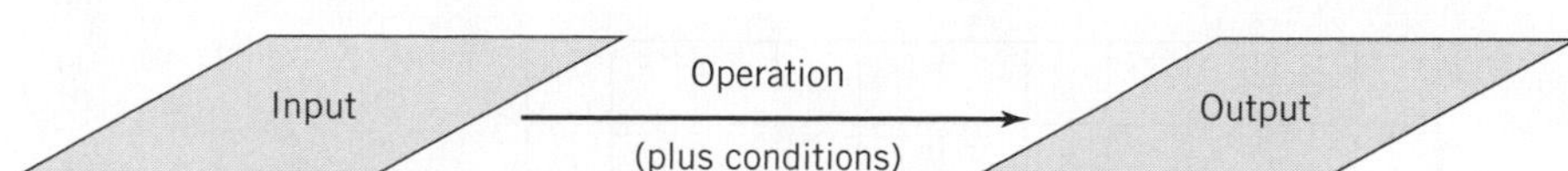

Figure 8.1 GIS flowchart. This is the general form of most GIS flowcharts. Although the specifics and symbols will change, virtually every GIS model begins with input, performs some operation (with or without conditions), and produces some output.

the flowchart idea more fully in Chapter 15, but the ideas of GIS analysis and GIS model flowcharting are so central to the analysis subsystem that many software packages now incorporate some form of flowcharting utility (cf. ESRI's Model Builder, ERDAS' GIS Modeler, and the flowcharting capabilities of IDRISI). Given the importance that GIS software vendors are placing on flowcharting and their recognition of its importance to systematizing our thinking and documenting our procedures, we should begin working with them immediately. By first flowcharting simple analysis tasks and becoming comfortable with the idea, it will be easier to use them for more complex modeling tasks.

All GIS tasks, no matter how simple, require us to perform at least one identifiable operation on at least one thematic layer as a model input to produce output. This provides us with the general form of a GIS model flowchart (Figure 8.1). In this simple example we have three primary components: (1) input (typically a map), (2) operation (a search, a measurement, or other GIS task), and (3) output. Assume the output is a map. Notice how the flowchart shows the objects, the operations acting on or creating them, and a direction of flow. This type of flowchart is called an **implementation flowchart** because it shows how the model actually works. As your modeling expertise grows you will need to create a flowchart that flows in the other direction first, so you can move from expected outcomes to the basic data elements upon which the model is built. This type of flowchart is called a **formulation flowchart** because it is used to design the modeling tasks. It is useful to begin modeling with this type of flowchart to help us organize our thinking. It is often easier in complex modeling to decide what our desired outcome is before we know what elements we need to achieve that output. For now we will examine implementation flowcharts to illustrate how the portions of a fully developed GIS model flowchart are created. Examples of simple implementation flowcharts will be included for selected analytical tasks as we encounter them so you can begin flowcharting now rather than waiting for later chapters. This general form is used for these examples, but the exact terminology is based on the command names and modeling approaches of your software.

GIS DATA QUERY

In a typical GIS database there are many maps, each containing a separate theme. Within each theme you will likely need to isolate numerous features for study. You will want to know which of the selected features occur most often, how often they occur, and where they are located. For example, you might be interested in the number of individual trees (point features) located inside your study area. Because these are point data, you are going to be counting

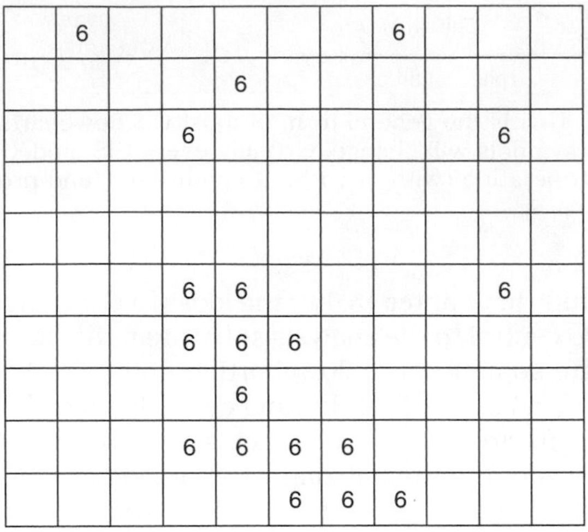

<table>
<tr><td></td><td>6</td><td></td><td></td><td></td><td></td><td></td><td>6</td><td></td><td></td><td></td></tr>
<tr><td></td><td></td><td></td><td></td><td>6</td><td></td><td></td><td></td><td></td><td></td><td></td></tr>
<tr><td></td><td></td><td></td><td>6</td><td></td><td></td><td></td><td></td><td></td><td>6</td><td></td></tr>
<tr><td></td><td></td><td></td><td></td><td></td><td></td><td></td><td></td><td></td><td></td><td></td></tr>
<tr><td></td><td></td><td></td><td></td><td></td><td></td><td></td><td></td><td></td><td></td><td></td></tr>
<tr><td></td><td></td><td></td><td>6</td><td>6</td><td></td><td></td><td></td><td></td><td>6</td><td></td></tr>
<tr><td></td><td></td><td></td><td>6</td><td>6</td><td>6</td><td></td><td></td><td></td><td></td><td></td></tr>
<tr><td></td><td></td><td></td><td></td><td>6</td><td></td><td></td><td></td><td></td><td></td><td></td></tr>
<tr><td></td><td></td><td></td><td>6</td><td>6</td><td>6</td><td>6</td><td></td><td></td><td></td><td></td></tr>
<tr><td></td><td></td><td></td><td></td><td></td><td>6</td><td>6</td><td>6</td><td></td><td></td><td></td></tr>
</table>

Figure 8.2 Finding raster attributes. Raster geographic information system (GIS) showing the isolation of points by attribute. Each grid cell is identified by a color or shading pattern that represents its unique attribute values. In this case the number 6 is used for illustation.

objects that conceptually occupy no spatial dimension. A raster GIS offers numerous ways to find these items, but the simplest way is to create a new layer that eliminates all unnecessary data. To do this you mask all other portions of the database through a simple reclassification process, classifying all the trees of the selected species as your target points (grid cells) and reclassifying everything else in your layer as background. Most raster GIS packages will then let you output the results in a table that allows you to count the target points and the background points (Figure 8.2, Table 8.1). From this we can tabulate percentages directly, making possible comparisons of one tree species with another. Of course, the trees are only located within the grid cell, not wholly occupying it, so the percentage area occupied by these target cells is not very accurate.

Locating point data accurately in raster GIS is impossible because the grid cells themselves occupy geographic space. Still, if you cannot specifically locate the points, you can identify the grid cells containing them. This is done using an onscreen cursor device that allows you to point to each grid cell individually. This usually results in a readout giving the row and column numbers, as well as the attribute codes for the item selected or pointed to. Even in raster, knowledge of the location of individual points is important. For example, you might be comparing the locations of trees to grid locations from remotely

TABLE 8.1 Listing of Grid Cells Chosen by Attribute Value Codes

Category	Attribute Value	Cells[a]
Not coded	0	225
Prime land	1	642
Urban	2	201
Federal land	3	188
Grazing	4	981
Water	5	64

[a]Number of cells of each type found in the database.

sensed data as a means of testing the sensor's ability to recognize the presence of trees at that location based on the electromagnetic properties returned to the sensor from that cell.

In raster systems, lines are strings of grid cells that touch one another either orthogonally or diagonally in Cartesian space. Polygons are groups of grid cells that are connected in much the same way or share attribute values (called **regions**). Because of the way attributes are linked in the simplest raster GIS software, to identify them as entities you must identify them by attribute, just as we did earlier when we isolated point data. And, as before, you can use a cursor to identify column and row locations and attributes for each location. Tabulation of the results will again show the amount of each category and their percentages. You need to keep in mind that for lines, percentages of the total database occupied by the various objects are most likely meaningless; for areas, the accuracy of the percentages is largely determined by the method used to encode the grid cell categories and by the grid cell resolution.

In vector systems, the process of finding points is performed by displaying all points in the cartographic portion of the database. If you need to find points with particular attributes (e.g., telephone poles, bird nesting sites, etc.), you will need to access the attribute database and its tables (Table 8.2). Most often you will perform a search that identifies all table locations that exhibit the appropriate codes. Because these are linked to the entities, you will be able to selectively isolate these specific types of point objects. Because vector data structures contain explicit spatial information, you can easily obtain the exact coordinates in projected space. These can be produced as tabular output or viewed directly on the screen by means of a pointing operation. In some cases you will point to individual items; in others you surround target objects with a graphic window. As with raster systems, you can also obtain attribute and coordinate information.

Because vector coordinates are explicit, you can obtain entity information about lines and polygons by selecting them individually. For example, you can

TABLE 8.2 Listing of Entities Chosen by Attribute Value in a Vector System[a]

Landuse Code	Name	Number of Polygons
100	Row crops	21
110	Fallow	18
120	Grains	65
130	Grazing	3
200	Residential	982
210	Commercial	124
220	Light industry	192
230	Heavy industry	54
Landuse Code	**Name**	**Polygon ID**
130	Grazing	25
130	Grazing	28
130	Grazing	29

[a]The lower half of this table shows a list of polygons selected from the top half. Notice that three polygons matched landuse code 130.

identify all points in the theme by displaying only points, or you can obtain a listing by accessing the attribute tables specific for points. This is true for lines of all kinds as well as for polygons. It is no difficult matter to identify all the polygons in a map, in addition to their attributes. To identify the absolute coordinates for lines and areas, you need multiple sets of coordinate pairs for each of the points that are used to identify all of the defining line segments. Again, this is done by pointing to individual objects or by surrounding them with a graphic window. And, as with point data, any line or polygonal object can also be selected on the basis of a database query of the attribute tables.

LOCATING AND IDENTIFYING SPATIAL OBJECTS

Although nearly any GIS can be used to find or locate objects, before we go any further we might ask why we need to be able to find and locate objects in our GIS. The twofold answer is (1) because it is a fundamental process comparable to traditional map reading and (2) because these tasks are fundamental components of more complex analyses. In the first case, it is important to be able to isolate, count, and locate objects because these activities give us a description of the overall complexity of our maps. The objects in the map represent features on the ground. Within the communication paradigm the primary purpose of the map is to display these objects. This facilitates the identification of the spatial relationships among objects in the map. The more objects occupy a given map space, the greater their density in the real world. It is important to know, for example, whether there are many houses (high-density housing) or just a few (low-density housing) in a particular area. These numbers often relate directly to population densities, which may, in turn, be vitally important in determining the risk of a population to hazardous materials spills, flood events, earthquakes, and other disasters, to cite just a few examples. Likewise, the numbers of plants or animals in an area may very well be related to the environmental health of the region. Or the number of road networks might provide important information for routing sales staff or truck delivery. In other words, a quantitative measure of the densities of these objects allows us to make direct, analytical comparisons to other areas or other variables in the same area.

The mere existence of objects and their patterns is often very important. Comparisons to other features on the landscape might be used to suggest spatial relationships among other spatial variables and even determine causes for their distributional patterns. For example, an enumeration of the number of houses may be useful for examining their relationships to the amount of land available for new housing. Areas on a map with low numbers of housing units can be compared to a map layer of available land for house construction through a process called **overlay**.

The simple enumeration of objects and their locations also lets you examine their relationships to more prominent objects in the same map. Landscape ecologists, for example, are interested in the relationships between small, isolated polygons (patches) to the overall background of the study area (Figure 8.3). By knowing how many patches occur in a study area, and how much area they

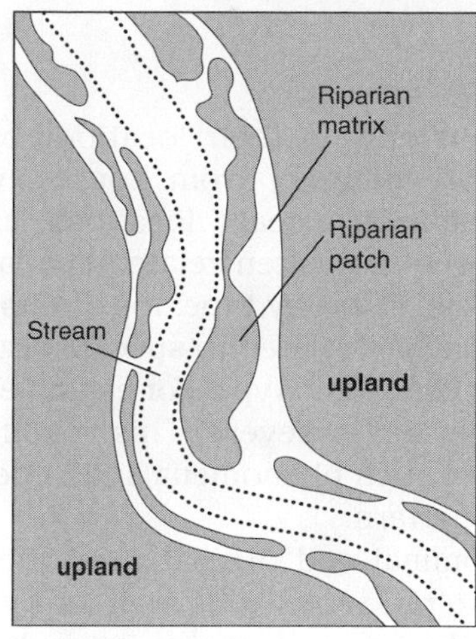

Figure 8.3 Elements of a river landscape showing upland regions and a patch and matrix configuration within the stream banks. Modern geographic information system (GIS) software can evaluate such maps to derive fragmentation, patchiness, and connectivity measures. To do this, the patches must be able to be isolated from the rest of the map elements.

occupy in the database landscape, ecologists obtain a measure of patchiness or fragmentation that is one indicator of overall landscape health. There are applications in economics as well. The locations of shopping centers, for example, may be highly fragmented, requiring additional roads and public utility services to afford access to isolated polygons of retail activity.

As implied by the last two examples, among the most important aspects of being able to find and locate objects on the map is the ability to make further measurements and comparisons. From absolute numbers we can make extensions to relative numbers. For example, we can move from the total number of houses, roads, or polygons of a particular land use to relative numbers of houses per unit area, the relative number of road miles per square mile, or the relative number of polygons of one land use type versus another. Each set of measurements or comparisons provides additional insights into our environment. All require the initial ability to isolate, count, and locate the individual entities within the database.

DEFINING SPATIAL CHARACTERISTICS

Being able to find point, line, and area entities on a map is of little value unless their search is purposeful. Most of the objects analysts find, count, and locate in coordinate space are selected not on the basis of entity type alone but rather on their real-world characteristics. Just as we would not simply record points as points, lines as lines, and areas as areas in a field survey, we likewise do not keep them stored as sterile entities in our GIS database. Instead, we note the types of point, line, and area objects, the amounts or values of each, and their categories, because our greatest interest is in their attributes or descriptors. For the same reason, we most often search for objects, count their numbers, and note their locations on the basis of these attributes. Let's examine individually the three entity types (points, lines, and areas) on the basis of how we would search a database for their attributes and the insights we might gain from doing so.

Point Attributes

Point objects, like all entities, differ not only in their locations but also in their attribute characteristics. Trees are different from houses, which are different from cars, which are different from disease locations, and so on. These differences provide us with different but often related spatial patterns of each group of objects. Point objects can differ by type; for example, we can isolate patients in a region that share diseases related to specific organ groups (Figure 8.4). These types may also be ranked within type. For example, digestive diseases could be ranked as mild, moderate, or severe. This would allow the user to identify patterns of distribution of such phenomena and to determine if there is some reasonable cause for this pattern.

We can classify point objects by nominal and ordinal categories; we can also select these objects by interval or ratio values as well as by type. For example, houses could be selected on the basis of market value: houses that are below $150,000, between $150,001 and $200,000, and so on. These attribute values must be stored in the database to permit their retrieval. Other examples might include the annual sales for individual businesses, the amount of nitrogen fixation for selected plants, and the shoulder heights of deer located through telemetry. These quantitative attributes allow us to select a wide range of groups or classes of each object depending on our needs.

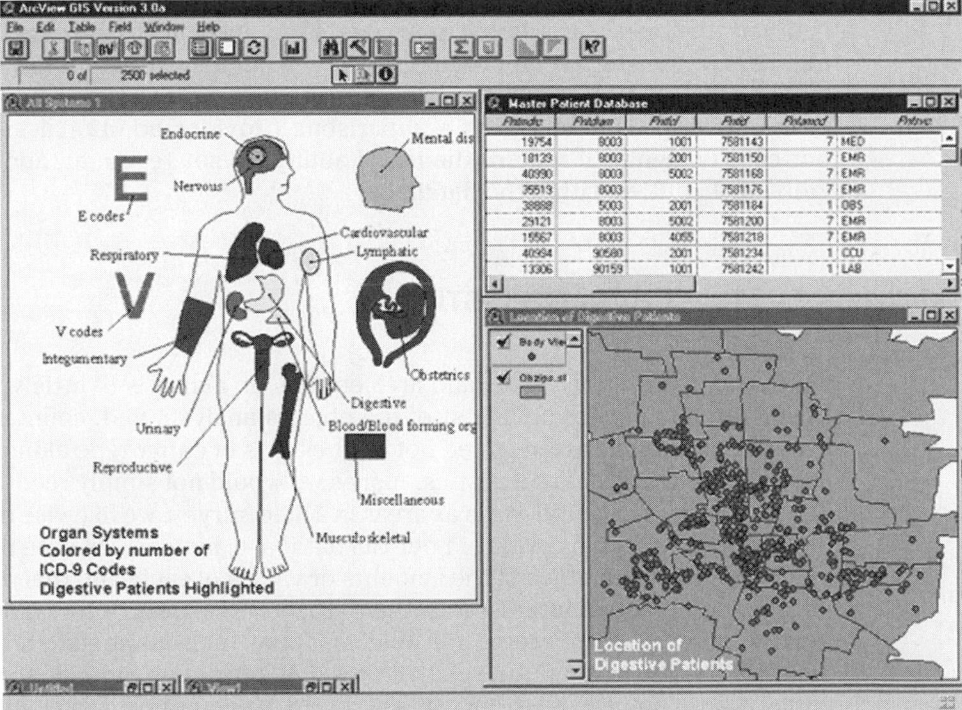

Figure 8.4 Body Viewer is a classic example of how point objects can be selected by type. In this case, selection of a graphic image of a human organ or system allows display of a linked table of disease types associated with those organs.

It is critical that a GIS be able to define each group or category separately and tabulate the results of a search. We also need to be able to produce a graphic of the selected classes of interest and omit all the others. Because GIS operates in a holistic cartographic paradigm, it should allow us to selectively retrieve and group data in any fashion we need.

Beyond isolating classes of data, the GIS must allow us to locate each item individually for each class and compare it to others of its own kind. That is, we need to be able to show the spatial relationships between individual items of a selected point object class so that we can later perform analytical operations to quantify those relationships. Such relationships might include proximity, density of distribution, or patterning (e.g., random, regular, or clustered).

If we need to show spatial relationships between like-kind or like-value objects, we eventually will need to compare them to other similar point objects of a different type or value. We will need to be able to illustrate and quantify the relationships, for example, between houses selling for $150,000 or less to those costing $150,001 and over. But even more, we might want to show the relationships among point objects of different types, say houses and streetlights. And if we need to compare the numbers and spatial locations of one group of point objects to another, we might need to show the relationships among linear and polygonal entities. For example, there may come a time when we will need to know the relationships between houses (points) and available paved roads, sewerage lines, water and electric utilities, parks, and shopping centers. Each of these other entity types will also need to be separable on the basis of type or magnitude.

Line Attributes

Recall that line objects are one-dimensional entities defined by two or more points with corresponding coordinate pairs. The line object can also contain nodes that are points specifically indicating either the beginning or the ending of a line, or some change in an attribute along the line. As with point objects, line entities can be identified by attributes with varying levels of data measurement (i.e., nominal, ordinal, interval, ratio, or scalar). Examples of line item types include railroads, streets, fault lines, fencerows, and streams. Each of these is separable from the others because all differ in kind (Figure 8.5). In other words, each different object should be able to be identified, retrieved, and located separately. And, as with points, they should be able to be tabulated and displayed separately to identify the unique pattern each exhibits on the landscape. Like point objects, lines should be separable based on ordinal rankings or some measure of magnitude. Highway types such as single lane, double lane, three lane, and interstate freeways are examples of line objects organized by ordinal ranks. These distinctly different road types can be compared only across the single spectrum of highway types; quantitative comparison to other nonhighway line features is not permitted. Another classic set of line entities exemplifying a measurable difference in magnitude might include lines representing the actual stream flow of different stream tributaries.

In some cases, a single line may experience a change in attribute type, rank, or magnitude at some point along its length. For example, a road may change

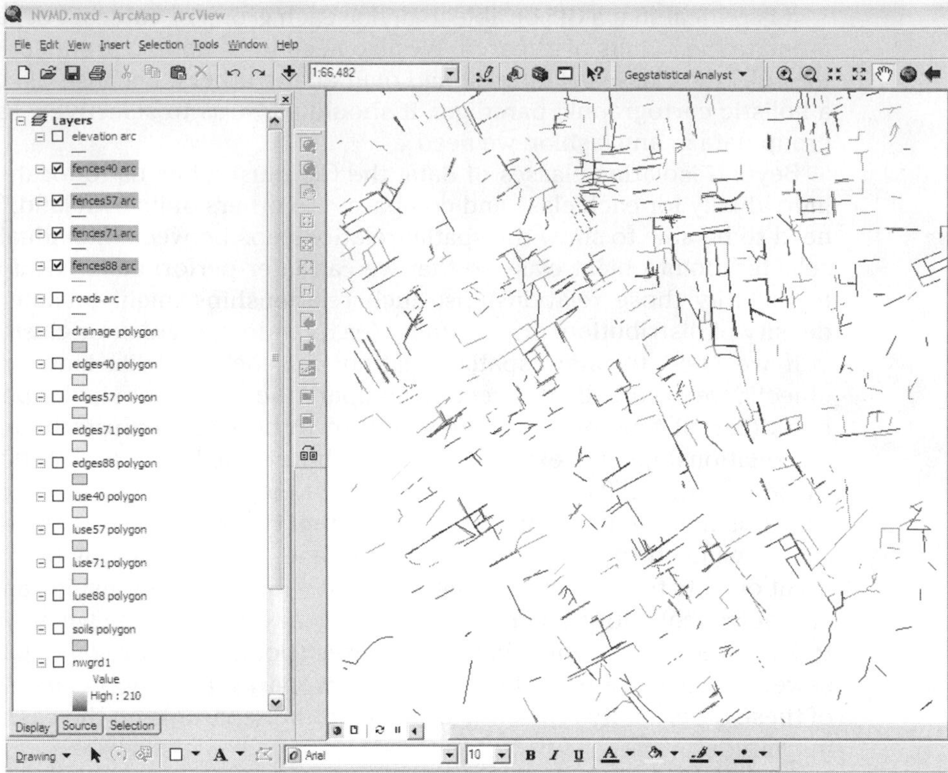

Figure 8.5 Selecting line objects on the basis of their attributes. Note how only the themes with fences, rather than the edges themes, are activated.

from single to double lane, a stream's flow may change to a higher value because of inputs from tributaries, or the measured traffic flow along a city street may change along its length where it connects to another street. By using nodes to indicate the changes, and by storing each segment between nodes with the appropriate attributes, we can identify each segment as a separate, identifiable entity. The attributes can subsequently be used to allow us to retrieve either whole lines or portions of a line.

Another line characteristic involves not just the line attributes but also a comparison of what falls on either side. For example, we may want to define a hedgerow or fencerow not on the basis of type of vegetation but rather on a comparison of the adjacent land covers on either side. We might want to be able to identify, for example, all hedgerows that have forest on one side and farm fields on the other, or all hedgerows that have only farm fields on both sides (DeMers et al., 1995). In raster systems, chores like this may be difficult, requiring us to perform elaborate categorical manipulations called **neighborhood functions**. In topological vector systems, a search could be performed to identify attributes for each bordering polygon. Because the topological relationships are explicitly encoded, a simpler method for both raster and vector is to explicitly encode the attributes for the line that relate to the neighboring polygons (e.g., a line might be coded as farm/urban, indicating its neighbors). With raster this would mean either creating a separate grid for the line entities to permit the attributes to be coded or, if your raster system employs a database management system, storing the attributes there as separate identifiers. Once again, you can see the importance of thoughtful planning before data entry, in this case with a view to easily answering questions you might ask.

To locate the lines, it is necessary to be able to identify all the coordinate pairs that make up the line in vector, or all the grid cell column and row values in raster. This adds three other retrieval factors that do not exist for point entities: length, orientation, and shape. Line entities may be straight with single orientations, or they may be jagged or sinuous with multiple orientations, like street or road patterns or meandering streams, where each straight-line segment indicates a unique orientation. Some line entities are simple, with a single line; others are complex, with a branching network that may form a hierarchy of lines, as in a branching stream network. All these are difficult for most raster GIS software to calculate, but vector GIS packages can easily perform length and azimuthal calculations. Length calculations require performing the distance-between-points calculations using the Pythagorean theorem if the data are in Cartesian coordinates, or great circle calculations for data that are projected (Robinson et al., 1995). Azimuthal calculations use the standard formulas for spherical trigonometry between each pair of coordinates (Robinson et al., 1995). Shape calculations for line entities are most often measures of **sinuosity**, or some ratio of the amount of overall length of a curvilinear line to its straight-line length. This merely requires two separate sets of calculations: one for the total length and one for the straight-line length. More complex analyses involving combinations of these calculations are possible but call for numerous sets of operations (Mark and Xia, 1994).

These three principal types of linear measure—length, orientation, and shape—can be defined as individual attributes for each line entity or for each separable portion of a line entity. For example, you may want to find all hedgerows that are longer than 80 meters, separate out all roads that have a north–south orientation, or isolate highly sinuous streams. To allow the lines or line portions to be selected, tabulated, and displayed separately, then, each should be separable with your GIS. In some cases the line entities may extend beyond the map boundaries rather than being wholly contained inside the map. If you want to be able to eliminate these, it might be useful to include an attribute indicating that the line is incomplete. Analyses that are not able to separate out incomplete line entities are likely to be inaccurate even if the data are aggregated. Most of these operations are designed to obtain overall or aggregate statistics for your map, such as the average length of a certain line entity in a map or the overall average orientation or sinuosity. At times, however, it may be necessary to combine operations to create regions of similarity in a map where the lengths, orientations, and sinuosities vary from portion to portion on the map.

Area Attributes

As with point and line entities, polygon entities can be defined, separated, and retrieved on the basis of category, class, or magnitude (Figures 8.6, 8.7). As before, each of these attributes needs to be explicitly stored in the database,

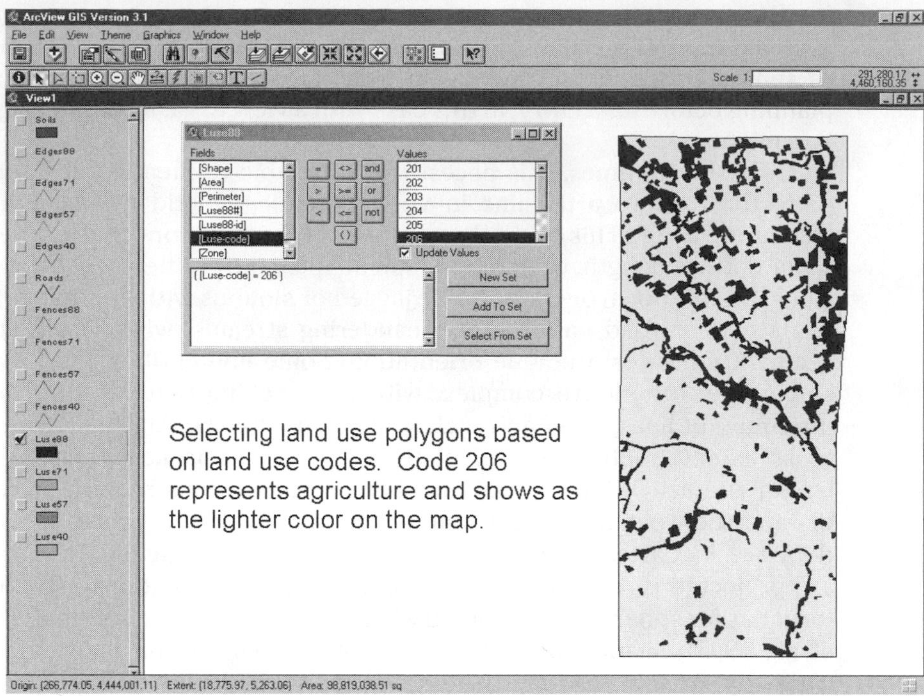

Figure 8.6 Selecting area objects on the basis of their attributes. In this case, the areas are selected by land use code.

whether as grid cell attributes or as vector polygon attributes. Isolation and retrieval are performed in exactly the same manner as for points and lines. But, like lines, areas have added dimensionality that permits more attributes to be examined.

Among the more useful attributes for polygons are their shapes. These shapes could be strictly Euclidean, that is, some variant of known geometric shapes with predefined properties, or they could be fractal, where the irregularity of the outside of a polygon is measured. Related to shape is a polygon's elongation, or the ratio between its long and short axes. Although not generally a built-in function of vector GIS, measuring the long and short axes and expressing these values as a ratio is a relatively simple matter. The coordinates for each of the polygon points can be used to identify the points that are the farthest in a particular direction. Elongation implies that one might also wish to determine the particular orientation of elongated polygons with respect to cardinal directions. Although measures of shape are not frequently invoked by the GIS community, there is growing interest in the relationships between shapes of both human and natural area features and their functioning (Forman

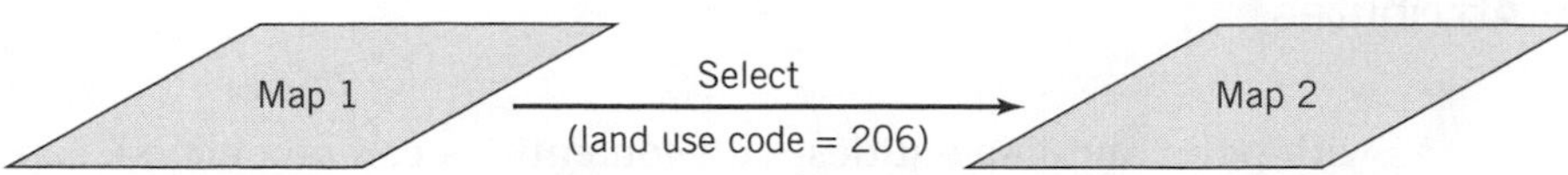

Figure 8.7 Attribute selection flowchart. A simple flowchart illustrating the use of land use codes to retrieve selected types of area attributes.

and Godron, 1987). At a minimum, some simple measures of shape would be useful to facilitate the isolation, selection, tabulation, and display of polygons of a given category of shape and orientation.

Another measurable attribute of polygons is their area. In raster, size is determined by counting the number of grid cells of a given category. These categories do not need to be restricted to a single polygon; rather, they simply indicate the number of grid cells with a given value. Often a reclassification process requires the ability to isolate these polygons, and then their area can more readily be found by a simple count of grid cells. Measures of area in raster are not going to be particularly accurate because of the quantized nature of the grid cell. In vector, the perimeter of the polygonal lines is easily calculated by adding the line lengths, whereas the area is calculated much as would be done manually, by a series of length-times-width calculations for portions of each polygon. The more complex the polygon, the more calculations must be performed. But in general, any commercial GIS package should be able to provide reasonably rapid area calculations. To isolate polygons on the basis of area or perimeter, one need only select a set of category criteria, place the appropriately sized polygons in each, and perform a simple query. Polygon area, like line length, is most often performed on each polygon for each category and is then averaged for subsequent tabulation. This indicates the relationship of polygon area to total area of the database occupied by that category of polygon.

Two other measures regarding areas are related to internal integrity of the map layer. The first is **contiguity**, a measure of the wholeness or amount of perforation of a polygon (Berry and Tomlin, 1984). A large polygon that contains many holes demonstrates less contiguity than one with only a few holes or none at all. Analyses of contiguity may prove useful for modeling animal habitats or forest fire hazards. By determining the amount of contiguity and then classing the results into groups, the analyst can easily retrieve and display these groupings or use them for further modeling.

The second measurable attribute that could prove useful for polygonal features is **homogeneity**, which may or may not be defined as a single polygon (Berry and Tomlin, 1984). Homogeneity measures how much map area is directly in contact with polygonal features with identical attributes. For example, two polygons that have identical attributes may touch each other only along a small portion. In this respect homogeneity is very similar to a simple measure of area. However, integrity can also involve the opposite of homogeneity—internal **heterogeneity**. One might, for example, group polygons on the basis of some heterogeneous mix of attributes rather than on their similarities. Other homogeneity measures could include minimum values, maximum values, averages, totals, and even diversity of attributes. Each of these can be grouped, resulting in a set of polygons with shared attributes. We could, for example, group all forest polygons that have 12 or more species of trees as a measure of forest diversity. This could prove useful for creating game reserves for animal species that require a high level of forest diversity to survive. Or we could group all polygons with that degree of human ethnic similarity—a lack of diversity in residential areas. We could then select all the polygons that contained fewer than three or four ethnic groups to perhaps indicate some measure of ethnic segregation in a city.

Both raster and vector GIS typically have some capability to perform this type of analysis. Once completed, we then have the capability of finding and retrieving polygons based on these attributes. The software should allow us to isolate, retrieve, and output the results individually or to combine them with other attribute characteristics. When we begin looking at GIS classification, we will return to all these techniques as methods of reclassifying our existing map attributes.

WORKING WITH HIGHER-LEVEL OBJECTS

Thus far, we have worked with points, lines, and areas based either on readily available descriptive characteristics or on attributes that are measurable through sometimes simple, sometimes complex methods. All of these entities, however, have certain attributes that set them apart from the rest. Some attributes are an artifact of encoding—for example, nodes that are encoded during the digitizing process. Others may have to be determined—for example, **centroids** indicating the center of an area. We call these "higher-level objects" for lack of a better term, simply because of their uniqueness and usefulness for later GIS analysis. We will separate these higher-level objects into points, lines, and areas and examine each individually.

Higher-Level Point Objects

Two primary types of higher-level point objects are centroids and nodes. A centroid is most commonly defined as the point that occurs at the exact geographic center of an area or polygon (Figure 8.8a). Its calculation is simple for

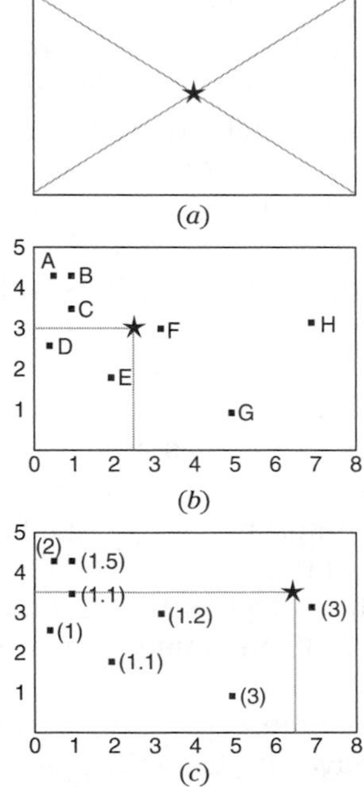

Figure 8.8 Types of centroid. (*a*) Simple centroid, (*b*) center-of-gravity centroid (mean center), and (*c*) weighted mean center, where numbers in parentheses indicate assigned weights, *f* values in Table 8.4 (note how high weights for points *G* and *H* of (*b*) pull the mean center to the right).

rudimentary polygonal shapes such as rectangles; when the polygons become quite complex, the complexity of the calculations needed to perform the task increases proportionately. Raster GIS is not well suited for this procedure. In many cases, even vector GIS does not include this calculation as a stand-alone function. Simple or geographic centroids in vector are calculated by a rule called the **trapezoidal rule**, which separates the polygon into a number of trapezoids. Then each trapezoid's centroid, or central coordinate, is calculated and its weighted average calculated (Figure 8.9). The centroid may be important if you are attempting to produce a surface map from samples taken from within areas. For example, if you were to produce an **isarithm** (defined as a line of equal value) or fishnet map of the population of the United States, but the data were sampled at the county level, you would have to place a centroid in each of the approximately 300 counties. The centroid would act as a point location, as if the data were actually calculated at this point. Then, through interpolation, the isarithms or surfaces could be calculated on the basis of these point locations.

Centroids can also be placed at the center of a distribution of some phenomena instead of at the absolute geographic center of a polygon (Clarke, 1990). A classic example of the use of this technique illustrates the center of distribution of the population of the United States through time (Figure 8.10), where each point represents an evaluation of a separate decade of population. This location, called the **mean center** or the **center of gravity**, requires us to separately average the X and the Y coordinates for all points in the map (McGrew and Monroe, 1993; Muehrcke and Muehrcke, 1992) (Table 8.3, Figure 8.8b). The final result is a single pair of values representing the central point of this distribution.

What if the individual points were not equally weighted? For example, if the points indicated both the locations and volume of shopping done at a selection of stores, we could place our center on the basis of this additional weighting factor. The procedure, called the **weighted mean center**, requires us to multiply each X and each Y by a weighting factor (amount of shopping in this case), sum them up, and then divide by the number of points (McGrew and Monroe, 1993) (Table 8.4, Figure 8.8c). The result is a single pair of X and Y coordinates indicating the mean center of the distribution, modified by the weighting factor.

Such calculations are often applied to large-scale market analyses and economic placement analyses to define the center of the market. Having done this, a market analyst might choose to select areas near the market center for

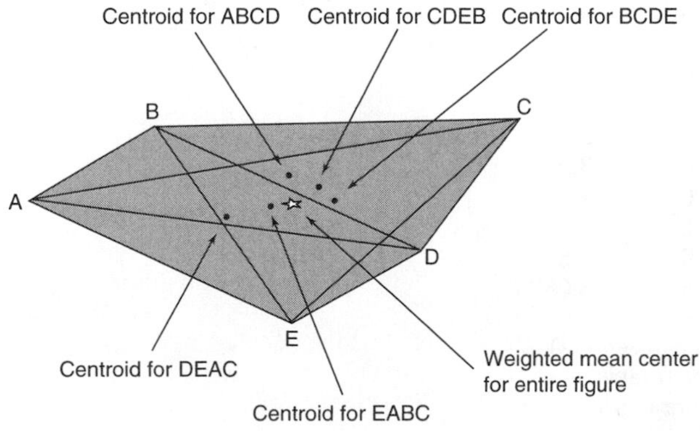

Figure 8.9 Trapezoid rule. Diagram of the trapezoids defined in an irregular polygon. Each overlapping trapezoid has a calculated mean center from which the weighted mean center can further be determined. This weighted mean center is the centroid of the polygon.

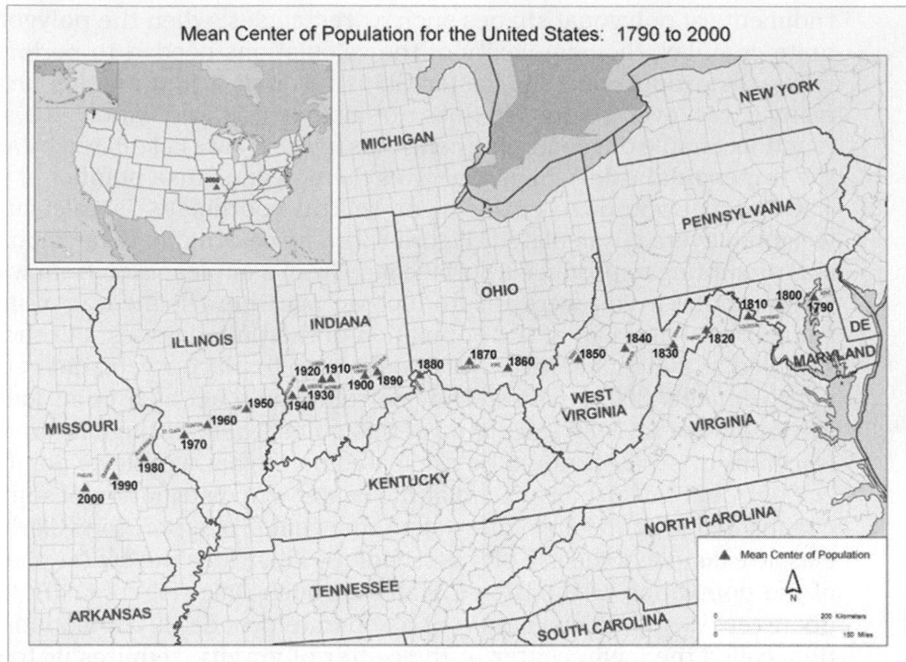

Figure 8.10 Mean center of population of the United States by decade. Each mean center is based on calculations of the center of gravity of population for each decade of population. Modified from US Census Bureau.

locating a new shopping center or other type of business. There are, of course, many other reasons for selecting either simple centroids or center-of-gravity centers; examples include studies of animal foraging activity (Koeppl et al., 1985) and analyses of the change in movement of population centers through time as a measure of large-scale migration (McGrew and Monroe, 1993).

TABLE 8.3 Table of Values Set Up to Allow Calculation on the Center of Gravity[a]

Point	*X*	*Y*
A	0.5	4.5
B	1.0	4.5
C	1.0	3.5
D	0.5	2.5
E	2.0	2.0
F	3.0	3.0
G	5.0	1.0
H	7.0	3.5
Total	20.0	24.5
Mean center	2.5	3.0625

[a]"Centers of gravity" (mean centers) are found by dividing the total values for *X* and *Y* by the number of points.

TABLE 8.4 Table of Values Set Up to Allow Calculations of the Weighted Mean Center on the Basis of the Weights Assigned to Each Set of *X* and *Y* Coordinates

Point	*X*	*Y*	*f*	*fX*	*fY*
A	4.5	2.0	0.5	2.25	1.0
B	1.0	1.5	4.5	6.75	1.5
C	1.0	3.5	1.1	1.1	3.85
D	0.5	2.5	1.0	0.5	2.5
E	2.0	2.0	1.1	2.2	2.2
F	3.0	3.0	1.2	3.6	3.6
G	5.0	1.0	3.0	15.0	3.0
H	7.0	3.5	3.0	21.0	0.5
Total				52.4	28.5
Weighted mean center				6.55	3.512875

Our second type of higher-level point object, the node in vector data models, is significant not as an individual but as a specific locator along line and area entities. Because nodes indicate a change in attribute, the ability to identify them is vital to many attribute selection and manipulation procedures. Nodes are generally encoded explicitly during input and should be easily separated or identified through simple search procedures. There will be difficulty only when a node has been improperly coded as a simple point rather than as a node. In such circumstances the ability to isolate line segments defined by nodes will be compromised. This is yet another illustration of the importance of good planning and careful execution during the input phase of the GIS.

One other situation that is covered in later chapters needs brief mention here: the use of point distributions to identify areas. For example, where closely packed points are found to exist, their area is distinctly different from sparsely packed regions. You may, for example, find a great many weeds in particular parts of agricultural fields, indicating that these areas are different. What remains to be determined is the source of the difference: a former disturbance, a lack of pesticide, or something inherently different about the soil or how it is manipulated.

Other pattern characteristics such as uniformity of distribution of point objects, randomness, or another identifiable pattern can also be used to define areas as specific **communities**, or areas sharing distributional patterns (Figure 8.11). Defining areas on the basis of point distributional patterns is not a strong part of most GIS systems, but it can often be calculated in both raster and vector systems. We'll return to this topic in Chapter 10.

Higher-Level Line Objects

There are three different types of higher-level line objects. The first are called **borders**, and are recognized by abrupt changes in polygon attributes as these lines are crossed. In other words, borders are particularly important because of their locations relative to adjacent polygons. Take a simple example of a national

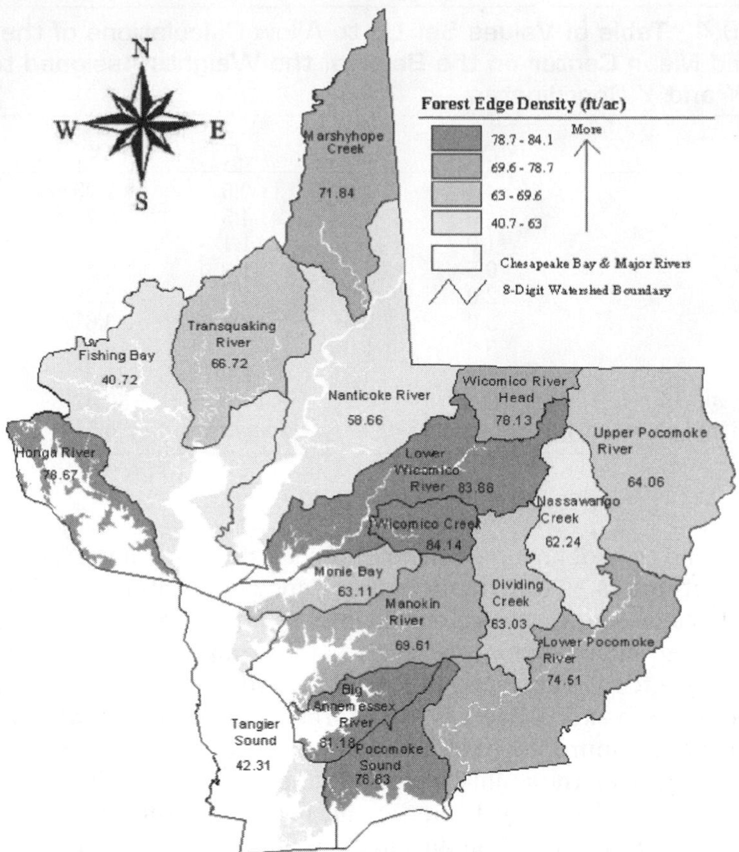

Figure 8.11 Communities based on object density. Map illustrating different communities of points (trees) based on density.

border between the United States and Canada. The line serving as the border between these two countries should allow one to identify all the states in the United States as fundamentally separate from all the provinces and territories of Canada on the north. Although this may seem obvious, if the line separating these two areas on the map is not provided with attribute information clearly indicating its status as an international border, you may have to force the GIS to separate the United States from Canada rather than simply performing a search on the basis of the neighbors of the borderline.

Lines can also be considered higher-level objects when they are located in a particular way relative to other lines. In many cases lines are not simply indications of locations of linear objects or of boundaries between polygons, but rather are connected by nodes to form **networks** (Figure 8.12). Networks can most appropriately be defined as a set of interconnected line entities whose attributes share some common theme primarily related to flow. Networks allow us to model many flow types, from automobile and rail traffic, commodities, and even animal species along corridors. In all these cases the software needs to be able to operate on networks, which means that the lines must share the attributes necessary for analyzing these flows (speed limits, frictions, etc.). Raster GIS is not well adapted to handling networks because there is no way to

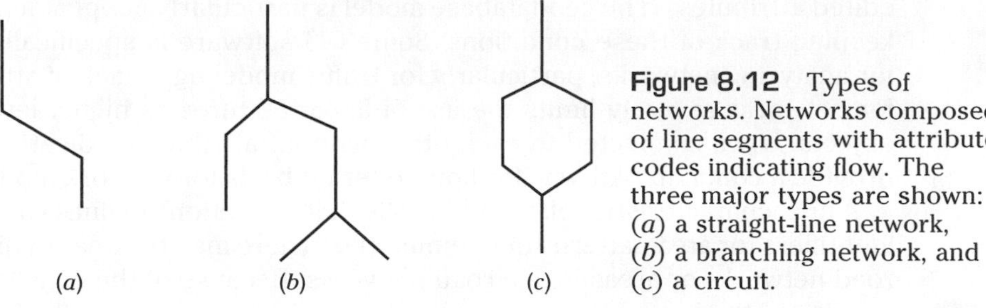

Figure 8.12 Types of networks. Networks composed of line segments with attribute codes indicating flow. The three major types are shown: (*a*) a straight-line network, (*b*) a branching network, and (*c*) a circuit.

define them explicitly except to assign specific attribute values to the grid cells. Still, this can be done, although most GIS professionals will defer to vector GIS for working with networks.

Networks come in three major forms: straight lines, as one might find in a major interstate highway (Figure 8.12a); branching trees, as one might expect when looking at stream networks (Figure 8.12b); and **circuits** (Figure 8.12c), as one might find in street patterns whose lines lead directly back to the starting point (Muehrcke and Muehrcke, 1992). In addition, all these network types can be defined as directed or undirected. In a **directed network**, the flows are allowed to move in a single direction only (Figure 8.13a). Streams, for example, will flow down slope and will not, under normal circumstances, flow in the opposite direction. Likewise, one-way streets limit the flow of traffic to a single direction. In the event that one link in the network intersects another at an angle, there may be a change in flow direction, or it may be necessary to restrict the locations at which turns can be made in passing from one link to another. At the intersection of a two-way street and a one-way street, for example, you will not be allowed to turn from the two-way street into the oncoming flow of one-way traffic. But in networks called **undirected networks** (Figure 8.13b), the flows can go back and forth along the network in either direction.

Because networks are capable of modeling either directed or undirected flows, and because some network links will be connected to some types of links but not to others (e.g., a road crossing over another road), all these characteristic attributes must be explicitly encoded either during data entry or later (as

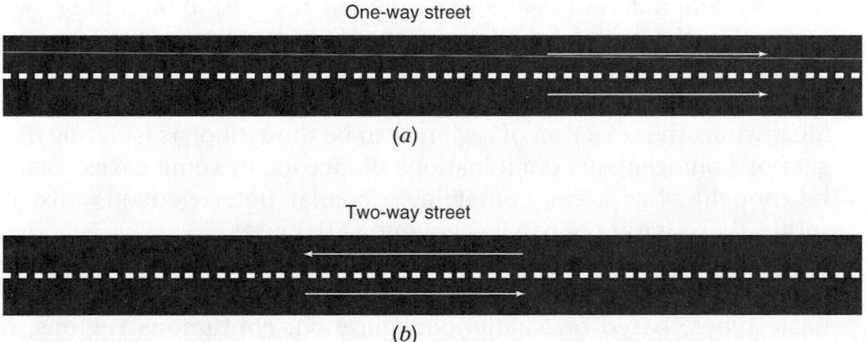

Figure 8.13 Directed versus undirected networks. Directed networks (*a*) restrict the flow to a single direction, whereas undirected networks (*b*) allow flow in both directions.

edited attributes). The geodatabase model is particularly adept at modeling and keeping track of these conditions. Some GIS software is specifically designed for analyzing networks, particularly for traffic modeling. A lack of attribute data for networks severely limits the use of linear features as higher-level network objects. Lines connected to each other without attributes indicating that they provide a common pathway for flows offer no basis for network modeling.

A final characteristic related to lines is that collections of lines can also act as identifiers for area patterns or communities. There may be areas of high-density road networks or areas low in road networks. Because of the density or lack of density of the line objects, we can define these areas as communities sharing these characteristics. Additional patterning, such as regularity or randomness, can be used to identify communities of these line features. Most GIS software, however, does not contain algorithms specifically designed for this process. Most often you will need to use alternative software or modify the existing software through the use of macro languages and other tool kits that may be obtainable from the vendor under certain conditions.

Higher-Level Area Objects

Areas can also occur as higher-level objects. The polygons themselves can be used to define regions of similar geographic attributes. In fact, among the more important aspects of geographic research today is the definition of regions: areas of the earth that exemplify some unity of describable attributes. For example, political regions are defined by national boundaries, ethnic regions by a similarity of origin, biogeographic regions on the basis of species composition, ecoregionalizations based on potential use, and hydrological regions based on availability and utility of water. Inside the GIS, these regional definitions can be based on the attributes defining each polygon or set of polygons. We might, for example, be able to define a forest region within our GIS by selecting all the polygons that show forest as the major vegetation component. We will have to know beforehand which regions we are looking for and how they are defined. Because defining regions is a major endeavor in itself, the chances are good that simply selecting the appropriate polygons or sets of grid cells will not suffice to produce definitions. Instead, we will likely combine several sets of attributes from several different layers to define our regions. In fact, the power to define regions on the basis of a large variety of characteristic attributes put together in as many ways as we need is one of the nicest features of the GIS. For this reason, we can define regions on the basis of what we intend to do with the data. Meanwhile, the selection of regions can be thought of as isolating homogeneous sets or homogeneous combinations of factors. In some cases regions can also be thought of as areas containing a similar heterogeneous mix of attributes rather than sharing only a few common attributes.

Formal regions differ not only in their attributes and in the way the attributes define them, but also in the way they are configured in space. There are three basic types, based on spatial configuration: **contiguous regions**, **fragmented regions**, and **perforated regions** (Figure 8.14). Contiguous regions are wholly contained in a single polygon. Although a contiguous region is wholly contained, its attributes can be defined as homogeneous or as some heterogeneous mix.

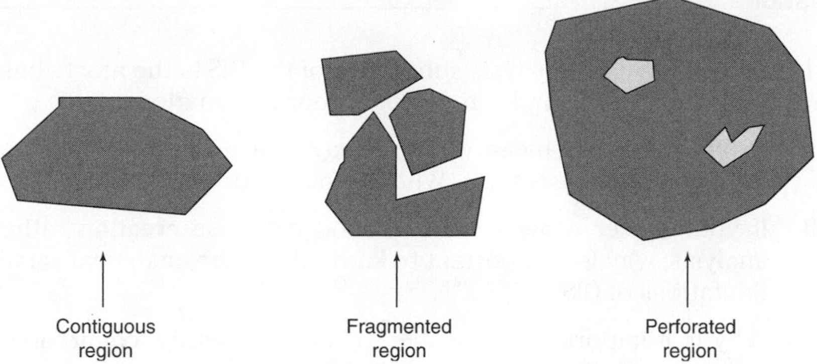

Figure 8.14 Types of regions. Three major types of regions: contiguous, fragmented, and perforated.

Fragmented regions, on the other hand, share some commonality of attributes, but they are composed of numerous polygons separated by intervening space whose attributes are different than the polygons. For example, one could define a scattered set of forest polygons as a forest region. There are no limitations to the separation of polygons as long as the similarity of attributes is maintained. Likewise, for perforated regions, the defining criterion for the region remains the same—homogeneity of attributes or of the attribute mix. In perforated regions, however, the uniform polygon is interspersed with smaller polygons that do not share the attributes of the surrounding polygon. As such, the region is defined as the surrounding matrix, whereas the smaller internal polygons are said to be the perforations. Clearly, there is a possible relationship between perforated and fragmented regions. If the smaller polygons contained in the perforated region are found to share common attributes, they too may be considered to be a region and can easily be separated out from the background region.

All the objects we have identified, whether simple or higher order, must be identifiable by the user so that they can be subsequently analyzed. Each must be able to be isolated, separately tabulated, and displayed. The ability to perform analysis is strongly linked to the ability of the GIS to do this. In vector data models, and in raster data models that are linked to a database management system, most often these objects are selected through a formal search of the attribute tables. For simple raster model systems, the separation is once again relegated to some form of reclassification process.

Terms

borders
center of gravity
centroid
circuits
communities
contiguity
contiguous regions
directed networks
formulation flowchart
fragmented regions
heterogeneity
homogeneity
implementation flowchart
isarithm
mean center
neighborhood functions
networks
overlay
perforated regions
regions
sinuosity
trapezoidal rule
undirected networks
weighted mean center

Review Questions

1. Explain why the analysis subsystem of the GIS is the most abused subsystem, and give examples to support your explanation.
2. Explain what we mean when we say that a GIS is an incomplete set of geographic analysis tools. What accounts for this?
3. If your career goals revolve around database creation rather than GIS analysis, why is it important to know about the analytical capabilities and limitations of GIS?
4. Why is it important to be able to isolate, identify, count, and separately tabulate and display individual items?
5. Describe the process of isolating, counting, and identifying point objects on the basis of their attributes within a raster and a vector GIS. Why do we need to use attributes to find these objects?
6. What are the differences between finding point objects and finding line and area objects?
7. What measurable attributes can be used to search for lines? Areas?
8. What are higher-level objects? How do they differ from simple objects?
9. Give some examples of higher-level objects for points. Do the same for lines and areas.
10. What are centroids? How are they found in a vector GIS? What are the different types of centroids? How do they differ in how they are found in a vector GIS?
11. What are networks? How do they differ from simple lines? What forms can networks take? What needs to be done to identify networks, as opposed to simple lines, in a GIS?
12. What is the difference between a directed and an undirected network?
13. What is a region? What are the differences and similarities among contiguous, fragmented, and perforated regions?
14. Give some examples of how collections of points, lines, and areas might be considered to be communities.

References

Berry, J.K., and C.D. Tomlin, 1984. *Geographic Information Analysis Workshop Workbook*. New Haven, CT: Yale School of Forestry.

Clarke, K.C., 1990. *Analytical and Computer Cartography*. Englewood Cliffs, NJ: Prentice-Hall.

DeMers, M.N., R.E.J. Boerner, J.W. Simpson, A. Silva, F. Artigas, and L.A. Berns, 1995. "Fencerows, Edges, and Implications of Changing Connectivity Illustrated by Two Contiguous Ohio Landscapes." *Conservation Biology*, 9(5):1159–1168.

Forman, R.T.T., and M. Godron, 1987. *Landscape Ecology*. New York: John Wiley & Sons.

Koeppl, J.W., G.W. Korch, N.A. Slade, and J.P. Airoldi, 1985. "Robust Statistics for Spatial Analysis: The Center of Activity." *Occasional Papers,* 115:1–14, Museum of Natural History, University of Kansas, Lawrence.

Mark, D.M., and F.F. Xia, 1994. "Determining Spatial Relations Between Lines and Regions in ARC/INFO Using the 9-Intersection Model." In *Proceedings of the 14th Annual ESRI User Conference*, pp. 1207–1213.

McGrew, J.C., and C.B. Monroe, 1993. *Statistical Problem Solving in Geography*. Dubuque, IA: Wm. C. Brown Publishers.

Muehrcke, P.C., and J.O. Muehrcke, 1992. *Map Use: Reading, Analysis, and Interpretation*. Madison, WI: J.P. Publications.

Robinson, A.H., J.L. Morrison, P.C. Muehrcke, A.J. Kimerling, and S.C. Guptill, 1995. *Elements of Cartography*, 6th ed. New York: John Wiley & Sons.

CHAPTER 9

Measurement

Once we find, count, and describe objects, our next step is to measure them. Remember that some of the attributes we used to retrieve entities involved measures of length, sinuosity, perimeter, and area. We also noted that collections of objects involve characterizing distances between and among each other. These attributes are most often calculated rather than just encoded during input. These calculations result in more robust attributes for each theme that can later be compared within a single theme or to those of other themes. Distances between objects can be measured as simple Euclidean distances, while in other cases we define them incrementally, adding the distance of one leg of our journey to another to produce a total distance. In still other cases we measure distance to search for better paths, less arduous terrain, seeking always the easiest, least-cost distances. These least-cost distances may be from point to point or from a single point to all other points in our terrain. In some cases we measure distance based on a need to traverse or to avoid barriers, and we can include some measure of the impact of the relative difficulty of traversing surfaces.

LEARNING OBJECTIVES

When you are finished with this chapter you should be able to:

1. Describe the process of measuring both straight and sinuous lines in vector and in raster, and discuss the advantages and disadvantages of each process.
2. Explain what impact elevation will have on actual travel distance and describe how the planimetric distance can be adjusted to more accurately reflect distance up and down slope.
3. Explain the process and the purpose for measuring polygons to determine long and short axes, perimeters, and areas. Provide a concrete application of this measure.
4. Explain some of the simpler measures of shape for lines and areas, and discuss why they might be important.

5. Understand the concepts of spatial integrity and boundary configuration.
6. Describe and calculate measures of spatial integrity and state what these tell us about the real-world objects we are analyzing.
7. Explain the process of measuring both isotropic and functional distances among objects, through friction surfaces, and around barriers.
8. In a raster system, calculate both isotropic and functional distances.
9. Explain the methods and potential problems associated with assigning friction and impedance values for modeling functional distance.

MEASURING LENGTH

Because points have zero dimensionality, there is no appropriate measure for them aside from attribute magnitude values, which are assigned to them and stored as grid cell attributes or saved in an attribute table linked to the entities themselves. Although these point magnitude values are important for analysis, they require no formal calculations, so we can eliminate them from the current discussion.

The single dimension of linear objects allows us to create searchable attributes based on measurements along their length. In vector, when we wish to measure the distance between two point locations on a globe, we use spherical geometry and trigonometric formulas to calculate great circle distance. Within a Cartesian reference system, the traditional distance equation, based on the Pythagorean theorem, takes advantage of the explicit X and Y coordinate systems for vector distance measurements. Through digitizing, each line segment is a straight line and its distance can therefore be calculated on the basis of the distance equation. By adding each of these segment distances, we can easily determine the total length of any line.

One often overlooked problem with calculating line distance is that of lines whose lengths extend over a range of elevational distances. For example, a map may show that the road distance from one point to another is somewhat less than what an automobile odometer might show. This typically happens when the vehicle is traveling up and down topographic slopes, thus adding road log miles to the overall distance (Figure 9.1). If the change in elevation is essentially linear, the additional mileage can be calculated with simple trigonometry. Because most topographic surfaces change in a nonlinear fashion, the calculation becomes more complicated. Because each line segment is a straight line, the Pythagorean theorem can be applied to each if the beginning and ending coordinates also have elevation values assigned to them. If not, interpolation may be required to define the missing elevation points and then the distances can be calculated. As before, we add the segment distances to obtain the total distance.

Calculating line lengths in raster is a matter of adding the number of grid cells together to achieve a total. Let's begin by working with a straight-line entity in grid format that occurs as a set of vertical or horizontal grid cells. Knowing the resolution of each grid cell, generally assumed to be from side to side (**orthogonal**) rather than on the diagonal, we need only add the number of grid cells and multiply by the grid cell resolution. If a vertical line is composed of

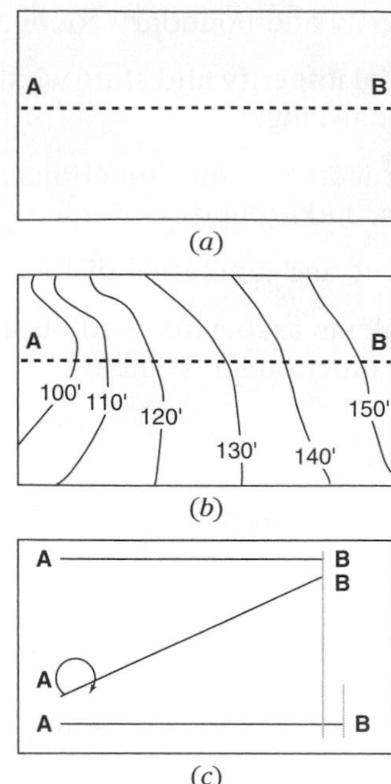

Figure 9.1 Relationship between map distance (*a*) and road log distance (*c*) as influenced by changes in elevation (*b*).

15 grid cells, each with a resolution of 50 meters, our line length is calculated by multiplying 15 grid cells by 50 meters to obtain a total line length of 750 meters. The identical process would be performed for a horizontal line.

Many lines are diagonal or have diagonal components along their length. What do we do when the grid cells in a line are connected to each other along a corner? Most modern raster systems are able to calculate the diagonal distance between each cell in a diagonal line of grid cells through the use of simple trigonometry (Environmental Systems Research Institute, 1993). Because a diagonal through a square grid cell produces a right triangle, the right-angle sides of which are identical to the grid cell resolution, the hypotenuse is calculated using the distance formula based on the Pythagorean theorem. You should note that the calculation of the hypotenuse will always give you a value approximating 1.414 times that of the horizontal or vertical distances ($\sqrt{1^2 + 1^2}$). Therefore, it is a simple task to multiply the resolution of each diagonally connected grid cell by 1.414 to obtain the correct value for distance. Depending on the resolution and the irregularity of the line entity's path, it is possible for whole loops to be represented by a single grid cell, in which case the length of the object will be underrepresented (Figure 9.2). For this reason, it is best, whenever possible, to use a vector data structure if your analysis relies heavily on measurements of linear objects.

The calculation of length along a vector line entity is only slightly computationally more exhaustive than simply adding the number of grid cells, as in raster, but the results are decidedly more accurate, as is the representation of the line itself. For each straight-line segment in the linear object, the software will have stored a set of coordinate pairs. The distance between members of each coordinate pair can be calculated through the distance theorem. Adding line-segment lengths produces a relatively accurate measure of the total or accumulated line length. Remember that the vector representation of

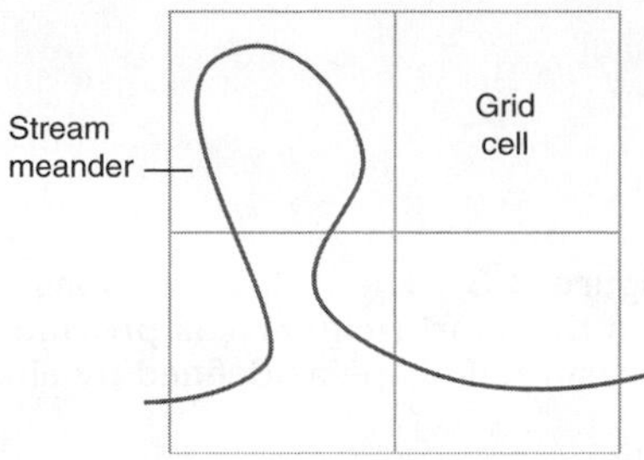

Figure 9.2 Subresolution line object. A highly sinuous linear object, such as a meandering stream, may have whole meanders self-contained within a single grid cell. This indicates the limitation of measuring length with raster data structures.

linear entities also employs a form of sampling, where changes in direction are represented as separate straight-line segments: the more line segments, the more accurately the linear object will be represented by the data structure, and the more accurate the measures of total line distance will be. This again reminds us of the importance of good map preparation and careful data input.

MEASURING POLYGONS

Polygonal features possess two dimensions: length and width. Because of the added dimensionality of polygons, we have more measurements that we can make upon them. We can, for example, measure the length of the short or long axis of a polygon, calculate the lengths of its perimeter, or measure its area. We know these measures can be used as search criteria for individual polygons, which can then be used for subsequent analysis.

Measuring Polygon Lengths

Because the orientation of the polygon is often linked to an underlying process, the orientation of polygonal features will at times be important to the GIS user (Muehrcke and Muehrcke, 1992). For example, asymmetrical forest patches oriented in a particular manner may trigger migrating birds to land (Baker, 1989; Forman and Godron, 1981; LaGro, 1991; O'Neill et al., 1988). Or a glacial geologist examining the movement history of glaciers may wish to know whether polygons representing certain glacial features have a particular orientation. Conceptually, the idea of polygon orientation is a matter of determining the direction of its longest axis. Most raster GIS software, however, lacks techniques for easily ascertaining the long axis. And because raster uses relative locations of grid cells, it is no easy task to determine orientation either.

In vector, the solution involves calculating the lengths of each pair of opposing polygon vertices. A comparison of these vertices will then show which line is the longest. That line is the major axis of the polygon, and its angular direction is calculated with spherical trigonometry (Robinson et al., 1995).

At times the GIS analyst will be searching not for orientation but rather for the relationship between the major and minor axes. The latter gives a simple measure of shape that can also be used as subsequent search criteria (Figure 9.3). Suppose an ecologist wants to study small mammalian species that

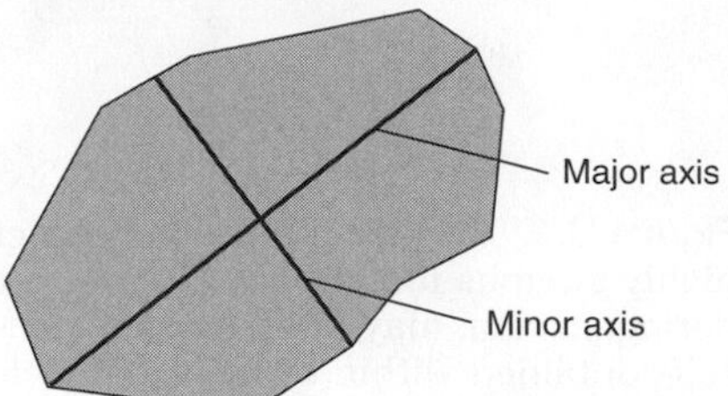

Figure 9.3 The ratio of the long (major) axis and the short (minor) axis provides a simple measure of shape as defined by elongation.

prefer to locate in long, skinny forest areas next to agricultural areas (Turner, 1991). This shows that being able to calculate the long-to-short axis ratio is useful. Its initial calculation in vector GIS is performed as before, but first distances of all the opposing vertices must be found. Then, by comparing their lengths, we can identify the shortest and the longest axes.

When measuring long and short axes for polygons, it is significantly easier to work with convex polygons than those that are not totally convex. If there are concave portions, or if the polygons are highly irregular, the measurements become considerably more difficult to describe and the results are less useful. One approach, called the **least convex hull**, allows us to sort of shrinkwrap the polygon so that it essentially resembles a totally convex polygon. These calculations, although useful, are beyond the intent of this introductory work.

Measuring Perimeters of Polygons

Calculating the perimeter of a polygon is a matter of calculating the distance of each bounding line segment using the distance formula and then adding these individual values. Once again, raster data structures are not particularly well suited for this task, but, like measuring distances themselves, they can be made to produce a perimeter. To do so, we must separately identify and reclassify each grid cell located at the outside perimeter of the collection of grid cells that make up a polygon to produce a map of just these perimeter-forming grid cells. Quite often this task calls for extensive interaction with the software. Finally, however, the perimeter is defined by the sum total of grid cells, multiplied by the grid cell resolution (Figure 9.4). As before, the more complex the polygon, the more grid cells will be at a diagonal to their neighbors and the less accurate your results will be.

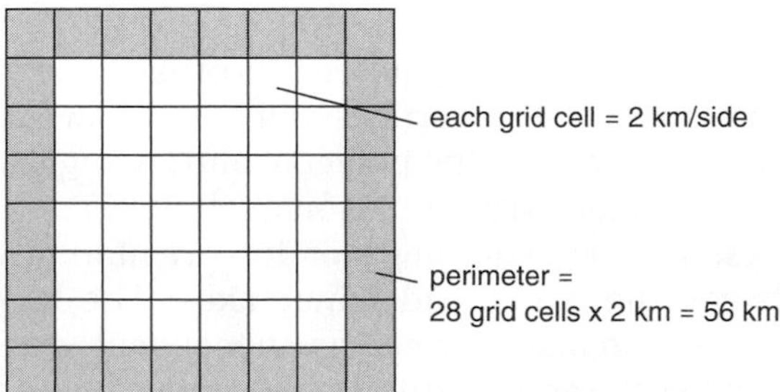

Figure 9.4 Finding perimeter in raster. By adding the perimeter grid cells and multiplying that number by the resolution, we obtain a reasonable estimate of perimeter.

Calculating Areas of Polygonal Features

A frequently used measure in GIS is that of polygonal area. Area gives us some quantitative measure of the amount of each classified area feature. Finding the area for a region in raster involves selecting the grid cells that share common attributes (our region) and counting the number of grid cells that the region occupies. Each grid cell's area (resolution) is then multiplied by the number of grid cells. Typically you just tabulate the data in your grid layer and read the number of grid cells sharing the assigned attributes from the resulting table (Figure 9.5). The table will also give you a percentage of grid cells in the database for that region, compared to the overall database. Other ratios can be performed by comparing the numbers of grid cells of one region to those of another.

If, on the other hand, you do not want to measure the grid cells of fragmented regions, the problem is a little more difficult. If you want to know only the area of a particular polygonal area—whether or not it constitutes a contiguous region—you need to isolate each polygon separately by selecting its grid cells and reclassifying them so that their attribute values are unique to the grid layer. There are raster functions that can isolate such areas by size (number of grid cells), providing an easy way to select the polygon you want to measure. If you have many polygons of approximately the same size within your grid layer, however, you will need to isolate them either by their row and column locations or by individually recoding the grid cells of your target area using available editing functions. Either way, this simple process requires forethought and planning.

In vector, the target polygons are selected on the basis of either coordinate locations or attribute values in the attribute tables. Then the area of the polygon is calculated. For simple Euclidean forms, such as rectangles, triangles, circles, parallelograms, and trapezoids, the equations are standard. It is only when the polygons are very complex that the computations become more difficult. The most common solution is to divide complex polygons into shapes that can be easily measured with readily available formulas (Clarke, 1990) (Figure 9.6). In many vector GIS, the areas of polygons are calculated during input. As the units in each set of line segments are digitized, the software determines the simple geometric shapes produced as the process continues, then calculates the areas of these shapes. Finally, the totals are computed to obtain the total polygonal area. Because the polygonal areas can be computed individually, their values are passed to the attribute tables immediately upon data input. Therefore, the area of a polygon can be obtained by selecting it from the attribute tables.

Frequently there is a need to combine the values of area and perimeter in a ratio. This relationship, called the **perimeter/area ratio**, provides a

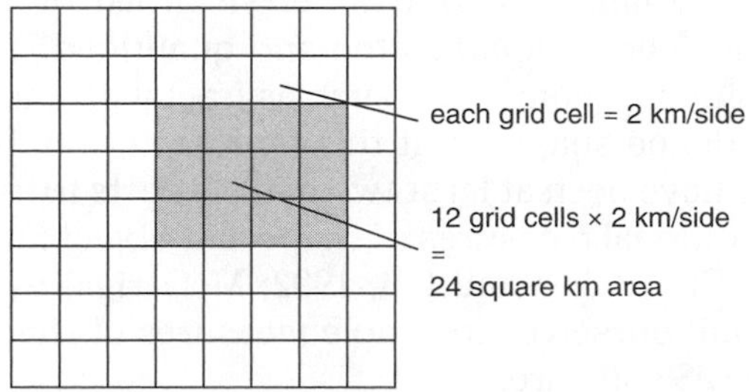

Figure 9.5 Finding area in raster. By adding the grid cells and multiplying that number by the area of each grid cell, we obtain a reasonable estimate of area.

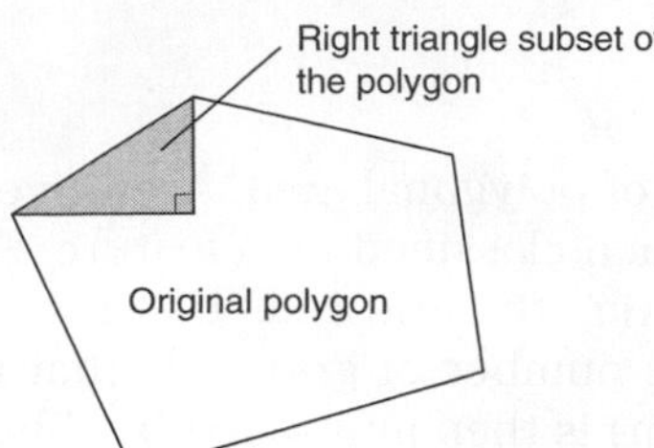

Figure 9.6 The area of odd-shaped polygons can be calculated by dividing them into smaller objects whose areas are easily calculated. Shown here is a right triangle as the first portion of the polygon whose area is to be calculated.

measure of the complexity of each polygon. Thus, a circular object will have the smallest perimeter/area ratio. By contrast, long, skinny polygons have larger perimeter/area ratios because their perimeters are so large and their areas comparatively small. Numerous natural resource-related operations require evaluation of this ratio. For example, the smaller the perimeter/area ratio of a forest stand, the more likely the species composition will be dominated by interior species (animals that prefer interiors, not edges). Or, if you are trying to develop lakefront property, the higher the perimeter/area ratio, the more beach area you can offer to prospective lot buyers. Thus, the higher the perimeter/area ratio, the more money you will make developing beachfront property.

MEASURING SHAPE

There is a close relationship between shape and such measures as perimeter and area for polygons and even length for linear objects. Often the shapes of polygons or some measure of the sinuosity of linear features will provide insights into the relationships between objects and their environment (Boyce and Clark, 1964; Lee and Sallee, 1970). For example, the sinuosity of a stream is related to such functions as stream sediment load, slope, and the amount of water flowing through the stream. In turn, these functional relationships have much to do with whether a stream is aggrading (building up sediment load), at grade (in balance with stream inputs and outputs), or degrading (down cutting). Hydrologists, geomorphologists, and other environmental scientists find these values useful in overall regional analysis. In polygons, the relationship between perimeter and area has much to do with the functioning of anthropogenic and natural features. Therefore, it is important to have at least a basic understanding of what these simple shape measures are and how they might be performed in a GIS.

While few measures of shape are readily available in GIS, there are some relatively sophisticated ones in the literature (Moellering and Rayner, 1982). Traditional Euclidean geometry limits us to a rather small list of identifiable shapes whose dimensions have been documented and quantified. There is an emerging interest in nontraditional geometries, such as fractal geometry, which uses imaginary numbers to define shapes. But this approach, too, is atypical. Interestingly, however, there have been at least two major efforts to incorporate such measures, as well as additional measures of shape and object interactions, into both raster and vector GIS (Baker and Cai, 1992; McGarigal et al., 2002). For the time being, we will limit ourselves to simple measures of shape that are commonly available in most GIS software.

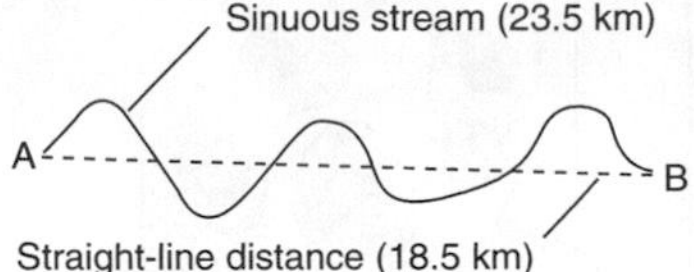

Figure 9.7 One simple measure of sinuosity is a ratio of the straight-line distance to the true distance. If the stream is perfectly straight (no sinuosity), its ratio would be 1.0. In this case, the ratio is 18.5/23.5, or approximately 0.787.

Measuring Sinuosity

There are two simple measures of **sinuosity** that can be used as a measure of linear shape. The first is the ratio of the actual linear distance (in the numerator) and the straight-line distance (in the denominator) (Figure 9.7). The larger the value, the more sinuous the line; the closer the value is to 1, the less sinuous the line. However, there is often a need to know more about the shape of a curve along a sinuous linear object. Very sharp curves in roads, for example, are likely to cause accidents. And in streams, sharp curves lead to high levels of erosion along the outside of the formation and more sedimentation along the inside. For this reason, it is useful to be able to find the radius of each curve in the line. To do this, we assume that the curves are essentially circular (Strahler, 1975). Then we fit a semicircle to each curve and measure their radii. When streams are represented as polygonal shapes rather than as lines, a ratio of the radius to the width of the stream gives yet another useful measure of shape.

Measuring Polygon Shape

There are two fundamental aspects of measuring polygon shape: The first, **spatial integrity**, is based on the idea of perforated and fragmented regions; the second is a measure of boundary configuration. Spatial integrity measures the amount of perforation in a perforated region. A common measure of spatial integrity is called the Euler (pronounced "oiler") function and is named after the Swiss mathematician (Berry, 1993). Many animals—for example, many bird species—prefer large, uninterrupted patches or polygons of a particular type of land cover, whereas others, like deer, seek out large forest areas interrupted by patches of grassland. A patch of forest that is unbroken is a contiguous region; if it completely surrounds smaller polygons, it is perforated; and if it is completely separated from similar patches by some intervening type, we call it fragmented. The **Euler function** is a numerical measure of the degree of fragmentation as well as the amount of perforation. Let's examine how it works.

Figure 9.8 shows three possible configurations of polygons. The Euler function describes these functions with a single number, called the **Euler number**. Numerically, the Euler number is defined by the following simple equation:

$$\text{Euler number} = (\#\text{holes}) - (\#\text{fragments} - 1)$$

where

#holes is the number of self-contained polygon perforations in the outside polygon

#fragments is the number of polygons in the fragmented region

(*a*) Euler = 4

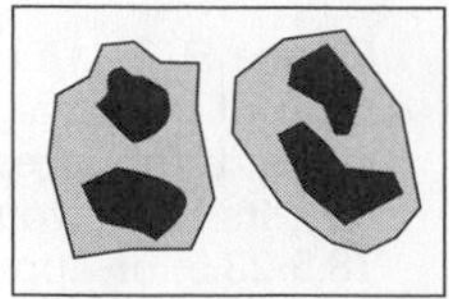

(*b*) Euler = 3

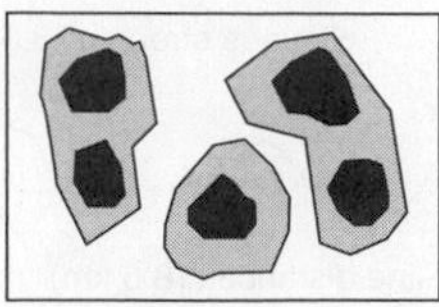

(*c*) Euler = 3

Figure 9.8 Example of Euler numbers. Three different land configurations and their associated Euler numbers: (*a*) four holes in a single contiguous region; (*b*) two fragments with two holes in each; (*c*) three fragments, two each with a pair of perforations and the third with a single perforation. It is important to note that the Euler numbers for (*b*) and (*c*) are identical even though the configurations are different.

In Figure 9.8a, there are four holes or perforations in a single contiguous region. By inserting these numbers into the equation, we get:

$$\begin{aligned}\text{Euler number} &= (4) - (1 - 1)\\ &= 4 - 0\\ &= 4\end{aligned}$$

In the second example (Figure 9.8b), we find two fragments, each with two perforations or holes. Again substituting in our formula, we find that our Euler number is $4 - (2 - 1)$, or $4 - 1 = 3$. The third example (Figure 9.8c) shows that with three fragments and a total of five holes, our equation gives $5 - (3 - 1)$, or 3, exactly the same as in Figure 9.8b. This shows that although the Euler number gives a measure of spatial integrity, care must be taken in the explanation of the results. You might want to examine a number of cases to see what results you get for, say, an area with three fragments but no holes: $0 - (3 - 1) = -2$. You should see, on the basis of the last example, that at times the numbers will not be positive. In addition, if you tabulate some of your results, you may begin to see the relationships between Euler numbers for different combinations of holes and fragments (Table 9.1). This table will prove useful in determining the twin meanings of any given Euler number.

A somewhat more sophisticated measure of integrity, called **lacunarity**, is a counterpart to fractals that describes the texture of a fractal area; in simple terms, it describes the size distribution of holes. Large gaps or holes yield high lacunarity; small gaps or holes yield low lacunarity. Lacunarity measures have been applied to landscape analysis (Plotnick et al., 1993, 1996) and have also proven useful in urban analysis to provide a measure of human interaction within cities (Wu and Sui, 2001).

The second group of polygonal shape measures—those related to boundary configuration—are quite numerous. Among the more prevalent are those (1) based on the axial ratio, (2) based solely on perimeter, (3) based on perimeter and area, (4) based only on areas, (5) those based on areas and areal lengths, and (6) measuring circularity mean side and variance of side. The formulas for these, which can be found in Chapter 5 of Davis (1986), can be referred to as your interests in shape analysis develop and your GIS skills improve. Some of the more esoteric measures are available in commercial GIS, whereas others

TABLE 9.1 Matrix of Euler Numbers for Different Hole-Fragment Combinations[a]

					#fragments				
#holes	**1**	**2**	**3**	**4**	**5**	**6**	**7**	**8**	**9**
1	1	0	−1	−2	−3	−4	−5	−6	−7
2	2	1	0	−1	−2	−3	−4	−5	−6
3	3	2	1	0	−1	−2	−3	−4	−5
4	4	3	2	1	0	−1	−2	−3	−4
5	5	4	3	2	1	0	−1	−2	−3
6	6	5	4	3	2	1	0	−1	−2
7	7	6	5	4	3	2	1	0	−1
8	8	7	6	5	4	3	2	1	0
9	9	8	7	6	5	4	3	2	1

[a]Note the number of configurations that can produce the same Euler number. Note also the mirroring of Euler numbers on either side of the diagonal made with the zero Euler values.

are available in geostatistical software or as macros for commercial GIS (Baker and Cai, 1992; McGarigal and Marks, et al. 2002).

Most of these measures are strongly related to perimeter/area ratio. In fact, the perimeter/area ratio itself can be considered to be a measure of polygonal shape. However, this doesn't necessarily describe the actual geometry of the object. For this we most often compare the polygonal shapes we encounter to more familiar geometries we can easily describe (Muehrcke and Muehrcke, 1992). We might, for example, want to compare the polygonal shapes to such geometric forms as parallelograms, trapezoids, and triangles. However, these shapes come in a wide array of forms, each one of which would have to be described separately. And, of course, the simplest, most compact, and most easily defined shape is the circle. For this reason, the basic method of measuring shape is to compare it to the circle.

Because we use the circle as the comparative shape, we can say that this measure of shape is also a measure of the **convexity** or **concavity** of the polygon. A circle has perfect convexity because no portion of its surface is concave or indented. This is why the circle is the most compact geometric shape and explains why we compare other shapes to it. All other geometric shapes have more perimeter than a circle. To compare our polygonal geometric forms to that of a circle is essentially the same as examining the amount of convexity of the polygon versus that of the circle. The general form of the convexity formula in vector GIS is

$$CI = \frac{kP}{A}$$

where

CI = convexity index
k = a constant
P = perimeter
A = area

This is a perimeter/area ratio for each object, multiplied by the constant. The constant is based partly on the size of the circle that would **inscribe** the irregular polygon. In addition, it is designed to provide a range of positive values from 1 to 99, where 100 indicates 100 percent similarity to a circle. Stated differently, a 1 is as far from a circle as can be measured, and a 99 is as close to a circle as a shape can get without actually being a circle. A perfect circle thus has a value of 100.

In raster, the formula is based on the exact same concept, but the area is now recorded as the number of cells and its square root is used to provide the same 1–99 range of similarity values. Therefore, the general form of the formula for calculating convexity in a raster GIS is

$$CI \frac{P}{\sqrt{\# \, cells}}$$

where

CI = convexity index
P = perimeter
$\sqrt{\# \, cells}$ = area in raster format

As before, a value of 1 is as far from a circular or totally convex shape as can be measured, and 99 is as close to perfect convexity as a shape can get without being a circle. Of course, in raster it is physically impossible to get a perfect circle.

Convexity measures are very important to a wide variety of both scientific and real-world applications. Denizman (2003) used the circularity index to examine karst formations in Florida, and lunar history can be interpreted by the circularity of its craters (Ronca and Salisbury, 1996). Many creatures, including humans, like the security of a circular shape, with its readily defended perimeter (hence the term's use in the military). Other interior species take advantage of the lack of edge that offers protection from competition for needed resources like food and shelter. Yet some species—for example, some small rodents—really like edges. They tend to use them as base areas from which to forage in the open along field crops, while the forest remains accessible for escape and shelter.

This last scenario suggests another measure of boundary configuration, called **edginess** (Berry, 1993), that uses a device called a **roving window**. Also known as a **filter**, a roving window is a matrix of numbers of a preselected size that can be moved across a raster GIS or digital remote-sensing database. The idea is to move the window across the database to either examine what is there or to modify the grid cell values. Although these uses of filters could more closely be associated with classification, we will look at their use for edginess here as a measure of boundary configuration.

Let's assume that we are using a 3 × 3 filter (a 9-cell window), which we will move along the respective boundaries of two types of polygons in a raster-based map (Figure 9.9). We assign digital values to all the cells inside the filter: 1 for edge cells and 0 if they are not. The amount of edginess is obtained simply by counting the number of grid cells that are 1s, or share the same attribute values as the polygon. The more 1s we get, the less edge we have, and the more interior. Thus, the value of 7 in Figure 9.9 illustrates very little edginess: in the first portion of the matrix covered, nearly all the grid cells are connected to each other. Likewise, a 9 would be total interior, with all grid cells connected and no edge. By contrast, the edginess value of 2 in Figure 9.9 indicates that a small number of the grid cells exist in the second portion of the map over which

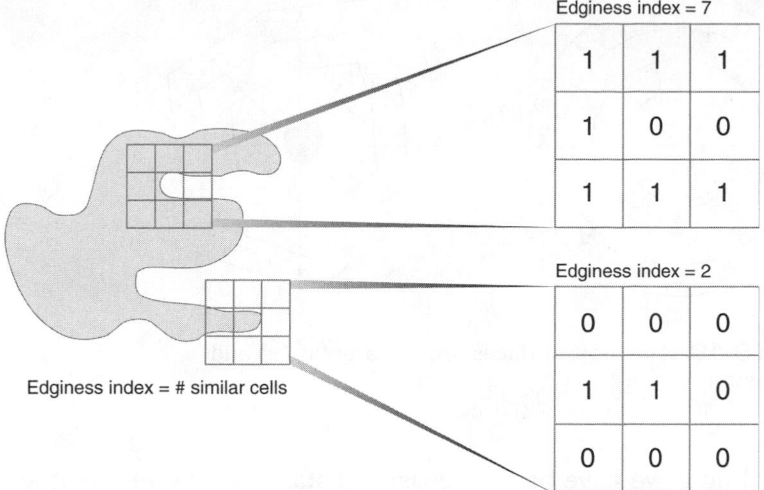

Figure 9.9 Raster measure of edginess. A moving filter is used to determine the edginess of a polygonal edge in raster. Note how much more edge there is as the number of neighboring polygons with a single attribute goes down. *Source:* Modified from J.K. Berry, *Beyond Mapping: Issues and Concepts in GIS*. Fort Collins, CO, GIS World, Inc., © 1993.

we have fitted the filter; the remainder are essentially background or matrix grid cells (0s). The 2 indicates a small, skinny protrusion into the matrix. Or stated differently, the lower filter covers a great deal of edge.

MEASURING DISTANCE

The distance along and between features is important not only in later analyses but also more immediately because it provides a means of estimating travel to, from, and around them (**travelsheds**). Distance can be measured in simple terms based on the absolute physical distance between places on the map. It can also be measured to include costs incurred while traveling rugged terrain, the cost of travel specifically along networks rather than off-road, or difficulties involving barriers that restrict or prevent movement. These latter approaches to distance are collectively known as **functional distance**.

Euclidean Distance

Calculating simple, or **Euclidean distance**, is the same as measuring length of lines and polygons. In raster, the method is to add the column and row distances in grid cell units and then multiply the number of grid cells by their grid cell resolutions—whether orthogonal or diagonal—to convert to standard distance measures. You might remember that the diagonal distance is 1.414 times the orthogonal distance (the hypotenuse of a right triangle for each grid cell).

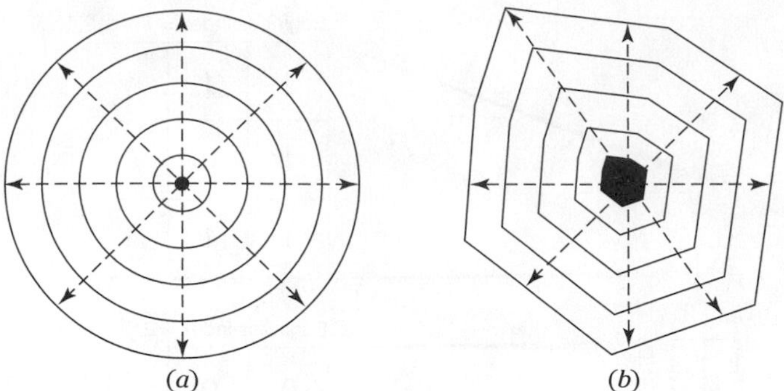

Figure 9.10 Isotropic surfaces around a point (*a*) and a polygon (*b*).

Until now, we have been discussing distances between point locations in a thematic map or along the grid cells of a linear object. However, this limits our perception of distance. We may, for example, want to know the distance between a single point location on a map and all other possible locations. In raster this is done by producing a series of concentric rings, each one grid cell in diameter, around the starting grid cell. Each ring is essentially one grid cell away from its preceding ring. The result of such a distance measure, assuming the surface is completely uniform, is called an **isotropic surface**—a surface that is completely symmetrical (Figure 9.10a). An isotropic surface map from a single starting point to all other locations is essentially a travelshed showing distances to every other location in the database. Its advantage over a calculation of point-to-point distance is apparent when there are many of these to calculate: the isotropic surface has already calculated them.

Travelsheds don't have to be measured from a single point. We may wish to measure the distance from a linear (e.g., distance from a river) or polygonal object (e.g., distance from a sewage lagoon) to all other locations on the map. The GIS operates as before, but begins its measurement at the margins of the line or polygon and moves outward in all directions. The result is another isotropic surface, but with an apparent line or hole in the middle where the starting line or polygon was located (Figure 9.10b). Figure 9.11 illustrates a simple flowchart of the process of measuring distance to produce a travelshed.

In vector, calculating distances between points along a line employs the Pythagorean theorem for each connected line segment and then adds them to obtain a total Euclidean distance measure. The same operations are used to calculate distances between any two point objects that are not connected by line segments. However, the points must actually exist on the map. Because

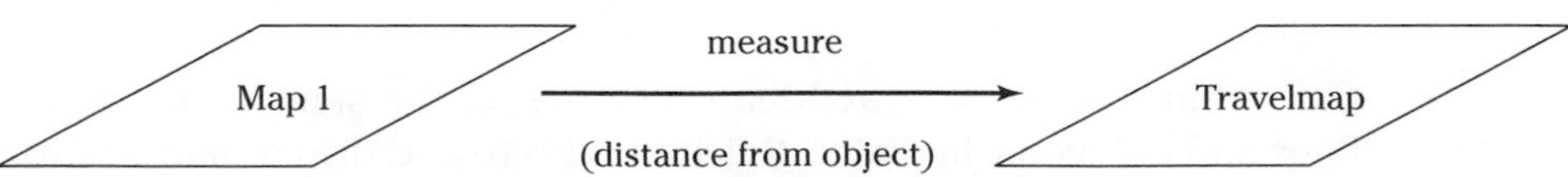

Figure 9.11 Distance flowchart. Notice how the distance operation is modified based on its source or starting place. This source might be a point, a line, or a polygonal object. The resulting travelmap based on measuring from a starting place to all other places on the map shows the travelshed.

vector data structures do not explicitly define the intervening spaces between objects, no calculations can be performed within these spaces unless objects are placed there.

The foregoing limitation also applies to developing isotropic surfaces, whether from points to all other locations or from polygons to all other locations. To perform such calculations requires the use of the TIN model described in Chapter 4. Distances can be calculated in the TIN data model, but at considerable computational expense. GIS users who work with surface models most often defer to the use of a raster system, especially when it is necessary to measure distances along surfaces.

Functional Distance

While measuring Euclidean distance is essential, the chance that one will encounter a totally isotropic surface is practically nonexistent because it is a theoretical construct. Still, we often ignore the surface conditions to simplify our surface measurement. There are many situations, however, where surface conditions cannot be ignored. Such conditions typically involve the measurement of non-Euclidean distance based on some functional variable such as time, energy, or gas consumption—thus the name functional distance. The implication here is that movement across nonisotropic surfaces incurs a cost.

There are two basic concepts we need to consider when looking at functional distance: **friction surfaces** and **barriers**. Both involve some form of **impedance**, or friction, **value**, which is a measurable surface condition that imposes the cost of movement. In friction surfaces, the impedance or friction value is generally considered to exist to some varying degree throughout a map area. Topography, itself a continuous variable, may impose a friction value based both on surface roughness and the variable cost of moving up or down hill (Figure 9.12). This

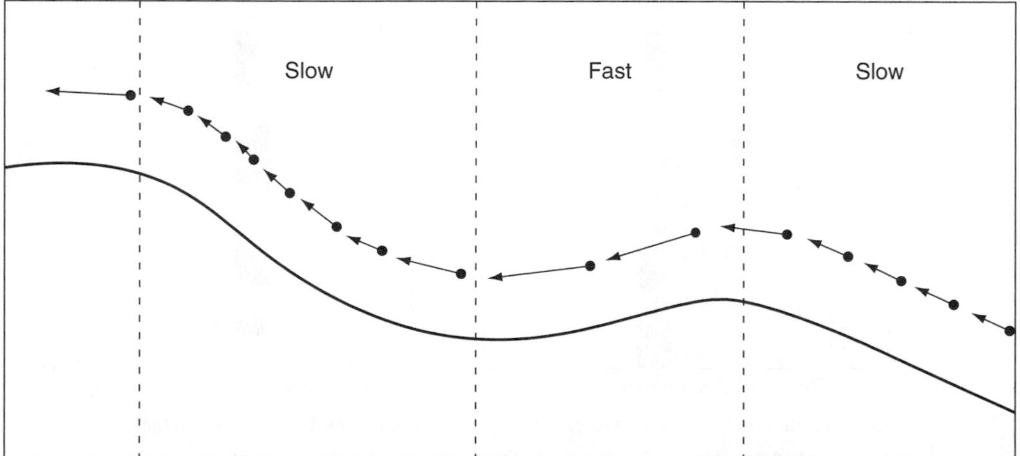

Figure 9.12 Topographic friction surface. Notice how the movement (dashed lines) get closer together as they move upslope and further apart as they move downslope. This shows how topography itself can impose a cost to movement.

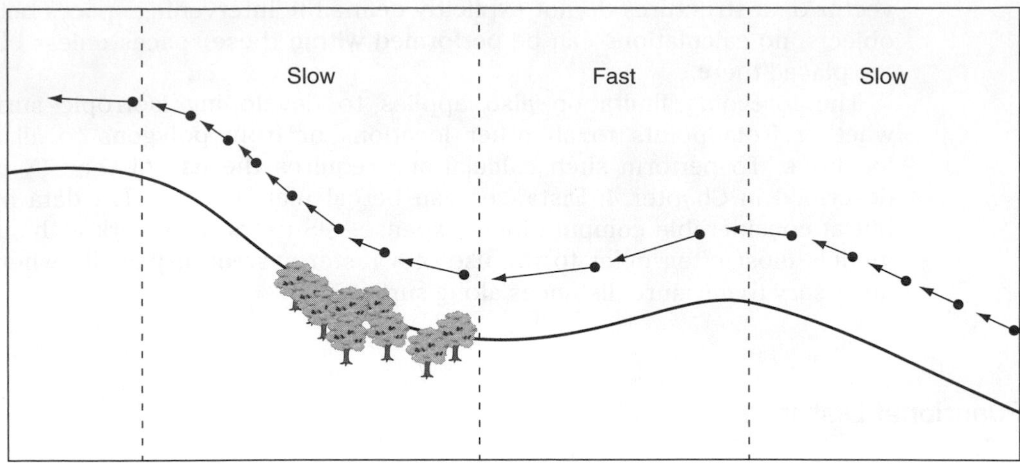

Figure 9.13 Vegatative friction surface. Notice how the existence of forest vegetation slows the movement (closely spaced dashed lines), while the absence of forest vegetation results in faster movement.

might be combined with still another variable, such as vegetation (Figure 9.13), to create a more complex friction surface.

In other cases rather isolated conditions or discrete obstacles called barriers interfere with movement across a surface. Barriers come in two types, absolute and relative. **Absolute barriers** (Figure 9.14), for example cliffs, fenced areas, and lakes, prevent or deflect movement. Although one normally characterizes absolute barriers as impermeable (thus preventing movement) (Figure 9.14a), they may be permeable with points of access through which travel is permitted (Figure 9.14b), such as bridges over streams. In such cases, movement across the barrier is restricted to these access points, producing a bottleneck that

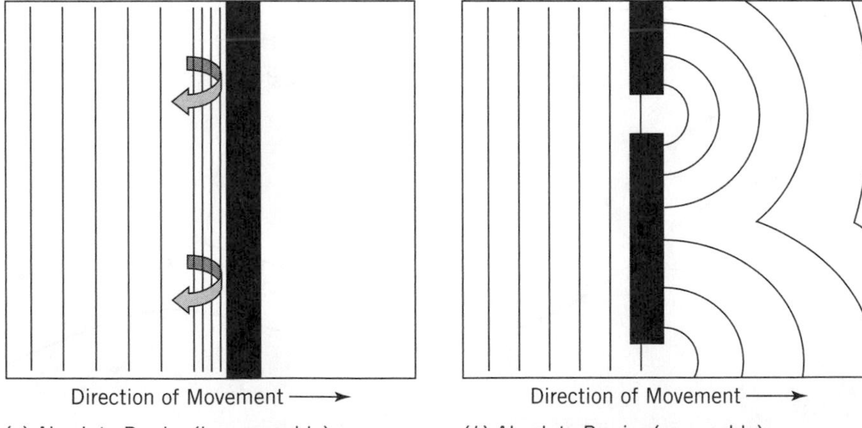

Figure 9.14 Absolute barriers. Notice how absolute barriers can be impermeable (*a*) where they stop or deflect movement, and permeable (*b*) that results in restricted access such as bridges across a river.

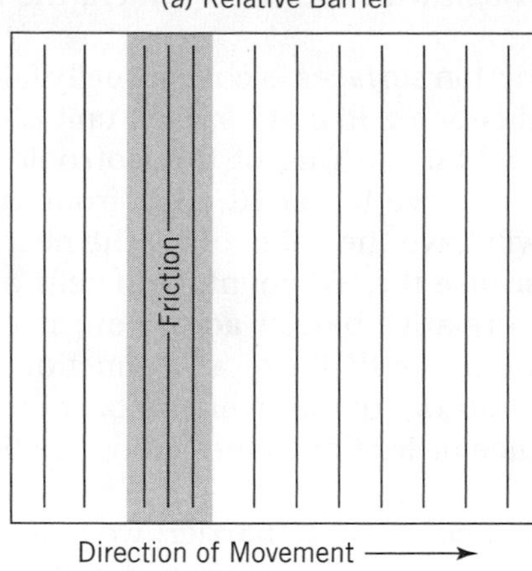

Figure 9.15 Relative barriers. Much like friction surfaces, except in their discrete spatial occurrence, relative barriers incur a cost for movement from one place to another.

results in a radial pattern of movement on its down-movement side. This is not unlike the pattern achieved as perpendicular onshore waves encounter the inlet to a bay, where the water that gets through produces a series of concentric waves inside the bay.

Relative barriers (Figure 9.15), like their less discrete friction surface counterparts, incur a cost for movement. Such barriers might be exemplified by narrow ridges of hilly terrain, shallow streams that are passable for off-road vehicles, or patches of forest that restrict but do not prevent completely the movement of animal herds. This results in either slower movement or the expenditure of additional energy—imagine the slower movement of large mammals through a swamp as opposed to rapid travel on an open prairie.

Whether you are employing friction surfaces or barriers, it is common practice to employ both an original map layer without friction or barriers and another layer with friction or barriers. In such cases you are modifying your measurement of Euclidean distance by the friction (or barrier) map layers; such movement across a topographic surface or through forests will modify the speed of movement. Figure 9.16 illustrates a simplified flowchart characterizing such a process. In all such cases you need to mathematically model the friction

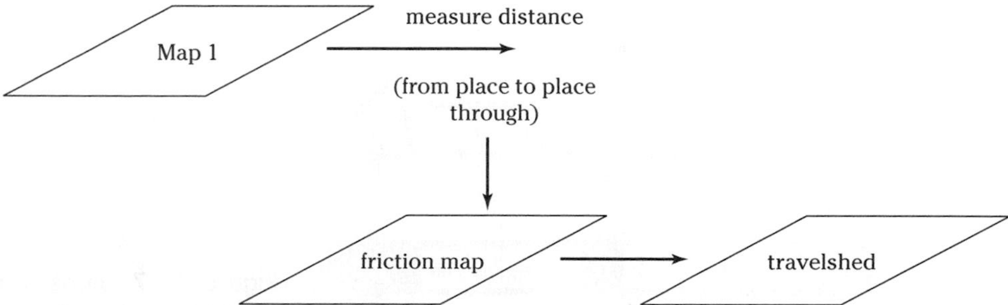

Figure 9.16 Flowchart illustrating distance measure through a friction map. The result of such a process is generally known as a travelshed.

surfaces or barriers themselves through which, or over which, the movement will take place.

Raster modeling of barriers and friction surfaces is conceptually fairly simple. If on an isotropic surface the GIS adds one grid cell for each unit of travel, the resulting map will show that each additional ring of the isotropic surface is recorded as 1 + the row before it. So if we travel 10 units from our starting point or polygon, the outside ring will have the value of 10, the next inner ring will have the value 9, and so on, because the GIS counts grid cells as it moves along the surface. What if we have a relative barrier across our map from top to bottom? We would want to show travel distance as a functional distance from left to right across the map—and of course over the barrier. To create the barrier, we would place impedance or friction values along the line (higher values represent higher friction).

As we move across our map and encounter the barrier, we count grid cells one by one until we reach the barrier edge. Then we add a grid cell value for each additional grid cell movement, plus one for each number encoded as friction. We cannot move forward in planimetric distance until the friction values have been counted. For example, if we used the impedance value of 5 for our barrier and began counting from left to right, it would continue to increase in successive grid cells by a value of 1 until it reached the barrier. It then must add three additional counts before it can continue. Such a barrier is considered to be a relative barrier because once it has counted five additional numbers, it will continue to move along, adding one number to each additional grid cell (Figure 9.17a). In many cases, such relative barriers or friction surfaces may vary in value and extend across the entire surface of the map. For example, we might want to move through a map of land cover containing five or six different land cover types, each with its own friction value for travel. Most often this is done by reclassifying the land use attributes to reflect their friction value and naming this new layer as the friction grid (Figure 9.17a).

To make the barrier an absolute barrier, we assign it a friction weighting factor sufficiently high that it will exceed either some maximum number designated in the GIS program or what would normally be the maximum number of grid cells in the map in the absence of friction. So if we have a raster grid that is 100 grid cells wide, we might put a barrier impedance value of 1,000 to ensure that the process of moving and counting the grid cells stops before it passes the barrier (Figure 9.17b).

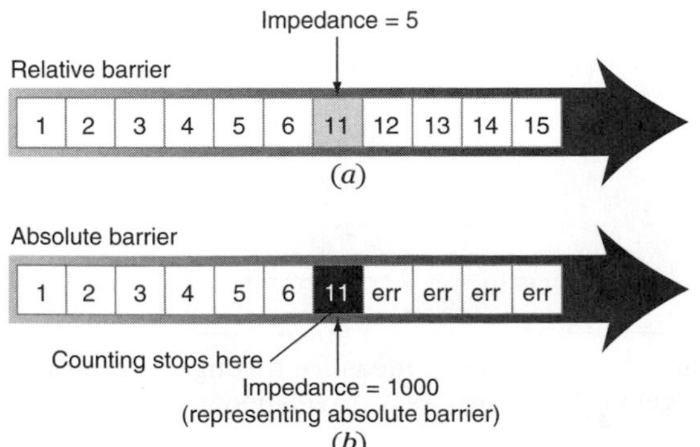

Figure 9.17 Impedance. Simple accrued values of grid cell counts as movement encounters a relative barrier (*a*) and an absolute barrier (*b*).

Although conceptually assigning friction or impedance values is simple, operationally it causes some difficulty. Just how much impedance does a forest cause to deer populations as they travel through it? How much more gasoline is used in moving up a 15 percent inclined surface than in moving over a flat surface? How much longer does it take a large mammal to swim a stream 100 meters wide than it would to walk across a bridge 100 meters long? Many similar types of questions must be answered in some quantifiable manner before we can create our barriers and friction surfaces. Most often the answers do not have supporting empirical data and depend on variables and criteria for which measurements are impossible to obtain.

Although these vagaries exist for many situations that require functional distance measurements, there are, in many cases, absolutes that allow us to set upper and lower limits. These limits give us a basis for creating a scale that is situation specific. In short, this is a perfect opportunity to employ the scalar measurement level. Although based on judgment and/or experience, the scalar values allow us to quantify the otherwise unquantifiable. This is also an opportunity to employ alternative mathematics such as Fuzzy Set Theory (Zadeh, 1965), where there is a range of set memberships between 0 (not a member) and 1 (fully a member)—in our case 0 membership implies no friction or impedance and 1 represents absolute impedance. Whatever method is employed, caution is advisable in assigning these values to ensure that there is a reasonable explanation for the system used and that some justifiable comparison can be made among the friction surfaces. Above all, the results of analysis of distance based on barriers and friction surfaces should be viewed with some trepidation, especially if they are to be used for decision making.

Before we discuss the mechanics of raster approaches to non-Euclidean and functional distance, we must examine two additional characteristics of distance. Distance can be viewed not only as Euclidean or non-Euclidean or as isotropic or functional, but also as **incremental distance** or **cumulative distance**. Incremental distance simply measures each unit, or increment, traveled, adding the second to the amount of distance traveled in the first, and so on. Each successive increment is added simply as a measure of distance, much like our procedure for working with the isotropic surface. In other words, the incremental distance measures the shortest path between two places without adding any friction or impedance values along the way. If incremental distance is measured for an entire surface, the result is a **shortest-path surface**; if the technique is restricted to lines or arcs (or linear arrangements of grid cells), we have a shortest path as a linear product rather than as a surface.

However, as you saw with friction and impedance, there is a continuous increase in cost as we encounter surfaces or obstacles. The purpose of calculating functional distance on a friction surface (e.g., a topographic surface) is to find the **least-cost distance** or shortest path between two points in a map layer. Of course, we may also want to calculate the **least-cost surface** for movement from one point location to all other locations. Let's take a look at these cases individually.

For this discussion, let us assume that our surface is a real topographic surface and that the cost is related to the change in elevation from one grid cell to another. Conceptually, we are placing a drop of water at the top of our topographic surface and watching as it streams or drains down the topographic surface to lower elevations, finding the path of least resistance. Indeed, many

GIS software packages use such terms as *drain* or *stream* as commands. To create a least-cost distance (as opposed to a **shortest-path distance**) in a 5 × 5 raster matrix, we begin at the top grid cell and perform a search of the eight nearest grid cells to evaluate which has the lowest friction or impedance value (in this case, lowest elevation relative to the starting point). This grid cell, which is assigned a numerical value or a pointer to indicate that it has been selected, then becomes the starting point for the next iteration in a search of its eight neighbor cells. The process continues until the lowest elevation value in the database has been selected. What we have then is a selected route from the top of our hill to the bottom that incurs the least amount of effort.

Let's continue with our simple visual example. We want to find the easiest path down the mountain represented by the 5 × 5 matrix of Figure 9.18a. Starting at the top (column 5, row 5), we examine our neighbors (there are only three because we are at the corner of our coverage) to determine which of the connected grid cells is lowest relative to (5,5). We note that grid cell (4,4) has the lowest value, and we move there next. Proceeding to grid cell (3,3), we again perform a search of all neighbor cells, again choosing to move to the lowest available grid cell value (2,2). When we have reached the bottom of the hill at location (1,1), the trail left behind is our least-cost path. An analog version of this search is shown in Figure 9.18b.

Although the least-cost route often suffices for point-to-point distances, it may prove useful to examine all possible routes from a particular place as we

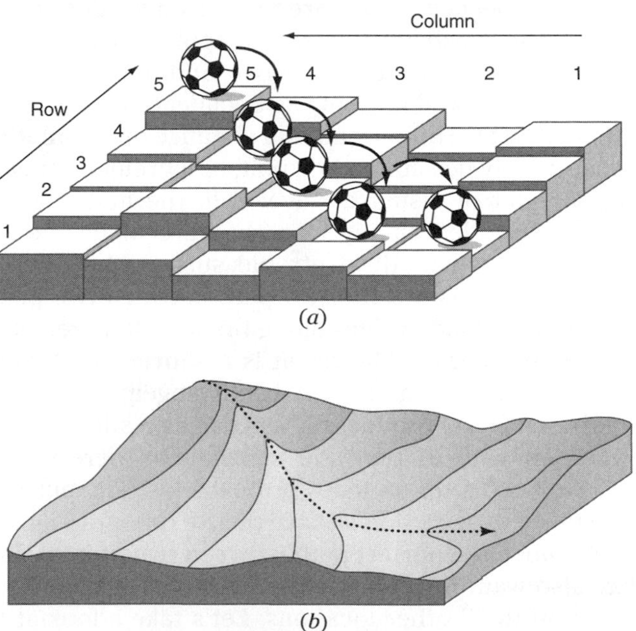

Figure 9.18 Least-cost path. (*a*) Calculation of a least-cost path requires the computer to compare the starting grid cell to its immediate neighbors. A soccer ball bouncing down the steps (grid cells) of a raster surface is used to illustrate the idea. (*b*) The same function as it might be performed by a stream flowing down a mountainside.

1	1	3	1
1	3	1	1
3	1	1	1
1	1	1	1

(*a*)

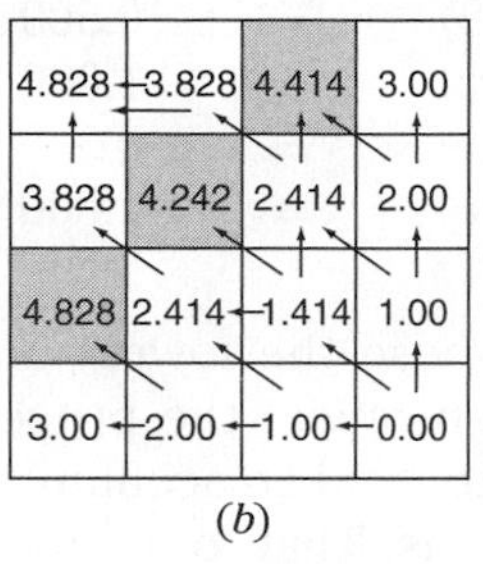

(*b*)

Figure 9.19 Accumulated distance. Accumulated cost surface shown by indicating (*a*) the friction coverage and (*b*) the calculated values. Note how the calculation of diagonal grid cells uses a value of 1.414 as the distance rather than 1.0 for orthogonal grid cells. The shaded grid cells are those whose friction value is 3. To accumulate distance, the previous value must be added to the calculated value. *Source:* Modified from J.K. Berry, *Beyond Mapping: Issues and Concepts in GIS*. Fort Collins, CO, GIS World, Inc., © 1993.

accumulate a cost for moving uphill or downhill through a friction surface. In this case, our friction values might be envisioned as topographic impedances, or friction impedances encountered during a trip through different types of land cover. Take a look at Figure 9.19. Here we have two grids—one indicating the friction or impedance factors and another indicating the calculated values of accumulated distance (weighted by the impedance factors). The process of computation is slightly more sophisticated than the shortest-path model we just examined. In this case, we see that the program doesn't simply search for the lowest value, but rather calculates a value for each adjacent cell based on both Euclidean distance and the impedance values. This example uses rational (decimal) numbers; diagonal distances between cells are measured when necessary.

We begin the process in the grid cell at column 0, row 0. The program then identifies all bounding grid cells and notes the friction factors (if any) for each of these bounding cells. The distance is calculated in half-step increments by multiplying each occupied grid cell (including the starting grid cell) by its friction factor and then by a value indicating its width (1 for orthogonal grid cells and 1.414 for each diagonal grid cell). Because we are moving from the center of the starting grid cell to the center of the next, we multiply each by 0.5 to indicate the half-step increments. Thus, our formula for each step of half a grid cell is (Berry, 1993)

$$0.5\,(\text{grid distance} \times \text{friction factor})$$

We then must add this to the previous friction value. So for a movement starting at column 4, row 4 and moving to grid cells (3,3), (2,2), and (3,4), our resulting calculations and values are

3,3	0.5 (1.414 × 1.00)	=	0.707
	0.5 (1.414 × 1.00)	=	0.707
	Subtotal	=	1.414
	+ Previous value	=	0
	Total	=	1.414
2,2	0.5 (1.414 × 1.00)	=	0.707
	0.5 (1.414 × 3.00)	=	2.121

	Subtotal	=	2.828
	+ Previous value	=	1.414
	Total	=	4.242
3,4	0.5 (1.000 × 1.00)	=	0.500
	0.5 (1.000 × 1.00)	=	0.500
	Subtotal	=	1.000
	+ Previous value	=	0
	Total	=	1.000

The next step is to choose the lowest accumulated distance for each of the adjacent cells and then repeat the process. The friction distances for all steps are added together to yield an accumulated distance. As you can see, the process can be quite tedious. That, of course, is why we use computers. The result is a least-cost surface, rather than a least-cost path, and we can choose the least-cost path to and from anywhere on the coverage. This is done by simply selecting the adjacent cells with the lowest cost values.

In cases of vector coordinates describing points, lines, or areas available for calculating distances, a series of modifications of the standard Pythagorean theorem can be used to calculate the distances. First, recall the original formula for Euclidean or straight-line distance between any two points:

$$d_{ij} = \sqrt{(X_i - X_j)^2 + (Y_i - Y_j)^2}$$

In this equation, the distance between points *i* and *j* is the square root of the sum of the square of the differences between the *X* and the *Y* coordinates.

If, however, we are unable to perform a straight-line distance—in other words, if there is an obstruction requiring us to deviate from a straight line—we can generalize the formula to a non-Euclidean form:

$$d_{ij} = [(X_i - X_j)^2 + (Y_i - Y_j)^{k1/k}]$$

where the squares and the square root sign have all been replaced by *k*, a variable representing any of a range of possible values (McGrew and Monroe, 1993). Instead of squaring the differences between the *X* and the *Y* distances, we raise them to the kth power. And where the sums of the *X* and *Y* coordinate values were under the square root sign, we now raise them to the $1/k^{\text{th}}$ power (keeping in mind that the 1/2 power is the same as square root).

By inspecting the preceding equation, you will see that by replacing each variable *k* with 2, you return to the original equation for calculating Euclidean distance. So the generalized equation works just as well for straight-line distances as it does for non-Euclidean distances. Let us assume, for example, that we are trying to find the distance between two points in the borough of Manhattan, where dozens of tall buildings and square blocks restrict our movement in square units. This measure, called the Manhattan distance, describes a rather restricted movement and changes our *k* factor to 1. Given that value, our formula now converts to the following:

$$d_{if} = |X_i - X_j| + |Y_i - Y_j|$$

As you can see, the calculation of any distance in a vector database can be calculated by simply modifying the *k* factor to fit our needs (Figure 9.20).

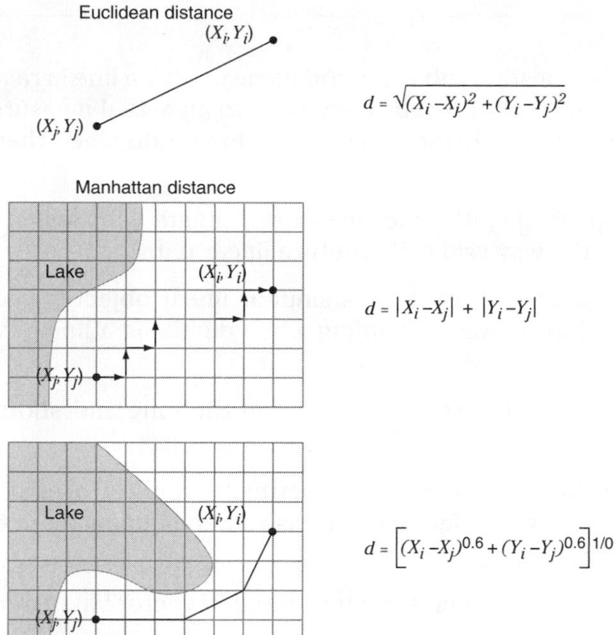

Figure 9.20 Vector distance measures. Two methods of calculating non-Euclidean distance in vector; each is a direct modification of the Pythagorean theorem formula. The exponent or *k* value is the only part of the equation that needs to be modified.

A *k* factor of 1.5, for example, simulates the distance modified by a combination of Manhattan and Euclidean. Even friction surfaces can be estimated through the use of these metrics by modifying the *k* factor. Take, for example, the use of a *k* factor of 0.6, which allows us to find the shortest distance around a barrier, such as a lake.

Although these metrics are very useful, we will at times want to perform distance measures for networks. We have the capability to find shortest paths and least-cost paths and to perform operations employing frictional impedance values as well as barriers. However, these are more closely related to connectivity than to distance, so we will postpone their discussion until Chapter 13.

Terms

absolute barrier
barrier
concavity
convexity
cumulative distance
edginess
Euclidean distance
Euler function
Euler number
filter
friction surface
functional distance
impedance value
incremental distance
inscribe
isotropic surface
lacunarity
least convex hull
least-cost distance
least-cost surface
orthogonal
perimeter/area ratio
relative barrier
roving window
shortest-path distance
shortest-path surface
sinuosity
spatial integrity
travelsheds

Review Questions

1. Describe the simplest possible method of measuring a line in raster format. How can a result so obtained be converted to an actual measure of length when the grid cells are orthogonal? What do you do when they are lined up diagonally?
2. What are the possible pitfalls for measuring length with raster, especially with regard to the way grid cells capture linear data?
3. How do we measure the length of a sinuous linear object in vector? What is the relationship between sampling when digitizing a line in vector and the length of the line as it is measured?
4. Why would we be interested in measuring the long and short axes of a polygon?
5. Why is the relationship between the perimeter and the area of a polygon important? How might differences in this ratio be important in the natural world? In the anthropogenic world?
6. How is the perimeter of a polygon measured in vector? In raster?
7. How is the area of a polygon measured in vector? In raster?
8. Given the relationship between feature shapes and function, why are there so few operational methods of shape analysis in GIS?
9. Describe two methods of measuring sinuosity of linear objects. Give two examples of such measures that might be useful to the GIS user.
10. What is spatial integrity? How can we measure it? Why is a given Euler number not a definitive measure of the number of holes and the number of polygons?
11. Describe how polygon boundary configuration might be calculated. What is a measure of convexity? Why is the circle the most frequently used geometric object against which polygonal shape is compared?
12. How can a roving window be used as a measure of polygon boundary configuration in raster?
13. How is Euclidian distance from a point to all other points in a coverage measured in raster? What is an isotropic surface, and what is its significance?
14. What is a friction surface? A barrier? What are the two types of barriers? What is the relationship between a relative barrier and a friction surface?
15. Describe a simple raster method of measuring distance through a friction surface.
16. What is the difference between incremental and accumulated distance? How can we measure the path of least resistance (best path) through a raster coverage that includes surface data? What is a more accurate method of measuring accumulated distance (best path surface) through a raster GIS than simply adding up the friction numbers?

17. How can we modify the Pythagorean theorem to account for Manhattan distance, paths around barriers, and paths through friction surfaces in vector?

18. What are the potential problems of assigning friction or impedance values for modeling distance?

References

Baker, W.L., 1989. "A Review of Models of Landscape Change." *Landscape Ecology*, 2:111–133.

Baker, W.L., and Y. Cai, 1992. "The r. le Programs for Multiscale Analysis of Landscape Structure Using the GRASS Geographical Information System." *Landscape Ecology*, 7(4):291–301.

Berry, J.K., 1993. *Beyond Mapping: Concepts, Algorithms, and Issues in GIS*. Fort Collins, CO: GIS World.

Boyce, R.R., and W.A.V. Clark, 1964. "The Concept of Shape in Geography." *Geographical Review*, 54:561–572.

Clarke, K.C., 1990. *Analytical and Computer Cartography*. Englewood Cliffs, NJ: Prentice-Hall.

Davis, J.C., 1986. *Statistics and Data Analysis in Geology*, 2nd ed. New York: John Wiley & Sons.

Denizman, C., 2003. "Morphometric and Spatial Distribution Parameters of Karstic Depressions, Lower Suwannee River Basin, Florida," *Journal of Cave and Karst Studies*, 65(1):29–35.

Environmental Systems Research Institute, 1993. *Learning GIS: The ARC/INFO Method*. Redlands, CA: ESRI.

Forman, R.T.T., and M. Godron, 1981. "Patches and Structural Components for Landscape Ecology." *BioScience*, 31:733–740.

LaGro, J.A., 1991. "Assessing Patch Shape in Landscape Mosaics." *Photogrammetric Engineering and Remote Sensing*, 57:285–293.

Lee, D.R., and G.T. Sallee, 1970. "A Method of Measuring Shape." *Geographical Review*, 60(4):555–563.

McGarigal, K., S.A. Cushman, M.C. Neel, and E. Ene, 2002. FRAGSTATS: Spatial Pattern Analysis Program for Categorical Maps. Computer software program produced by the authors at the University of Massachusetts, Amherst. Available at www.umass.edu/landeco/research/fragstats/fragstats.html (last visited 10/23/07).

McGrew, J.C., and C.B. Monroe, 1993. *Statistical Problem Solving in Geography*. Dubuque, IA: Wm C. Brown Publishers.

Moellering, H.M., and J.N. Rayner, 1982. "The Dual Axis Fourier Shape Analysis of Closed Cartographic Forms." *Cartographic Journal*, 19(1):53–59.

Muehrcke, P.C., and J.O. Muehrcke, 1992. Map Use: *Reading, Analysis and Interpretation*. Madison, WI: J.P. Publications.

O'Neill, R.V. et al., 1988. "Indices of Landscape Pattern." *Landscape Ecology*, 1(3):153–162.

Plotnick, R.E., R.H. Gardner, and R.V. O'Neill, 1993. "Lacunarity Indices as Measures of Landscape Texture." *Landscape Ecology*, 8:201–211.

Plotnick, R.E., R.H. Gardner, W.W. Hargrove, K. Prestegaard, and M. Perlmutter, 1996. "Lacunarity Analysis: A General Technique for the Analysis of Spatial Patterns." *Physical Review E*, 53:5461–5468.

Robinson, A.H., J.L. Morrison, P.C. Muehrcke, A.J. Kimerling, and S.C. Guptill, 1995. *Elements of Cartography*, 6th ed. New York: John Wiley & Sons.

Ronca, L.B., and J.W. Salisbury, 1996. "Lunar History as Suggested by the Circularity Index of Lunar Craters," *Icarus*, 5(1-6):130–138.

Strahler, A.N., 1975. *Exercises in Physical Geography*, 2nd ed. New York: John Wiley & Sons.

Turner, M.G., 1991. "Landscape Ecology: The Effect of Pattern on Process." *Annual Review of Ecology and Systematics*, 20:171–197.

Wu, X.B., and D.Z. Sui, 2001. "An Initial Exploration of a Lacunarity-Based Segregation Measure." *Environment and Planning B*, 28 (3):433–446.

Zadeh, L.A., 1965. "Fuzzy Sets," *Information and Control*, 8:338–353.

CHAPTER 10

Classification

Classification of the earth's features is an old problem that has engaged the minds of natural scientists of many types, each looking at the earth through different eyes and using a different vocabulary and a different intellectual filter. Although the concept of classifying earth's features may seem quite simple, the practical aspects of the task are far from straightforward. How we classify what we see will be strongly reflected in how we analyze data, our proposed application, and what our conclusions mean to that application. The process cannot be taken lightly, for poor classification will often lead to poor spatial decision making.

Let's say we have a digital database digitized from old land cover maps. You might wish to compare these maps to more modern land cover maps. For this to happen it is absolutely necessary that the class types and numbers match exactly for any comparisons to take place. In addition, the description of what the classes mean must also match. For example, wetland as defined in the old maps as areas supporting waterfowl will not necessarily correspond to a modern definition of wetland as the opposite of upland. These differences must be reconciled for map comparison to be effective.

As the foregoing suggests, how we classify things has an impact on how we view them, their scale or resolution, and how we subsequently manipulate the data for analysis. Our classification may also influence the instruments we choose to employ: whether ground based or deployed from aircraft (aerial photography) or a satellite. The classifications may be simple, based on a single observed criterion such as ground cover, or complex, based on multiple sets of criteria (elevation, rainfall, ecological functioning, etc.). Some classifications can be created by combining many measurements of many different themes. The way we classify can change dramatically for a single set of themes, based entirely on what questions we ask.

In this chapter you will become acquainted with the general problem of land classification. We will briefly examine some well-known classification systems and some basic GIS-compatible methods of classification. We will also look at how classifications themselves can be decomposed or combined with data from other layers to reclassify what we originally had.

LEARNING OBJECTIVES

When you are finished with this chapter you should be able to:

1. Understand the importance of thematic map classification as afforded by new technology.
2. Be familiar with some of the common thematic map classification systems available, especially for land cover and land use maps.
3. Know how to aggregate data in both vector and raster to achieve new classification systems.
4. Be able to reclassify data based on the manipulation of nominal, ordinal, interval, ratio, and scalar data.
5. Understand the types of filters available and know how to use both static and roving window filters for reclassification.
6. Have a thorough understanding of what neighborhoods are, what types are available, and how total and targeted neighborhood analysis can achieve new classifications.
7. Be able to describe the many ways neighborhoods can be created in GIS for two-dimensional map data.
8. Have a working knowledge of map algebra and explain how it differs from matrix algebra.
9. Understand the differences among local, focal, zonal, block, and global operations and be able to calculate their values and apply the techniques to modeling tasks.
10. Diagram and describe the different types of buffers.
11. Describe how buffers are used for creating neighborhoods, and list and explain the four basic methods for determining the buffer distance.

CLASSIFICATION PRINCIPLES

The tendency for humans to want to impose a classification framework has long been extended to the analysis of earth's surface features. The centuries-old history of cartography shows how people have grouped features by physical type (land versus water), by political subdivision, and by human endeavor (inhabited versus uninhabited). For centuries, these classifications were placed on maps with little formal discussion about how they should be organized or what impact they might have on decisions. Only when cartographers began to recognize the importance of map classification as a means of decision making and scientific inquiry did the problems of area classification become a topic of academic inquiry (Sauer, 1921). Although resource managers and land planners began to show interest in the impact of land classification in the early 1900s, an increased urgency emerged with the advent of satellite remote sensing, digital cartography, and GIS. Because of the abilities of these systems to obtain and manipulate vast amounts of spatial data, conditions were right for the development of new and better classifications of the earth's features. Increasing pressures on our natural and economic resources make it more important than ever to view existing systems of classification with a measured sense of cynicism because of the potential limitations these classification **filters** impose on our interpretations and decisions.

Land classification depends on the types of objects we are going to group. There are separate classifications for vegetation (Küchler, 1956), soils (Soil Survey Staff, 1975), geological formations, wetlands (Cowardin et al., 1979; Klemas et al., 1993), agriculture, land use, and land cover (Anderson et al., 1976; U.S. Geological Survey, 1992). These classifications may be simple, such as a classification of vegetation for mapping based strictly on the plants present, or they may be more functional, such as mapping ecosystems rather than vegetation alone (U.S. Department of Agriculture, 1993; Walter and Box, 1976; Omernick, 2004; Bailey et al., 1994). Variations in classification are dictated by scale when, for example, vegetation is mapped not for small regions but for the entire earth (United Nations Educational, Scientific, and Cultural Organization [UNESCO], 1973). Or the system might be specifically oriented toward the mandates of a particular organization (Klemas et al., 1993; Wilen, 1990). Other classifications are dictated more by the technology used to obtain the raw data, such as satellite remotely sensed data (Anderson et al., 1976). Yet others are designed specifically to address decisions based on factor interactions, as exemplified by maps of biophysical units, land capability, or land suitability, which have strong similarities but reflect subtle differences in the manner in which the classifications are applied. Biophysical units use a combination of spatially arrayed co-occurrences of biological and physical phenomena related, for example, to soils and topography to evaluate the viability of the land units to support natural systems. Land capability most often applies specifically to the ability of the soil to support housing, septic systems, wildlife, agriculture, and other major categories. Land suitability maps areas on the basis of not only soils but also other biological and physical properties, and they are used primarily to evaluate the utility of the respective regions to support a variety of human activities.

Collectively, all of these classifications and thousands of others have one thing in common: they all have an intended purpose and specific users. The end users of some classified data sets, such as land suitability, are very specific, while others will have a broader audience. The UNESCO (1973) vegetation classification system was designed to unify vegetation classifications around the world, to permit anyone interested in vegetation, anywhere on earth, to talk about plants using the same basic system. Although this broad approach to classification may be practical as a general standard, its usefulness as a decision tool is restricted both from a perspective of scale, but more importantly because it is a single, static classification. With today's GIS, we can store raw data and classify them to meet changing needs. This is the most important point about classification in the GIS environment: the more closely your classification can be made to fit user needs, the more useful it becomes. A GIS provides a wide variety of ways to classify and reclassify stored attribute data to achieve this result. The GIS operator either displays the existing classification (a storage and retrieval operation) or manipulates the existing attributes to create a classification more appropriate to the questions being asked. In the remainder of the chapter, then, we will refer to the function as "reclassification."

ELEMENTS OF RECLASSIFICATION

Frequently map entities together with their attributes are input to a GIS under a specific set of criteria. You have seen how points and lines can be reclassified by recoding the attributes in their tables or by recoding the grid cell values to produce new point or line themes. This way the users change the attributes, not the entities. The process is much the same when working on polygonal features with raster systems: by selecting the attributes for the areas in which you are interested, you merely change the numerical codes or attribute names for the grid cells.

GIS reclassification is simplest in raster. Whether focusing on individual point locations or groups (neighborhoods) of cells, the process is essentially one of changing the attribute codes or manipulating the numerical grid cell values. Although simple to perform, care should be taken to avoid such obvious mistakes as multiplying grid cell numbers representing nominal categories by those representing ordinal, interval, ratio, or scalar categories. Such multiplication may result in sometimes silly results. Manipulating grid cells that contain 0s or "no data" can also result in surprising results (DeMers, 2002). A simple example of raster reclassification might be if we had a map with three types of grid cell values, one each for corn, wheat, and row crops. We might have cell values of 2, 4, and 7 respectively. To simplify our map, we could reclassify corn and wheat (2 and 4) as grain crops with a value of 1 and leave the row crops as a separate classification.

In vector, reclassifying areas requires us to change both entities and attributes. We first remove any lines that separate adjacent classes that are going to be combined. This is called a **line dissolve** operation because it selects a particular line entity and dissolves or eliminates it. Then the attributes associated with the combined polygons are rewritten for the new map as a single new attribute. Let's take a very simple example featuring only two polygons, one for wheat and the other for corn (Figure 10.1). Our purpose is to produce a single category called *grain* by "dissolving" (removing) the line separating the two original categories. We place the new grain category in our attribute table and assign it explicitly to the new, larger polygon. We now have a new map with a single category. Most real applications will be far more complex, requiring multiple line dissolves and attribute changes, but the process is identical.

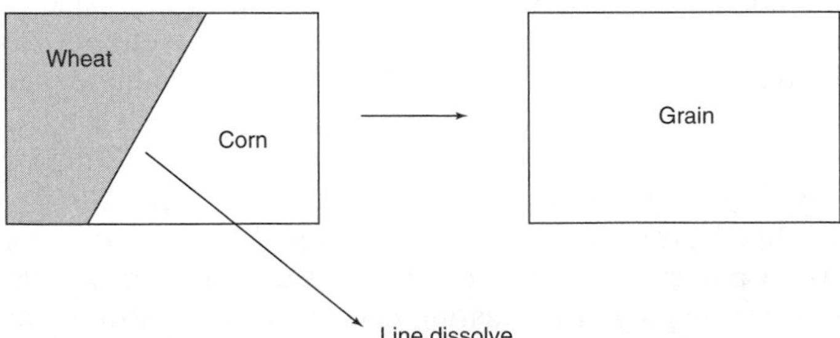

Figure 10.1 Reclassification and line dissolve. Reclassification through data aggregation: the regrouping of corn and wheat into a larger class called grain.

In both our raster and the vector examples, an interesting pattern emerged. We ended with fewer categories than we had originally. This result, called **aggregation**, is a useful, common, and simple reclassification. Imagine trying to reverse the process—separating a large single category polygon into two parts: one wheat, the other corn. Unless you could compare this information to another map, it would be impossible to find the line separating them. There are more complex methods that allow separating out more detail from coarse polygonal information. These methods require the use of two or more map layers compared with each other through a process called **overlay**, which we will cover in Chapter 14.

Thus far our reclassification was based on nominal data only. The same aggregation process can apply to other levels of data measurement as well. One way is to create range categories of data, often called **range-graded classifications**. We simply recode raw data on the basis of where they fit within the range classes. As we did with nominal data, we recode our grid cells or perform a line dissolve and attribute change in vector. We can also perform such operations as ordering the grid cell or polygon values, inverting them, or using a process called **constrained math**, which permits the values of the target polygons to be operated on (multiplied, divided, etc.) by another number. But the process is fundamentally the same as what we did with nominal data.

All the methods so far have one thing in common: the reclassification process is directed toward renaming the objects on the basis of the attribute values at their own location. It is a local or worm's eye view, and the operations are called **local** in the modeling language called **map algebra** (Tomlin, 1990). In local operations each collection of grid cells or each polygon is seen as a distinct individual entity, and the classification is constrained to the immediate target area. Reclassification based on **locational information** is only one of four basic conditions that include **position**, **size**, and **shape** information. But these concepts do not separate out as individual techniques. Instead, they are often combined to provide a wide array of reclassification techniques. We will look at these, starting with a general description of **neighborhood functions** (based on the idea of a bird's-eye view), then proceed to examine several types including filters, two- and three-dimensional neighborhoods, and buffers. As you read along, notice the frequent similarities among the techniques as they are applied under different circumstances.

NEIGHBORHOOD FUNCTIONS

Reclassification based on locational attributes is very useful, but it limits us to the attributes within each object or feature. It would be nice if we could classify features based on a bird's-eye view. In other words, if we could fly over an area, we would be able to find out not only that a particular feature exists but where it is relative to its surroundings. Such procedures, called neighborhood functions, characterize each target object based on attributes it shares with its neighborhood. This is not unlike how we as residents classify our own neighborhoods, each of which shares particular characteristics. These characteristics might be political, economic, religious, ethnic, or others, but the concept is identical to how neighborhood functions are applied.

Neighborhood functions can operate on both two- and three-dimensional objects. They can also be separated into either **static neighborhood functions**, where the analysis takes place all at once for the selected target area, or **roving window neighborhood functions** (in raster), where the analysis takes place within the framework of a window that moves across the coverage. You saw an example of using a windowing function in Chapter 9 to characterize the edginess of a polygonal boundary.

ROVING WINDOWS: FILTERS

The example in Chapter 9 where we used a specified window to reclassify cells for a calculation of edginess is part of a larger group of operations called windowing functions. They are also referred to as filters, especially when the window itself contains numbers against which the grid cells are compared. These can be used to reclassify whole raster grids based on a wide range of alternative filter values and combinations of values. This technique is used quite frequently in image processing (Lillesand and Kiefer, 1995), but it has equal utility in raster GIS operations. In particular, the filter is used, much as in remote sensing, to isolate edges or linear features (an approach called **high-pass filter**), to emphasize trends by eliminating small pockets of unusual values (often called **low-pass filter**), or even to give a measure of orientation or objects in an image (**directional filter**). Although these filters can be statically placed over a target neighborhood to provide a neighborhood characterizing result, most often they are applied as roving windows to characterize an entire coverage (Figure 10.2).

The high-pass filter (Figure 10.3) is designed to separate out detail in a raster image that may be obscured by nearby grid cells that contain relatively similar attribute values. In remote sensing, these values most likely represent electromagnetic reflectance. However, we can use almost any surface-related data. Let's say that we are interested in finding small topographic ridges in our raster database. With each grid cell containing a value for elevation, we want to highlight the differences between the slightly higher values of a ridge and the slightly lower values surrounding this feature. The standard method for performing a high-pass filter is to create a 3×3 filter (a kernel) with the weighting value of 9 at the center grid cell (2, 2), and a weight of -1 for the

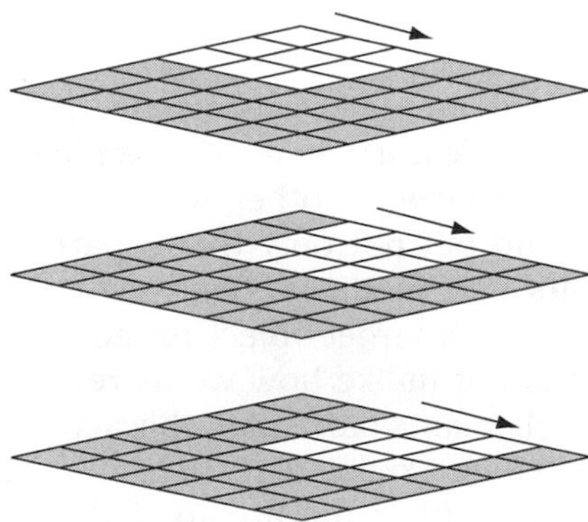

Figure 10.2 Roving window. A window of cells is placed at a starting point where the calculations are performed and the output cell value is assigned. Then the filter is moved one cell at a time across the entire raster coverage in serpentine fashion until it is finished.

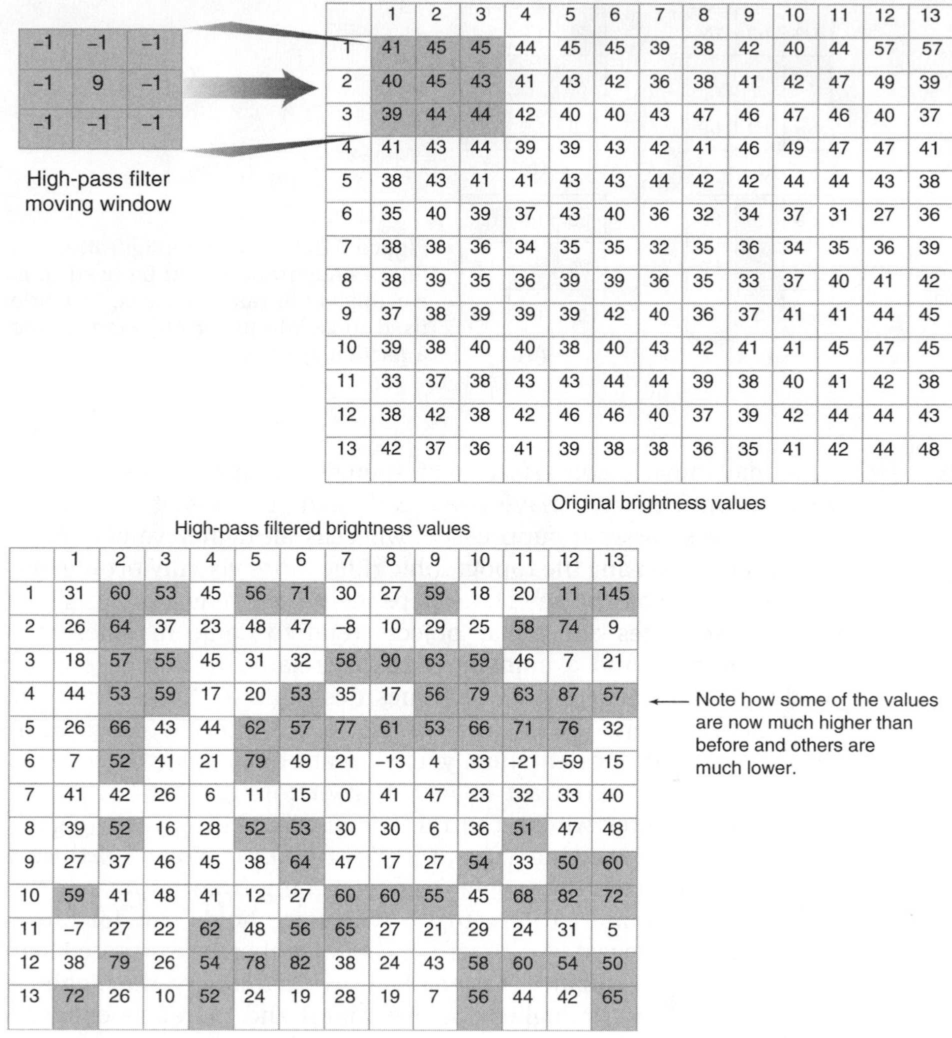

-1	-1	-1
-1	9	-1
-1	-1	-1

High-pass filter moving window

Original brightness values

	1	2	3	4	5	6	7	8	9	10	11	12	13
1	41	45	45	44	45	45	39	38	42	40	44	57	57
2	40	45	43	41	43	42	36	38	41	42	47	49	39
3	39	44	44	42	40	40	43	47	46	47	46	40	37
4	41	43	44	39	39	43	42	41	46	49	47	47	41
5	38	43	41	41	43	43	44	42	42	44	44	43	38
6	35	40	39	37	43	40	36	32	34	37	31	27	36
7	38	38	36	34	35	35	32	35	36	34	35	36	39
8	38	39	35	36	39	39	36	35	33	37	40	41	42
9	37	38	39	39	39	42	40	36	37	41	41	44	45
10	39	38	40	40	38	40	43	42	41	41	45	47	45
11	33	37	38	43	43	44	44	39	38	40	41	42	38
12	38	42	38	42	46	46	40	37	39	42	44	44	43
13	42	37	36	41	39	38	38	36	35	41	42	44	48

High-pass filtered brightness values

	1	2	3	4	5	6	7	8	9	10	11	12	13
1	31	60	53	45	56	71	30	27	59	18	20	11	145
2	26	64	37	23	48	47	–8	10	29	25	58	74	9
3	18	57	55	45	31	32	58	90	63	59	46	7	21
4	44	53	59	17	20	53	35	17	56	79	63	87	57
5	26	66	43	44	62	57	77	61	53	66	71	76	32
6	7	52	41	21	79	49	21	–13	4	33	–21	–59	15
7	41	42	26	6	11	15	0	41	47	23	32	33	40
8	39	52	16	28	52	53	30	30	6	36	51	47	48
9	27	37	46	45	38	64	47	17	27	54	33	50	60
10	59	41	48	41	12	27	60	60	55	45	68	82	72
11	–7	27	22	62	48	56	65	27	21	29	24	31	5
12	38	79	26	54	78	82	38	24	43	58	60	54	50
13	72	26	10	52	24	19	28	19	7	56	44	42	65

Note how some of the values are now much higher than before and others are much lower.

Figure 10.3 High-pass filter. Operation of a high-pass filter showing the use of a 3 × 3 matrix of numbers (the filter) designed to enhance the higher values and suppress the lower values. *Source:* Robinson et al., *Elements of Cartography*, 6th ed., John Wiley & Sons, New York, © 1995, modified from Figure 12.16, page 214. Used with permission.

remaining grid cells. This filter is placed over each 3 × 3 matrix of grid cells in your grid, and the members of each corresponding grid cell pair are multiplied. That is, the elevation value of the center grid cell in your topographic layer is multiplied by 9, and all the remaining values are multiplied by −1. Next, these nine newly created values are summed to obtain the final, high-pass value that is assigned to the center grid cell. In other words, this single operation of nine grid cells operated on by the filter of nine grid cells produces a single value that is placed at the center of the new coverage.

Once this first operation is finished, the filter is moved one grid cell to the right, so that the central cell in the topographic coverage is now (3, 2). The calculations are repeated as before, resulting in a new high-pass value for column 3, row 2. The filter is moved and the calculations are repeated one cell

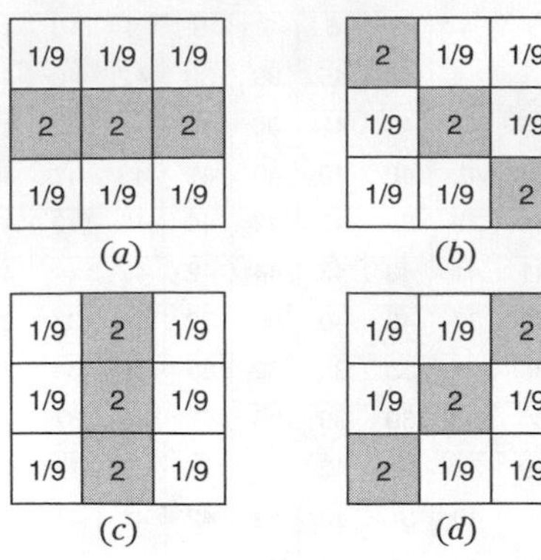

Figure 10.4 Filter configurations. A variety of filter configurations can be used to identify linear objects in raster images. The orientations are used to isolate linear objects that are oriented in a particular way.

at a time, moving from left to right, then down, then right to left—in serpentine fashion until all cells have been evaluated. The result will be that all lower values are somewhat suppressed, whereas all higher values are enhanced or made larger, making the topographic ridge more visually prominent. It can then be reclassified as you see fit, perhaps identifying it as a ridge.

Because edges and linear objects tend to come in different orientations (e.g., you may find a number of ridges related to some geological cycle of folding and erosion), it is sometimes desirable to "direct" your filter toward a particular orientation. If, for example, you are trying to highlight ridges that have an east–west orientation, you will want to enhance these features, leaving the others as low-contrast objects in your topographic coverage. To do this, you create a filter whose central row of numbers is high (e.g., 2) while the numbers in the remainder are 1/9 (Figure 10.4a). For a northwest–southeast bias, you will have the higher numbers going from upper left to lower right on your spatial filter (Figure 10.4b). The same holds true for north–south and northeast–southwest bias, where the high numbers are oriented in the direction you are trying to enhance (Figures 10.4c, d). In this way, the higher numbers will be multiplied by the higher positive values and therefore enhanced, or made larger, and the remainder will be suppressed.

All of these previous examples involve enhancing features in a grid. If you wish to suppress rather than enhance the higher values to remove meaningless topographic peaks or pits, thus obtaining a grid more closely representative of the overall topographic trend, you can use a filter kernel that mutes these unusually high or low values. The most common filter used for this low-pass operation is a 3×3 filter whose numbers are all the same, commonly 1/9, because there are nine grid cells in the filter (Figure 10.5). Because the idea now is to make all the numbers closer together in value, you move the filter as before, one grid cell at a time through the entire image, but instead of adding the values to obtain the new number for the central grid cell, you average them. Your result for each pass will be an average number for the grid cells under the filter.

Although the normal method of filtering in raster databases employs the 3×3 filter, and even though set number combinations are programmed into standard software applications, there is no need to stick to either convention. Most software that contains the filtering capability will allow some flexibility in the size of the filter and the numbers used. Smaller filters are more often used for edge enhancement, whereas larger ones are more common for low-pass filtering

1/9	1/9	1/9
1/9	1/9	1/9
1/9	1/9	1/9

(*a*)

42	38	71
68	89	72
70	59	85

(*b*)

	66	

(*c*)

Figure 10.5 Low-pass filter. Multiplying each cell in the filter (*a*) by each cell in the original data (*b*) and totaling assigns the resultant cell value (*c*) to the central cell in the output matrix (i.e., $1/9 \cdot 42 + 1/9 \cdot 38 + 1/9 \cdot 71 + 1/9 \cdot 68 + 1/9 \cdot 89 + 1/9 \cdot 72 + 1/9 \cdot 70 + 1/9 \cdot 59 + 1/9 \cdot 85 = 594/9 = 66$).

or averaging. The larger the filter size, the more averaged the numbers will become because of the larger set of numbers used for averaging. It often takes experimentation to decide on the utility of abandoning the default filter size and number combinations. These operations are much the same as those used in your desktop graphics software to remove scratches in scanned photographs.

STATIC NEIGHBORHOOD FUNCTIONS

Suppose that we are interested in the identity of a region's polygons within a certain neighborhood or distance. For example, say we are studying the spread of innovative farming practices to see whether a pattern of innovation diffusion emerges and are interested in neighborhoods of no-till agriculture. First we retrieve a layer showing only locations in which no-till agriculture is being performed, and then we select a distance value as an estimate of where imitation is most likely to be observed (either adjacent or some distance beyond that). The GIS takes this distance value and spreads outward until it reaches its maximum search radius, all the while adding up the number of no-till agriculture grid cells (essentially measuring the area of groups of polygons). The result will be groups (neighborhoods) of fields that indicate an adoption of this new practice.

In the previous example the GIS reclassified the original no-till region so that all the polygons or grid cells that fall within the specified distance of one another are assigned the same attribute class. Each of these neighborhoods can subsequently be reclassified by measuring its size (i.e., large, medium, and small groups of no-till agriculture). We might conclude that the farmers in the larger groups of no-till neighborhoods are more apt to talk to one another, or that the farms themselves are larger, or perhaps that these farmers knew some people at the nearby agricultural college. In any event, the values suggest that there is either a different mechanism for larger as opposed to smaller neighborhoods sharing these ideas, or some other functional reason for the difference in sizes of these clumps. The interpretation of the causes of these results will generally have to be further verified, but the neighborhood function of the GIS allowed us to recognize that differences exist.

In the preceding characterization of neighborhoods, we were looking at a single attribute (no-till agriculture) to produce our groups or clumps based on a specified distance. Many times, however, we are less interested in groups of identical polygons or grid cells and more interested in defining the similarities and differences within a specified neighborhood distance. As an example, let's say that we want to determine the average age of people in a specified region on the basis of census block data. By selecting a search radius, just as we did before, the software looks at the attributes for all the different census block polygons or grid cells, and then performs a numerical average of these values. The result is a new map for average age.

But average is not the only thing we can use to define new neighborhoods. Perhaps we are looking at a particular species of animal known to be especially attracted to highly varied land cover types. The more diverse the cover types, the more these animals like the neighborhood, which provides a wide range of possible places to nest, forage, and take shelter from the sun and from predators. We may also know that this species needs a particular amount of land area in which to operate. We convert this known amount of area to a search radius and begin as before. In this case, the software looks at all the different land cover patterns within the search radius and counts them, returning the number of cover types as a measure of spatial or landscape diversity (Figure 10.6). Areas with the highest diversity would be the most likely place for the target species to nest. If we were trying to introduce this species into a region, we would most likely choose the areas of highest diversity.

If we can perform an average on the basis of our surrounding neighborhood polygons or grid cells, we can also perform a wide range of other calculations. We might want to know some maximum value for a neighborhood—for example, the maximum number of crimes for the neighborhood for a given year (Figure 10.7). Or we might want to know the minimum value, such as the lowest price for

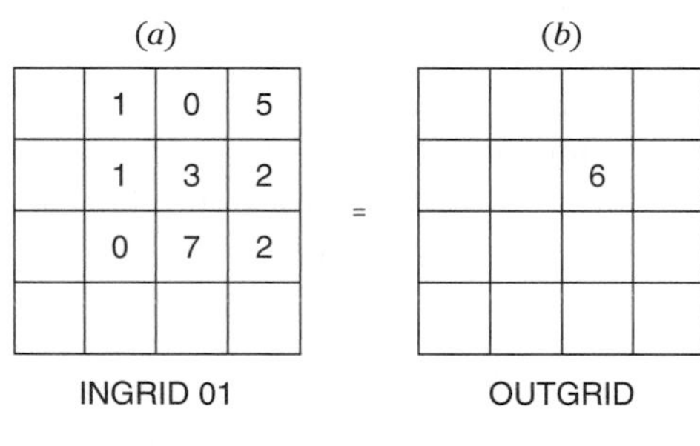

Figure 10.6 Neighborhood function based on diversity measure. By examining the surrounding grid cells (*a*), this focal function counts the number of land uses in the neighborhood (3 × 3 cell matrix) and returns that number (3) to a single cell (*b*).

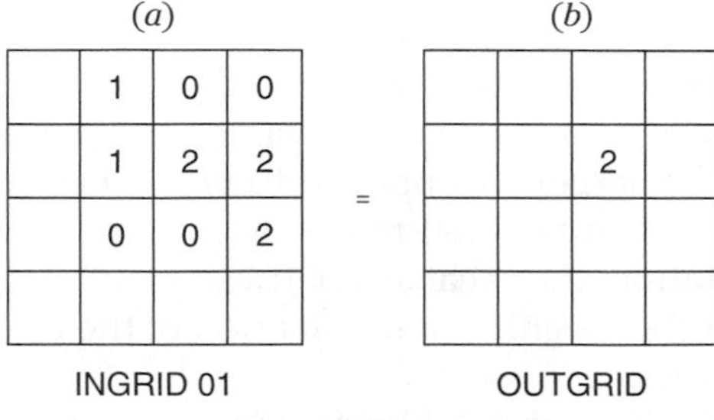

Figure 10.7 Neighborhood function based on maximum measure. By comparing surrounding grid cells (*a*), this focal function returns the maximum value that occurs in its neighbor cells (3 × 3 cell matrix) and returns that number to a single cell (*b*).

homes in the neighborhood to determine whether we could afford to live there. Other measures include total counts of all types in the neighborhood; the median, highest, and lowest frequencies; deviations from a central point to the mean of the surrounding values; and even the proportion of the neighborhood that shares attributes common to the central cell.

The measures we have identified in the last two paragraphs can operate in many different ways. If, for example, they return the value from analysis to the central point or target cell (a **focal function**), that value (average, diversity, median, etc.) of the neighborhood is assigned to the central location in the neighborhood for a new grid. We also do not necessarily need to assign a target distance at all, using instead a roving window approach to characterize the entire coverage on the basis of the same measures as before (average, diversity, median, etc.). In this case, the output values will not be returned to a central location but will be assigned throughout the new coverage, giving us an idea of the trends from one part of the coverage to another. We might also decide to create a roving window function in which the results are applied to the entire grid under consideration (e.g., 3 × 3 blocks or 9 grid cells), and the filter is moved a block at a time rather than once cell at a time. These roving window functions are called **block functions** (Figure 10.8).

In some cases we can characterize a neighborhood on the basis of the values within a single map, or we can use a separate layer for target cells and another for the neighborhood measure. For example, if we want to know the neighborhood diversity for known bird nest locations, we would choose the bird locations as target cells and another map of land cover as our measure of neighborhood. In addition, measures of neighborhoods such as average, minimum, and maximum are comparisons of the surrounding polygons with each other, whereas other cases (e.g., measures of deviation from the average, the proportion of a neighborhood that contains the same value as the target location) require the target location to be compared with the surroundings.

But the ability of the GIS to measure size and shape offers other ways of characterizing neighborhoods. A measure of size, for example, is often combined with a measure of the clumps of a single polygonal attribute value. Most often this information is used to rank or order the results of analytical techniques that group data together into localized or regionalized clumps. The

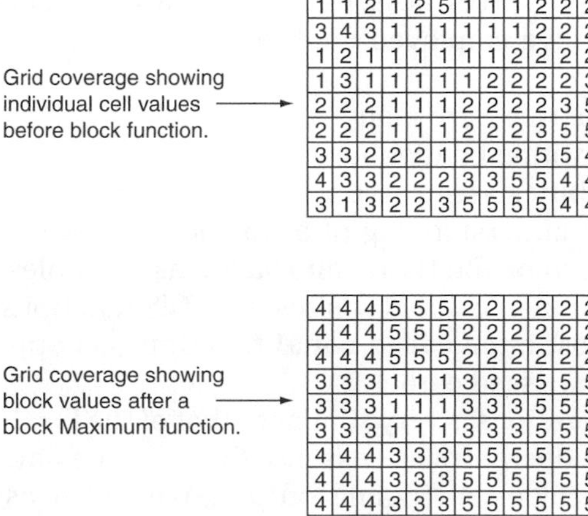

Figure 10.8 Block function. In some cases, we want to convert whole groups of cells on the basis of the neighborhood values. This block function uses a maximum value for the surrounding neighborhood cells, thus defining the 3 × 3 cell matrix on the basis of the highest value in the neighborhood. Notice that the blocks do not overlap.

sizes of clumps may be very important in our analysis. Ecologists, for example, are aware of the sizes of range required by wolves and other large predators (Forman and Godron, 1987; Ortega-Huerta and Medley, 1999). They are also familiar with the minimum requirements of individual forest stands to support a diversity of animal species or even to continue to exist as a forest patch. In some cases there may be a need to mix a neighborhood measure of diversity (where higher diversity indicates better habitat) with a measure of size (where larger areas are preferred). Many other combinations can also be employed. Let's take a look at an example, and while you read about this one, think of some others that might be of interest to you.

Jaguars are large predatory cats that require different types of habitat, especially jungle vegetation, for cover, and running water, because a large portion of their diet consists of fish. For the sake of simplicity, let us say that jaguars require approximately 200 square miles of territory in which to live and only two types of habitat—jungle forest and stream corridor. Let us further assume that researchers have indicated a requirement for territory that is at least 1 part stream corridor to 10 parts forest. We can identify the ratio between target points along the stream from a stream's coverage and the amount of surrounding (neighborhood) forest in the vegetation coverage that also includes stream corridor as a category. We can reclassify the vegetation coverage into stream corridor and forest vegetation to simplify our calculations. Then by using the neighborhood function that computes the ratio of the amount of neighboring area having the same category as the target location, we can identify the neighborhoods that have at least one-tenth stream corridor. Finally, now that we have created neighborhoods that suit the habitat-type needs of the jaguar, we can consider size. Neighborhoods that have at least 200 square miles of area with a stream corridor/forest ratio of 1:10 will be ideal sites for jaguar habitat.

Suppose, however, that the habitat neighborhood revealed by our analysis looks somewhat like an hourglass, having a narrow middle with the stream running through it. The areas outside our forest and stream corridor habitat are inhabited by humans, a species jaguars generally avoid. There may not be enough stream corridor habitat in this bottleneck to support the animals' activities. That is, as far as the jaguars are concerned, the two ends of the hourglass may be functionally separate habitat areas, neither of which is large enough for survival purposes. Using the ability of the GIS to measure distances across polygonal objects, we can perform an analysis that locates and classifies regions based on some value of narrowness, usually the narrowest portion of the polygon's shape. Thus we have employed another measure of neighborhood, this one based on shape.

As you have already seen, there are many ways to reclassify neighborhoods using values available from within contiguous regions. However, we can extend the concept to include attributes associated with fragmented or discontinuous regions as well. For a clear understanding of a fragmented region, one might easily use the islands of Japan or the Hawaiian islands as examples of noncontiguous areas that share common characteristics. Our GIS functions applied to these types of regions is most often called **zonal functions**, as opposed to the focal and block functions discussed earlier.

The idea behind zonal operations is very similar to that of the focal operations. The available mathematical operators are essentially the same and the results are typically stored as a single-cell (point) output. We begin as before by

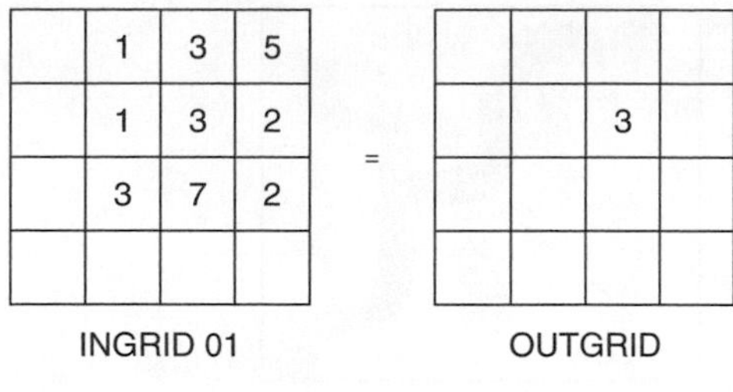

Figure 10.9 Zonal neighborhood function based on average. Using the average of numerical values of a fragmented region, this function returns that value to a single cell. Unlike the focal function, the value is based on zones rather than immediate neighbors.

selecting a target location (most of a grid cell) upon which we will operate. Instead of selecting an adjacent neighborhood, however, we begin by classifying a region based on some measurable characteristic. We might characterize our fragmented region as a collection of forested patches that might be found appealing to some bird species if other conditions are satisfactory, so the bird will evaluate the forest patches to determine if an appropriate nesting site is available at our target cell.

We examine the characteristics of all portions of the fragmented region to determine, for example, that there is a selected average tree density. This is just like what we did before using the focal function, except we now "average" the values for the extended neighborhood (i.e., the fragments) and then pass the value on to the output target cell (Figure 10.9). Although similar in approach, this zonal method allows for characterization of single-point locations based on often widely separated fragments of a region that share common attributes.

BUFFERS

Another very common method of reclassification, based this time on distance, is called **buffering**. A buffer is a polygon created through reclassification at a specified distance from a point, line, or area. Because it is based on the location, shape, and orientation of an existing object, we could easily have classified buffering as a method of reclassification based on position. However, a buffer can be more than just a measured distance from any other two-dimensional object; it can also be closely related and even controlled to some degree by the presence of friction surfaces, topography, barriers, and so on. That is, although buffering is based on position, it has other substantial components as well. An area surrounding a stream that indicates something about the stream corridor offers an example of a buffer. In Figure 10.10 the buffer was produced by reclassifying the area on either side to differentiate it from the amorphous background. Although the figure shows the stream plus the buffer, normally the buffer is generated as a separate feature and is often stored as a separate coverage as well. To produce this buffer merely requires that a specified distance be measured in all directions from the target object. You have seen how a GIS performs a distance measure in both raster and vector; creating the buffer is merely an extension of that procedure. But because of the utility of the technique and the frequency with which it is applied,

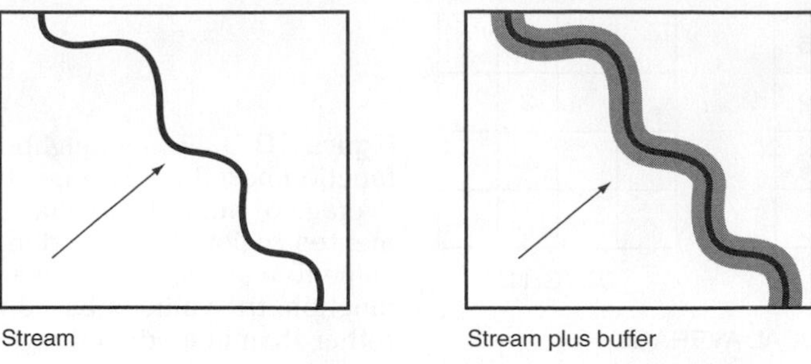

Figure 10.10 Line buffers. Stream and its buffer based solely on a distance measure selected by the user.

most GIS software includes separate routines that allow the buffers to be produced.

Buffering is a matter of measuring a distance from another object, whether a point, a line, or another polygon. In the case of point features we measure out a uniform distance (in raster or vector) in all directions from that point (Figure 10.11). In some cases, we may want to produce a buffer around a linear feature, as in Figure 10.10. Or, returning to Figure 10.11, we could begin with an area and measure a set distance from its outer perimeter. There may even be a requirement for a buffer around a second buffer around still another buffer to produce a **doughnut buffer**, so called because of its resemblance to the pastry (also shown in Figure 10.11). This procedure is relatively simple in raster because each buffer is simply a number of grid cell distances beyond the existing feature or the previously created buffer.

In vector data structures, however, we must explicitly encode the topological information for each polygon we produce. In particular, we are required to provide topological information about the connections between polygons by explicitly placing nodes at the beginning and end of each bordering line segment.

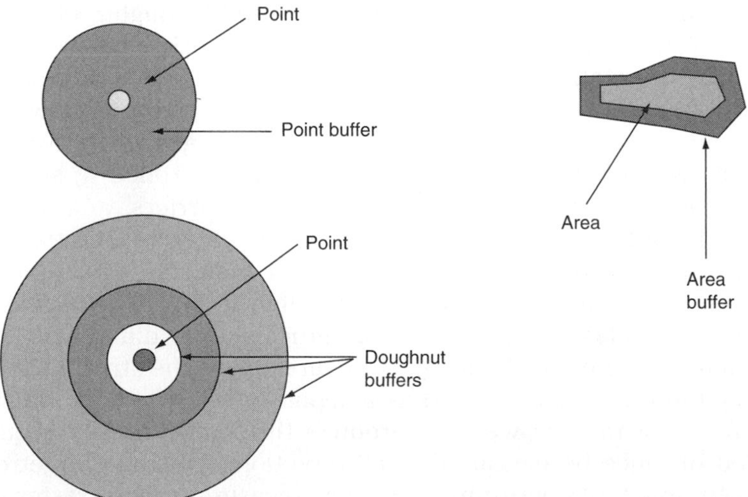

Figure 10.11 Types of buffers. Point, area, and doughnut buffers.

A doughnut buffering procedure attempts to create an island polygon that is not connected explicitly to the neighboring polygon. The difficulty of producing a doughnut buffer in vector is largely a function of the data model used, but by carefully following the software instructions, you can generally obtain one. You may want to experiment on a test database before you try to produce a doughnut buffer on a real database.

Although you have seen that buffers can be extended outward from a point, line, or polygon, they are not limited to that direction. In fact, buffers can be directed to extend both outward and inward at the same time (Figure 10.12a). In addition, where real-world buffers such as setbacks are used—for example, the setback from the center of a suburban street to some portion of a homeowner's property that is available to the city for lighting and utility work—buffers can easily be designed using a GIS (Figure 10.12b).

Now you have seen that buffers are essentially distances measured in any or all directions from point, line, and area objects. But we have not mentioned one of the most pressing problems related to creating buffers—that of distance. How big should the buffer be? A buffer is not worth much if you have no purposeful guidelines as to how large it should be. While circumstances and settings differ, and the objects around which buffers are created change, the basic concept of buffer size is inextricably linked to its purpose. Based on this idea we can create a general set of rules for buffer size decisions.

Some buffers are designed to indicate that around a given entity, for an unknown and perhaps unknowable distance, lies a region that needs to be protected, studied, guarded, or otherwise afforded special treatment. This scenario is not uncommon. Many real-world buffers are just as arbitrary as their digital counterparts. Yet in both cases, there is usually an underlying but often nonarticulated heuristic behind our choice of a particular arbitrary distance. Contractors frequently build buffers around construction sites to protect passersby from falling debris and heavy machinery. The boundaries of areas believed to be contaminated by poison gas, nuclear disaster, and hazardous materials spills are generally imposed by governmental or law enforcement agencies. But like the areas cordoned off around spaceships in the 1950s science fiction movies, quite often these danger zones are only guesses—that is, **arbitrary buffers**. Most often the decisions are based on guesses or anecdotal information of unknown source. They do share one characteristic, however, and it illustrates a time-honored approach—when in doubt, err on the side of safety. In other words, all these arbitrary buffers seem to be larger than they actually need to be. You probably notice this when you

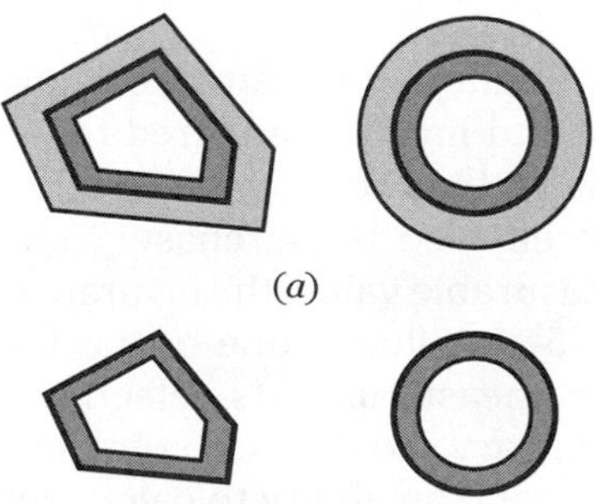

Figure 10.12 Bidirectional buffers and setbacks. Although buffers can be extended on either direction of the object (*a*), they can also be directed toward the center of polygonal features to model setbacks (*b*).

are forced to walk or drive around a construction buffer, or when barriers surrounding a relatively small parking lot pothole force you to find another place to park. The extra distance that inconveniences the public is the added margin of safety for an arbitrary buffer.

Buffer distances can also be based on nearly any of the measures and reclassification procedures you have seen so far, whether two-dimensional or three-dimensional. For example, we could create a second type of buffer based on functional rather than Euclidean distance from the target object. This would be a **causative buffer**—one based on a priori knowledge about the area within which the buffer is produced. Let's say, for example, that we are putting a buffer along a stream to indicate the potential for contaminating soil on either side of the stream. We know, however, that the soils on one side of the stream are heavily clay rich, whereas the soils on the other side are richer in sand. Because contaminants will pass through sand faster than through clay, the buffer must be based on the frictional or impedance quality of the clay soil. The result will be a buffer smaller in Euclidean diameter on the clay side and larger on the sand side, reflecting the difference in permeability of the soils. The use of friction surfaces and barriers is common when producing buffers, because they offer some justification for the buffer width we choose. However, as you remember from our discussion of friction surfaces, they tend to be poorly understood. Thus, a buffer assigned on the basis of a friction or barrier impedance value may be of little more use than an arbitrary buffer of simple design.

A buffer can also be based on visibility measures. In that case, the buffer is selected not on the basis of an arbitrary value or on a poorly defined friction value, but rather on a definable, measurable value—a **measurable buffer**. This is the third basic type of buffer we can employ. The measurements are not arbitrary but are quite accurate, being based solely on measurable phenomena. A common approach is to combine the second and third methods of buffering, relying on a measurable phenomenon whose interaction with the buffer area is poorly understood. For example, we know that trees work quite well as filters of contaminating materials along stream corridors. We know also that the more trees, the more effective the filter will be. So we could measure the density of vegetation along a stream corridor and then reclassify it to produce a number of neighborhoods, each representing a different class of tree density. Next, we could use these different tree densities as surrogates for friction surfaces, which we would then employ to modify the amount of buffering. However, we are not absolutely sure how directly the tree density is related to the amount of movement of contamination through the vegetation and into the streams. Thus, we have some measurable values, and they can be logically applied, but we don't know the exact relationships among them. This is often the case when buffers are employed.

When the fourth and final method of buffering is employed, however, legal or otherwise mandated measures exist and must be adhered to—a **mandated buffer**. For example, if you build a home within the 100-year flood zone in some communities, you are not likely to be allowed to purchase flood insurance. Although the 100-year flood zone is a measurable value, the insurance companies could just as easily have selected the 75-year flood zone or the 150-year flood zone. In other words, the value itself is measurable; its selection as opposed to other potentially useful values is arbitrary. To develop a buffer of this kind usually requires a database of the terrain and an ability to calculate the volume of water that would fill the stream floodplain if a flood of a size not likely to occur more than once in a century were actually to occur.

But other mandated buffers can be employed. We are told how close to a fire hydrant we can park, and how much of our frontage property actually belongs to the local community. Building codes specify distances around utility areas and between buildings; conservation organizations obtain easements on properties; legal ownership buffers are established around rail lines and power corridors, and so on. You might be able to think of a number of others. In each case, there is a legal reason for placing the buffer a set distance from other objects.

No matter which of the four basic buffer methods (arbitrary, causative, measurable, and mandated) is used, there is always the possibility that the buffer will not be the same along the entire length of linear objects or on all sides of a polygon. Such differences, exemplified by our different buffer widths along the stream based on soils, create a general class of buffers called variable buffers (Figure 10.13). The variable buffer can be based on friction, barriers, or any of our other neighborhood functions. It can be selected arbitrarily or on the basis of a measurable component of the landscape, or it can be mandated by law. In each case, special procedures must be invoked in the creation of the buffers. For vector, the nodes between line segments can most often be used to dictate buffer differences along the line segments. In raster, the grid cells need to be coded selectively so that a buffer for each group of grid cells can be established; these buffers most likely will be combined later into a larger buffer map layer.

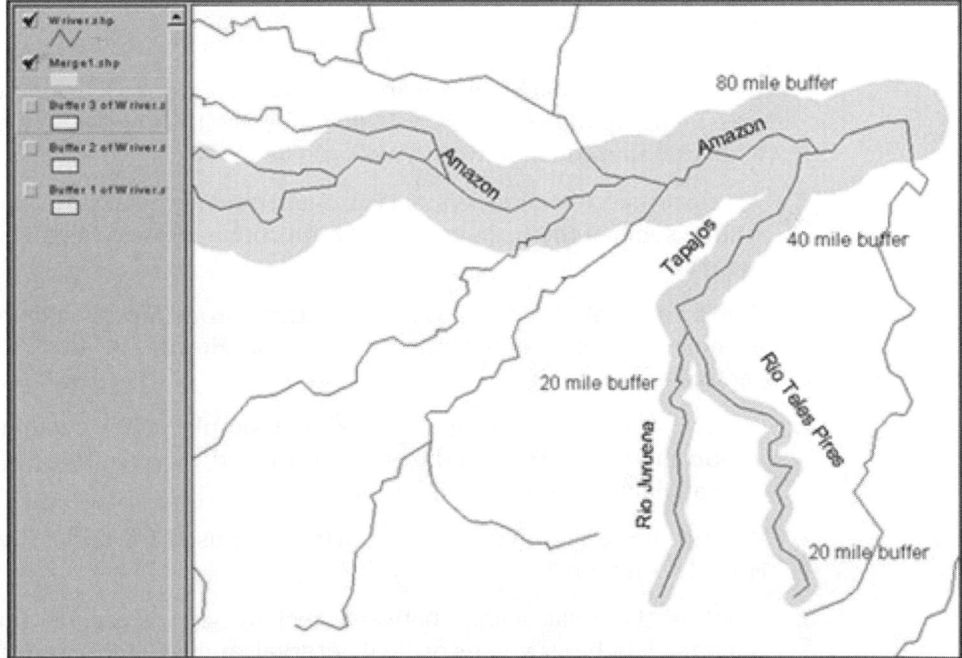

Figure 10.13 Variable buffer. Here you can see that three different buffer sizes are used to indicate the difference between trunk streams and tributary streams. Each of these different stream types has a different code, allowing different sized buffers to be created.

Buffers are useful methods of reclassifying the landscape and are common features in many GIS analyses. The fundamental problem with buffers is that they frequently require us to know more about the interactions of our landscape's elements than we do. You should always attempt to overcome this obstacle by seeking out the best available knowledge about each situation before proceeding. The more you know, the more confident you will be about the distance you have buffered. If you have a legally mandated buffer distance, you need only be sure that your buffer corresponds to that mandate. But if your knowledge about the interactions of your landscape elements is weak, or if you have no knowledge on which to base selection of the buffer distance, your best bet is to make the buffer larger rather than smaller.

Terms

aggregation
arbitrary buffer
block function
buffering
causative buffer
constrained math
directional filter
doughnut buffer
filters
focal function
high-pass filter
line dissolve
local
locational information
low-pass filter
mandated buffer
map algebra
measurable buffer
neighborhood functions
overlay
position information
range-graded classifications
roving window neighborhood functions
shape information
size information
static neighborhood functions
zonal function

Review Questions

1. What two technological advances have heightened the importance of classifying spatial data? Why is it important that we view classification with a measured amount of cynicism?

2. Why are classification systems for earth-related data so varied? After stating some of the reasons, identify the common theme that runs through them all.

3. Give an example of a simple reclassification procedure that aggregates two or more classes together into a single class. How is this done in raster? In vector?

4. When one is reclassifying areas using a simple raster system, why is it important to keep track of the order in which the grid cells are reclassified? Give an illustrative example.

5. What is a line dissolve operation? How is it used for reclassifying areas? How does it work?

6. What is the relationship between reclassifying areas on the basis of nominal data types versus ordinal, interval, and ratio data types? What are range-graded data? How are they used in the reclassification process?

7. What are filters? How are they used for reclassification processes? What's the difference between static and roving window filters?

8. What is the difference between a high-pass and a low-pass filter? Describe how both work and give their respective purposes. What is a directionally biased filter? How does it work to reclassify features?
9. What are neighborhoods? Describe how neighborhood conditions can be used to classify a target grid cell or polygon.
10. Define the following types of map algebra operations: *local, focal, zonal*, and *block*. Provide a diagram illustrating each of these and label its parts, including the input values and output grid locations and values.
11. Describe a process of defining neighborhoods on the basis of a single attribute. Provide some concrete examples of how this could be applied in decision making.
12. How could we use such mathematical procedures as average, maximum, and diversity to define a neighborhood? Create a table that shows a number of possible math-based measures of neighborhoods and an example of how each could be used. Describe how any one of these (e.g., diversity) actually works. Diagram the process.
13. Give examples of neighborhoods defined by a combination of possible methods (size, shape, diversity, etc.). Explain how this will be done: what steps are taken, how they are to be performed, and in what order. If you have access to a GIS, you may want to create a simple database to try this out.
14. What are buffers? How are simple buffers based only on distance calculated? What is a doughnut buffer? Why is it so difficult to execute in vector GIS?
15. Describe how buffers can be produced on the basis of functional distance, neighborhood classifications, and so on. What is a variable buffer? How is this done in vector? In raster? When would it be used?
16. What are the four basic methods of determining buffer distance? Describe them.

References

Anderson, J.R., E.E. Hardy, J.T. Roach, and R.E. Witmer, 1976. *A Land Use and Land Cover Classification for Use with Remote Sensor Data*. Geological Survey Professional Paper 964, U.S. Geological Survey. Washington, DC: U.S. Government Printing Office.

Bailey, R.G., P.E. Avers, T. King, and W.H. McNab, Eds., 1994. *Ecoregions and Subregions of the United States* (map). Washington, DC: USDA Forest Service. 1:7,500,000. With supplementary table of map unit descriptions, compiled and edited by W.H. McNab and R.G. Bailey.

Cowardin, L.M., V. Carter, F.C. Golet, E.T. LaRoe, 1979. *Classification of Wetlands and Deepwater Habitats of the United States*. U.S. Department of the Interior, Fish and Wildlife Service, Washington, DC, 131pp.

DeMers, M.N., 2002. *GIS Modeling in Raster*. New York: John Wiley & Sons.

Forman, R.T.T., and M. Godron, 1987. *Landscape Ecology*. New York: John Wiley & Sons.

Klemas, V.V., J.E. Dobson, R.L. Ferguson, and K.D. Haddad, 1993. "A Coastal Land Cover Classification System for the NOAA Coastwatch Change Analysis Project." *Journal of Coastal Research*, 9(3):862–872.

Küchler, A.W., 1956. "Classification and Purpose in Vegetation Maps." *Geographical Review*, 46:155–167.

Lillesand, T.M., and R.W. Kiefer, 1995. *Remote Sensing and Image Interpretation*. New York: John Wiley & Sons.

Omernik, James M., 2004. "Perspectives on the Nature and Definition of Ecological Regions," *Environmental Management*, 34 (Supplement 1):s27–s38.

Ortega-Huerta, M.A., and K.E. Medley, 1999. "Landscape Analysis of Jaguar (Panthera Onca) Habitat Using Sighting Records in the Sierra de Taumalipas, Mexico," *Environmental Conservation*, 26(4):257–269.

Sauer, C.O., 1921. "The Problem of Land Classification." *Annals, Association of American Geographers*, 11:3–16.

Soil Survey Staff, 1975. *Soil Taxonomy: A Basic System of Soil Classification for Making and Interpreting Soil Surveys*. U.S. Department of Agriculture, Soil Conservation Service, Agriculture Handbook No. 436. Washington, DC: U.S. Government Printing Office.

Tomlin, C.D., 1990. *Geographic Information Systems and Cartographic Modeling*. Englewood Cliffs, NJ: Prentice-Hall.

United Nations Educational, Scientific, and Cultural Organization, 1973. *International Classification and Mapping of Vegetation*. Paris: UNESCO.

U.S. Department of Agriculture, Forest Service, 1993. *National Hierarchical Framework of Ecological Units*. Washington, DC: U.S. Government Printing Office.

U.S. Geological Survey, National Mapping Program, 1992. "Standards for Digital Line Graphs for Land Use and Land Cover, Technical Instructions, Referral ST0–1–2." Washington, DC: U.S. Government Printing Office.

Walter, H., and E. Box, 1976. "Global Classification of Natural Terrestrial Ecosystems." *Vegetation*, 32:75–81.

Wilen, B.O., 1990. "The U.S. Fish and Wildlife Service's National Wetlands Inventory." *U.S. Fish and Wildlife Service, Biological Report*, 90(18):9–19.

CHAPTER 11

Statistical Surfaces

Thus far we have concentrated on point, line, and area features, but we have made frequent reference to surface features as well. You've seen how the latter can be modeled in the computer environment, and have just found that they can be used to modify our ideas of what a neighborhood is and how it operates. Surfaces are used frequently to model the friction or impedance that we encounter as we travel. They can be grouped based on their shapes and orientation as well as the effects these have on visibility and on movement of fluids, and can be classified into recognizable landforms like valleys, hills, and watersheds. But the surfaces we will encounter in GIS are not always topographic. There are surfaces based on economics, population, barometric pressure, temperature, precipitation, and a host of other quantitative information. Our geographic filter can distinguish surfaces that are **continuous** or **discrete**, **smooth** or **rough**, physical or cultural. In other words, our definition of surfaces must expand to include data of any type that either exist or can be assumed to exist as changing values throughout an area.

The topographic surface is the most recognizable of the statistical surfaces, and throughout this chapter you will see frequent references to topography as the classic type. Topographic features are developed because at locations throughout the areas they occupy there are measurable objects, in this case elevation values. Because topographic elevation values occur everywhere, we say that the surface is a continuous surface. However, as we try to record this information with a view to producing reasonable and quantifiable descriptions, we find a major dilemma, not unlike that associated with trying to describe trees in a forest or blades of grass in a prairie—there is so much data that we simply cannot record the complete set. And so, as before, we must produce a meaningful sample of the elevation values from which we can reconstruct the essence of the topography. There is a strong similarity between the properties we sample for topographic surfaces—like barometric pressure, temperature, and humidity—and other continuous surfaces. These also occur everywhere, but we cannot record data for them at every location. Instead, we select locations to represent the distribution. In this chapter we will learn how this is done to produce the most reliable results.

But although continuous surfaces present a particular set of problems for the GIS professional, many other spatial variables do not occur continuously everywhere; rather, they are found as discrete objects at specific locations.

Because they occur with a very high frequency, or because we want to record them for very large regions, however, these discrete data must also be sampled. In some cases, as with continuous data, we will take samples at specific point locations, whereas in other cases we must collect data for whole areas at a time. For example, we would surely find as we travel across the United States that there are differences in human population numbers for each county. Rather than taking the time to tour each county, locating individual people at every stop, we simply add up the number of people for each of the 3,000 or so counties and use them as point samples. In both cases, we can produce maps that resemble the contours of a topographic map by assuming that they occur everywhere. In this way we can produce either a contour-like map or a fishnet map that models the trend or form of the distribution. And although we know that the objects we measure are not continuous, it is useful to employ these techniques for the sake of ease in communicating the shapes and patterns of distribution. In addition, we will see how these techniques can be used to predict distributions in places we have not sampled.

LEARNING OBJECTIVES

When you are finished with this chapter you should be able to:

1. Define and describe the statistical surface and provide concrete examples of how the concept applies to such data as elevation, population, temperature, and disease occurrences.
2. Explain and provide concrete examples of continuous and discrete statistical surfaces.
3. Describe how statistical surfaces can be represented using isarithms, and tell how the method of sampling data changes how statistical surfaces are represented.
4. Explain how statistical surfaces can be represented by regular and irregular lattices and how these can be converted to other forms, such as discrete altitude matrices, digital elevation models, and TIN models.
5. Explain the problems of control points for creating reliable statistical surfaces, including number and placement of such control points. Describe the differences between control point placement for isometric and isoplethic surface maps.
6. Describe and provide a concrete example of the use of trend surface analysis.
7. Explain and perform the process of spatial sampling for surface data.
8. Describe the differences between isometric and isoplethic isarithms and evaluate the advantages and disadvantages of each for particular spatial data.
9. Explain how a continuous surface is represented in raster format.
10. Understand and describe the different methods of interpolation, their uses, and their advantages and disadvantages.

11. Explain the four factors to be considered in any form of interpolation and understand how problems of each type can be solved.
12. Describe the concept of spatial autocorrelation and explain its role in exact forms of interpolation. In answering this question provide a sample semivariogram and label it.
13. Diagram and explain the idea of selecting a study area larger than the project area when interpolation of surfaces is involved in the analysis.

WHAT ARE SURFACES?

Surfaces are features that contain height Z values distributed throughout an area defined by sets of X and Y coordinate pairs. The Z values need not be restricted to elevation values. Instead, any measurable values (i.e., in ordinal, interval, and ratio data scales) that can be thought of as occurring throughout a definable area can be regarded as comprising a surface. The general term is **statistical surface**, because the Z values constitute a statistical representation of the magnitude of the features under consideration (Robinson et al., 1995). Perhaps because the statistical surface extends our geographic filter to include such values as population, salary, animal densities, and barometric pressure, it is often considered to be among the most important cartographic concepts.

Behind the statistical surface is the idea that all the measured values (Z values) either occur continuously across the area of interest or can be assumed for the sake of mapping and modeling to occur continuously across it. Statistical surfaces that employ data that occur to some degree at every location inside the study area, such as elevation data, are called continuous (see Figure 2.3). Those that occur only as individuals but with some difference in numbers per unit area, such as the number of houses per square mile in each neighborhood, are called discrete (see Figure 2.3). Conceptualizing the statistical surface for these two data types can be tricky, so we will examine them separately.

"Continuous" surface data are said to have an infinite number of possible points that can occur at any location in the area. That is, it is possible to obtain a measurement, no matter how small, for this variable anywhere within the area in question. We know, for example, that air temperature occurs everywhere. At every location, we could obtain a measure of the temperature with a thermometer. If we did this, we would see that the values change gradually from place to place in a continuous stream. In some cases the temperature changes slowly, perhaps less than a tenth of a degree per 100 meters. We say that these continuous data demonstrate a smooth surface, with little change in statistical information per unit distance. At times, however—when we cross the space between two totally different air masses along a weather front, for example—the temperature values change very abruptly. We call the surface produced by rapidly changing values a rough surface because there is a major change in the statistical data with small changes in distance. In elevational surfaces this is sometimes referred to as rise over reach.

The advantage of continuous over discrete data is the certainty that there are plenty of data to work with. Remember that they are considered to have an infinite number of data points from which we can make our measurements, which poses two closely related problems. First, it is physically impossible, even with the most advanced scientific instruments, to measure an infinite number of locations. Second, even if we could measure everywhere, no computer is capable of handling an infinite amount of data. So defining a continuous surface on the basis of an infinite number of data points must be replaced by a method of modeling that allows us to use samples taken at critical locations. These samples will serve to represent the most significant representative changes in surface as a simplified representation.

SURFACE MAPPING

The statistical surface is most often displayed in analog form as an **isarithmic map**. This is certainly the most common approach for continuous types of surface data. You can envision isarithmic mapping as a series of lines draped over the statistical surface. Each line, called a **contour line** when applied to topography, represents all points that occur at the same elevation. The general term for lines connecting points of equal statistical value is **isarithm**. Figure 11.1 shows a portion of a digital elevation model of the Colorado Rocky Mountains that illustrates the concept of isarithms, which are especially prominent in the upper left corner of the model.

Viewed from the top, isarithms appear as a series of semiparallel lines. If they surround a topographic form, such as a hill, they will be closed and will encircle

Figure 11.1 Digital elevation model (DEM) of a portion of the Colorado Rocky Mountains. Note how the isarithms appearing in the upper left portion of the map are unevenly spaced horizontally to adjust for their vertical sampling.

the feature; if not, they continue until they hit the edges of the map. They allow us to observe specific patterns and shapes based on the changes in elevation topography encountered. We see, for example, that isarithms that cross stream valleys tend to look like the letter *V* pointing uphill or upstream. Steep slopes have more isarithms spaced closer together horizontally than gently sloping surfaces. You can now see how the traditional contour lines can be used to represent neighborhoods of slope as well: steep slopes with closely spaced contour lines; gentle slopes with widely spaced contour lines (Figure 11.2).

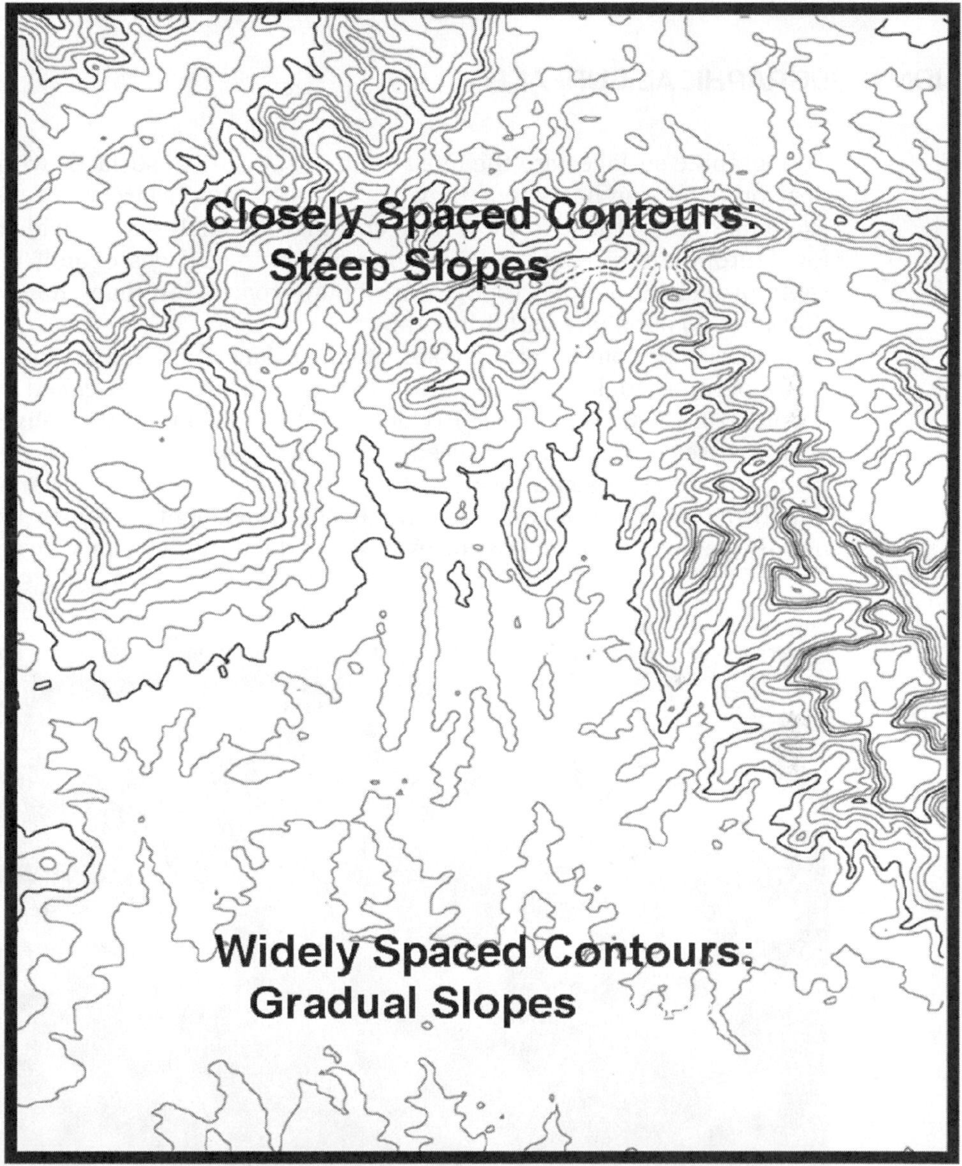

Figure 11.2 Contour spacing and surface configuration. Steep slopes are shown by closely spaced isarithms and gentle slopes are shown by widely spaced isarithms. The distance between lines (the contour interval or class interval) is based on elevation differences, not horizontal differences.

We can also observe another important aspect of these isarithms. It is no accident that the vertical distance between isarithms is the same for any two pairs of lines. The vertical distance between lines was preselected to allow the reader to more easily understand what the lines represent. Each contour map has its own vertical distance between isarithms, depending on whether there is a rapid or a slow change in elevation. This preselected value, called the **contour interval**, is used to divide or quantize the change in Z value by choosing a uniform amount of elevational change. Thus, the person viewing the map knows that each contour line indicates a set difference in elevation.

NONTOPOGRAPHICAL SURFACES

Topographic surfaces are the most common statistical surfaces with which we deal on a regular basis, but they are one of a much larger set. We need to examine some of these to allow us to consider the analyses available to surfaces for nontopographic data. Some nontopographic surfaces exhibit many of the same properties as their topographic counterparts, but many have properties that they do not share with topography.

Examples of nontopographic statistical surfaces that share most properties of topography include temperature (whose isarithms are called isotherms), barometric pressure (isobars) (Figure 11.3), and humidity (isohyets). These surfaces are continuous, measured on the ratio scale, and are typically modeled using surface representations. Another nontopographic surface variable is time (isochrones), which allow us to model the movement of things like species, floods, glaciers, populations, or even ideas. Like the previous three examples

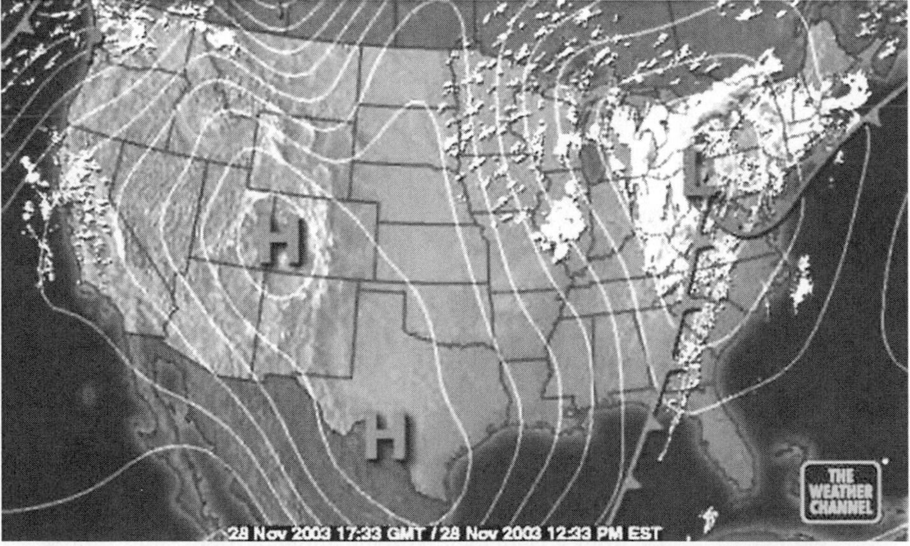

Figure 11.3 Nontopographic statistical surface. This weather channel map shows lines of equal barometric pressure (isobars) and is an example of a continuous, nontopographical statistical surface.

(temperature, pressure, humidity), these represent continuous, interval, or ratio-scale data, although they are not directly measurable on the earth itself.

Surface data that do not resemble topographic data but can be represented in many of the same ways include such things as family income, crime levels, housing costs, and cancer rates. The primary difference between these and our previous examples is that they do not occur continuously across the earth. In short, they are not continuous. They are, however, measured on the interval or ratio scale. Examples such as surveys of the level of people's interest in sports, value of scenic areas, or importance of college degrees are noncontinuous because the people from whom the data are collected do not occur continuously across the earth and are more apt to be measured at scalar measurement scales. In short, they fit two characteristics of the definition of statistical surfaces in that they are "assumed" to be continuous and are measured at the interval or ratio scale.

SAMPLING THE STATISTICAL SURFACE

Because continuous surfaces have an infinite number of possible *Z* values, it is necessary to sample them for input to the GIS. Discrete surfaces can also have many data points requiring sampling. Because of this there are two major ways to obtain samples of *Z* surface data. The first employs selected point locations to sample, in which case the isarithmic map produced is said to be **isometric**. This is the method most commonly applied to continuous surfaces such as elevational, barometric pressure, temperature, and other natural surface data. However, we may also wish to work with discrete surfaces whose data are aggregated by small areas such as population by county. Although we know that these data are discrete, we can treat them as if they were continuous. By assuming that each of these areas represents a point sample, we can produce the same kind of isarithmic map that would result if the data had been gathered at point locations. This type of isarithmic map, called **isoplethic**, requires us to determine where within the areas we should place the points. You have already studied how centroids or centers of gravity can be calculated and used in determining such placement (Chapter 8).

In these isoplethic maps, we choose the data collection point from within each area by employing either the center of area or the weighted mean center. But in isometric mapping we have to decide where we are going to select the data sampling points in the first place. There are two general approaches. The first employs a **regular sampling procedure** or **grid**, often called a **regular lattice** sampling. The points are later connected to form a triangular lattice **structure** much like trellises used in gardens to support climbing plants (Figure 11.4a). The regular lattice sampling approach has the advantage of simplicity in that there is no need to decide where the points are to be placed. A set identical distance in both the *X* and *Y* direction is selected, and the points are measured where the *X* and *Y* lines cross. This method incorrectly assumes that by sampling regularly throughout a surface, we can adequately represent both smooth and rough surface features. A rough surface implies a rapid change in *Z* value with distance. We say that there is more information content per unit area because every change in elevation is considered to be a piece of information about the surface. The

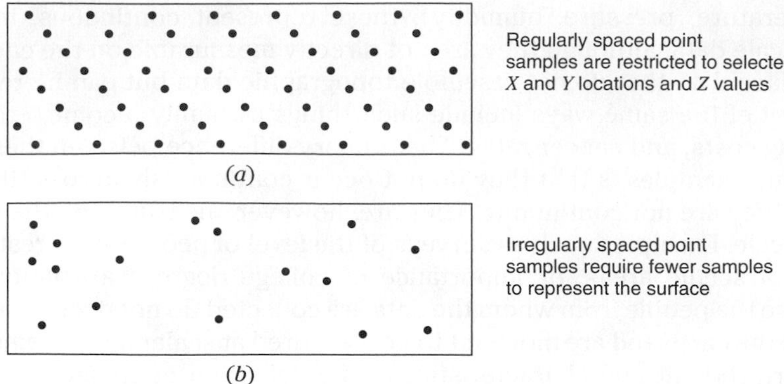
Regularly spaced point samples are restricted to selected X and Y locations and Z values
(a)
Irregularly spaced point samples require fewer samples to represent the surface
(b)

Figure 11.4 The lattice. (*a*) Regular and (*b*) irregular lattices of points. The points can be sample locations themselves or points as they are represented in the GIS database.

more information contained in the surface, the more need there is for sampling. Because of this we will obtain a better understanding of surface shape if we use an alternative method of sampling called the **irregular lattice** (Figure 11.4b).

In an irregular lattice we can determine the amount of surface sampling on the basis of prior knowledge of how rugged or smooth it is. We are not restricted in either the *X* or the *Y* direction or by the number of point samples we can make within any given area. One might assume that, in the absence of restrictions, we are likely to make more samples using this method. This is not necessarily the case. Most often we can obtain a good model of the surface with fewer data samples because we can reduce the number of samples for relatively smooth surfaces well below the number that would be needed for a regular lattice. So while we increase sample density for areas with high information content, we can decrease the sample density for areas with less surface information. Increased sample density for smooth surfaces adds no relevant information; it just adds unnecessary time and effort as well as increased data storage.

THE DEM

Now that you have procedures for sampling surface data, you need to understand how a surface could be represented inside the computer, both in raster and in vector. In Chapter 5 you learned about the TIN model, the basic vector data structure for representing surfaces in the computer. The TIN model is one of a number of methods of storing *Z*-value information, creating a group of products collectively called **digital elevation models (DEMs)**. Such methods are based either on mathematical models or on image models (including TIN) designed to more closely approximate how they are normally sampled in the field or represented on the paper map (Mark, 1978). Although mathematical models are quite useful, the currently available DEMs are most often based on image models. We will therefore focus on image models and leave the more advanced student to pursue the mathematical model representations. Image models come in two general types: those based on points and those based on lines.

Image models of surfaces based on lines are nearly the graphical equivalent of the traditional method of isarithmic mapping. In many cases such models are produced by scanning or digitizing existing contour lines or other isarithms. The purpose is to extract the form of the surface from the lines that most commonly depict or describe that form. Once input, the data are stored either as line entities (each line with a common elevation attribute, and each line segment connected by points) or as polygons of a particular elevational value. Because it is not particularly efficient to calculate slopes and aspects and to produce shaded relief output from such data models, it is more common to convert them to point models, treating each point connecting each line segment as a sample location with an individual elevation value. This produces a **discrete altitude matrix**.

The discrete altitude matrix is a point image method that represents the surface by a number of points, each containing a single elevational value. Much like the methods of sampling topography in the field, the altitude matrix uses a number of data points sampled from stereo pairs of aerial photographs, usually employing an analytical stereoplotter that locates and measures elevational data based on the displacement of the two photographs (Kelly et al., 1977). The analyst can use a regular lattice or an irregular lattice, but because the regular lattice oversamples where the topographic information is minimal and undersamples where it is abundant, the irregular lattice is preferred. Progressive sampling using automatic scanning stereoplotters that increase the number of samples in highly changing topographic regions offers efficient execution (Markarovic, 1973).

Irregular lattices can be converted to TIN modes under two different approaches. The first is to use the data points in the irregular lattice itself as the basis for the triangular facets in the TIN. This requires little additional input, instead allowing the GIS to create the triangular facets. Alternatively, some form of interpolation among data points can be used. The product of such an interpolation approach is a set of predicted data points that can be used to create the TIN facets. Although interpolation allows the addition of data points to the altitude matrix and, ultimately, to the TIN, predicted values are not as accurate as measured data points. Thus, any model produced from such a technique has added error. Even the order in which DEM data points are selected for insertion into the TIN can have an impact on the overall quality of the spatial model (Little and Shi, 2003).

DEMs are readily available for many parts of the world as altitude matrices with grids of 63.5 meters obtained from 1:250,000 scale topographic maps; they are becoming available in larger scale formats such as can be obtained from 1:25,000 maps and aerial photography (Appendix A). Among the advantages of using DEMs derived from interpolation techniques that create a regular matrix is ease of input to raster GIS. DEMs based entirely on an irregular lattice and input to a raster GIS will have to be interpolated.

INTERPOLATION

Because *Z* values for continuous surfaces are samples, whether they are topographic, economic, demographic, or climatic, we must be able to create models that depict with relative accuracy the features we observe. For example, we need to be able to convert from point elevation samples to a display that uses isarithmic lines to represent the surface form. We must also be able to produce visual representations of other types, such as fishnet maps and shaded relief maps. And, of course, we need to have the additional capability to calculate slopes, aspects, and cross sections and to predict unknown elevations for objects that occur at places for which elevational data do not exist. Interpolation provides much of what is needed to perform these operations.

The process of interpolation is conceptually very simple, but it requires an a priori assumption. Before we begin working with surface data, let's look at a mathematical examination of interpolation based on the idea of a mathematical progression (progressions are numbers that occur in some identifiable sequence). The mathematical sequence

$$1 \quad 2 \quad 3 \quad 4 \quad 5 \quad 6 \quad 7 \quad 8 \quad 9 \quad 10$$

is based on starting at 1 and adding 1 at each successive step. It is called a linear or arithmetic sequence because it increases by the same amount each time, and each increment is developed by simple arithmetic. Other arithmetic series could be

$$10 \quad 20 \quad 30 \quad 40 \quad 50 \quad 60 \quad 70 \quad 80 \quad 90 \quad 100$$

which uses 10 as the additive factor, or

$$1{,}000 \quad 900 \quad 800 \quad 700 \quad 600 \quad 500 \quad 400 \quad 300 \quad 200 \quad 100$$

in which 100 is subtracted at each step. Still other arithmetic series may be created using multiplication or division.

Linear Interpolation

Within these simple series we can easily identify the a priori assumption mentioned earlier that the numbers at each successive step can be determined by a simple mathematical procedure. If we can discern the mathematical procedure, we can also insert any missing values. So, for example, if we are told that the following simple series is missing two values:

$$30 \quad 40 \quad 60 \quad 70 \quad 80 \quad 100$$

we can infer that the series is based on addition and that the missing values lie between 40 and 60 and between 80 and 100. Because this is a linear series, we can see that the missing numbers are 50 $(40 + 10)$ and 90 $(80 + 10)$. This is nearly identical to **linear interpolation**, the method of assigning values between points of known elevation spread over an area.

Take the simple example illustrated in Figure 11.5. We are looking at a single line transect of data points that range between 100 feet and 150 feet in elevation. If we assume that the surface changes in a linear fashion, just as in a simple series, and we have a linear progression, it is obvious that four numbers, spaced equal distances apart, can be interpolated between 100 feet and 150 feet. By segmenting the distances between these two points into five equal units, we

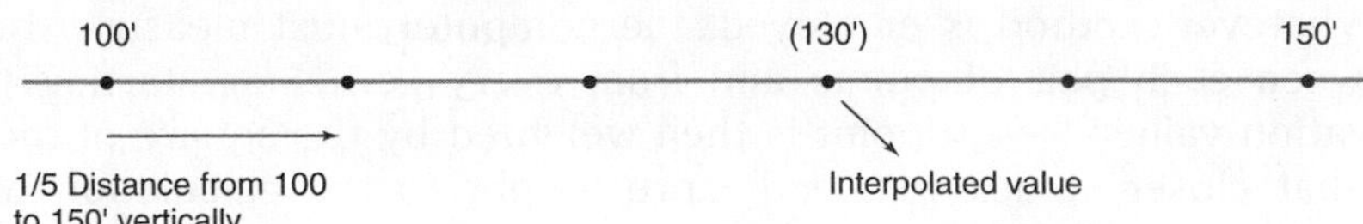

Figure 11.5 Linear interpolation. Predicting missing values between 100 and 150 feet assuming a linear arithmetic sequence.

can treat the distances as surrogates for changes in elevation. Therefore, at each unnumbered segment we need only insert a 10-foot elevation to obtain the missing values. If we do this for an entire area rather than just for a single line transect, we can obtain values for all of the 10-foot contour interval locations. By drawing smooth lines to connect these segments, we can create contours of 100, 110, 120, 130, 140, and 150 feet. In other words, we are able to create an isarithmic map (contour map in this case) that allows us to visualize the elevational features.

Thus far we have been assuming that surfaces change linearly, but this rarely if ever happens. In some cases, the series is more logarithmic; in others it is predictable only for small portions of the surface. Under such circumstances, the results of a linear interpolation method will not accurately depict the surface. And there are other needs for surface information that may require us to determine the overall trend in surface, rather than trying for the most accurate depiction. Some of these techniques can get complicated mathematically, and it is not the purpose of this book to examine them in painful detail. Instead, we will stick to the conceptual level in examining a few of the nonlinear methods of interpolation to get an idea of how they might best be used in GIS.

Methods of Nonlinear Interpolation

Nonlinear interpolation techniques eliminate the assumption of linearity called for in linear methods. We will look at three basic types of nonlinear interpolation methods: weighting, trend surfaces, and Kriging. There are advanced texts that cover the many additional approaches in great depth (Burrough and McDonnell, 1998; Davis, 1986). Here we consider only the general types.

Weighting methods assume that the closer together elevation sample points are, the more likely they are to be associated to one another. With the exception of cliffs and other drastic changes in elevation, we observe that elevations within a meter of where we stand are fairly close to our elevation. As we move further away the chances of this are lessened. And at a certain point, for example, 1,000 meters or more, the chance that elevation values would be similar is virtually nonexistent. Because close values are more closely related than farther ones, we should place more emphasis (more weight) on nearby values than on distant ones. So to accurately predict the topography, we need to select points within a reasonable neighborhood that demonstrate this surface similarity. This is done by a number of search techniques, including defining the neighborhood by a predefined distance from each point, predetermining the number of sample data points, or selecting a certain number of points in quadrants or even octants (i.e., one point for each quadrant is used for the interpolation) (Clarke, 1990).

Whatever method is employed, the computer must measure the distance between each pair of points and from every kernel or starting point. The elevation value at each point is then weighted by the square of the distance, so that closer values will lend more weight to the calculation of the new elevation than closer distances (Figure 11.6). There are many modifications of this approach. Some reduce the amount of distance calculations by employing a "learned search" approach (Hodgson, 1989); others modify the distance by weighting factors other than the square (e.g., cubed or higher powers); and some include barriers that simulate coastlines, cliffs, or other impassable features likely to affect the output of an interpolation (Shepard, 1968). The barrier method is especially useful in the development of surface models that can account for these local objects. And as in the case of barriers for functional distance modeling, the interpolation cannot pass through the barrier in its search for neighboring weights and distances.

Under some circumstances we are more interested in general trends in Z surface rather than in the exact modeling of individual undulations and minor surface changes. For example, we might want to know the general trend in population across a country to support demographic research, or whether a buried seam of coal trends toward the surface to indicate how much overburden needs to be removed for surface mining. The most common approach to this type of surface characterization is called **trend surface**. As in the weighting method, in trend surfaces we use sets of points identified within a specified region. The region is based on any of the methods already discussed for weighted methods. Within each region, a surface of best fit is applied on the basis of mathematical equations such as **polynomials** or **splines** (piecewise polynomials). These equations are best described as nonlinear progressions that approximate curves or other forms of nonlinear series. To develop the trend, we examine each of the values in the region and fit it to the mathematical

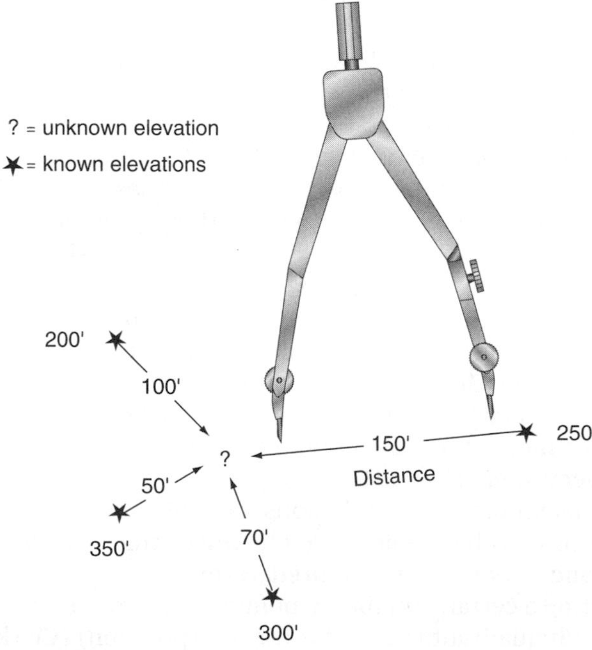

Figure 11.6 Distance-weighted interpolation method. Note how the closer Z values have more weight than those far away. For example, the missing value is more likely to be closer to 350 feet in elevation because of its proximity to that elevation.

equation. Based on the equation of best fit, a single value is estimated and assigned to the kernel. As you have seen, the process continues for other target points or kernels, and the trend surface can then be extended for the entire layer.

The number assigned to the target cell or kernel may be a simple average of the overall surface values within the region, or it may be weighted on the basis of the trend direction. Trend surfaces can be relatively flat, showing an overall trend for the entire layer, or they can be relatively complex. The complexity of equation used will determine the amount of undulation in the surface. The simpler the trend surface, the lower the degree it is said to have. For example, a first-degree trend surface employing a first-order polynomial will show a single plane that slopes across the layer—that is, it trends in a single direction. If it trends in two directions (second-order trend surface), it is said to be a second-degree trend surface (Figure 11.7). If it trends in three directions (third-order trend surface), it is a third-degree trend surface.

Our final method of interpolation, known as **Kriging**, optimizes the interpolation procedure on the basis of the statistical nature of the surface (Oliver and Oliver, 1990). Kriging uses the idea of the regionalized variable (Blais and Carlier, 1967; Matheron, 1967), which varies from place to place with some apparent continuity, but cannot be modeled with a single smooth mathematical equation. As it turns out, many topographic surfaces fit this description, as do changes in the grade of ores, variations in soil qualities, and even a number of vegetative variables. Kriging treats each of these surfaces as if it were composed of three separate values. The first, called the **drift** or structure of the surface, treats the surface as a general trend in a particular direction. Next, Kriging assumes that there will be small variations from this general trend, such as small peaks and depressions in the overall surface that are random but still related to one

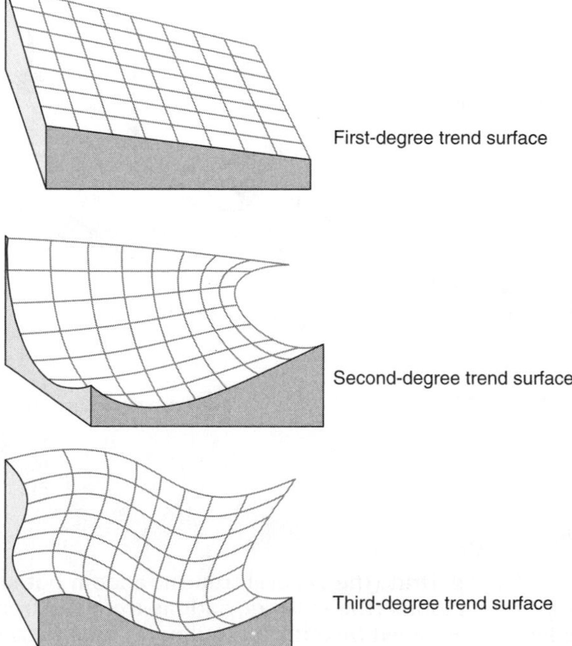

Figure 11.7 Degrees of trend surface. First-, second-, and third-degree trend surfaces based on the complexity of the polynomial equation used to represent the surface.

another spatially (we say they are spatially autocorrelated). Finally, we have **random noise** that is neither associated with the overall trend nor spatially autocorrelated. Clarke (1990) aptly illustrated this set of values by means of an analogy: if we are hiking up a mountain, the topography changes in an upward direction between the starting point and the summit; this is the drift. But along the way, we find local drops denting the surface and accompanied by random but correlated elevations. Also along the way we find boulders that must be stepped over, which can be thought of as elevation noise because they are not directly related to the underlying surface structure causing the elevational change in the first place (Figure 11.8).

Now that we have three different variables to work with, each must be evaluated individually. The drift is estimated using a mathematical equation that most closely resembles the overall surface trend. An expected, elevational distance is measured with the use of a statistical graphing technique called the **semivariogram** (Figure 11.9), which plots the distance between samples, called the **lag**, on the horizontal axis; the vertical axis gives the **semivariance**, defined as half the variance (square of the standard deviation) between each elevational value and each of its neighbors. Thus, the semivariance is a measure of the interdependency of the elevational values based on how close together they are. We then place a curve of best fit through the data points to approximate their locations (giving us a measure of the spatially correlated random component). If you look closely at the semivariogram, you will notice that when the distance between samples is small, the semivariance is also small. This means that the elevational values are very similar and are therefore highly dependent on one another because of their close spatial proximity. As the distance (lag) between points increases, there is a rapid increase in the semivariance, meaning that

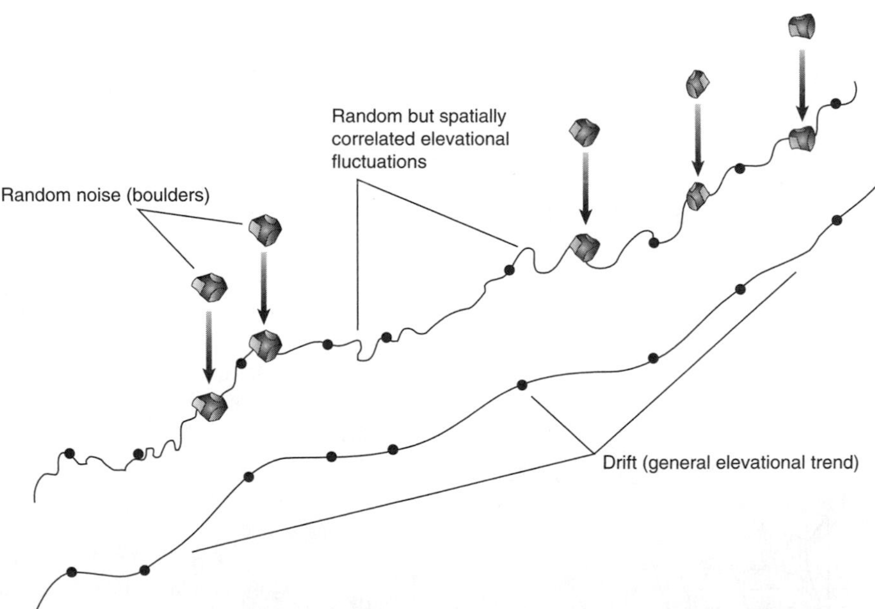

Figure 11.8 Elements of Kriging. Drift (the general trend), random but spatially correlated elevational fluctuations (the small deviations from the trend), and random noise (boulders) exemplified by a hiking trail up a mountainside.

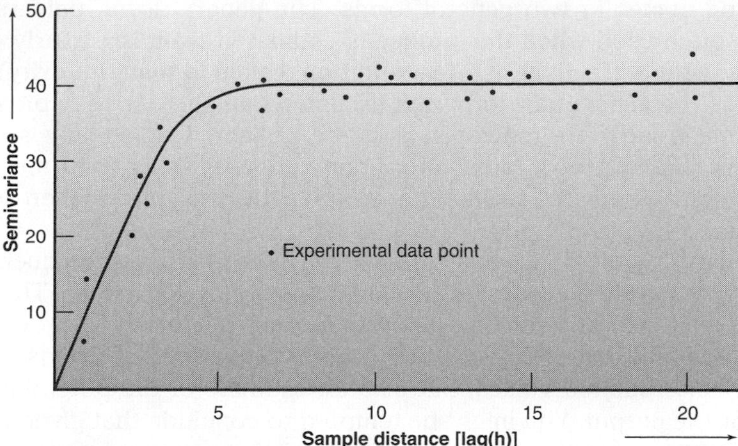

Figure 11.9 Example of a semivariogram showing the relationship between sample points (dots) and the fitted line. Note how within the lag the *Z* values are related to one another, and beyond the lag there is no relationship at all because the dots are too far apart. *Source:* Adapted from P.A. Burrough, *Principles of Geographical Information Systems for Land Resources Assessment*, Oxford University Press, Oxford Science Publications, Monographs on Soil and Resources Survey, No. 12, © 1986; Figure 8.11.

the spatial dependency of values drops rapidly. Eventually, a critical value of lag known as the **range** occurs, at which point the variance levels off and stays essentially flat. Within the range (i.e., from 0 lag to the point at which the curve levels off), the closer together the samples, the more similar their elevation values. Beyond the range, the distance between sites makes no difference; the sites are totally unrelated at any of these larger distances. This information gives us a measure of what neighborhood needs to be applied (e.g., with weighted interpolation techniques) to encompass all points whose elevational values are going to be related.

A third feature of importance in the semivariogram is that the fitted curve does not run directly through the origin. Both mathematically and conceptually, if you have no distance between samples, there should be no variance because the items would essentially be the same point sample. But remember, the curve is an estimate of the locations. The difference between an expected variance value of 0 at 0 lag and the predicted positive value is an estimate of the residual, spatially uncorrelated "noise" variance and is called the **nugget variance**. As Burrough and McDonnell (1998) put it, this nugget variance "combines the residual variations of measurement errors together with spatial variations that occur over distances much shorter than the sample spacing, and that consequently cannot be resolved."

Now that we have defined the three important components of the regionalized variable with the semivariogram, we can determine the appropriate weights to perform interpolation for local regions. Unlike the weighting methods employed before, however, the weights for interpolation within neighborhoods are chosen to minimize the estimation variance for any linear combination of elevation samples. This variance can be obtained directly from the model that created the semivariogram in the first place.

Kriging comes in two general forms. The general form, universal Kriging, is most often used when the surface is estimated from irregularly distributed samples where trends exist (a condition called **nonstationarity**). Punctate Kriging is the elementary form and assumes that the data exhibit stationarity (lack of a trend), are isotropic, and are collected at equally spaced point locations (Davis, 1986). Most often, punctate Kriging is used specifically for finding point estimates from other point estimates rather than for defining surfaces.

Because Kriging is an exact method of interpolation, it frequently gives a relatively accurate measure of the elevations of missing values. This exactness often comes at a cost in time and computing resources, however. Even so, Kriging has another advantage over other interpolation methods—it provides not only interpolated values, but also an estimate of the potential amount of error for the output. You might be tempted to conclude that given its ability to supply an error measure and the exactness of the computed surface, it might be best to use this method all the time. However, this is not the case. When there is a great deal of local noise due to measurement error or large variations in elevation between sample points, it is difficult for this technique to produce a semivariogram curve. Under such circumstances, the results of Kriging are not likely to be substantially better than those of any other nonlinear method.

Regardless of which interpolation technique you use, most GIS software will provide a number of approaches for your use. In TIN models, the process of interpolation is most easily performed by isolating individual points and their associated elevational values and converting them to an altitude point matrix. From that point matrix, the interpolation procedures can easily operate on the selected algorithms, just as described. Actually, the TIN model itself can also perform interpolation (McCullagh and Ross, 1980), but the method is somewhat more involved, and we leave such techniques for an advanced GIS course. In raster grids with missing values, the grid cells that contain elevational values are treated like points located inside each grid cell. Then the distance between grid cells can be measured as described earlier, and the elevational values are interpolated as before. If your GIS does not provide the desired interpolation method, there is usually a method for converting the point matrix to a form compatible with specialty interpolation software. They can be converted back to your original software for later processing. For some basic overviews of interpolation methods, you might consult Lam (1983) and Flowerdew and Green (1992).

Problems of Interpolation

You have now seen a number of methods of interpolation, some more exact than others. When performing any of them, however, four factors need to be considered:

1. The number of control points
2. The location of control points
3. The problem of saddle points
4. The area containing data points

Generally the more sample points we have, the more accurate the interpolation will be. However, there is a limit to the number of samples that can be made for any surface. Eventually one reaches a point of diminishing returns, where having more points doesn't substantially improve the quality of the output and may result in increased computational load and data volume. In some cases, too much data will tend to produce unusual results, because clusters of points in areas where the data are easy to collect are likely to yield a surface representation that is unevenly generalized and therefore unevenly accurate. In other words, having more data points does not always improve accuracy. Figure 11.10 shows that at a certain number of points, the accuracy actually falls.

The number of control points is frequently a function of the nature of the surface: the more complex the surface, the more data points you need. But for important features such as depressions or stream valleys, we should also place more data points to be sure of capturing the necessary detail. Additionally, although the location of sample points relative to one another has an impact on the accuracy of the interpolation routine, the relationship is not perfectly linear (Figure 11.11).

The problem of sample placement is more severe when we consider interpolation from data collected by area to produce an isoplethic map. When the polygons are relatively evenly distributed, it is easiest to use the centroid-of-cell method to locate the sample data points. The center-of-gravity method is most applicable when the sample polygons are either clustered or unevenly distributed. With either method, however, there is a chance that the center will occur outside the sample polygon, especially if the polygons are of unusual shape. When this occurs, the easiest solution is to pull the centroid or center of gravity just inside the polygon at its nearest possible location. This will probably have to be done interactively within the GIS environment.

The **saddle-point problem**, sometimes called the alternative choice problem, arises when both members of one pair of diagonally opposite Z values forming

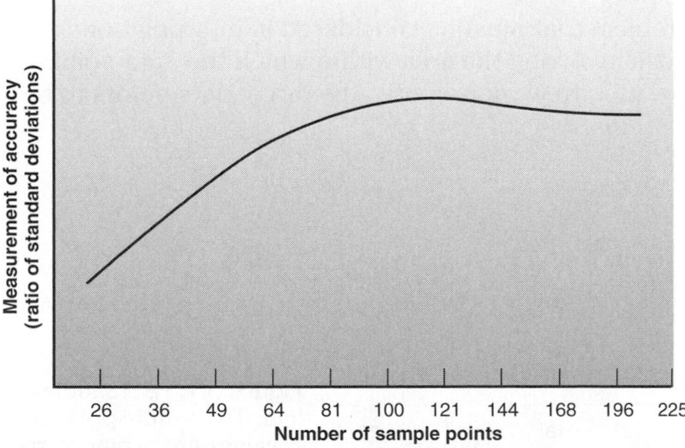

Figure 11.10 Accuracy of isarithmic map versus number of data points. Characteristic curve of a hypothetical relationship between the number of points and the accuracy of an isarithmic map. *Source:* A.H. Robinson et al., *Elements of Cartography*, 6th ed., John Wiley & Sons, New York, © 1995. Adapted from Figure 26.29, page 513. Used with permission.

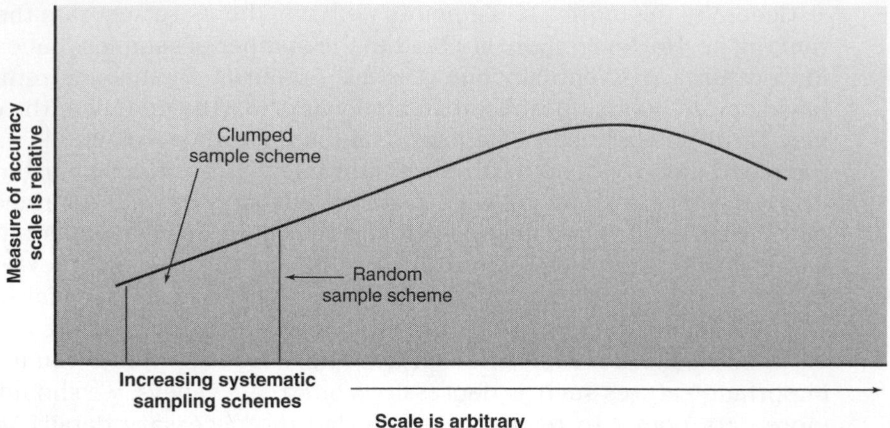

Figure 11.11 Sample point scatter and isarithm accuracy. Characteristic curve of a hypothetical relationship between the distance between data points and the accuracy of an isarithmic map. *Source:* A. H. Robinson et al., *Elements of Cartography*, 6th ed., John Wiley & Sons, New York, © 1995. Adapted from Figure 26.30, page 513. Used with permission.

the corners of a rectangle are located below, and both members of the second pair lie above, the value the interpolation algorithm is attempting to solve (Figure 11.12). This generally occurs only in linear interpolation, because any distance weighting will likely solve the problem. When it does occur, however, the computer software is presented with two probable solutions to the contour line location (Figure 11.12). A simple way to handle this problem is to average the interpolated values produced from the diagonally placed control points, and then place this average value at the center of the diagonal. Then the interpolation can proceed with a single solution to the problem because it now has additional information with which to calculate the interpolated values (Figure 11.13).

The final problem that must be considered in interpolation is a common one in GIS operations involving the area within which the data points are collected. For the interpolation to work properly, the target data points to be interpolated

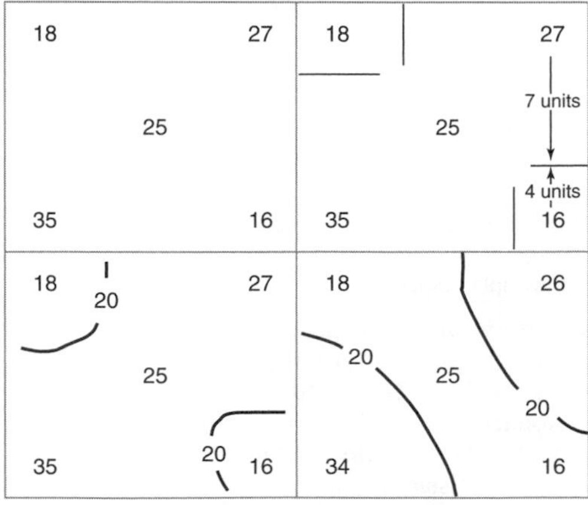

Figure 11.12 Saddle-point problem. The saddle-point problem, where *Z* values are arranged rectangularly. Note the two possible solutions to the same interpolation problem. *Source:* A.H. Robinson et al., *Elements of Cartography*, 6th ed., John Wiley & Sons, New York, © 1995. Adapted from Figure 26.27, page 512. Used with permission.

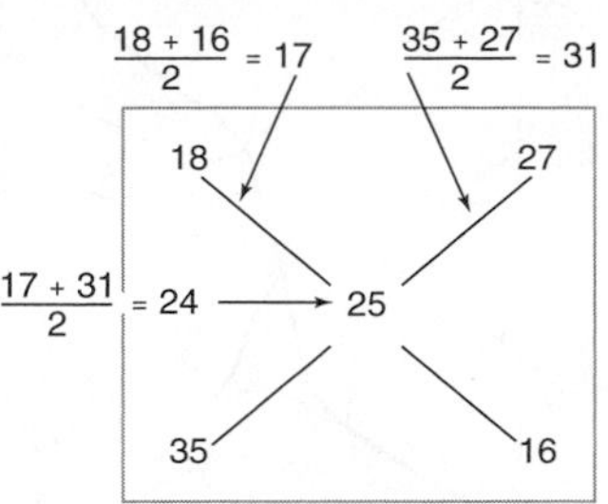

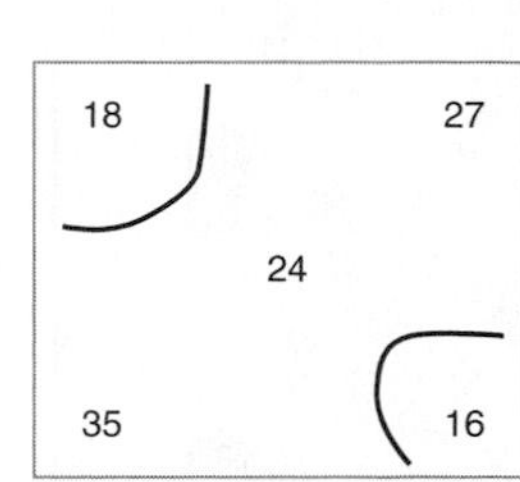

Figure 11.13 Solution to saddle-point problem. The solution uses an average value placed at the exact center. *Source:* A.H. Robinson et al., *Elements of Cartography*, 6th ed., John Wiley & Sons, New York, © 1995. Adapted from Figure 26.28, page 512. Used with permission.

must have control points on all sides. If, as is often the case, we select a study area for analysis and use that same study area to perform interpolation, we will soon need to interpolate points near the margins of the study area. When we approach the map margin, we see that the interpolation routine is faced with control points on two or three sides of our unknown elevation points because the map border precludes any data points beyond the margin. As you have seen, best interpolation results are obtained when we are able to search a neighborhood in all directions for selection of control points and determination of weights. In the absence of these surrounding data points, the algorithm will use whatever is available, biasing the results away from the border.

As an example, let's try to perform interpolations along the left margin of a map. We will assume that this margin cuts directly through a hill that rises from the center of the map toward the left margin. The adjacent map sheet indicates that the hill continues to climb for some distance beyond the left margin of our study area, but because there are no data points to the left of the map border, the algorithm will search for its neighbors to the right, above, and below the point to be estimated, and the interpolated values will be influenced in these three directions. That is, the matrix generated by the algorithm probably will use elevational values that are lower than those that in fact occur on the adjacent map sheet. Therefore, when the interpolation is performed, it will tend to underestimate the missing elevations, producing a surface map that is of little use at the margins. Any slope calculations, azimuthal calculations, line-of-sight, least-cost path, or surface calculations will be in error near the margins as well.

The solution to this problem is to extend the borders of the elevation layer beyond the boundaries of the initial study area (Figure 11.14). In doing so we create a surface layer whose outside margins are larger than the original study area, and then perform the interpolation to produce the necessary surface configuration. Once the interpolation is complete, the outside margins of the original study area can be used as a sort of cookie cutter to shave off the edges (of the elevation model). The idea is to operate on the larger elevational layer to develop the final topographic surface layer that will be compared to other layers. Determining the extent of the surface layer needed to produce reasonable interpolation results is difficult. It would be useful to perform some type of neighborhood analysis, perhaps like that used in Kriging, to determine how far apart data points must be to ensure that they do not influence their neighboring elevational values. However, in practice, this is time consuming and impractical. It is best to err on the conservative side by expanding the area sufficiently to ensure good calculations. Most often, extending the margins

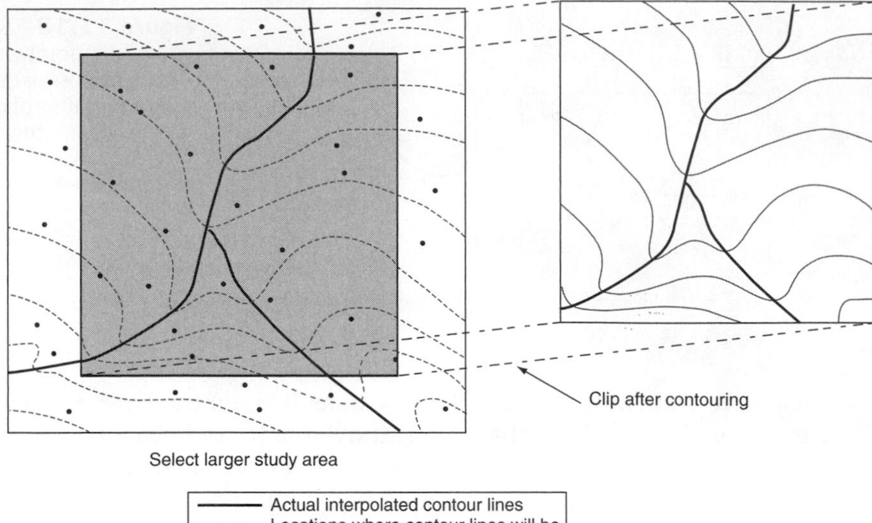

Figure 11.14 Avoiding interpolation error at map margins. Solution to the problem of missing data for interpolation at the margins of a study area: We extend the study area and perform interpolation or other surface procedures, and then the completed map can be clipped and subjected to operations by additional layers, with accurate results.

10 percent on all sides will guarantee good results, but the complexity of the surface must be factored into this decision. The more rapidly elevation changes along the margins, the wider the margins should be to compensate for the impacts the neighboring control points will have on the calculation of missing values when weighting or Kriging techniques are used. No standard guidelines exist here—you are left to use your own experience and firsthand knowledge of your data. Any additional border will increase the quality of your analysis, and is always preferred to no additional borders at all.

Terms

continuous
contour interval
contour line
digitel elevation model (DEM)
discrete
discrete altitude matrix
drift
irregular lattice
isarithm
isarithmic map
isometric
isoplethic map
Kriging
lag
linear interpolation
nonlinear interpolation
nonstationarity
nugget variance
polynomials
random noise
range
regular grid
regular lattice
regular sampling procedure
rough surface
saddle-point problem
semivariance
semivariogram
smooth surface
splines
statistical surface
structure
trend surface
weighting methods

Review Questions

1. Define a statistical surface. Describe a nontopographic statistical surface. What is the difference between a continuous and a discrete statistical surface?
2. What is the difference between a smooth and rough statistical surface? What is the importance of each in terms of point sampling to represent the surface configuration?
3. What is an isarithm? How do the inferences we can make from closely spaced isarithms differ from what widely spaced isarithms tell us? What is the specific term assigned to isarithms that measure topographic elevation?
4. What is a contour interval? Why must a contour interval remain the same throughout a map?
5. What is the difference between an isarithmic map and an isoplethic map? What assumptions must be made before we can use isolines for isoplethic maps?
6. What is the difference between a regular and an irregular lattice in terms of sampling for surface data? What are the advantages and disadvantages of each?
7. What are DEMs? What is the relationship between a discrete altitude matrix and a TIN model?
8. Describe and diagram the method of representing continuous surface data with a raster data model. Why is it important to know where in each grid cell the actual point locations for topographic data are intended to lie?
9. Describe and diagram linear interpolation. What are the major drawbacks to linear interpolation?
10. Why might we be more inclined to use nonlinear interpolation instead of linear interpolation? What are the three basic types of nonlinear interpolation routine?
11. Describe and diagram the use of weighted methods of nonlinear interpolation. Why might we want to include barriers in weighted interpolation methods?
12. What are trend surfaces? When might we want to use this interpolation approach rather than weighted methods? Give an example showing its utility.
13. What is the difference between Kriging and weighted interpolation methods? What is a semivariogram, and what does it tell us about the drift, random spatially correlated variations in the surface, and random noise?
14. Create a diagram of a semivariogram based on hypothetical data. Explain the relationship between lag and semivariance. What can we say about data within and beyond the range? What does this tell us about selecting the neighborhood for interpolation?

15. What is the difference between universal and punctate Kriging? When is Kriging not likely to give better results than any other form of interpolation?

16. Give some concrete examples of the use of interpolation beyond simply creating an isarithmic map.

17. What four things do you need to look out for in any interpolation routine? Describe the problem of sample placement in isoplethic mapping. What is the saddle-point problem? How can it be rectified?

18. Describe the problem of no data points at the map margin as it applies to interpolation results. How can such inadequate results be avoided?

References

Blais, R.A., and P.A. Carlier, 1967. "Applications of Geostatistics in Ore Evaluation." *Ore Reserve Estimation and Grade Control*, 9:41–68.

Burrough, P.A. and R.A. McDonnell, 1998. *Principles of Geographical Information Systems* 2nd ed. New York: Oxford University Press.

Clarke, K.C., 1990. *Analytical and Computer Cartography*. Englewood Cliffs, NJ: Prentice-Hall.

Davis, J.C., 1986. *Statistics and Data Analysis in Geology*, 2nd ed. New York: John Wiley & Sons.

Flowerdew, R., and M. Green, 1992. "Developments in Areal Interpolation Methods and GIS." *Annals of Regional Science*, 26:67–78.

Hodgson, M.E., 1989. "Searching Methods for Rapid Grid Interpolation." *Professional Geographer*, 41(1):51–61.

Kelly, R.E., P.R.H. McConnell, and S.J. Midenberger, 1977. "The Gestalt Photomapping System." *Photogrammetric Engineering and Remote Sensing*, 43(11):1407–1417.

Lam, N.S., 1983. "Spatial Interpolation Methods: A Review." *American Cartographer*, 10:129–149.

Little, J.J. and P. Shi, 2003. "Ordering Points for Incremental TIN Construction from DEMs," *GeoInformatica* 7(1):33–53.

Mark, D.M., 1978. "Concepts of Data Structure for Digital Terrain Models." In *Proceedings of the DTM Symposium*, American Society of Photogrammetry, American Congress on Survey and Mapping, St. Louis, MO, pp. 24–31.

Markarovic, B., 1973. "Progressive Sampling for Digital Terrain Models." *ITCJ*, 1873–3: 397–416.

Matheron, G., 1967. "Principles of Geostatistics." *Economic Geology*, 58:1246–1266.

McCullagh, M.J., and C.G. Ross, 1980. "Delaunay Triangulation of a Random Data Set for Isarithmic Mapping." *Cartographic Journal*, 17(2):93–99.

Oliver, M.A., and R.W. Oliver, 1990. "Kriging: A Method of Interpolation for Geographic Information Systems." *International Journal of Geographical Information Systems*, 4(3):313–332.

Robinson, A., J. Morrison, P. Muehrke, A. Kimmerling, and S. Guptill, 1995. *Elements of Cartography* (6th Edition). New York: Wiley.

Shepard, D., 1968. "A Two-Dimensional Interpolation Function for Irregularly Spaced Data." In *Proceedings of the Twenty-Third National Conference of the Association for Computing Machinery*, pp. 517–524.

CHAPTER 12

Terrain Analysis

In the last chapter we examined the nature and interpolation of surfaces. In doing so we used topographic surfaces as the classic example of surfaces found in GIS databases. Over the years a substantial set of analytical techniques has evolved to analyze the topographic surfaces. These techniques, collectively called **geostatistics**, include methods of interpolation, but have become far more advanced than predicting missing elevation values. These techniques are grouped as **terrain analysis**. Among the more common of the surface analytics include the ability to group terrain into categories of slices of elevation (e.g., treeline analysis), **slopes** and **aspects** for subsequent incorporation into models applied to such endeavors as siting solar facilities and wind farms, evaluating plant and animal habitat, fire and landslide hazards analysis, windthrow evaluation, and providing options for residential development. Another often used surface analysis technique is the classification of areas of visibility, called **viewsheds**, from a point or a series of points. This allows for determining appropriate placement of military resources, privacy for residential housing, and scenic views for recreational facilities. In some circumstances the viewshed analysis can be modified to evaluate sound (**soundshed**) to allow for reduction in highway noise, stealthy movement of military equipment, or to protect wildlife whose mating might be affected by noise volumes.

Terrain analysis techniques are available for evaluating the shape or form of landforms, both as cross-sectional diagrams and as volumetric shapes. The tradition of separating portions of elevational surfaces into specific landforms based on shape analysis is more commonly an academic rather than an application-driven endeavor, but no less important in the geostatistical toolkit. The very nature of landforms is important in relationship to understanding soils and vegetative characteristics which, in turn, are powerful environmental analysis tools. Moreover, geostatistics include the capacity to determine volumes of overburden in mining circumstances (i.e., cut and fill evaluation), calculate volumes of slump blocks, or the amount of water likely to fill basins upon damming a stream. A wide variety of additional surface hydrological analyses are also dependent on the nature and shapes of landforms upon which they exist, including their surface area. The following pages will give a brief introduction to some of these techniques.

LEARNING OBJECTIVES

When you are finished with this chapter you should be able to:

1. Define geostatistics and terrain analysis, explain the similarity and differences in these terms, and provide general examples of the application of terrain analysis.
2. Describe and provide concrete examples of *Z*-surface slicing and treeline analysis.
3. Describe and provide concrete implementations and real-world applications of slope and aspect analysis.
4. Understand how visibility and intervisibility analysis work to produce viewsheds.
5. Explain the methods and application of soundshed analysis.
6. Explain the use of terrain analysis to describe the shapes and volumes of landforms and provide concrete examples of their application.
7. Explain how raster GIS calculates volumes both when the comparative surface is flat and when it is variable.
8. Explain and provide examples of surface hydrological analysis, including watershed delineation and stream network analysis, based on terrain.

TERRAIN RECLASSIFICATION

Characteristics of surfaces that can be used for describing neighborhoods include zones of elevation (slicing the *Z* surface), steepness of slope, azimuth or orientation, shape or form, intervisibility, and volume. To one degree or another, all can be performed with both raster and vector GIS, again depending on the sophistication of the software. In many cases, as with reclassifying neighborhoods with two-dimensional layer data, these characteristics may very well be used in combination. We will look at each one individually and finish with a discussion of how they might be combined for more complex analyses.

Elevation Zones

You have seen that a common method of displaying surface information makes use of isarithms, and that isarithms are commonly drawn at set intervals (a contour interval in topographic maps). The interval chosen allows us to depict the shape of the surface, but we always assume that the elevational values between the contour lines exhibit some continuously changing value because the surface itself is continuous. We also assume that the interval was selected to best display the surface features. Most GIS software, whether raster or vector, offers the ability to change the contour interval or even to convert the area under each individual contour interval to a flat surface, eliminating the need to assume a continuous surface. For simplicity we will call this general group of techniques **slicing**, and we will envision their execution as applying a sharp knife horizontally through the surface.

Slicing can be a matter of selecting a different isoplethic class interval set to permit us to look at surface features differently (Figure 12.1). We might, for example, want to make the vertical distance between isopleths larger, to bring out overall shapes of features without great amounts of confusing detail. This useful visualization technique gives an impression of trends without necessitating the actual computation of a trend surface for the area. Alternatively, to see more of the detail in the surface, we can decrease the vertical distance between the isopleths. Of course, this latter approach assumes that the sample data are sufficiently detailed in the first place.

A more common application of slicing could more appropriately be called a neighborhood function. This approach assumes that by placing contour lines at individual intervals, we have effectively reduced a continuous surface to a discrete surface more closely resembling a stair-step feature. Why would we want to degrade our data from continuous to discrete? Consider the following example.

You are trying to define appropriate land uses for a large portion of land on the basis of a combination of soil capabilities and elevation classes. Elevation classes have been chosen rather than slope classes partly because most soil surveys incorporate slope into the soil capability classes and partly because agricultural advisers know that some plants grow better at certain elevations than at others. On the basis of each crop's elevational requirements (let's say there are five crops), you can divide the surface into five elevational zones

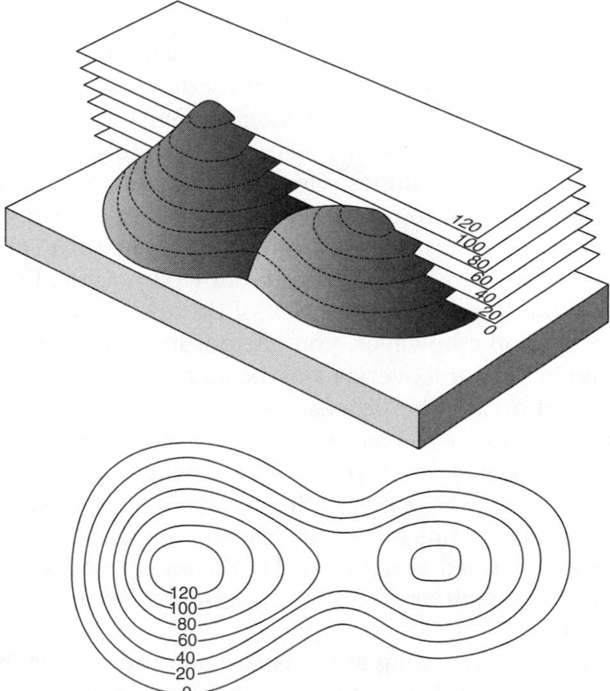

Figure 12.1 Slicing the *Z* surface. Slicing through a *Z* surface to produce different contour intervals.

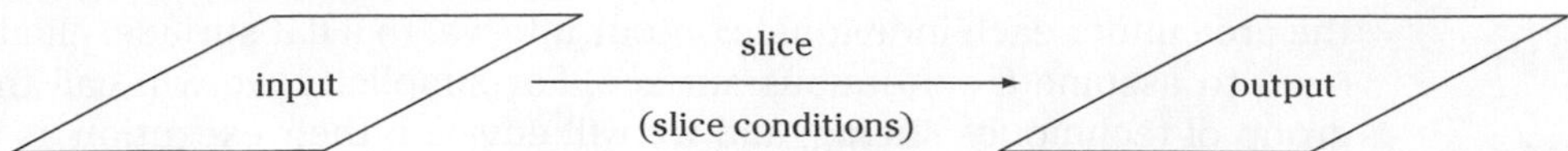

Figure 12.2 Flowchart showing the process of reclassification of topographic data into elevation zones by slicing.

by slicing through it in four places. Once you have performed the slices, the groups can be reclassified by their influence on the five crops. So you now have five groups of areas, based on elevational classes as they affect the five crops, and called "crop1elevations," "crop2elevations," and so on. Thus, you have converted the ratio data of the surface to nominal data on the basis of the interactions between crops and elevations. These can now be combined with other layers to make decisions about agroecosystem requirements for particular crops (Nautiyal, et al., 2002–2003). Beyond agricultural applications, this slicing approach—sometimes called treeline analysis (a simplification)—is used to analyze the habitat requirements of both plants (Camarero, et al., 2006; Heiri, et al., 2006) and animals (Finch, 1989; Rickart, 2001). The simple flowchart in Figure 12.2 shows how slicing can be applied to habitat analysis. The analysis of elevational zones has been used to evaluate the habitat limitations of many types of organisms, including some that are rather obscure (Aubrey, et al., 2005), making this capability of GIS highly relevant to ecologists, biogeographers, and environmental planners.

Slope Analysis

Our previous analysis of terrain involved examining zones of elevation, but these zones can occur gradually or rapidly. The evaluation of the relative rapidity of elevational change or slope adds another dimension to our modeling capabilities. If you are planning to build a mountain cabin, you will want to know where the flattest portions of the terrain are to avoid unnecessary construction costs due to steep slopes. Or, if you intend to cut some timber along a mountainside, you want to use the slopes to roll the freshly cut pieces downhill (don't try this at home). Or maybe you are planning a ski resort and want to put up three different levels of slopes: Snowplow Only, Extra Padding Needed, and Call Out the Rescue Squad. In all of these scenarios, you need to know something about slope; however, you are not interested in the shortest route down but rather in a general overview of the locations of steep, moderately steep, and relatively flat areas. The process requires knowing the relationship between the horizontal distance (measured in vector or raster) and the vertical change in elevation from bottom to top. A common way of expressing slope is rise over reach (rise/reach), where rise is the change in elevation and reach is the horizontal distance. To do this in vector, you will need a data model similar to the TIN model we discussed in Chapter 11. Raster, of course, can readily manage this, although some minor errors due to grid cell quantization of space must be compensated for.

The general method of calculating slope is to compute a surface of best fit through neighboring points and measure the change in elevation per unit of

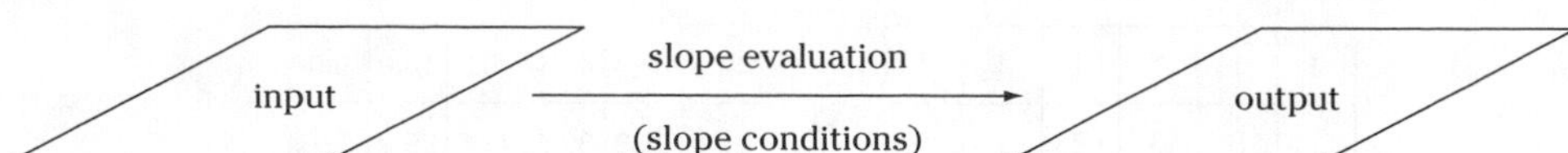

Figure 12.3 Flowchart showing the process of reclassification of topographic data into slope levels. In this case the result is two categories of slope (goodslope and badslope), based on whether or not the slopes are greater than 25 percent.

distance (Clarke, 1990). Specifically, the GIS calculates the rise/reach throughout the entire layer, generating a set of slope categories. If we wish fewer slope categories than are actually developed, we can reclassify the set at will. Let's say we are trying to find slopes that are less than 25 percent (e.g., 25 meters' change in elevation over 100 meters in horizontal distance) to define areas possibly suitable as sites for a cabin. In vector, using a TIN-like data model, we can perform any number of interpolation algorithms to provide what essentially amounts to a contour map. To determine slope, the software compares the vertical distance between the vertices of each of the TIN facets with the respective horizontal coordinates. Because the TIN stores these calculated values in its attribute tables, the calculations are already precalculated. Each facet value can be retrieved, and the slopes grouped or classed as those with less than 25 percent slope and those with 25 percent or more slope. We can then rename these as "badslope" (>25% slope = too steep for building), and "goodslope" (<25% slope = acceptable for building). We have reclassified neighborhoods on the basis of slope to facilitate decision making in the single area of suitability for building (Figure 12.3).

Both simple and complex methods of reclassifying neighborhoods based solely on slope can also be performed in raster GIS. The simplest method is to use a search of the eight immediate neighbor cells of each target cell. This is most often done by looking at all grid cells in the database and examining their neighbor cells, so that the slope values for the entire layer can be performed. The software fits a plane through the eight immediate neighbor cells by finding either the greatest slope value for the neighborhood of grid cells or an average slope. For each group of cells, the software uses the grid cell resolution as the measure of distance and compares the attribute values (elevations) from the central cell to all the surrounding cells (Figure 12.4). If, for example, we are building a ski resort and we want slopes no greater than 15 percent for the use of inexperienced or novice skiers, between 16 and 25 percent for talented amateurs, and between 26 and 45 percent by professionals, we can select these three classes through a simple reclassification process. We would identify all neighborhoods whose maximum (or average) slopes are between 0 and 15 percent, 16 and 25 percent, and 26 and 45 percent. All slope values greater than 45 percent would be called "badslopes" because they are too steep for skiing.

Aspect Analysis

Slope is inextricably linked to our next terrain feature (aspect), because slopes are by definition oriented in a particular direction. Without a slope, there is no topographic aspect. You have seen that azimuth or orientation can be useful for

INGRID

1	1	1	1	1	1
1	3	3	2	1	
1	1	3	2	2	2
1	2	2	2	2	2
1	1	1	2	2	2
1	1	1	1	1	2

=

OUTGRID

19.5	38.3	41.5	29.2	10.0	0.0
26.6	38.3	32.5	41.5	29.2	
21.6	43.6	21.6	28.8	14.0	10.0
14.0	28.2	28.2	14.0	0.0	0.0
10.0	21.6	29.2	29.2	21.6	10.0
0.0	0.0	10.0	21.6	29.2	14.0

SLOPE (degree or %)

Figure 12.4 Characterizing surface shape in raster. Raster and vector output for characterizing a neighborhood based on form. In vector, the output is a cross-sectional profile, whereas in raster the result is a layer that shows the relationships between the target cell and its two neighbors. The two neighbors are selected as vertical, horizontal, or diagonal neighbor cells.

classifying neighborhoods of 2-D features. The same can be said for 3-D terrain features. There are numerous applications of this technique. For example, biogeographers and ecologists are aware that there is generally a noticeable difference between the vegetation on slopes that face north and slopes that face south (Brown and Gibson, 1983). A major reason for this differential includes the availability of sunlight to green plants, but our interest in the phenomenon is that GIS can separate out north- versus south-facing slopes for comparison to related layers such as soil and vegetation. Another example of the utility of knowing slope orientation is found when we try to build wind generators relative to the prevailing winds as they move through the terrain, or to locate solar facilities that require slopes oriented toward the sun. Geologists frequently want to know the prevailing slopes of fault blocks or exposed folds to understand underlying subsurface processes. A grower may want to place an orchard on

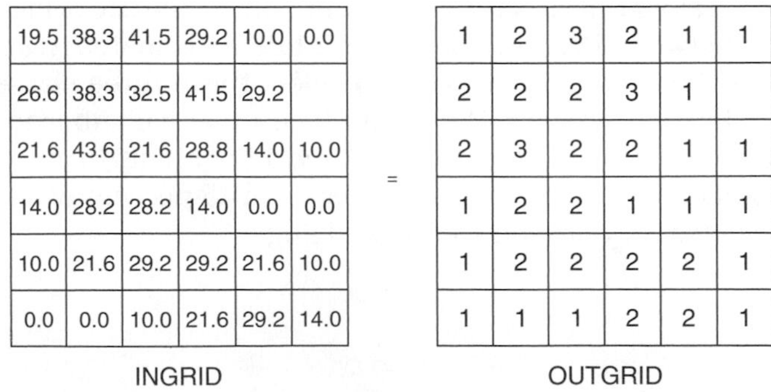

INGRID

19.5	38.3	41.5	29.2	10.0	0.0
26.6	38.3	32.5	41.5	29.2	
21.6	43.6	21.6	28.8	14.0	10.0
14.0	28.2	28.2	14.0	0.0	0.0
10.0	21.6	29.2	29.2	21.6	10.0
0.0	0.0	10.0	21.6	29.2	14.0

=

OUTGRID

1	2	3	2	1	1
2	2	2	3	1	
2	3	2	2	1	1
1	2	2	1	1	1
1	2	2	2	2	1
1	1	1	2	2	1

RECLASS (SLOPE)

Figure 12.5 Aspect analysis. On the left is an unclassified topographic surface. On the right is the same surface classified as north-facing and south-facing.

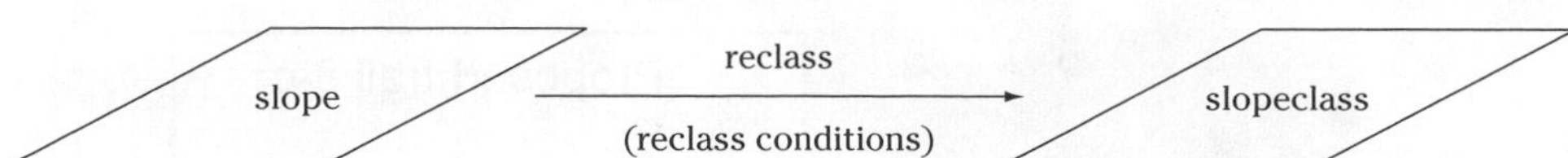

Figure 12.6 Flowchart showing how topographic data can be classified based on north versus south aspects.

the sunny side of a hill to be able to take advantage of the maximum amount of sunshine. All these determinations and many more can be performed through the classification of sloping surfaces based on their aspect.

In a vector GIS TIN-based model, the slope and aspect values are explicitly part of the model itself. Each TIN facet has a specific slope and aspect. The aspect is defined as the compass direction associated, and is linked to each sloping surface (Evans, 1980), thus requiring no additional calculation. We can group these aspects into classes. For example, if we are biogeographers interested in only north- and south-facing slopes, we can create three separate classes based on criteria selected earlier. Thus, north-facing slopes could be classed within a single neighborhood if they range in azimuth from 345 degrees to 15 degrees, and slopes facing south could be classed into a single neighborhood if the range is between 165 degrees and 195 degrees. This means that the classes have a 30-degree range, or 15 degrees on either side of the cardinal direction (Figure 12.5). All of the remaining aspect values can then be classed as "badaspect" because they are not required for analysis (Figure 12.6).

In raster, the most common method is to perform a roving window analysis throughout the entire layer, repeatedly comparing a target cell to its neighbors. In this case, the surface fitted through the nine-cell matrix evaluates either an average or a maximum direction by looking at the high and low elevation values within the matrix. If, for example, the highest value is located at the top center of the matrix and the lowest at the bottom center, the solution for this portion of the matrix is that the general aspect is south (assuming that your layer is oriented north and south). Or you may find that the highest and lowest grid cell elevation values are located in the upper left and lower right, respectively. This will give you a southeast aspect to your sloping surface. The results of raster analysis of aspect can come either in degrees (0–360) or in a simpler set of vector values resembling the Freeman chain codes, where, for example, north, south, east, and west would be 0, 2, 4, and 6, respectively, and northeast, southeast, southwest, and northwest would be 1, 3, 5, and 7, respectively. The actual numbers and methods will depend on the software.

Shape or Form

Another useful method of reclassifying statistical surfaces is to provide a measure of their form. We have already discussed 2-D shape methods, which are applicable for planimetric views of surface features. In addition to these, a simple and useful method of visualizing surface form is to produce a **cross-sectional profile** of the surface. In analog this is done by transferring each elevational value to a sheet of graph paper, where the horizontal position is identical to the lines between the data points and the vertical axis is scaled to some **vertical**

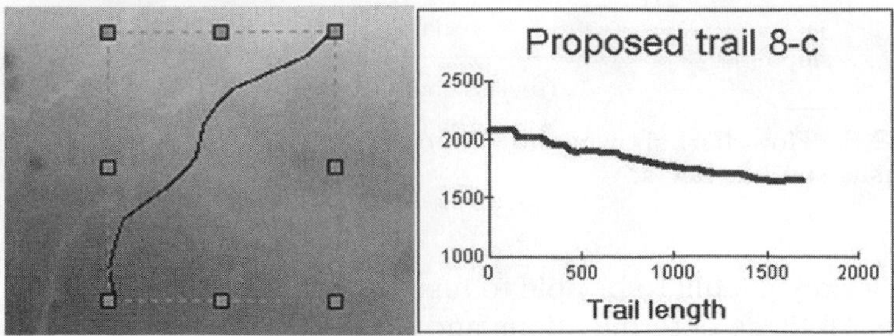

Figure 12.7 Profile as a measure of landform shape. Note how the transect line is not a straight line.

exaggeration (usually a whole-number power) of the original surface elevation values. The process is easy to perform in a vector system using a TIN model, where the line is drawn along some portion of the layer (it need not be a straight line) (Figure 12.7). The software then generates a profile identical to that produced with analog maps in the manual method. Keep in mind, however, that this visual is not a layer, and the results are used merely as a means of interpreting what is there. We know, for example, that *V* shapes in profiles are most often associated with stream valleys and that *U* shapes indicate glacial troughs. The more you know about the surface processes you are working with, the better your interpretation of the output is likely to be.

Cross-sectional analyses are designed to produce neighborhoods based on changes in surface value that can be interpreted by the user to represent specific features. Thus, ridges, channels, peaks, watersheds, and so on may need to be identified as specific topographic features for later analysis. Watersheds, for example, are defined as all areas that drain into a stream network (Band, 1986, 1989; Douglas, 1986). More than simply draining into the stream channel, watersheds tend to function ecologically as single, uniform regions. Ecologists, hydrologists, engineers, pollution (White et al., 1992) and flood control experts, and many other specialists need to be able to define these areas precisely. We have only seen the simplest method, but other more sophisticated methods have been applied to just this type of problem. As your skills improve, you may want to read up on these (Band, 1986, 1989; Douglas, 1986; White et al., 1992).

A very important set of surface shape analysis tools relate to streams and their related landform features. We can delineate watersheds and extract their stream networks (Tarboton, et al., 1991) to determine their complexity (called bifurcation ratio). Before we can use GIS to delineate watersheds, we must first be able to explicitly define the elevational characteristics that the software will use for that delineation. In this context we define a watershed as the upslope area draining water or other fluids (including pollutants) to a common location or outlet. As you might guess, this is explicitly defined by elevation: all points in a watershed must flow from higher elevations on the margins to the lowest elevations at the streams (Figure 12.8). Sometimes called stream basins, catchments, subwatersheds, or contributing areas, they are typically delineated with a DEM. Generally a raster-based function, the GIS determines either flow **accumulation threshold** or **pour points** (Figure 12.9) (low points at which the water flows into the basin) to determine flow direction. When threshold values are used, the pour points are the junctions (branchings) of the stream network.

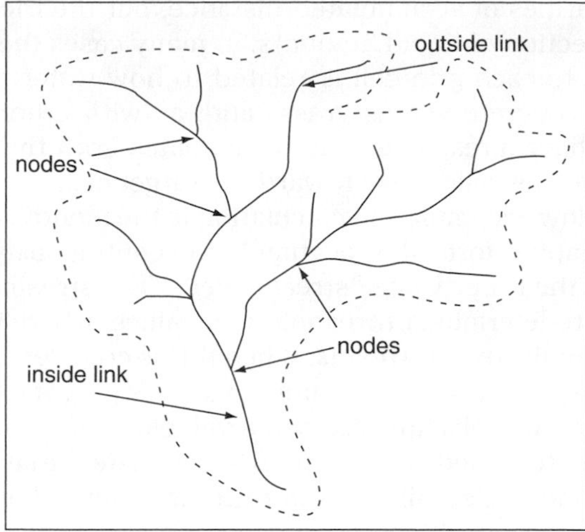

Figure 12.8 A watershed, sometimes called a stream basin, is composed of subbasins, a trunk stream, tributaries, and pour points. All water in a watershed drains internally toward the pour points.

A flow accumulation grid will be developed, as well as a group of cells that constitutes the stream (the threshold).

The first step is to determine flow direction by comparing the elevation values of each grid cell with its neighbors. For simplicity we will assume there are no small variations (sinks) disrupting our analysis (a **depressionless surface**). These are correctable, but are beyond our introductory discussion. The lowest values will be defined as the stream network grid cells, and the highest will define the outside boundaries of our watershed. Where these stream grid cells connect to form branches of the network, we say that these grid cells are pour points—points where each subbasin's water ultimately flows.

A primary purpose of the delineation of the watershed is a determination of the accumulation of fluid downslope in a basin. This is called **flow accumulation**

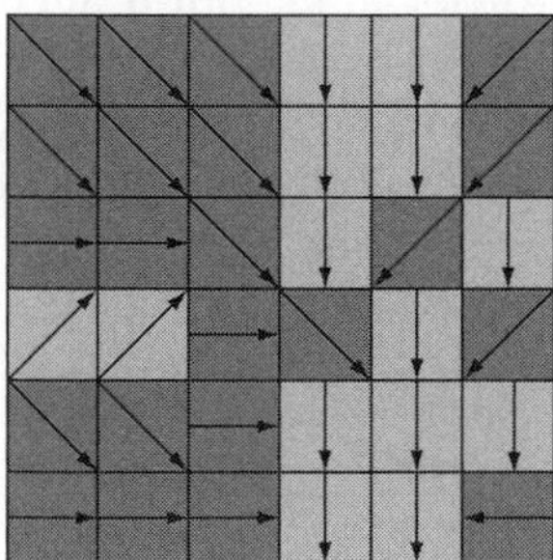

0	0	0	0	0	0
0	1	1	2	2	0
0	3	7	5	4	0
0	0	0	20	0	1
0	0	0	1	24	0
0	2	4	7	35	2

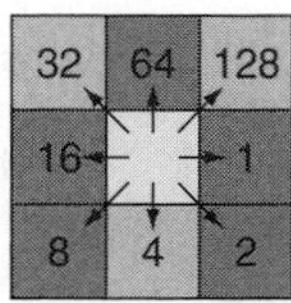

Direction Coding

Figure 12.9 Raster method of determining flow accumulation. Each grid cell's contribution is added until the highest values are reached at the stream (center).

and shares strong similarities of accumulated distance, but the friction values are replaced by the collection of liquid amounts. In many cases the amount of flow or weight assigned to each grid cell is related to how water moves over different surfaces (e.g., concrete versus grass) and, as with other functional distance measures you have already seen, it accumulates both the cell values as well as their weights as the cells move toward the target cells.

A direct result of the flow accumulation calculation is the identification of the stream network. In its simplest form, flow accumulation continuously adds cells that ultimately flow into the lower value "stream" cells. The stream cell values are selected based on predetermined threshold call values, where the highest accumulated values will indicate the streams. When this is complete, the stream cells are typically assigned (reclassed) a common value (e.g., a value of 1), and the background or nonstream cells are assigned a value of "no data." Once the streams themselves are determined, it is common to calculate the **stream order**. Stream ordering (sometimes called **bifurcation ratio**) is a method of classifying streams based on their relative positions and connectivity within the network. In short, it is a method of classifying the numbers and configurations of tributaries in a stream network. These can be inferred from stream order in which, for example, first-order streams are characterized by overland flow of water and don't have upstream concentrated flow. A direct result of this characteristic is that they are most susceptible to nonpoint source pollution. There are several methods of calculating stream order; the two most commonly applied are the Strahler (1957) and the Shreve (1966) methods (Figure 12.10). Both begin by assigning a value of 1 to the most "upstream" stream segment or exterior link. That is where the similarities end. For the Strahler method, probably the most common method, the stream orders increase as stream segments of the same order value intersect. So as two first-order streams come together, they form "second-order" streams; two second-order streams linking form "third-order" streams; and so on. If dissimilar stream order segments link, they retain their own unique order number and, as such, do not account for all network links and are sensitive to the addition and deletion of stream links.

The Shreve method also starts at the first-order links, but it accounts for all links regardless of stream order number. All interior links are additive where, for example, the intersection of first- and second-order streams results in a third-order link, and the intersection of a second- and third-order stream results in a fifth-order link. Because, unlike the Strahler method, the numbers are additive, they are referred to as **stream magnitudes** rather than stream orders. Before determining stream order, the method must be determined.

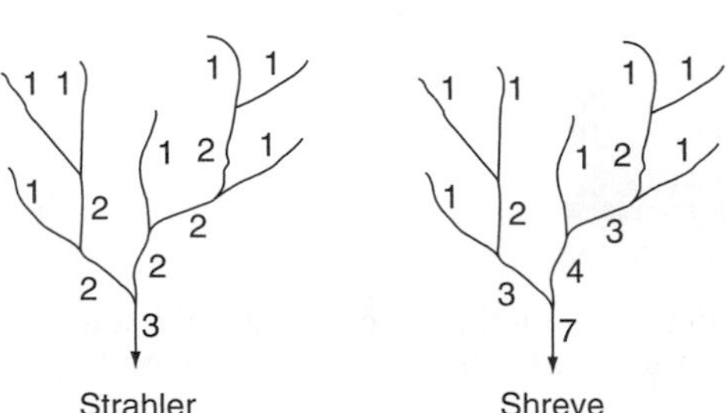

Figure 12.10 Methods of stream ordering. Strahler's method (left) orders streams based on connections of identical stream orders, while Shreve's method (right) creates stream magnitudes based on adding the stream orders as they come together, regardless of whether they are identical orders. The Strahler method is the most common, while Shreve's method accounts for all possible tributary linkages.

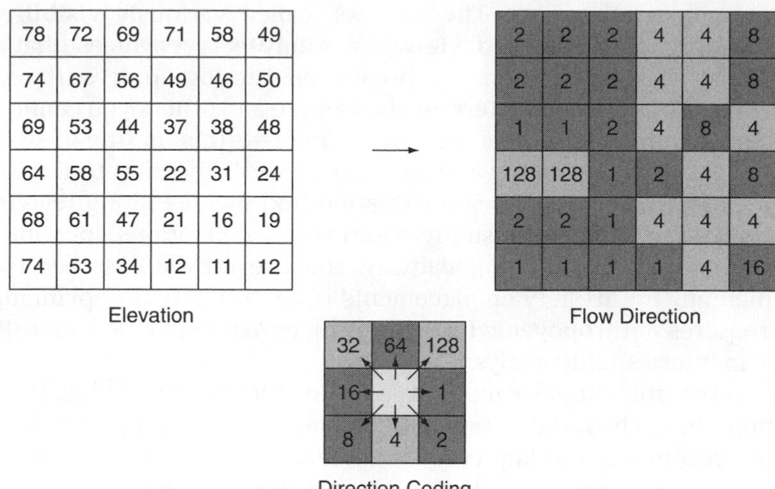

Figure 12.11 Raster calculation of flow direction. The elevation value of the central cell is compared to the eight surrounding elevation values.

You have already seen how flow accumulation is modeled and how it is used in connection with delineation and characterization of watersheds and stream networks. A direct outcome of this type of analysis is the determination of direction flow. This is mentioned here because if we generalize our definition of fluid to include wind and all the associated things that are affected by wind (e.g., plant seeds and spores, diseases, dust, sparks and fire, etc.), we can easily imagine how these same tools can be used for the movement of fire and the spread of vectors and plant species. We can also easily correlate the flows related to elevation to other types of materials, such as mud (mudflows), debris (debrisflows), snow (avalanches), and large chunks of the earth (landslides and slumps). All of these can easily be modeled by assigning weights and appropriate values to existing elevational data.

The general model for flow direction is contained in the simple formula (change in *Z*-value / distance * 100). On a 3×3 set of grid cells, each target (central) cell's elevation (*Z*) value is compared to its neighbor cells, where distance is measured as 1 for orthogonal cells and 1.414 for diagonal cells (Figure 12.11). Most software will enlarge the size of the neighborhood as necessary if no differences exist. Recall our previous discussion of functional distance, and you can see the direct correlation to flow direction calculation.

VIEWSHED ANALYSIS

You have seen the impact that topography can have on the movement of fluids, particulates, and even solids, and have examined the capabilities of the GIS software to model these situations. Topographic configuration also has a profound impact on the ability to see portions of our landscape where intervening objects or landscape features interfere with our field of view. Most professional GIS software packages have the capability to determine and model such relationships. The process, called variously **visibility analysis**, **intervisibility analysis**, and **viewshed analysis**, recognizes that if you are located at a particular point or points on a topographic surface, there are locations or portions of the terrain you can see and others you cannot see. When comparing individual points we usually refer to this as **line-of-sight analysis**, where a straight line is drawn from an observer's location to the location of a target object. The many uses for this method include siting television, radio, and cellular telephone transmitters and receiving stations, locating towers for observing forest fires, routing highways that are not visible to nearby residents, and planning for artillery emplacements (Clarke, 1990). Any planning objective that requires anthropogenic features to be either visible or concealed will find utility in intervisibility analysis.

In vector, the simplest method is to connect a viewing location (observer location) to each possible target in the layer. Next you perform **ray tracing**; that is, you follow the line (ray) from each target point back to the starting point, looking for elevations that are higher (Clarke, 1990). Higher points would obscure the observer's view of what is behind it (Figure 12.12a). There are many possible ways to determine intervisibility in vector; the large number of calculations involved makes the ray tracing method a simpler and more useful

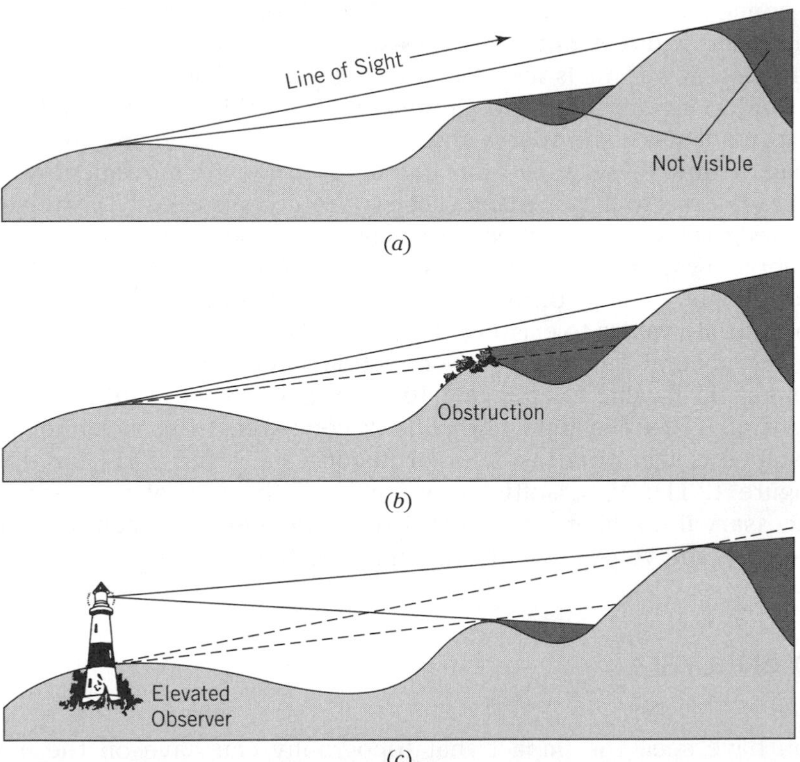

Figure 12.12 Line-of-sight ray tracing. Point to point visibility analysis employs a process called ray tracing. The general method (*a*) can also be modified to include the addition of obstructions like trees (*b*), or can include the possibility of an elevated observer (*c*), such as observing out terrain from a lighthouse.

technique, although perhaps less accurate. Intervisibility performed in vector requires the use of a TIN data model in which the surface is defined by the triangular vertices. A number of algorithms for this process have been provided for vector data structures, including TIN (DeFloriani et al., 1986; Sutherland et al., 1974). Our purpose here is not to go into these algorithms but rather to show the concept, so let's take a quick look at a possible application of intervisibility in vector to see what it does and how it might be applied.

Suppose you are planning to build a home in the foothills of a mountain range, and you want to be able to see as much terrain as possible from your front porch. You have limited your candidate locations to three. For each location, identified in your GIS terrain layer (TIN model), the software looks out in all directions at the vertices of the model and identifies them. It then retrieves the elevation values for all these points. Next it compares these elevations to the elevation of the potential building site. All areas that are higher in elevation (and all those beyond those locations) are not visible to you and must be reclassified as invisible; all the remaining areas are visible. The resulting polygon map shows you how much area is visible (Figures 12.13, 12.14) for each layer tested. Because areas of visibility rather than points are the final output, they are typically called viewsheds and the approach is called viewshed analysis rather than visibility or intervisibility.

Raster methods of viewshed analysis operate in much the same way, but they are less elegant and more computationally expensive. The process begins by defining a viewer cell as a separate layer against which the elevation layer will be tested. Starting at the location of the viewer cell, the software evaluates the

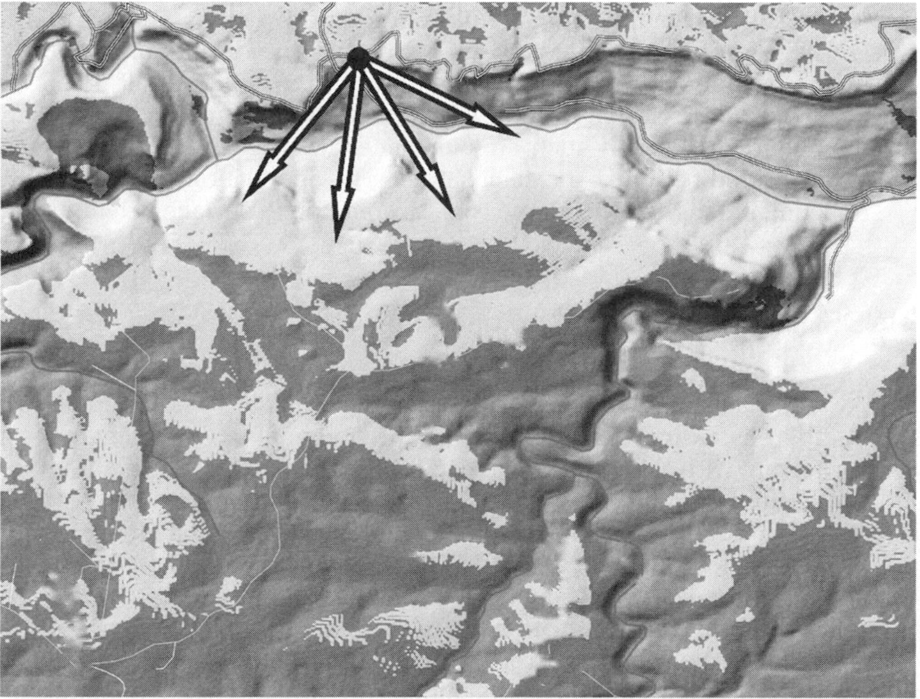

Figure 12.13 When visibility analysis involves evaluating multiple lines of sight, the result is a set of visible polygons called viewsheds.

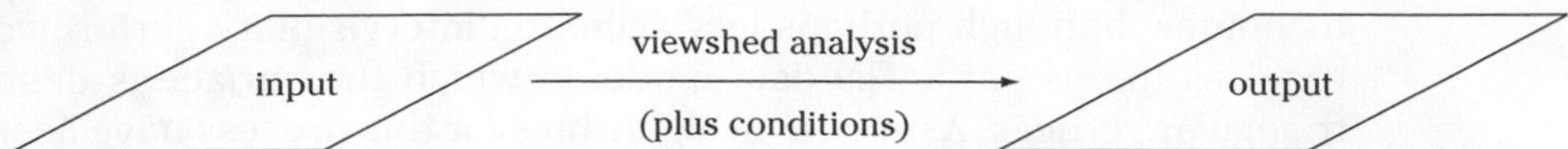

Figure 12.14 Viewshed flowchart. A sample viewshed analysis flowchart for a mountain cabin. Notice that the flowchart requires at least two operations (the original viewshed operation) plus a reclassification process specific to the needs of the mountain cabin.

elevation (in the second layer) that corresponds to that location. Then it moves out in all directions, one grid cell at a time, comparing the elevation values of each new grid cell it encounters with the elevation value of the viewer grid cell. Each time it encounters a grid cell with an elevation value higher than the viewer grid cell elevation, it reclassifies it as not visible and codes it to indicate a nonvisible grid cell. Each time it encounters a grid cell with an elevation value lower than the viewer elevation, it places a different code, indicating that the area represented by the cell is visible from the viewer's location. This, of course, is the simplest way to perform this method. There are others, and each yields different but computationally valid results (Anderson, 1982; Dozier et al., 1981). You might want to check your software's documentation to evaluate the utility in your work of one type over another (Fisher, 1993).

Most applications of viewshed analysis are based solely on topographic surfaces, but in some cases the topographic surface will have forest cover with known individual heights or grouped heights associated with the trees. To perform intervisibility where the heights of these or other obstructing objects are known, the elevation layer values must include the obstruction heights (Figure 12.12b). These can be added in both vector and raster, usually by means of a mathematically based (addition) combination of the two themes. Where topography itself is not important, the heights of the obstructions can be used alone to determine the degree of visibility of all points compared to those of the viewer. In some cases the viewer is actually higher than the terrain (Figure 12.12c). In such circumstances, the height of the viewer's object can be added to adjust the calculations. The applications of viewshed analysis are endless, and the technique is quite common in surface neighborhood analysis.

SOUNDSHED ANALYSIS

You have seen terrain analysis functions linked to flows and to visibility, but there is an increasing interest in the potential of employing our knowledge of terrain with the vast resources of the acoustic engineering literature. Common settings for the evaluation of noise include planning for airports and urban areas (Moreno-Jiménez and Hogdart, 2003; De Klujver and Stoter, 2003) to abate the volume of noise due to transportation. Other sources of noise that might be examined include those from military testing, mining, construction, and a wide array of other anthropogenic sources. Because unwanted noise reduces the value of recreational facilities, residential neighborhoods, and even the workplace, an ability to employ GIS to either design abatement strategies or to locate facilities in such a manner as to reduce their impact would be useful. Moreover, biologists and acoustic scientists have found that noise can have profound impacts on the health of humans, wildlife, and domestic stock (Dufour, 2005).

These effects are often related to the elevations in which either the target or the source of the noise is located. We know, for example, that if we are in the line of sight of a high-volume highway that we are likely to experience much higher volumes of road noise than if we are behind a substantial hill. Unlike light, however, sound can travel substantially around or over obstacles resulting in noise that, while less than for areas facing the source, may still be unacceptably loud. Sound also has a property of being modified differently based on the texture, structure, orientation, and material composition of obstacles it encounters. These factors might prove useful in designing sound barriers near noisy activities. Without examining the detail of how one might use terrain and terrain-based objects, it is sufficient to think about the problem as one of a modified viewshed analysis with some additional factors affecting how the model is set up. Software has been developed specifically for mapping noise (e.g., SoundPlan www.soundplan.com; last visited 1/2/2007) and more packages are forthcoming. These are also likely to provide data models and output in forms that are compatible with nonacoustic types of GIS software.

CUT AND FILL

So far we have been treating surfaces as if they were true 3-D objects, but we have not been analyzing the third dimension of volume adequately. There are many circumstances when we wish to know, for example, how much overburden there is in cubic yards or cubic meters because each cubic quantity costs money to remove. We may also want to know how much water a basin developed behind a dam is capable of holding. In the building trades we often need to estimate how much earth needs to be transported in to build up a pad for building a home or other structure. All of these and many more are part of the GIS toolkit (most often raster) called **cut-and-fill analysis**.

In short, cut-and-fill analysis summarizes the volumes of change between two surfaces, although the surfaces might actually be the same surface at different times (e.g., before and after filling, or before or after flooding). Determining cut-and-fill analysis in vector, while possible, is complicated and time consuming, so the analysis is most often performed in raster. Let's take a look at how this is accomplished.

We'll begin with a simple, fictional backyard pool example. Suppose our backyard pool has vertical sides and covers an area of 16 square meters (4 × 4 meters). The pool is uniformly 2 meters deep across the entire surface area. By multiplying the depth (2 meters) by the area (16 square meters), we obtain the overall volume of the pool: 32 cubic meters. Now let's turn this into grid cells. Let's say we are going to represent our pool with 10 centimeter grid cells (10 grid cells per meter). If this is the case, we would have a 40 × 40 grid cell area (1,600 grid cells), where each grid cell represents a depth of 2 meters. Given that each of these 1,600 grid cells represents 2 meters of water, then each also represents 2 × 0.1 meters × 0.1 meters, or 0.02 cubic meters of water. So we can multiply each grid cell's volume by the number of grid cells (1,600) and we get 32 cubic meters, just as we did before.

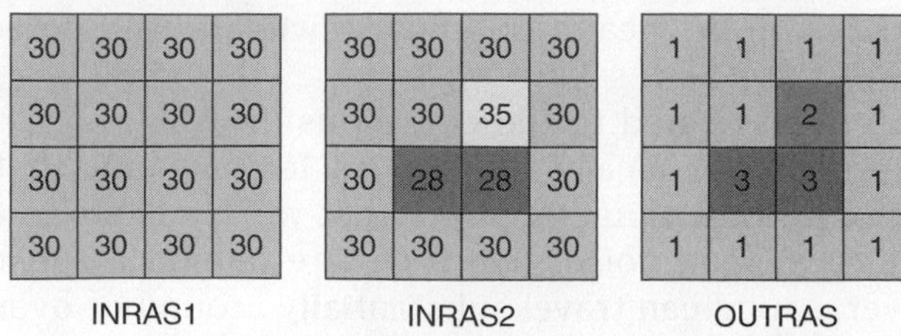

30	30	30	30
30	30	30	30
30	30	30	30
30	30	30	30

30	30	30	30
30	30	35	30
30	28	28	30
30	30	30	30

1	1	1	1
1	1	2	1
1	3	3	1
1	1	1	1

Figure 12.15 Raster representation of volumes allows easy calculation. By multiplying the elevation of each column of grid cells by the area of each grid cell, one obtains the volume for each column. Adding all of these columnar volumes produces a total volume for the body in question.

But many things that are measured across a large area are not uniformly distributed. What if our pool had two different depths? For example, one half of the pool might only be 1 meter deep as opposed to 2 meters deep. To calculate this we could examine each half set of grid cells separately. This would result in the deep half of the pool with a volume of 16 cubic meters and the shallow half with a volume of 8 cubic meters, with a total of 24 cubic meters (Figure 12.15). For more complex configurations we merely need to consider the volume for each separate columnar group of grid cells and add them to achieve a total.

This volume calculation is the same for a hill or a lake in that the comparative surface is essentially flat, thus simplifying the problem somewhat. But if your purpose is to determine how much ore there is in a subterranean deposit, both the upper and lower surfaces will be irregular. Still, you can use the same approach as before, except that you can no longer assume that one of the surfaces is flat. By subtracting the upper elevational values from the lower, you obtain a variable height value. Then, by noting the area for each grid cell column, you multiply by the depth and add the column totals to compute the total volume.

Terms

accumulation threshold
aspect
bifurcation ratio
cross-sectional profile
cut-and-fill analysis
depressionless surface
flow accumulation
geostatistics
intervisibility analysis
line-of-sight analysis
pour points
ray tracing
slicing
slope
soundshed
stream magnitude
stream order
terrain analysis
vertical exaggeration
viewshed
viewshed analysis
visibility analysis

Review Questions

1. Describe the relationships between the terms *geostatistics* and *terrain analysis*.
2. What is slope? How is it implemented in a vector GIS? In raster? How can slope be used to define neighborhoods?

3. What is aspect? How is it used to create neighborhoods? Give an example of using aspect to define a neighborhood, describing both how it is done and what the output will tell you about your data.
4. Create a simple flowchart illustrating the process of converting raw elevational data into classes of slope and/or aspect. Be sure to include the reclassification step after the classes are created.
5. Provide a diagram illustrating how the shape or form of a surface (as a cross-section) is depicted.
6. Create an illustration of how the cross-section can be applied to a line that curves and wanders around the topographic surface.
7. Speculate on how cross-section lines can be used to define neighborhoods of terrain features (i.e., landforms).
8. Define, diagram, and label a watershed. Explain why the characteristics of the labeled parts are important to watershed delineation.
9. Describe the process of watershed delineation.
10. Diagram how flow direction is calculated in raster.
11. Create a simple diagram showing a branching stream network with at least eight tributary streams. Make a second copy of this diagram. Demonstrate your command of stream ordering by manually labeling the stream branches according to Strahler's stream order and Shreve's stream magnitude methods.
12. Describe what the differences are between the Strahler and the Shreve methods used to calculate stream network configuration. Describe any advantages or disadvantages you might see with either or both.
13. What is a viewshed? What is a line of sight? What is the difference?
14. Assuming the observer is at ground level and the target object is also at ground level, create a ray tracing diagram showing how line-of-sight analysis is achieved on an undulating terrain. Now add obstructions on the landscape and repeat the procedure. Finally, put the observer in a lighthouse and diagram it again.
15. Speculate on how obstacles and observer platform elevations can be accounted for in line-of-sight analysis.
16. What is a soundshed? How does it differ from a viewshed? What factors must be considered to perform soundshed analysis that might not be particularly important to viewshed analysis?
17. Speculate on some real-world applications of soundshed analysis based on what you have read in the text, seen in the literature, or found on the Internet.
18. What is *Z*-value slicing? Other than changing the class interval (e.g., the contour interval), how can slicing be used for creating neighborhoods? Give an example from biology. Give another example from agriculture.

19. Create a simple set of data representing grid cell resolutions and upper and lower depths for a subterranean ore body. Calculate the volume of the ore body.

References

Anderson, D.P., 1982. "Hidden Line Elimination in Projected Grid Surfaces." *ACM Transactions, Graphics*, 1(4):274–291.

Aubrey, S., F. Magnin, V. Bonnet, and R. Preece, 2005. "Multi-Scale Altitudinal Patterns in Species Richness of Land Snail Communities in Southeastern France," *Journal of Biogeography*, 32(6):985–998.

Band, L.E., 1986. "Topographic Partitioning of Watersheds with Digital Elevation Models." *Water Resources Research*, 22(1):15–24.

Band, L.E., 1989. "Spatial Aggregation of Complex Terrain." *Geographical Analysis*, 21(4):279–293.

Brown, J.H., and A.G. Gibson, 1983. *Biogeography*. St. Louis, MO: C.V. Mosby.

Camarero, J., E. Gutiérrez, and M. Fortin, 2006. "Spatial Patterns of Plant Richness Across Treeline Ecotones in the Pyrenees Reveal Different Locations for Richness and Tree Cover Boundaries," *Global Ecology & Biogeography*, 15(2):182–191.

Clarke, K.C., 1990. *Analytical and Computer Cartography*. Englewood Cliffs, NJ: Prentice-Hall.

DeFloriani, L., B. Falcidieno, and C. Pienovi, 1986. "A Visibility-Based Model for Terrain Features." In *Proceedings of the Second International Symposium on Spatial Data Handling*. International Geographical Union Commission on Geographical Data Sensing and Processing the International Cartographic Association, Seattle, WA, July 5–10, pp. 235–250.

De Klujver, H. and J. Stoter, 2003. "Noise Mapping and GIS: Optimizing Quality and Efficiency of Noise Effect Studies," *Computers, Environment and Urban Systems*, 27(1):85–102.

Douglas, D.H., 1986. "Experiments to Locate Ridges and Channels to Create a New Type of Digital Elevation Model." *Cartographica*, 23(4):29–61.

Dozier, J., J. Bruno, and P. Downey, 1981. "A Faster Solution to the Horizon Problem." *Computers and Geosciences*, 7(2):145–151.

Dufour, P.A., 2005. "Effects of Noise on Wildlife and Other Animals, Review of Research Since 1971." *Ecological Abstracts*, 1980, 106 pp.

Evans, I.S., 1980. "An Integrated System of Terrain Analysis and Slope Mapping." *Zeitschrift fur geomorphologie* (supplement), 36:274–295.

Finch, D.M., 1989. "Habitat Use and Habitat Overlap of Riparian Birds in Three Elevational Zones." *Ecology*, 70(4):866–880.

Fisher, P.F., 1993. "Algorithm and Implementation Uncertainty in Viewshed Analysis." *International Journal of Geographical Information Systems*, 7(4):331–347.

Heiri, C., H. Bugmann, W. Tinner, O. Heiri, and H. Lischke, 2006. "A Model-Based Reconstruction of Holocene Treeline Dynamics in the Central Swiss Alps." *Journal of Ecology*, 94:206–216.

Moreno-Jiménez, A., and H.L. Hogdart, 2003. "Modeling a Single Type of Environmental Impact from an Obnoxious Transport Activity: Implementing Locational Analysis with GIS." *Environment and Planning*, A 35(5):931–946.

Nautiyal, S., R.K. Maikhuri, K.S. Rao, R.L. Semwal, and K.G. Saxena, 2002–2003. "Agroecosystem Function Around A Himalayan Biosphere Reserve." *Journal of Environmental Systems*, 29(1):71–100.

Rickart, Eric A., 2001. "Elevational Diversity Gradients, Biogeography and the Structure of Montane Mammal Communities in the Intermountain Region of North America." *Global Ecology and Biogeography*, 10(1):77–100.

Shreve, R.L., 1966. "Statistical Law of Stream Number," *Journal of Geology.* 74:17–37.

Strahler, A.N., 1957. "Quantitative Analysis of Watershed Geomorphology." *Transactions of the American Geophysical Union.* Vol. 8. Number 6, pp. 913–920.

Sutherland, I.E., R.F. Sproull, and R.A. Schumacker, 1974. "A Characterization of Ten Hidden-Surface Algorithms." *Computing Surveys*, 6(1):1–55.

Tarboton, D.G., R.L. Bras, I. Rodriguez-Iturbe, 1991. "On the Extraction of Channel Networks from Digital Elevation Data." *Hydrological Processes*, 5:81–100.

White, D.A., R.A. Smith, C.V. Price, R.B. Alexander, and K.W. Robinson, 1992. "A Spatial Model to Aggregate Point-Source and Nonpoint-Source Water-Quality Data for Large Areas." *Computers and Geosciences*, 18(8):1055–1073.

CHAPTER 13

Spatial Arrangement

Thus far we have focused on characterizing the objects we have seen. But to evaluate our environment effectively, we also need to know the distributional relationships among the individual items we see as well as the properties of their intervening spaces. Rather than looking at the amount of space occupied by a single object or how the object's space is configured (i.e., its shape), we now concern ourselves with the overall arrangement of objects in space. Such arrangements can be characterized by how many occur in a particular area and by how they are positioned or distributed in that area of geographic space. In this chapter we will examine the distance relationships between objects themselves and their relationships to the overall size of the area they occupy.

Such distributional patterns occur for a wide array of natural and anthropogenic objects and settings. We know, for example, that some human population distributions are widely dispersed in sparsely populated rural areas of the earth, while others are more concentrated in larger collections we call cities. Plants and animals may occur individually in evenly distributed arrangements, or they too may form more concentrated groups. Even physical features such as sediment types and landscape features—streams and hills, mountains and valleys—may occur as widely spaced individuals or as larger groups. Anthropogenic objects like roads, fences, and houses can occur in patterns as well. As your geographic filter increases, you will see still more. Observing that there are differences in **spatial arrangements** of objects allows you to ask questions about what these patterns are, how they can be classified, and how they might provide insights into the processes that made them.

With time we see that the observed patterns change. Regions that once appeared to show diffused and scattered patterns may now show evidence of coalescence. Objects that were once randomly organized in space may now occur in **regular**, repeating patterns. Some distributional patterns may show growth and expansion, whereas others may be shrinking or vanishing. Patches or areas may merge; single linear objects may begin to connect to form networks; stabilized dunes may begin moving and diffusing. In all these cases time is an important component in our understanding of spatial arrangement. And we soon wonder what the processes are that cause the change from one arrangement to another. We can ask many questions about direction of change, driving forces, upper and lower limits of driving forces and many more.

LEARNING OBJECTIVES

When you are done with this chapter you should be able to:

1. Clarify what is meant by arrangement of objects and evaluate the importance of such knowledge.
2. Explain what density is and provide examples of high- and low-density distributions.
3. Define a uniform distribution. Tell what makes a uniform distribution regular versus random. Define a clustered distribution.
4. State the significance of regular, random, and clustered distributions in terms of the potential processes that caused them. Give some examples of each distributional pattern and give a hypothesis that might describe the causes.
5. Explain what nearest neighbor analysis tells us about the distribution of objects.
6. Perform a nearest neighbor analysis on point objects and explain the results.
7. Give the purpose of Thiessen polygons. Diagram and describe how they are created.
8. Give three concrete, real-world examples of the use of distance or adjacency as a method of retrieval of cartographic features.
9. Describe the process of performing a nearest neighbor statistic on line patterns and state what this quantity tells us about the distribution of our lines. Describe some situations in which the nearest neighbor statistic for lines might give misleading results.
10. Describe the use of line intersect methods for analyzing line pattern distributions. Define a random walk.
11. Define a vector resultant and state what it tells us about the patterns of linear objects. Define resultant length and discuss how it compares to resultant force problems in physics.
12. Define the mean resultant length, explain how it differs from the raw resultant length, and tell when we would use it. Answer questions such as the following: What would a large mean resultant length tell us? What is the difference between circular variance and mean resultant length?
13. Describe how we can adjust our measurements for mean direction, mean resultant length, and circular variance to account for orientations that can be measured in either of two different directions.
14. Describe how the gamma index is performed, spell out what it tells us about a given network, and know what a gamma value of 0.48 tell us in terms of the amount of connectedness in our network.

15. Tell what the alpha index is, how it is different from the gamma index, and how it is the same. Give some nonroad network examples of how both these indices might be useful.
16. Describe, in general terms, the gravity model. Explain how distance affects the interactions of point objects and how the magnitude of the nodes changes the interactions between them. Give an example of how the gravity model might be applied.
17. Describe the types of attributes that might need to be assigned to the nodes, and the arcs between nodes, in working with routing and allocation problems.

POINT, AREA, AND LINE ARRANGEMENTS

Spatial arrangement is the placement, ordering, concentration, connectedness, or **dispersion** of multiple objects within a confined geographic space. Until now, most of the analytical techniques we have examined have dealt either with individual objects or with collections of objects when they might be defined as regions or neighborhoods or conceived of as statistical surfaces. We have even touched very briefly on the possible **interactions** of objects, regions, surfaces, and neighborhoods with those of other map layers. However, most objects we encounter in a single map also have a definable spatial patterning that might indicate possible mechanisms for producing them.

Spatial arrangement generally refers to the simple cartographic display of spatially distributed objects. In such a "map as communication" mode, we say that the map shows us where the objects are located and illustrates the shape of this distribution. There is, however, more that can be used to describe the interactions of each individual object to its neighbors and the relationship of all these objects to the whole space in which they reside. If we can identify ways to measure these relationships, we can also find ways to isolate and understand the possible functional mechanisms that produce such patterns.

POINT PATTERNS

Perhaps the most common techniques for analyzing spatial distributions are applied to point patterns. Point objects can be individual trees, houses, animals, street lights, and even cities, depending on the scale (Figure 13.1a). As you will see later, point objects can be represented in linear or area form as well (Figure 13.1b).

The simplest measure of point patterns is **density**, the number of points per unit area, which provides us with a quantifiable descriptor of the numbers within the distribution. Population density, housing density, tree density, and so on are commonly used to provide a measure of the compactness of points. With that information, we can compare the densities to those of comparable point objects in other areas to compare or contrast the relative mechanisms operating in each. Or we might compare the points in the same location but at different times

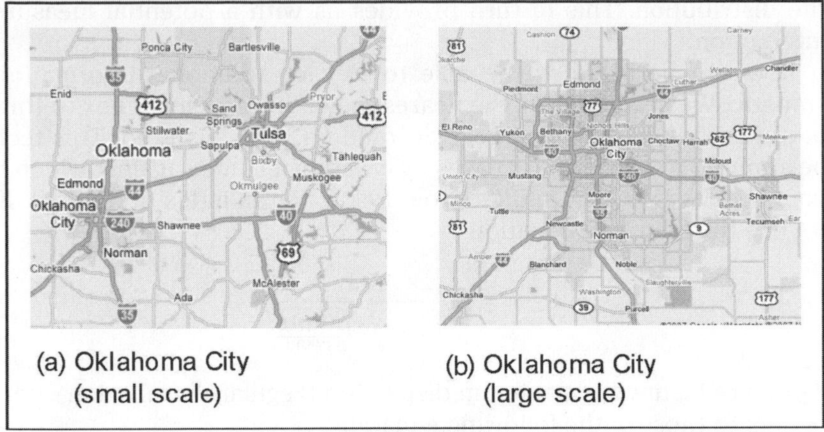

(a) Oklahoma City (small scale)

(b) Oklahoma City (large scale)

Figure 13.1 Point and area representation of cities. Area objects, in this case cities, can be viewed (*a*) as points or (*b*) as lines, depending on the scale at which they are presented. This shows the close association between the analytical techniques that can be applied.

to give us an idea of change in density through time. We might find, for example, that population density is increasing in urban areas through time, or that housing density is increasing, or that tree density decreases as the trees mature and compete for space and sunlight. Even this simple statistic, easily calculated in either vector or raster, can give us many useful insights into our data.

No matter what the overall density of the distribution, we may be interested in the form of the distribution in addition to the numbers per unit area. Point patterns are said to occur in one of four possible arrangements, each with a specific set of criteria. A pattern is **uniform** if the number of points per unit area in one small subarea is the same as the number per unit area in each other subarea. If the points occur on a grid, separated by exactly the same distance throughout the entire area, the uniform pattern is said to be regular, as you saw with a regular grid in Chapter 11 when sampling surface data points. In some cases, uniformly distributed points may occur in **random** locations scattered throughout the study area. Or, in other cases, the points are grouped in tight arrangements in a pattern called **clustered** (see Figure 2.8).

NEAREST NEIGHBOR ANALYSIS

Thus far, we have confined our characterization of point distribution patterns to the areas they occupy. However, it is also instructive to examine the locational relationships from one point to another. To do this, we most often rely on an alternate method of point pattern analysis, called **nearest neighbor analysis**, which is a common procedure for determining the distance of each point to its nearest neighbor and comparing that value to an average between-neighbor distance. Calculating this statistic involves first selecting each data point and determining its nearest neighbor and measuring each pair-wise distance. Then the software determines the mean of these nearest neighbor distances. The mean nearest neighbor distance is an index of the spacing between points in the distribution. This in turn provides us with a potential measure of object interaction.

It would be useful to be able to extract additional information from the statistic. We can compare the nearest neighbor spacing index to three possible patterns that might occur: regular, random, and clustered. This technique can be applied to each of these cases by creating an index against which to compare your results (McGrew and Monroe, 1993). For an index of random distribution, we use the following equation:

$$\frac{1}{2 * \sqrt{\frac{\#\text{points}}{\text{area}}}}$$

If you are testing for maximum dispersion (regular distribution), compare your results to those of the following equation:

$$\frac{1.07453}{\sqrt{\frac{\#\text{points}}{\text{area}}}}$$

Finally, for a test of maximum clustering, we assume the value of the divisor is 0. Any value you obtain for your index that is not 0 will be positive. A simple comparison of your average nearest neighbor distance to that of these three index values gives you an idea of where they fall along the continuum.

Let's examine how this works with the set of data in Table 13.1 and the associated plot (Figure 13.2). We have six data points, each with a single (X,Y) coordinate pair, all falling within an area of 25 square units. The average nearest neighborhood value is approximately 1.4 (average of the six nearest neighbor distances). For randomly distributed points, the index is found from:

$$\frac{1}{\sqrt{\frac{6}{25}}} = 1.02$$

Our average nearest neighbor of 1.4 is somewhat higher than the random value, 1.02, and much smaller than the dispersed value of 2.19 calculated from the

TABLE 13.1 Calculating Nearest Neighbor Distance[a]

Point	Coordinates *X*	*Y*	Nearest Neighbor	Nearest Neighbor Distance (NND)
A	0.7	1.0	B	1.6
B	1.25	3.0	C	1.4
C	2.5	3.7	D	1.3
D	3.3	2.75	C	1.3
E	4.0	4.0	C	1.34
F	3.8	1.0	D	1.5
Total				8.44
Average NND				1.4
Random Average NND				1.02

[a]Nearest neighbor distance calculated by using the Pythagorean theorem.

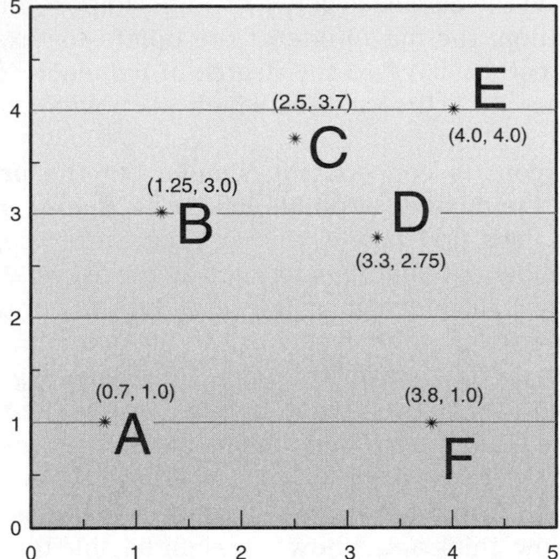

Figure 13.2 Nearest neighbor point coordinates. Locations of data points for determining nearest neighbor distance (see Table 13.1). Each point (e.g., point *A*) has an associated nearest neighbor (in this case, point *B*). The distances are measured using the Pythagorean theorem.

following:

$$\frac{1.07453}{\sqrt{\frac{6}{25}}} = 2.19$$

And, of course, a perfectly clustered distribution is 0, and our average nearest neighbor is much larger than that. So what we find is that our average nearest neighbor is slightly more dispersed than random, or somewhat between truly uniform and random. In other words, it is starting to take on a more regular configuration but as yet is still rather randomly distributed.

These values give only a measure of the patterning of these point features, and we might need to decide how we might later compare these to other map layers or simply use the statistic as a descriptor for further statistical analysis. The nearest neighbor statistic is an absolute measure; hence, it is not immediately amenable to comparison with nearest neighbor statistics for other point distributional patterns. The nearest neighbor index can be standardized to permit such comparisons (McGrew and Monroe, 1993), but the techniques are beyond an introductory course in GIS. In addition, there are other methods of determining the degree of clustering on the basis of other statistics (Davis, 1986; Griffiths, 1962, 1966; Ripley, 1981), but again, these are more detailed than we need for an introduction to the use of nearest neighbor analysis in GIS.

THIESSEN POLYGONS

Points can be organized into a more regional context without relying on the use of other map layers for comparison by means of **Thiessen polygons** (also referred to as **Dirichlet diagrams** and **Voronoi diagrams**). Beyond examining the arrangement of points, we can grow polygons around each point to illustrate the "region of influence" it might have relative to other points in the map.

For example, we assume that the distances between points impose an attraction on the neighbors. In addition, the magnitude of the point—for example, the size of a city—is often directly related to the degree of influence. To simplify this discussion, we restrict it to the case in which all points are of equal magnitude.

Creating Thiessen polygons is conceptually simple, but the process can become quite involved as the number of points increases. Before we examine how these are developed, let's first take a look at what these shapes are to represent. If we have a number of point objects, such as towns, we can imagine that each is surrounded by a single irregular polygon. But the polygon has an important property—all the space within it is closer to the point encircled than to any other. By contrast, every location outside each Thiessen polygon is closer to some other point than to the enclosed point. In other words, the boundaries of each polygon give to the point it surrounds the most compact possible area of influence. In most cases, each point in the map will have its own Thiessen polygon, sometimes called a **proximal region**, to show its exclusive area of influence (Clarke, 1990). Now, think about how we might be able to do this.

Let's take a small sample of points to show this (Figure 13.3). With five points as a starting place, we might conceptualize how the Thiessen polygons are derived by envisioning a bubble growing outward from each point, much like the bubbles formed from bubble pipes. If you look carefully, you can see that each of the interfaces between bubbles forms a straight line (if viewed orthogonally). This straight line is oriented perpendicular to a line that connects each pair of neighboring points. If you measure the distance between the two end points of the perpendicular line, you see that the distances are identical on either side of the line forming the interface. In other words, the edges of the polygons are formed by an equidistant (bisector) line perpendicular to the line connecting each pair of points. Algorithms for producing Thiessen polygons have been implemented in both CAC and GIS systems for decades, both in

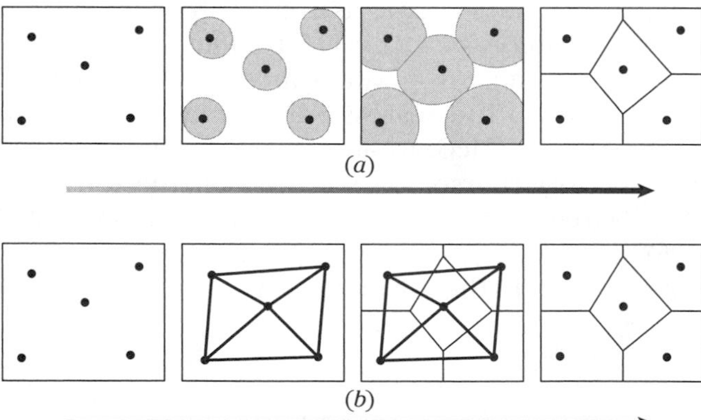

Figure 13.3 Creating Thiessen polygons. Construction of a Thiessen polygon from five points: (*a*) arrangement of the data points, (*b*) construction of the related Thiessen polygons. Note the similarity between the form of the Thiessen polygons and the pattern we might get from a bubble pipe. *Source:* Figure derived from Environmental Systems Research Institute, Inc. (ESRI) drawings.

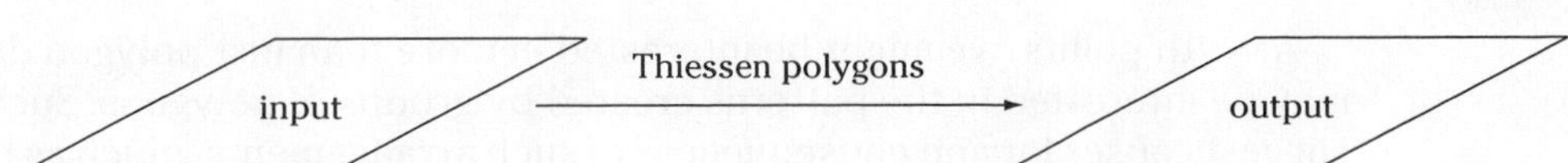

Figure 13.4 Flowchart illustrates the application of Thiessen polygon operations to identify proximal areas.

vector systems (Brassel and Reif, 1979) and even in quadtree data structures (Mark, 1987). The flowchart in Figure 13.4 shows how Thiessen polygons might be implemented in a GIS.

So now that you know how to create Thiessen polygons, what do you do with them? You can gain some initial ideas by looking at the reason for their development in the first place. The Thiessen polygon was named after the climatologist A.H. Thiessen, who wanted to be able to interpolate climate data from highly uneven distributions of weather station data. He was trying to describe and analyze point data with area-based symbols and analytical techniques. Thus, if we have only a few scattered points of data but we want to characterize regions on the basis of these points, we use Thiessen polygons to divide our space into a polygonal form. Because we assume that in each polygon the influence of the enclosed point is absolute, we can treat the data as a polygonal map.

Common uses of Thiessen polygons involve determining the influence of point data representing shopping centers, industries, or other economically based activities. If we modify the position of our perpendicular line on the basis of the size or other magnitude of each point, our Thiessen polygons become even more representative of the actual influence of industries or shopping on the surrounding space. With such information, the economic placement specialist might determine how much of a city's population (on the basis of proximity) will likely frequent a planned shopping center. One unique noneconomic example involves their use to detect spatial patterns of vegetation (Hutchings and Discombe, 1986). Actually, the use of this technique is likely to increase as the functional capabilities of the GIS become familiar to users.

AREA PATTERNS

So far we have focused on the distribution of points. If we change our scale somewhat, our points become areas, and their patterns can then be analyzed by similar methods. We can begin to analyze area patterns much as we did with points—by determining the density of polygons per unit area of our study area. In performing a density measure of polygons, however, we must first measure the area of each class of polygons in which we are interested. Then, knowing the area of each polygon, we add the area of all polygons with the same attributes to obtain a total area for that class. Then we divide that by the overall area of the map. This gives a percentage rather than a number per unit area as in point density. It is possible, of course, to calculate the number of polygons per unit area, but because of the potential for widely varying polygon sizes, this value has little value.

As with points, we might be interested in more than just polygon density: We may be interested in the patterns created by groups of polygons. Such patterns suggest causes for and consequences of such arrangements, much as they do for point patterns. But before we can examine the interactions of polygonal objects with one another, we must know something about how we can describe their patterns. As with point patterns, areas can be clustered, dispersed (regular), and randomly spaced relative to one another (see Figure 2.8). In addition, area patterns can be connected to one another or separated by some definable distance. Let's look at some methods of analyzing these pattern metrics.

DISTANCE AND ADJACENCY

Arrangement is inherently based on distance. Among the techniques commonly available for evaluating arrangement are those involving retrieval based on distance. For example, to evaluate if there are school yards within 1,000 meters of a highway, we query the software to select and highlight all school polygons within 1,000 meters of the highway. Or we might want to know if there are substantial numbers of at-risk people (e.g., elderly) within 1,000 meters of hazardous cargo routes on a highway. To answer this, we search for polygons representing rest homes or subsidized housing. Another scenario might include a request for all high-income neighborhoods within a certain distance of the local mall so the shop owners could perform a market analysis. All of these scenarios and many more merely require a query based on distance.

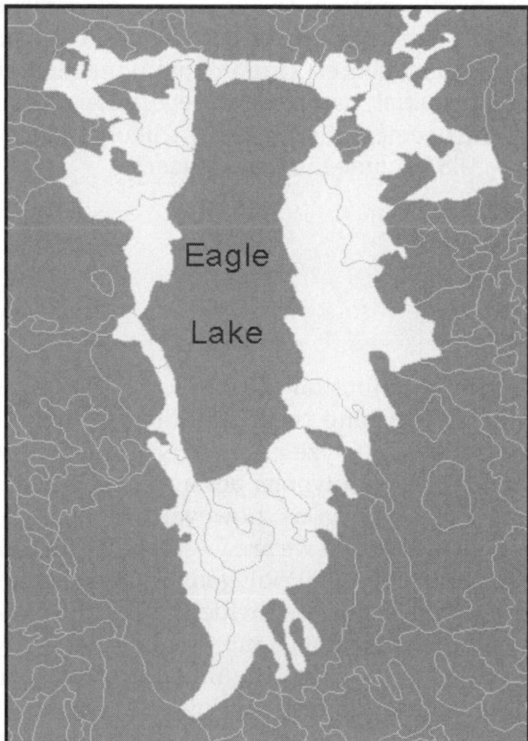

Figure 13.5 Adjacency. Notice how only the polygons that are touching Eagle Lake are highlighted. Unlike buffering, this technique only locates and selects polygons that are within a particular distance. In this case, the distance was set to zero so adjacent polygons would be selected.

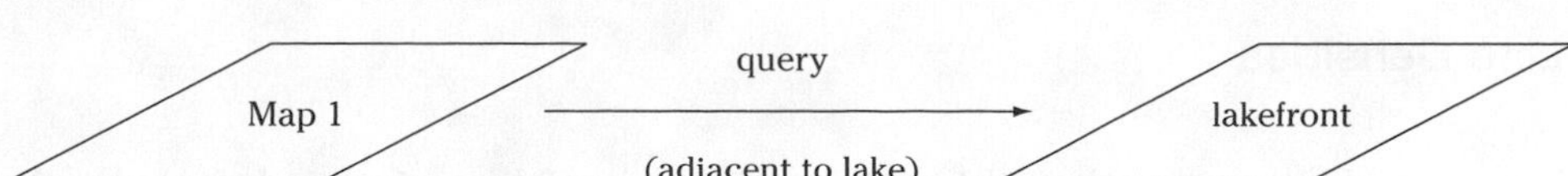

Figure 13.6 Adjacency flowchart. The query measures distance from the lake, but the distance is set to 0. Notice how the flowchart itself simplifies the notation.

Adjacency is a special case of this arrangement measure and the distance-based retrieval method that implements it. The most common form of this measure is to begin with a polygon or a number of polygons and to determine which of these are adjacent to other polygons. Let us say, for example, that you want to know which land parcels are adjacent to a lake (essentially a measure of lake front property). To do this we would select the lake first and query all polygons that are immediately connected (adjacent) to the lake (Figure 13.5). The simple flowchart in Figure 13.6 shows how this might be illustrated.

OTHER POLYGONAL ARRANGEMENT MEASURES

Analyses of polygon arrangements can become quite sophisticated, and software links to a GIS have provided opportunities to perform them (Baker and Cai, 1992). Among the most well known of such specialized software packages is FRAGSTATS (McGarigal and Marks, 1994), which was designed specifically to analyze polygons. Landscape ecologists have produced and use a large set of these techniques, usually treating polygons as patches, especially with respect to a larger, more uniform background (matrix) (Forman and Godron, 1986). Their techniques include measures of polygonal **isolation**, measures of **accessibility** (connectedness of polygons to lines), polygon interactions, and dispersion. Because many of these measures are borrowed from the literature in geography, biogeography, ecology, forestry, and other disciplines, the examples will be varied enough to give you a feel for how these additional measures might be used.

LINEAR PATTERNS

As with points and areas, lines occupy space and occur in particular configurations and patterns. We see many linear patterns but often fail to note them. Streets and highways form lines with particular definable patterns (networks) running between spatially arranged points called towns. We see fencelines, also occurring in particular configurations and numbers depending on field size, lot size, and shapes of the polygons they trace (Simpson et al., 1994). Striations on exposed bedrock show parallel lines indicating the dragging of rock below the glaciers as they scoured the land thousands of years ago. The mechanisms that caused each of these line patterns can best be understood if we first define the specific patterns and concentrations before us. Let's look at some of the more common measurements of line patterns available to the GIS professional.

Line Densities

Because of the added dimensionality of lines (as with polygons), especially lengths and orientations, linear analysis is somewhat more complicated than the study of points. Some researchers have tried to examine the distributions of line lengths (Aitchison and Brown, 1969), and some have looked at line spacing (Dacey, 1967; Miles, 1964), much akin to the analysis of nearest neighbors in point distributions (Davis, 1986). We examine these as well as the larger body of knowledge of line orientation in the next sections. Meanwhile, we need to introduce the simplest measure of line patterning—line density.

Like point and polygon arrangements, lines can occur with greater or lesser density. To measure this for points (0-D), we divide their number by the area. When we worked with polygons, however, we accounted for the area occupied by each polygon by adding all the areas and dividing this sum by the total map area. So, to measure line density (with one dimension), we divide the sum of all the line *lengths* by the map area, which will be in units such as meters per hectare or miles per square mile. Except for comparing with other values for different regions or for the same region at different time periods, there is not much we can do with this information. We need to know more about the distributional patterns of the lines, just as we did with points and polygons. Next, we examine the nearest neighbor statistic as it is applied to lines rather than points. Although the calculations are somewhat different than for points, the results provide a similarly useful characterization of line patterns.

Nearest Neighbors and Line Intercepts

You have seen how the nearest neighbor statistic is used to characterize point arrangements. The relative distribution of neighboring lines is determined in much the same way. We might choose a simple solution by selecting the center of each line and performing a nearest neighbor analysis on that point alone. Because the linear objects are often of different lengths, however, this procedure is not going to give us a true picture of the arrangement of the lines themselves. From a statistical standpoint, it is often considered useful to sample randomly. Following that precept, our first task in measuring a linear nearest neighbor is to select a random point on each line on the map (or on each line segment if the lines are not straight). Next we draw a line perpendicular between each nearest neighbor line and measure its distance (Davis, 1986) (Figure 13.7), and then we calculate the mean nearest neighbor distance for all the distances.

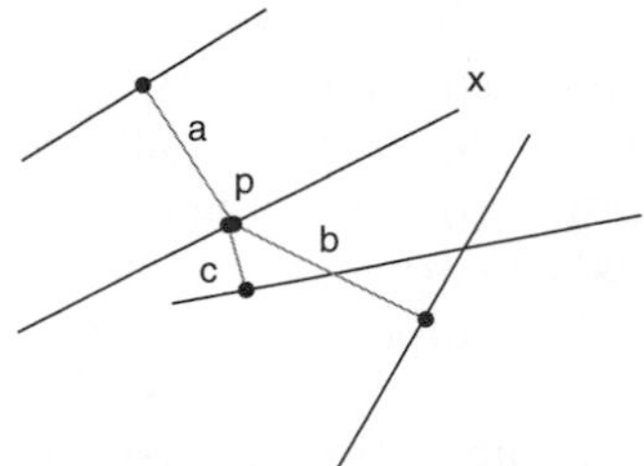

Figure 13.7 Nearest neighbor distances between lines. Finding the nearest neighbor between lines using a randomly selected point along each line as the sample location. *Source:* J.C. Davis, *Statistics and Data Analysis in Geology*, 2nd ed., John Wiley & Sons, New York, © 1986. Adapted from Figure 5.16, page 313. Used with permission.

As with all nearest neighbor statistics, we need to be able to test this value against a random distribution. Dacey (1967) determined values for expected nearest neighbor distance, expected variance, and standard error in a random distribution of lines. These values allow us to compare the expected and the observed and to generate a statistic against which to test our hypothesis of randomness (Davis, 1986).

This test will work for most line patterns, whether the features are straight lines or curved lines, but it has some limitations. If lines in your map are very sinuous, this approach is less than optimal. For the test to be useful, the lines should be at least 1.5 times the length of the mean distance between lines. If the number of lines on the map is small, the estimate of density used in nearest neighbor analysis should be adjusted by a weighting factor of $(n - 1)/n$, where n is the number of lines in the pattern. So rather than simply dividing the sum of the lengths by the area, we use the formula

$$\frac{(n-1)L}{nA}$$

where

L is the sum of the lengths
A is the area

This adjusted line density value will improve the quality of the nearest neighbor statistic.

Line intersect methods are alternative methods for analyzing line distributions. One approach is to convert the two-dimensional pattern into a one-dimensional sequence by drawing a transect line across the map and noting where the sample line intersects the map line objects. There are at least two basic methods of producing sample lines (Getis and Boots, 1978). The first is to randomly select a pair of coordinates and connect them with lines. A second is to draw a radius at a randomly chosen angle and starting point, then measure a random distance from the center, finally constructing a perpendicular to that radial line (Davis, 1986). Once the random test transect has been placed over the map lines, the distribution of intersection intervals can be tested with a **runs test** or other tests of sequences of data. An alternative to a single line is to use a zigzag line that crosses the map two or three times. The zigzag path (often called a **random walk**) will again produce a series of line intersects, the distances of which can be tested with any simple statistical test for sequences of data (Figure 13.8). If the resulting pattern distribution is not random, it suggests a nonrandom process as its cause.

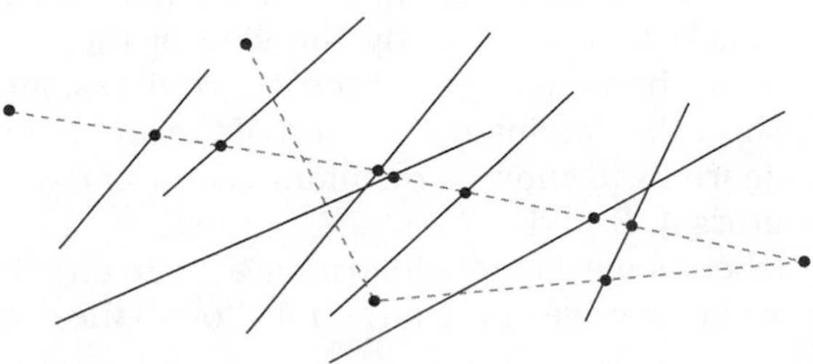

Figure 13.8 Random walk method to evaluate line patterns. Modification of the line intersect using a zigzag line called a random walk to identify sample points. *Source:* J.C. Davis, *Statistics and Data Analysis in Geology*, 2nd ed., John Wiley & Sons, New York, © 1986. Adapted from Figure 5.17, page 314. Used with permission.

DIRECTION AND CIRCULAR STATISTICS

Linear objects, and in some instances area objects, exhibit more than just distributional patterns on the landscape. Often we find that such features as sedimentary bedding planes, drumlins, glacial striations, water-deposited pebbles, glacial boulder trains, fencerows, street grids, and forest blowdowns exhibit a preferred orientation. These features and many more are oriented in a particular direction, often strongly related to a functional force. We know, for example, that tree shelterbelts in agricultural regions are often placed in directions opposite the prevailing winds, that trees felled from hurricanes give evidence of the wind directions occurring within the storm, and that glacial striations are indicative of the movement of the glacier. But when we analyze orientation, we see that if lines are oriented in, say, a predominantly north–south direction, we have two directions to choose from. If these linear objects are, for example, one-way streets, the orientation tells us the direction the street is facing, but doesn't tell us the direction of traffic flow within it. So, besides orientation we also need to consider directionality. We can also think of the distributions of linear objects as either two-dimensional (i.e., distributed purely in azimuthal directions) or as three-dimensional (i.e., distributed in an azimuthal direction combined with an angular direction in a sphere). Although both are important, we will concentrate on the simpler two-dimensional forms and leave the spherical directional and orientation measures for more advanced treatments (Davis, 1986).

In circular statistics data from maps of linear objects (or major axes for area objects) are plotted on a circular graph called a **rose diagram**, where each line starts at the center and is drawn in its correct direction. In some rose diagrams, the length of the line also indicates the magnitude (such as wind speed) or length of the feature (as in the lengths of hedgerows or fences). But although rose diagrams provide a useful framework for visual inspection of directions, the measurements themselves obtained directly from our map are much more useful for numerical analysis. Our first analytical technique is to determine the **resultant vector**.

Let's take an example of directional analysis in two dimensions by considering a large number of trees that were blown down by strong straight-line winds. Each blown-down tree might be plotted as a line feature on a map; the positions of the top and bottom of each tree would be recorded as well, giving us both the orientation and the direction of every downed tree (Figure 13.9). To deduce the overall direction of the winds on the basis of the felled trees, our first task is to analyze the vector resultant of all the blown-down trees.

The vector for each tree is defined by an angle (θ) (theta) measured from the base of a tree to its top. We multiply the X coordinate of each tree by the cosine of theta and each Y coordinate by the sine of theta. To find the resultant vector, we sum all these values for both coordinates, and resultant vector values X_r and Y_r show the dominant direction of the end points of all the trees in the blowdown. Figure 13.10 shows a resultant vector of R obtained from three values of θ, the vectors A, B, and C.

We may want to go further, however, and determine a mean direction $\overline{\theta}$ based on the resultant vector. As an average of the directions for all the downed trees,

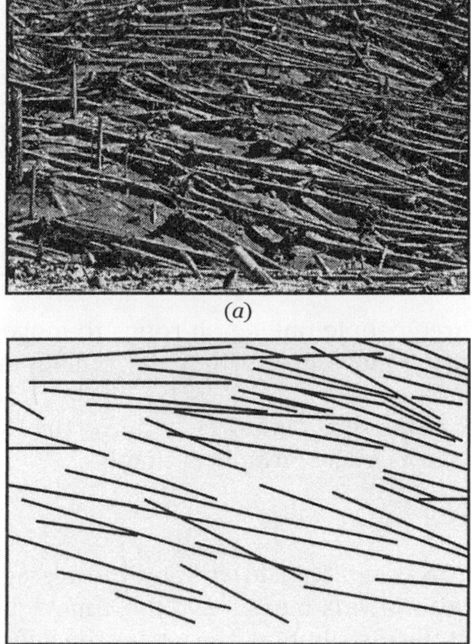

Figure 13.9 Line patterns of felled trees. A portion of a photograph of part of Mount St. Helens after the eruption (*a*). The photograph could be digitized to show the linear objects and their general trend (*b*).

it is analogous to the mean of any other set of data. The formula for mean is

$$\bar{\theta} = \tan^{-1}\left(\frac{Y_r}{X_r}\right)$$

Because the mean direction of our vectors depends not only on the dispersion of the trees but also on the number of observations, we can standardize these values by dividing the coordinates of each resultant vector by the number of line objects in the map. This allows us to compare two different areas. For example, by comparing the average vectors for downed trees from two study areas, we could determine whether the winds were generally coming from the same direction. The flowchart in (Figure 13.11) shows how the mean vector (sometimes called **linear directional mean**) might be applied in GIS.

As with any sample in which the mean serves as a measure of the central tendency of the distribution (in this case our orientation), we can use the average to develop other statistics that define the amount of spread away from the mean. Figure 13.12 shows two cases of three vectors being used to determine the resultant vector R. When the vectors are closely spaced, the resultant vector is long, whereas the widely dispersed vectors give a shorter resultant vector.

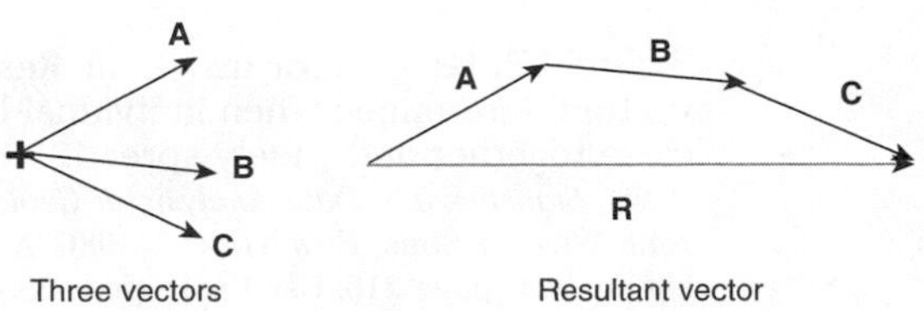

Figure 13.10 Resultant vector for three individual linear features. *Source:* J.C. Davis, *Statistics and Data Analysis in Geology*, 2nd ed., John Wiley & Sons, New York, © 1986. Adapted from Figure 5.20, page 317. Used with permission.

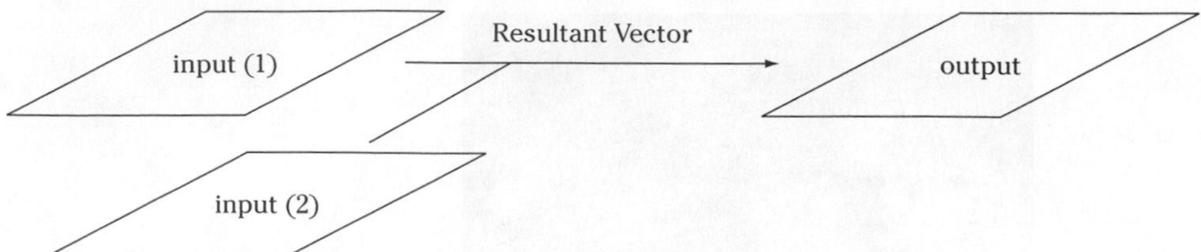

Figure 13.11 Flowchart illustrating the implementation of resultant vector (mean directional vector).

Think about this in terms of three people pulling on ropes to move an object. The closer the people are and the more nearly they are pulling in the same direction, the more force results.

We can determine the length of the resultant force by using the Pythagorean theorem on our resultants (X_r and Y_r). The formula is simply

$$R = \sqrt{X^2 + Y^2}$$

where R is the **resultant length**. Thus, not only do we know the average direction in which the trees lie, but we also have a measure of the compactness of the distribution. The more compact the distribution, the longer the line. Again, this is analogous to the three ropes pulling an object. The closer together the ropes (in other words, the more closely spaced the vectors), the stronger the force (or the greater the length of the resulting vector).

To compare the resultant length of our vector to that of another site, we again need to standardize the data. We find the **mean resultant length** $\overline{R}$ by dividing the resultant length R, found with the Pythagorean theorem, by the number of observations, n. This value will range from 0 to 1. The mean resultant length resembles variance in linear statistics inasmuch as it is a measure of spatial dispersion about a mean value. The value of mean resultant length, however, is expressed in the opposite direction; that is, large values indicate that the lines are closely spaced and small values mean they are more dispersed in direction. Thus, a large mean resultant value in our tree example would indicate that the winds were nearly perfectly undirectional, whereas a small value would suggest eddies in the overall wind pattern. If the inverse nature of the mean resultant length seems counterintuitive, you can use its complement, defined by 1 minus

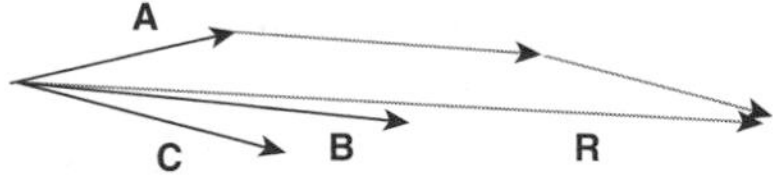

Closely spaced around a single direction

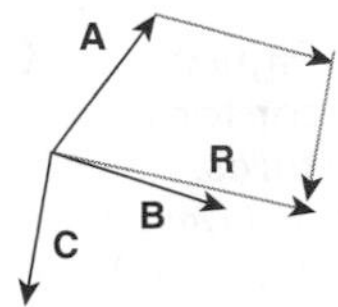

Widely dispersed vectors

Figure 13.12 Vector dispersal. Resultant vectors determined when individual lines are close together and widely spread. *Source:* J.C. Davis, *Statistics and Data Analysis in Geology*, 2nd ed., John Wiley & Sons, New York, © 1986. Adapted from Figure 5.21, page 318. Used with permission.

the mean resultant length. This value, called the **circular variance**, will give increasing values as the spread of vectors increases. Directional analogs to standard deviation, mode, and median also exist (cf. Gaile and Burt, 1980) and may prove useful in some circumstances.

One aspect of orientation for linear objects is that each possesses two possible (and opposite) directions. For example, an object might be oriented north-south. Fortunately, Krumbein (1939) found a simple way to resolve this problem. He observed that when he doubled the angles, no matter which direction was originally used to record the data, the same angle was recorded. Suppose that a fencerow is oriented northwest to southeast (that is, 315 degrees to 135 degrees). If we multiply each of these values by 2, we obtain the same number: $315 \times 2 = 630$ degrees (2,360 degrees $\times$ 270 degrees) and 135 degrees $\times\ 2 = 270$ degrees. The calculations for mean direction, mean resultant length, and circular variance will now be doubled. To get our true values for these measures, all we need to do is divide by 2.

These simple measures of directionality and dispersion can be tested for randomness (Batschelet, 1965; Gumbel et al., 1953) and for specific trends (Stephens, 1969) through standard processes of hypothesis testing. Other sets of directional data can be compared to those for our original map (Gaile and Burt, 1980; Mardia, 1972) but are beyond the scope of this book.

CONNECTIVITY OF LINEAR OBJECTS

An important aspect of the spatial arrangement of lines is their ability to form networks. Networks occur in many different forms and are both natural and anthropogenic. Roads and rail lines are networks that serve to move and transport people and materials from place to place. Telephone lines are networks to allow the movement of information and communication. Rivers and streams are networks for aquatic creatures, and even hedgerows may act as networks to allow the movement and migration of small mammals through a landscape. The list of systems that may be defined as networks is quite large. And although the density of such features and their orientation may be of interest to us, we need to be able to analyze the actual connections made by these features and the amount of **connectivity** provided from place to place. Connectivity is a measure of the complexity of a network. There are several devices for calculating this value (Haggett et al., 1977; Lowe and Moryadas, 1975; Sugihara, 1983; Taaffe and Gauthier, 1973). The two most common ones are the **gamma index** and the **alpha index**.

The gamma (γ) index compares the number of links, L, in a given network to the maximum possible number of links between nodes. To calculate this, we simply produce a ratio of the two. First we count the number of links (line segments between nodes) that actually exist in the map. As you might guess, this is very difficult in raster GIS but is easily accomplished in vector. Once we have the number of actual links, we need to determine the number of possible links in the map. This is not as difficult as it might appear, because limits to the number of links are determined entirely by the number of nodes. If, for example, we have three nodes present, only three links are possible (Figure 13.13), but if we add another node, we see that three additional links are possible, for a total of six (Forman

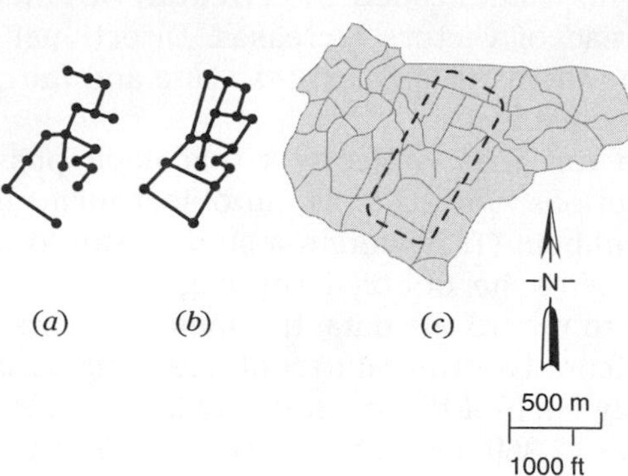

Figure 13.13 Gamma index. Two different networks of hedgerows based on the same set of nodes. The network on the left (*a*) is minimally connected and has no circuits, whereas the network on the right (*b*) has higher connectivity and has circuits that provide alternate routes for travel. (*c*) Shows the study area for each set of networks. *Source:* R.T.T. Forman and M. Godron, *Landscape Ecology*, John Wiley & Sons, New York, © 1986. Adapted from Figure 11.12, page 418. Used with permission.

and Godron, 1986). So if we assume that no new intersections (line crossings) are formed, the maximum number of possible links is increased by three each time (Figure 13.13). In other words, the maximum number of links, L_{max}, is always $3(V - 2)$, where V is the number of nodes. To find the gamma index, we simply divide the number of links, L, by the maximum number of links, L_{max}.

$$\gamma = \frac{L}{L_{\max}} = \frac{L}{3(V-2)}$$

The gamma index ranges from 0, where no nodes are connected, to 1.0, where all possible links between nodes are present (Forman and Godron, 1986). Figure 13.13 reveals the difference in connectivity for two different cases with 16 possible nodes. The case in Figure 13.13a shows 15 links among the 16 nodes. This gives us a connectivity of

$$\gamma = \frac{L}{3(V-2)} = \frac{15}{3(16-2)} = \frac{15}{42} = 0.36$$

whereas the case in Figure 13.13b has 20 links for the 16 nodes, yielding a connectivity of

$$\gamma = \frac{L}{3(V-2)} = \frac{20}{3(16-2)} = \frac{20}{42} = 0.48$$

So the first network is about one-third connected (36 percent), whereas the second is almost half connected (48 percent).

Figure 13.13 shows hedgerows, which might be used by small mammals as transportation corridors. The same arrangement of features could easily have been a road network. In the first network, the route from the starting point at R to the endpoint at S would be relatively long compared to that for the second network with its increased connectivity. The increased connectivity will result in improved movement.

However, connectivity is not enough to fully characterize networks. Recall our discussion of the formation circuits (loops) within networks. A good example of a circuit is an interstate highway beltway in the United States, a circular highway surrounding urban areas. As we travel across the United States, we must decide whether to travel through or around each city. If we choose to go around, thus avoiding the urban traffic, we have two possible routes: we can turn either left or right to go around the city counterclockwise or clockwise. Thus, the loop feature offers choices for travel, allowing us to select the shortest route to a destination on the other side of the city.

The index designed to measure **circuitry**, the degree to which nodes are connected by circuits of alternative routes, is called the alpha (α) index. Similar to the gamma index, this measure is a ratio of the existing number of circuits to the maximum possible number. We see that a network with no circuits present contains one link fewer than the number of nodes: $L = V - 1$. If you look again at Figure 13.13, you see that this is true where the first network (a) has 15 links and 16 nodes. This figure is minimally connected in that it has the fewest number of links possible given the number of nodes, but each is still connected to at least one link. By adding a link to this network, we create a circuit or loop. Thus when a circuit is present, $L > (V - 1)$. Therefore, the number of circuits present in the network (i.e., the number of links present minus the number of links in our minimally connected network) can be given by $(L - V) + 1$. Thus we know the actual number of the circuits in the network (Forman and Godron, 1986).

An alpha index for the amount of circuitry is given by the ratio of the actual number of circuits present to the maximum number possible in the network. Therefore, because the number of possible links is determined by $3(V - 2)$ and the minimum number of links in a minimally connected network is expressed as $V - 1$, we can determine the maximum possible number of circuits (by combining these two simple formulas). That is, we subtract the number of links in the minimally connected network, $V - 1$, from $3(V - 2)$, the maximum number of links resulting in $2V - 5$. From that we obtain the final formula for the alpha index:

$$\alpha = \frac{\text{actual number of circuits}}{\text{maximum number of circuits}} = \frac{(L - V) + 1}{2V - 5}$$

The values for the alpha index range between 0 (no circuits) and 1.0 (a maximum number of circuits).

If we again use the network from Figure 13.13, the alpha value for Figure 13.13a is

$$\alpha = \frac{(L - V) + 1}{2V - 5} = \frac{(15 - 16) + 1}{(2 \times 16) - 5} = \frac{0}{27} = 0$$

illustrating that there are no circuits present. And the alpha value for Figure 13.13b is

$$\alpha = \frac{(L - V) + 1}{2V - 5} = \frac{(20 - 16) + 1}{(2 \times 16) - 5} = \frac{5}{27} = 0.19$$

demonstrating a circuitry of 19 percent. So in Figure 13.13a there are no options for travel; movement from point R to point S has only a single option. For Figure 13.13b, on the other hand, we have numerous alternative routes, some longer than others, to get from R to S.

Because of the addition of links needed to produce circuits, we could consider the alpha index as an alternative method of measuring connectedness. However, because the alpha and gamma indices give us a different look at the overall network patterning, they might more appropriately be combined to provide an overall measure of the amount of **network complexity**. To perform the alpha or the gamma index in a GIS requires that the GIS be a vector system. This relationship between these two indices and the vector GIS model is particularly strong because, like the vector GIS model itself, the indices are based on topological properties. In this case, the primary topological foundations are **graph theory**, which, as with the DIME topological model, is concerned less with line lengths and shapes than with their degree of connection between them.

For modeling transportation we need to know more than just how connected the network is. We also need to know, for example, the lengths of the lines between nodes, the permitted travel direction, and their impedance values. In addition, there are numerous other simple indices, derived mostly from transportation and communication theory, for characterizing the connectivity of networks. We can, for example, calculate the linkage intensity per node or the number of alternate routes between nodes; we can determine which node has the most links (**central place**); and we can delineate functional (as opposed to formal) regions based on connectivity and accessibility (Haggett et al., 1977; Lowe and Moryadas, 1975). And all these could be combined with each other and with the measures of linear dispersion, arrangement, and orientation to give a more detailed view of networks.

GRAVITY MODEL

In discussing connectivity we assumed that all the nodes were of equal size or importance. This allowed us to concentrate on the degree to which the links connected these nodes or the number of circuits and alternate paths available. However, think about the nature of towns and cities. Large cities are obviously more likely to provide more opportunities for shopping, more access to the arts, and more occasions to attend professional sporting events. This is why we tend to go to cities for shopping and entertainment. But towns and cities are not the only pairs of nodes that can exhibit different amounts of attraction. Take, for example, a small pond versus a lake. The lake is likely to attract more waterfowl than will be found near the pond.

Regarding the two examples of contrasting nodes, note that we have defined a magnitude for each that we can use to separate them out in a hierarchical fashion. Both large cities and lakes have a greater likelihood of attracting activity, whether by ducks or by people, than do their smaller counterparts. The magnitude of the attraction can be thought of much as astronomers envision the gravitational attraction between two celestial bodies. The larger the body, the greater the gravitational attraction between itself and neighboring bodies. We know this is true because astronauts who landed on the lunar surface were experiencing a gravitational attraction of only about one-sixth that of earth.

Converting the concept of gravitational attraction to a two-dimensional setting provides the measure of the interaction between two nodes in a GIS map that we call the **gravity model**. In its simplest form, the gravity model takes the form of the following formula:

$$L_{ij} = K\frac{P_iP_j}{d^2}$$

where

L_{ij} is the interaction between nodes i and j
P_i is the magnitude of node i

P_j is the magnitude at node j
d is the distance between the two nodes
the constant K relates the equation to the types of objects being studied (population size, animals, etc.)

The values of P can be represented by a force of interaction between the nodes, such as the demand for products, the amount of retail sales for city shopping centers, or the amount of wetland habitat for waterfowl.

As with planetary gravity, the larger the node magnitudes (planet sizes), the greater their interaction potential. Conversely, the greater the distance between the two, the less their attraction. Recall our city shopping example. The larger the city, the more shopping it is likely to offer, but the farther away you are from a given city, the less likely you are to visit it regardless of the potential shopping opportunities.

Gravity models (sometimes called potentials models) are normally implemented in vector GIS software, although they are also implemented in raster systems. In both events there are quite a number of variations on the basic model, but they are best studied in more advanced texts on spatial analysis. And although most gravity models are used for economic placement analysis, there are quite a number of possible applications. Investigators have used gravity models to describe the flow of passengers by air and land between cities, the volume of telephone calls, flows of birds, and the seeds the birds disperse among woodlots (Buell et al., 1971; Carkin et al., 1978; McDonnell, 1984; McQuilkin, 1940; Whitcomb, 1977). As long as your concern is the movement of objects between nodes of different magnitude, the gravity model may be of use to you.

ROUTING AND ALLOCATION

Among the most useful applications of networks in GIS are the related tasks of **routing** and **allocation**. Routing involves finding the shortest path between any two (in the simplest case) nodes in a network (Figures 13.14, 13.15). Because the nodes can be assigned weighting factors, a route might be between one point and the nearest point with the highest weight (e.g., product demand) (Lupien et al., 1987).

Each link in the net can also be assigned an impedance value, much like a friction surface but imposed only on the line itself. The impedance value

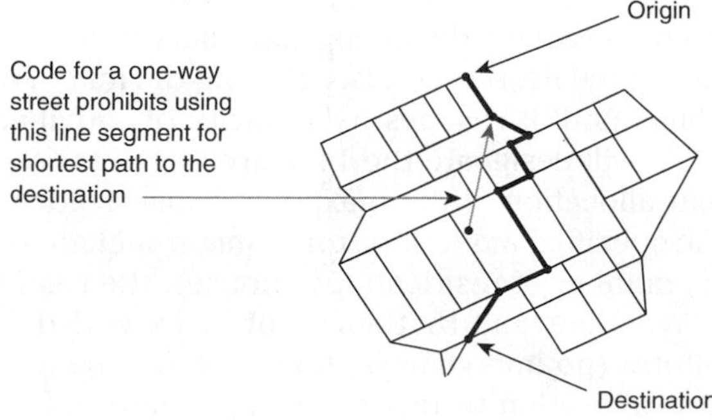

Figure 13.14 Shortest path along a network. A simple road network showing the solution for the shortest path algorithm. Notice the code entered along the diagonal road to indicate this one-way street is going in the wrong direction for it to be usable for the shortest path solution.

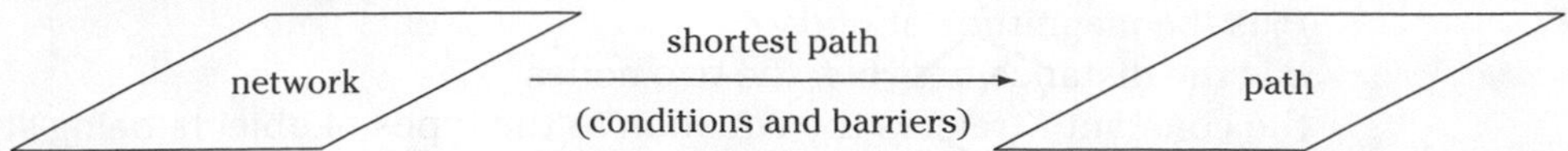

Figure 13.15 Flowchart illustrating the application of routing to find the shortest path.

might be related to a speed limit along a street. By using an accumulated distance, based on a combination of calculated distance and the impedance factor, the most *efficient* route can be determined, rather than just the shortest. Nodes can also be coded with stops (indicative of traffic signals or stop signs), turning impedances based on the difficulty of turning left or right at an intersection, and even barriers preventing movement and forcing traffic along another path. As with calculating surface distances, all of these measures require prior knowledge of the nature of the streets, intersections, and other nodes. Frequently the weights and impedances are somewhat arbitrary or are based on intuitive knowledge rather than on absolute certainty of how they will affect travel.

Routing is best performed in a vector system. Because of the close relationship of graph theory topology and the topological vector data model, the two perform well together. You should also be aware that given the many possible routes, especially where circuits are involved, there are likely to be a number of possible ways for finding your route. In short, the solutions to such problems may be approximations rather than exact solutions. This is particularly true with complex systems. As each route segment is taken, the network's configuration is effectively altered. This is called the **Traveling Salesman Problem**, and is a topic of **geocomputation** rather than introductory GIS.

Allocation is a process that can be used to define, for example, the location of a market, the areal extent of a water treatment center, or the boundaries of a series of fire station service areas. A network structure within a vector GIS is most often used. The idea here is that the capacity of a given service is distributed throughout the net. Each link in the network has a specified number of items that must be served. For example, each street segment contains a number of homes that would need to receive water service from the water treatment center. Or each house could be thought of as a fire-prone structure that may need the services of the nearby fire station (Lupien et al., 1987). In addition, each service or market center is reasonably capable of operating at a certain maximum capacity, so the maximum number of units that each service area can accommodate is compared to the number of service sites that occur along the network links. Say, for example, that a fire station can service only 100 homes with a reasonable level of safety. Beginning at the service location, the GIS will travel outward, following the nearest roads, counting the number of houses that occur along those links until it reaches its capacity of 100. If each street contains 10 houses, the GIS will designate the 10 nearest streets (each with 10 homes) to ensure efficient allocation of the maximum capacity for the service center (Figure 13.16). The specified amount of route might include the number of people to service (i.e., number of residents per house), the road miles that can be traveled, and a travel time limit that must not be exceeded. If our road net were completely uniform (no impedances, no stops, no changes in speed limit, etc.), calculating this allocation would be a simple matter of deciding on

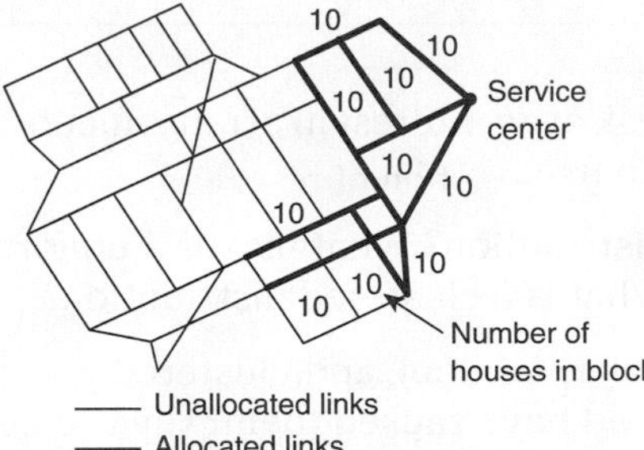

Figure 13.16 Allocations of networks. A simple road network showing how links containing 10 homes each are allocated to a service center (e.g., a fire station) that can service only 100 homes safely.

the criteria for allocation and pushing our boundary outward until they are met. For example, if we wanted to allocate the routes for newspaper carriers so that each person's vehicle would have to travel only a certain number of road miles, the software would simply add up the road miles as each route spreads outward from the starting point, until that value is reached, whereupon the links in the road that were allocated would be assigned attribute codes to correspond to the different carriers.

Most real allocation problems are quite complicated. In some cases our primary interest might be the number of people we can service in a given amount of time. If you were planning to allocate newspaper routes, you would need, for example, to be able to distribute newspapers to the subscribers by a specified time of day, say 9 A.M. And, of course, no road network is totally uniform. Some roads have higher speed limits than others, some have stop signs, some include turns that are slower than others, and so on. And, of course, you need to know that individual addresses are somehow connected to the streets you intend to service (the process of linking addresses to streets is called **address matching**). The importance of quality geocoding for effective address matching cannot be understated here. Most of the work of determining these allocations requires us to encode all the appropriate attributes, impedance values, turning impedances, and other variables. Once this has been done, the software is quite capable of providing effective results.

Terms

accessibility
address matching
allocation
alpha index
central place
circuitry
circular variance
clustered
connectivity
density
Dirichlet diagrams
dispersion
gamma index
geocomputation
graph theory
gravity model
interactions
isolation
line intersect methods
linear directional mean
mean resultant length
nearest neighbor analysis
network complexity
proximal region
random
random walk
regular
resultant length
resultant vector
rose diagram
routing
runs test
spatial arrangement
Thiessen polygons
Traveling Salesman Problem
uniform
Voronoi diagrams

Review Questions

1. What do we mean when we speak of an interest in arrangements of objects? Why is this knowledge useful to us?
2. How do we define a uniform distribution? What makes a uniform distribution regular versus random? What is a clustered distribution?
3. What is the significance of regular, random, and clustered distributions in terms of the processes that could have caused them? Give some examples of each distributional pattern and give a hypothesis that might describe the causes.
4. What does nearest neighbor analysis tell us about a distribution?
5. What is the purpose of Thiessen polygons? Diagram some examples and describe how they are created.
6. Explain, with examples, how the concept of adjacency and distance might be employed without highly sophisticated specialty software such as FRAGSTATS.
7. Describe the process of extracting a nearest neighbor statistic from line patterns. What does this quantity tell us about the distribution of our lines? Describe some situations in which the nearest neighbor statistic for lines might give misleading results.
8. Describe the use of line intersect methods for analyzing line pattern distributions. What is a random walk?
9. What is a vector resultant? What does it tell us about the patterns of linear objects? What is the resultant length? How does it compare to resultant force problems in physics?
10. What is the mean resultant length? How does it differ from the resultant length? When would we use it? What would a large mean resultant length tell us? What is the difference between circular variance and mean resultant length?
11. How can we adjust our measurements for mean direction, mean resultant length, and circular variance to account for orientations that can be measured in either of two directions?
12. Describe how the gamma index is performed. What does it tell us about our network? What does a gamma value of 0.48 tell us in terms of the amount of connectedness in our network?
13. What is the alpha index? How is it different from the gamma index? How is it the same? Give some nonroad network examples of how both of these indices might be useful.
14. Describe, in general terms, what the gravity model is. How does distance affect the interactions of point objects? How does the magnitude of the nodes change the interactions between them? Give an example of how the gravity model might be applied.

15. Describe the types of attribute that might need to be assigned to the nodes and the arcs between nodes in routing and allocation problems.

References

Aitchison, J., and J.A.C. Brown, 1969. *The Lognormal Distribution, with Special Reference to Its Uses in Economics*. Cambridge: Cambridge University Press.

Baker, W.L., and Y. Cai, 1992. "The r. le Programs for Multiscale Analysis of Landscape Structure Using the GRASS Geographical Information System." *Landscape Ecology*, 7(4):291–301.

Batschelet, E., 1965. "Statistical Methods for the Analysis of Problems in Animal Orientation and Certain Biological Rhythms." *American Institute of Biological Sciences Monograph*. Washington, DC: AIBS.

Brassel, K.E., and D. Reif, 1979. "A Procedure to Generate Thiessen Polygons." *Geographical Analysis,* 11(3):289–303.

Buell, M.F., H.F. Buell, J.A. Small, and T.G. Siccama, 1971. "Invasion of Trees in Secondary Succession on the New Jersey Piedmont." *Bulletin of the Torrey Botanical Club*, 98:67–74.

Carkin, R.E., J.F. Franklin, J. Booth, and C.E. Smith, 1978. "Seeding Habits of Upper-Slope Tree Species. IV. Seed Flight of Noble Fir and Pacific Silver Fir." U.S. Forest Service Research Note PNW–312. Portland, OR: USFS.

Clarke, K.C., 1990. *Analytical and Computer Cartography*. Englewood Cliffs, NJ: Prentice Hall.

Dacey, M.F., 1967. "Description of Line Patterns." *Northwestern Studies in Geography,* 13:277–287.

Davis, J.C., 1986. *Statistics and Data Analysis in Geology*, 2nd ed. New York: John Wiley & Sons.

Forman, R.T.T., and M. Godron, 1986. *Landscape Ecology*. New York: John Wiley & Sons.

Gaile, G.L., and J.E. Burt, 1980. "Directional Statistics: Concepts and Techniques in Modern Geography." *Geo Abstracts*, No. 25, University of East Anglia, Norwich, England.

Getis, A., and B. Boots, 1978. *Models of Spatial Processes: An Approach to the Study of Point, Line and Area Patterns*. Cambridge: Cambridge University Press.

Griffiths, J.C., 1962. "Frequency Distributions of Some Natural Resource Materials." *Pennsylvania State University, Mineral Industries Experiment Station Circular*, 63:174–198.

Griffiths, J.C., 1966. "Exploration for Natural Resources." *Journal of the Operations Research Society of America*, 14(2):189–209.

Gumbel, E.J., J.A. Greenwood, and D. Durand, 1953. "The Circular Normal Distribution: Tables and Theory." *Journal of the American Statistical Society*, 48:131–152.

Haggett, P., A.D. Cliff, and A. Frey, 1977. *Locational Analysis in Human Geography*. New York: John Wiley & Sons.

Hutchings, M.J., and R.J. Discombe, 1986. "The Detection of Spatial Pattern in Plant Populations," *Journal of Biogeography,* 13:225–236.

Krumbein, W.C., 1939. "Preferred Orientation of Pebbles in Sedimentary Deposits." *Journal of Geology*, 47:673–706.

Lowe, J.C., and S. Moryadas, 1975. *The Geography of Movement*. Boston: Houghton Mifflin.

Lupien, A.E., W.H. Moreland, and J. Dangermond, 1987. "Network Analysis in Geographic Information Systems," *Photogrammetric Engineering and Remote Sensing*, 53(10):1417–1421.

Mardia, K.V., 1972. *Statistics of Directional Data*. London: Academic Press Ltd.

Mark, D., 1987. "Recursive Algorithm for Determination of Proximal (Thiessen) Polygons in Any Metric Space." *Geographical Analysis*, 19(3):264–272.

McDonnell, M.J., 1984. "Interactions Between Landscape Elements: Dispersal of Bird-Disseminated Plants in Post-agricultural Landscapes." In *Proceedings of the First*

International Seminar on Methodology in Landscape Ecological Research and Planning, Vol. 2, J. Brandt and P. Agger, Eds. Roskilde, Denmark: Roskilde Universitetsforlag GeoRuc, pp. 47–58.

McGarigal, K., and B.J. Marks, 1994. "FRAGSTATS: Spatial Pattern Analysis Program for Quantifying Landscape Structure, Version Two." Forest Science Department, Oregon State University, Corvallis.

McGrew, J.C., and C.B. Monroe, 1993. *Statistical Problem Solving in Geography*, Dubuque, IA: Wm C. Brown.

McQuilkin, W.E., 1940. "The Natural Establishment of Pine in Abandoned Fields in the Piedmont Plateau Region." *Ecology*, 21:135–149.

Miles, R.E., 1964. "Random Polygons Determined by Lines in a Plane, I and II." *Proceedings of the National Academy of Sciences*, 52:901–907, 1157–1160.

Ripley, B.D., 1981. *Spatial Statistics*. New York: John Wiley & Sons.

Simpson, J.W., R.E.J. Boerner, M.N. DeMers, L.A. Berns, F.J. Artigas, and A. Silva, 1994. "48 Years of Landscape Change on Two Contiguous Ohio Landscapes." *Landscape Ecology*, 9(4):261–270.

Stephens, M.A., 1967. "Tests for the Dispersion and for the Model Vector of a Distribution on a Sphere." *Biometrika*, 54:211–223.

Stephens, M.A., 1969. "Tests for Randomness of Directions Against Two Circular Alternatives." *Journal of the American Statistical Association*, 64(325):280–289.

Sugihara, G., 1983. "Peeling Apart Nature." *Nature*, 304:94.

Taaffe, E.J., and H.J. Gauthier Jr., 1973. *Geography of Transportation*. Englewood Cliffs, NJ: Prentice Hall.

Whitcomb, R.F., 1977. "Island Biogeography and 'Habitat Islands' of Eastern Forest." *American Birds*, 31:3–5.

CHAPTER 14

Map Overlay

The process of map **overlay** requires both graphic and attribute comparisons. Some of the details can become quite complicated, especially the computer graphics associated with vector overlay. The following descriptions of the overlay process will give you a feel for how the computer accomplishes this task. Rather than detailing what is essentially computer graphics, this chapter imparts a general, conceptual understanding. You are encouraged to consult the numerous texts on computer graphics or to consider taking a course or two if your interests lie in this direction.

This chapter does not detail all possible methods of digitally combining themes. Instead, it focuses on a few general categories from which the detailed approaches can be selected. The two primary approaches to cartographic overlay involve formal logic based on forms of set theory and mathematical combinations of thematic content. Each provides a unique opportunity to simulate the real world by representing known or expected co-occurrences of spatial variables. Among the more important aspects of overlay operations is a knowledge of which of the map themes should be overlaid for specific types of operations. For example, if you know that patterns of map variables correspond spatially, you would overlay them to determine the degree to which this is true. Alternatively, you might use overlay operations to ascertain whether two variables are spatially correlated as a form of hypothesis testing. Once you have decided what to overlay, the next most important aspect is to decide which of the many logical and mathematical approaches is most appropriate for your application. While there are no recipe manuals for which approaches and which data sets are most appropriate, this chapter will give you some insights from which you can draw your own conclusions.

LEARNING OBJECTIVES

When you are done with this chapter you should be able to:

1. Understand the conceptual relationship between manual and automated overlay approaches.

2. Explain how the McHarg acetate overlay approach works, its relationship to digital cartographic overlay, and describe its advantages and disadvantages.
3. Describe the overall advantages of overlay as a GIS analytical technique, and the advantages and disadvantages of raster versus vector overlay approaches.
4. Explain the possible measurement level problems involved in mathematical overlay operations, particularly with mixed measurement level datasets.
5. Describe examples of the use of point-in-polygon and line-in-polygon overlay operations.
6. Describe examples of the use of polygon overlay and provide rudimentary flowcharts showing how specific overlay operations might work.
7. Provide detailed explanations of the different types of overlay in both raster and vector.
8. Understand the difference between graphical overlay operations and cartographic overlay operations.
9. Explain dasymetric mapping and describe how it might be used in GIS.
10. Understand the entity-related problems associated with vector overlay operations and be able to discuss the types of errors associated with them.
11. Describe the "weakest link hypothesis" and discuss its merits and its limitations, particularly with regard to the ability of a GIS to "weight" the overlay themes individually.

THE CARTOGRAPHIC OVERLAY

Among the most powerful and most frequently used features of the modern geographic information system is its ability to place the cartographic representation of thematic information of a selected theme over that of another. This process, commonly called map overlay, is so intuitive that its application long preceded the advent of modern electronic geographic information systems. A very early application of the analog overlay technique dates back to the days of the American Revolution. During the Battle of Yorktown in 1781, General George Washington engaged the services of French cartographer Louis-Alexandre Berthier, who employed map overlay to examine the relationship between Washington's troops and those of British General Charles Cornwallis (Wikle, 1991). This was certainly not the first use of the manual overlay operation, but you can see that the idea has been around literally for centuries.

The idea that spatial phenomena occur in much the same location as other spatial phenomena has long been recognized for some categories of mapped data, but certainly not all. Sauer (1925), for example, established a model for the interactions of general categories of earth-related data in his work on the morphology of landscape. His research established the existence of strong spatial correlations among human activity, landforms, and other physical parameters. Although Sauer did not formalize these correlations cartographically, it was clear that he saw connections among distributions of these phenomena on the surface of the earth. Later researchers formalized this design into a wide variety of approaches named, for example, sieve mapping (Tyrwhitt, 1950) and **biophysical mapping** (Hills et al., 1967).

Among the most influential practitioners of this early developing approach was Ian McHarg, whose work relating environmental phenomena spawned an entire school of thought among today's landscape architects, allowing considerably more work to be performed using the computer than was originally possible from field observation and single map cartography alone (Simpson, 1989). McHarg (1971) retained an analog method, applying clear acetate overlays of selected mapped environmental phenomena to allow him to evaluate the

SLOPE
ZONE 1 Areas with slopes in excess of 10%.
ZONE 2 Areas with slopes less than 10% but in excess of 2½%.
ZONE 3 Areas with slopes less than 2½%

SURFACE DRAINAGE
ZONE 1 Surface-water features—streams, lakes, and ponds.
ZONE 2 Natural drainage channels and areas of constricted drainage.
ZONE 3 Absence of surface water or pronounced drainage channels.

SOIL DRAINAGE
ZONE 1 Salt marshes, brackish marshes, swamps, and other low-lying areas with poor drainage.
ZONE 2 Areas with high-water table.
ZONE 3 Areas with good internal drainage.

BEDROCK FOUNDATION
ZONE 1 Areas identified as marshlands are the most obstructive to the highway; they have an extremely low compressive strength.
ZONE 2 The Cretaceous sediments: sands, clays, gravels; and shale.
ZONE 3 The most suitable foundation conditions are available on crystalline rocks: serpentine and diabase.

SOIL FOUNDATION
ZONE 1 Silts and clays are a major obstruction to the highway; they have poor stability and low compressive strength.
ZONE 2 Sandy loams and gravelly sandy to fine sandy loams.
ZONE 3 Gravelly sand or silt loams and gravelly to stony sandy loams.

SUSCEPTIBILITY TO EROSION
ZONE 1 All slopes in excess of 10% and gravelly sandy to fine sandy loam soils.
ZONE 2 Gravelly sand or silt loam soils and areas with slopes in excess of 2½% on gravelly to stony sandy loams.
ZONE 3 Other soils with finer texture and flat topography.

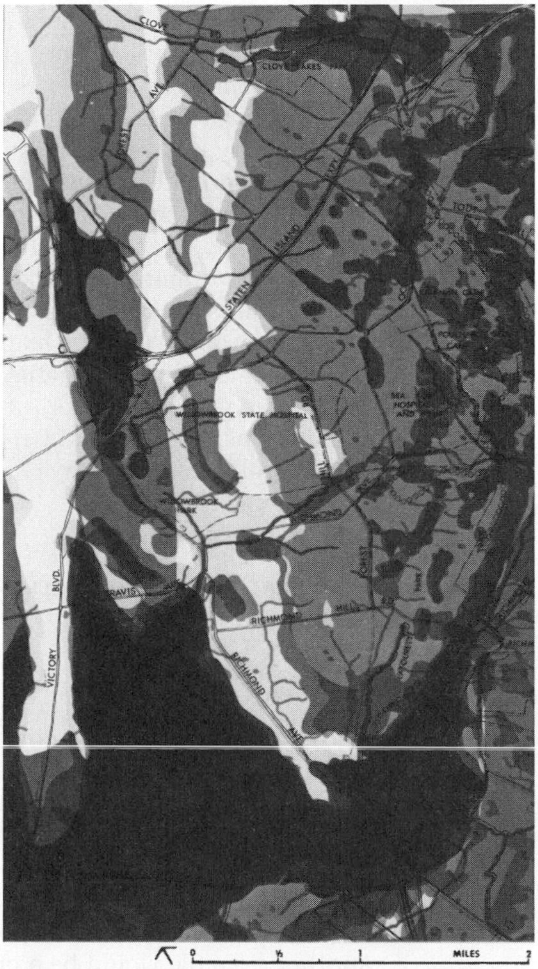

COMPOSITE: PHYSIOGRAPHIC OBSTRUCTIONS

Figure 14.1 Acetate overlay to determine environmental sensitivity. Example of Ian McHarg's use of manual acetate overlay operations to illustrate increased sensitivity as a number of categories of physical parameters overlap. This map shows the overlapping of slope, surface drainage, bedrock foundation, soil foundation, and susceptibility to erosion for the Richmond Parkway study area in New Jersey. *Source:* Ian McHarg, *Design with Nature*, John Wiley & Sons, New York, © 1995. Reprinted with permission.

environmental sensitivity of an area. Each environmental sensitivity factor was represented by gray tone shading on the acetate. As each of these acetate layers was placed over the others, those that corresponded spatially became darker—thus representing higher overall environmental sensitivity. The area was then remapped showing a gradation from higher to lower sensitivity, and decision makers could use the map product in assessing available alternatives (Steinitz et al., 1976) (Figure 14.1).

There is still much to learn about the nature, composition, relevance, and causal relationships among spatial phenomena. This subject comprises a large part of what geographers do. Its importance is seen in what is now known as geographic information science—essentially a subdiscipline of geography. Geographic information systems are now providing easily available map overlay procedures that may result in the development of new hypotheses, new theories, and new laws about these pattern similarities. Many specialists in both geography and other disciplines are learning more about corresponding spatial patterns that had, until recently, been unobserved, perhaps even unobservable without automation. As more disciplinary specialists learn about unique spatial patterns, we will have an increasingly robust set of rules about which layers can legitimately be overlaid and which ones can reasonably permit statements about cause and effect. Until then, we would be wise to stay alert, especially because the modern GIS provides dozens of methods for quickly and easily overlaying maps. This power to overlay can easily be translated into a large number of opportunities to make mistakes and to overlay either nonassociated or autocorrelated layers, leading to false conclusions. It can also increase our ability to knowingly lie with maps well beyond what was possible before GIS could be used to compare spatial phenomena (Monmonier, 1991). In the following pages, we will look closely at our options for overlay operations, consider our limitations, what we can reasonably overlay, when to overlay, and the potential problems of overlay.

POINT-IN-POLYGON AND LINE-IN-POLYGON OVERLAY

Traditionally we envision map overlay in one of two ways—analog or digital—and primarily as a method of comparing areas (polygons). This approach works well if all our data are polygonal. Often, however, point and line data must be compared either to each other or to the locations of polygons. Thus, a means of comparing these disparate cartographic forms might prove useful to the GIS practitioner. A wildlife biologist, for example, may wish to know if a particular species of bird (point data) has a preference for a particular type, size, or configuration of vegetated patch (polygon). The necessary operation is to overlay the point data on the polygonal data, called **point-in-polygon overlay**.

Another example would be a crime analyst trying to determine whether certain neighborhoods are experiencing an increase in incidence of purse snatching. By plotting historical crime event records of purse snatching and overlaying them on the neighborhood polygons, our crime analyst notices that a particular neighborhood (we'll call it Pleasantview) seems to have a concentration of this crime type despite being a relatively average neighborhood. Later, by mapping these same data on a monthly basis, he notices a concentration of purse snatching for the month of December. He also notices that this pattern seems to disappear starting in 2003—the same year that a new shopping mall was built in Pleasantview. As he compares a map of point locations of purse snatching with a detailed map of the Pleasantview neighborhood, he sees not only that a spatial relationship exists between these crime data and the Pleasantview polygon, but also that certain high-traffic streets leading to and from the mall and its parking lot seem to experience the most purse-snatching events in the neighborhood. This latter comparison shows the importance of certain lines (in this case, streets) occurring inside polygons (in this case, neighborhoods).

Our detective has demonstrated the strong relationship between points and polygons, and you can see how useful comparing these entities can be. He has also demonstrated the close relationship between analyzing single map variables and between-map comparison of those variables with other layers. In examining his maps, he discovered not only that there were more points in a certain polygon, but also that the points themselves were grouped in close proximity to themselves and to linear features in the area.

A third example will illustrate the possibility of correspondences between line phenomena and polygons, or a **line-in-polygon** procedure. Suppose you are an historical geographer in the Minneapolis–St. Paul metropolitan area. You are interested in the influences on city growth, and you know that the Twin Cities area has experienced several growth spurts. You also know that the architecture during different time periods is easily identifiable and quite distinctive. You first map the architectural period regions as separate polygons. Three major zones of urban expansion appear. Because of the somewhat linear nature of the polygons, you begin to suspect an underlying linear feature as the cause for the patterns.

You compile a series of maps based on the most prominent linear patterns in any urban environment—the transportation nets including rail lines, streetcar rail lines, and major highways. Next you overlay the patterns to view possible correspondences. Results show that the polygons with the earliest growth

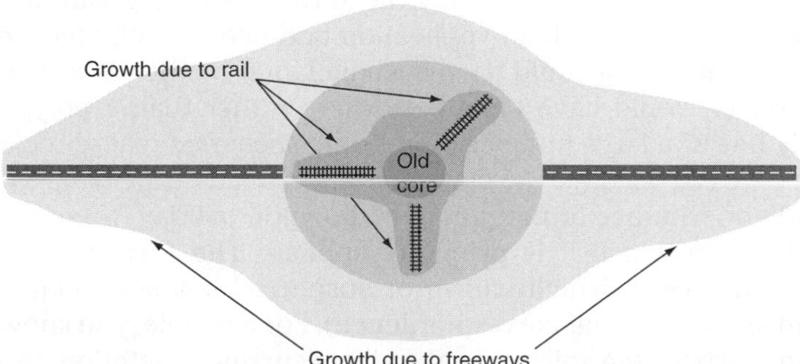

Figure 14.2 Line-in-polygon operation. Schematic showing how a line-in-polygon overlay for the Minneapolis–St. Paul area might appear. The periods of urban expansion are illustrated by polygons, whereas the lines within the polygons are illustrative of the strong spatial relationship between growth periods and major period transportation nets. *Source:* Adapted from John S. Adams, 1970. "Residential Structure of Midwestern Cities." *Annals, Association of American Geographers*, 60(1):37–62, Blackwell Publishers. Used with permission.

spurt also have the major rail lines that existed in the area at that time. You next overlay the streetcar lines and immediately see a similar pattern for the second growth spurt. Finally, after overlaying the major highways, primarily the interstate system begun in the 1950s, you see once again the strong spatial relationships between transportation networks and urban expansion (Adams, 1970) (Figure 14.2). These results allow the formulation of hypotheses relating urban expansion to transportation, enabling examination of their cause-and-effect relationships. They also clearly demonstrate the both practical and theoretical benefits of point-in-polygon and line-in-polygon overlay.

POLYGON OVERLAYS

As you have seen, the traditional approach to map overlay compares one set of polygons to another. The predominance of polygon overlay in GIS derives from the practical considerations and impetus for the development of the Canada Geographic Information System. Canada's needs were primarily to examine spatial relationships of large regions on an area-by-area basis. Still a major thrust among GIS practitioners, vendors have developed software with strong polygon overlay operations. Thus, there are many approaches to performing polygon overlay, each answering specific user needs.

WHY PERFORM AN OVERLAY?

First, you must decide why are you performing an overlay operation. Part of this question involves asking whether polygon overlay is needed at all. Many polygon overlay procedures involve a simple reclassification of existing data. If, for example, we have two maps, with one showing grain and another showing row crops, logically we could overlay these two maps to create a map of just agriculture. However, if such an application had been anticipated prior to map input, the two categories could have been included on the same map. This way a reclassification could have been performed rather than a polygon overlay. This is only one of a large number of possible scenarios where reclassification could eliminate the need for the more computationally expensive overlay.

To decide on whether or not you need polygon overlay, there are generally two conditions in which it is primarily applied. The first involves trying to evaluate the degree to which known or suspected spatially related variables correspond spatially (**areal correspondence**). For example, you know that there is a relationship between soils and naturally occurring vegetation. By overlaying a map of soils with one of natural vegetation, you can examine the degree of areal correspondence that actually exists. You might be trying to determine if any disturbance has changed this pattern of association. The overlay provides a visual as well as a quantifiable level of correspondence that might require subsequent analysis.

There are situations when we may suspect a possible areal correspondence, but have only anecdotal evidence to support our suspicions. This data exploration is an alternative, if infrequent, use of cartographic overlay. As with points and lines, we often make cursory observations of things being colocated. As you saw in the earlier crime example for points and lines, when performing on overlay of two suspected variables we can, as before, determine not only the existence of areal correspondence, but also the percentage of such correspondence. This allows us to use GIS as an effective spatial hypothesis testing toolkit.

TYPES OF MAP OVERLAY

Now that you know what situations are appropriate for overlay, let's examine the methods by which it is accomplished. The two basic categories are logic or "set theoretic based" and "mathematically based." Both approaches are implemented in both raster and vector systems, and include options for determining selective rules of combinations that determine specific sets of factors that are to be combined and how they will be reclassified.

Set theoretic-based approaches most often use Boolean search strategies and come in three basic varieties: union, intersection, and complement (Figure 14.3). In the union-based overlay method, all polygons found in all overlaid maps (or grids) are combined through the Boolean operator *and* (Figure 14.3a). We use this approach when we are interested in combining categories to make more inclusive classifications. For example, the McHarg acetate overlay might be thought of as a **union overlay** in that all categories are combined to produce a new map. To be more selective in what we want as output, we would use an **intersection overlay** that will result in only those categories that meet the criteria of all map layers (Figure 14.3b). By employing the Boolean operator *or*, this type of overlay might show us, for example, only those areas that have a particular soil from one layer, landform types from another map, and vegetation from a third. The result of this approach is a small subset of all three categories (where all three areas co-occur). We might want to take the opposite approach for our last example, where we want to identify particular soil, landform, and vegetation areas that do not correspond spatially. This approach employs the **complement overlay** method using the Boolean operator *not* to achieve the

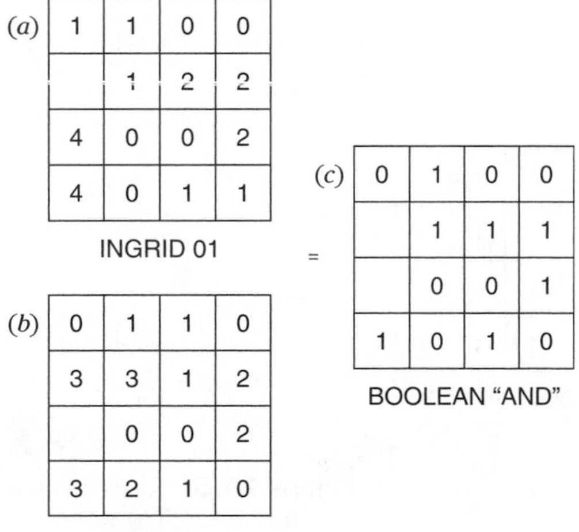

Figure 14.3 Set theoretic overlay. The three basic types of logical map overlay include (*a*) union, (*b*) intersection, and (*c*) complement.

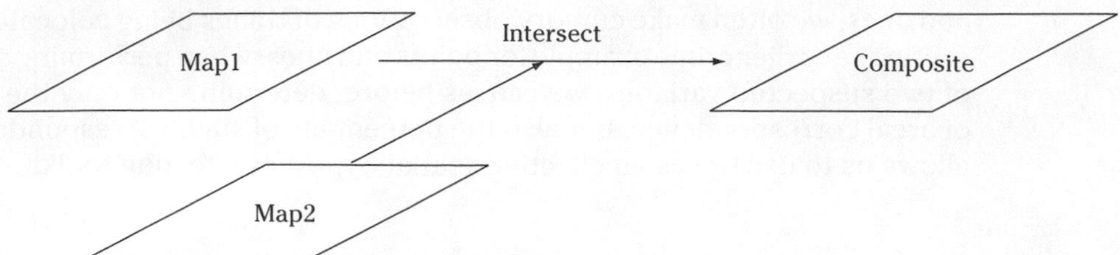

Figure 14.4 Flowchart illustrating the application of complement-based logical overlay. This method results in isolating those items not contained in the initial map layer.

results (Figure 14.3c). The flowchart in Figure 14.4 illustrates how set theoretic overlay might be applied in the complement case.

Boolean operators are easily implemented in raster through the map algebra modeling language. Figure 14.5 shows all three of the basic methods of logical overlay (union, intersection, and complement). The same Boolean operators—and, or, and not—are applied just as in the vector approach. One common variant of the complement method in raster, however, is called the "exclusive complement." This procedure considers nonzero values as true and zero as false. If the cell value on one map is true (1) and a comparable cell value is false (0), the program returns a value of 1. If both are either true or both are false the output is 0, and if one or both values are "NoData," the output is "NoData."

The second major type of overlay is mathematical and is most commonly performed in raster GIS using some variant of map algebra. This approach uses the numeric values stored in each grid cell as a means of comparison with two or more grids. A simple example is using addition to add the grid cell values of one grid to those of another to obtain a third (Figure 14.6). Other operators could include subtraction, division, multiplication, power, roots, absolute value, and pretty much any other standard mathematical procedure. Each of these methods of overlay always assumes that the positional locations of the grids are coregistered for each layer (grid): a property separating map algebra from

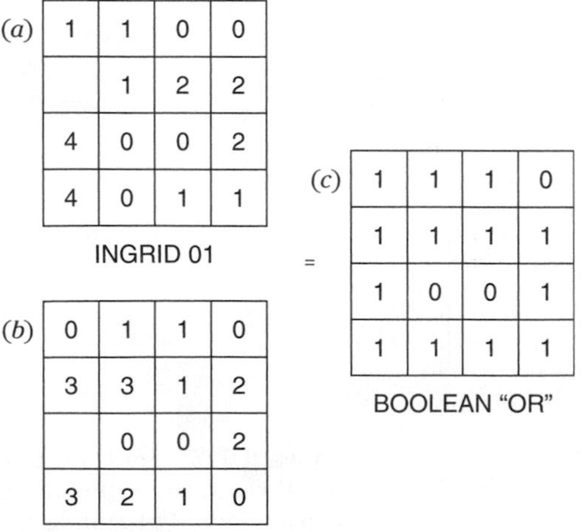

Figure 14.5 Raster versions of (*a*) union overlay, (*b*) intersection overlay, and (*c*) complement overlay.

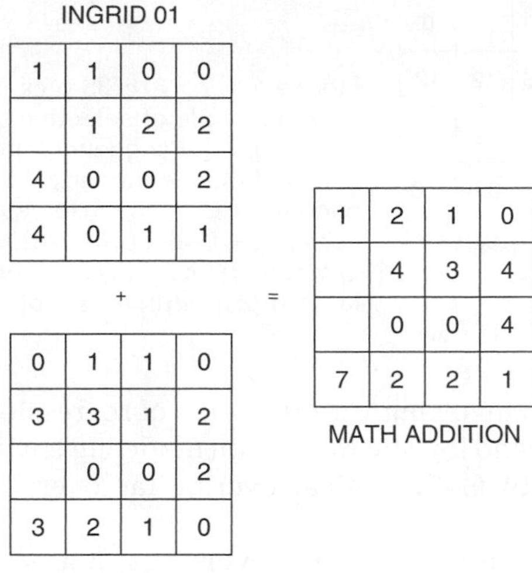

Figure 14.6 Grid cell mathematical overlay based on map algebra. This example shows adding grid cell values from each of two grids. Virtually all available mathematical operators can be applied to mathematical overlay.

matrix algebra. Even with this property the potential consequences of this type of overlay must be considered before its application. For example, if one of your grids contains cells with zeros, multiplication with nonzero values in a second grid will result in zeros (0 times anything equals 0). The same results from the use of "NoData" grid cells. This can either be used to purposely block out portions of your grid that you wish to ignore or it can create an unfortunate result from unplanned interactions. The same can be said of the use of division-based mathematical overlay. Division could result in remainders (that in many systems will be truncated), and can also produce errors if you attempt to divide one map with another that has zeros (division by 0 error).

These problems abound with the use of raster overlay, but are often not as potentially detrimental as those involving the improper interaction of grid cells coded in different data measurement levels. Let us assume that you have a grid with numbers representing nominal land use categories (e.g., 1 = residential, 2 = agriculture, 3 = abandoned, and 4 = industrial). Another grid shows ratio scale elevation values (e.g., 100 feet to 1,500 feet). In trying to determine the relationship between elevation zones and land use type, you multiply these two grids together. Two problems arise from this. First, there are a number of circumstances in which the value of, say, 300 could be achieved (e.g., 3 × 100, 2 × 150, 1 × 300). This confusion makes an interpretation of the results difficult at best. But more importantly, these same numbers, while they often look visually compelling and quite logical, produce some unusual categories. For example, 3 × 100 is essentially "abandoned feet," while 2 × 150 is "agriculture feet," and 1 × 300 is "residential feet." The software will allow the user to make these types of mistakes. This is another reason to review your data levels before performing mathematical overlay.

Still, despite the potential for erroneous results, the mathematical overlay is among the most powerful overlay modeling tools available. In cases where deterministic relationships exist among variables, this approach allows us to employ these relationships in a spatial model. Examples include nonpoint source pollution modeling, soil loss equations, and ecological systems models where inputs are directly linked to known outcomes. Other uses include statistical

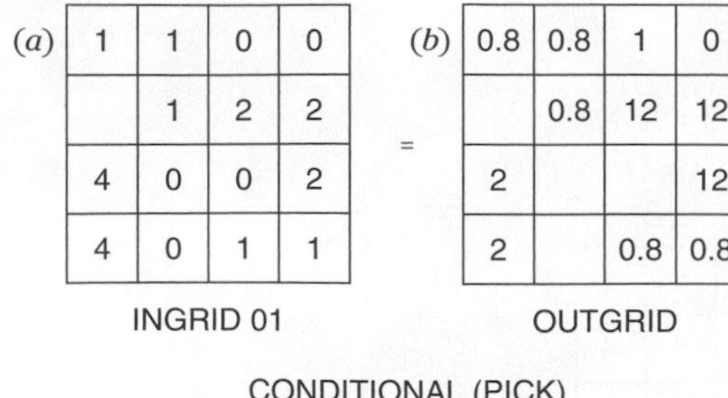

Figure 14.7 ArcGIS pick operation as an example of selective overlay. Ordinarily the sequence of operations is based on the numerical sequences encountered in the grid cells themselves (*a*). Conceptually this can be envisioned as multiple layers (each layer representing a set of rules) (*b*).

predictive modeling approaches involving various forms of regression models, including nonlinear multivariate and logistic regression modeling environments. In short, the potential power of mathematical overlay far overshadows its potential for problems.

Another general category of overlay, **selective overlay**, is not necessarily a unique category but rather an extension of the techniques we have discussed thus far. In many cases we will have a set of rules that allow us to decide which factors we might want to use to cover other factors as we compare two layers. This approach, sometimes called **rules-of-combination overlay** (Chrisman, 1996) (or **conditional overlay**) allows us to use exclusionary rules, weightings, mathematical manipulations; local, focal, zonal, and block functions; and Boolean logic in appropriate combinations. Such a set of rules might look something like the following map algebra example from ESRI's ArcGIS:

pick (<grid > *x*), <expression, . . . expression>

In this software the grid is the name of the input grid to be evaluated by the expressions. The order of the expressions determines which of the possible expressions will be applied to the grid cell values in the input grid. So, if a grid cell value is 1, it will apply the first expression. If it is 2, it will apply the second expression, and so on. Conceptually this is the same as using a separate grid for each of these conditional expressions (Figure 14.7). Other approaches for conditional overlay vary with software but include an ability to select which grid cell values for an input grid will be compared (through whatever method you select) to which other grids. This requires your software to allow for conditional program flow statements through some form of macro language (Figure 14.8).

Among the most useful qualities of mathematical overlay is that not only the grid cells and their values can be defined, but the grid layers themselves

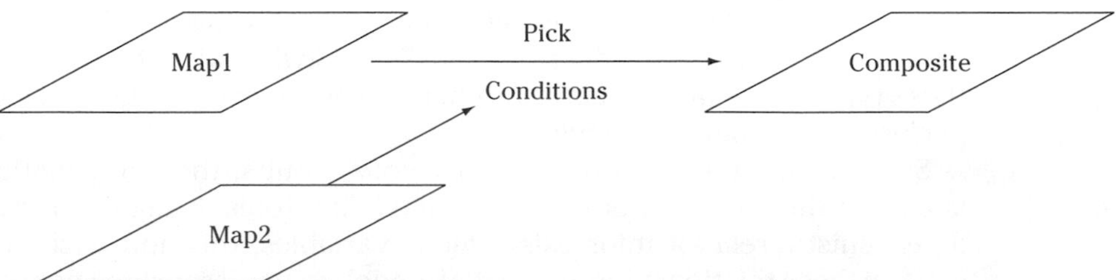

Figure 14.8 Flowchart illustrating selective overlay operations illustrated in Figure 14.7.

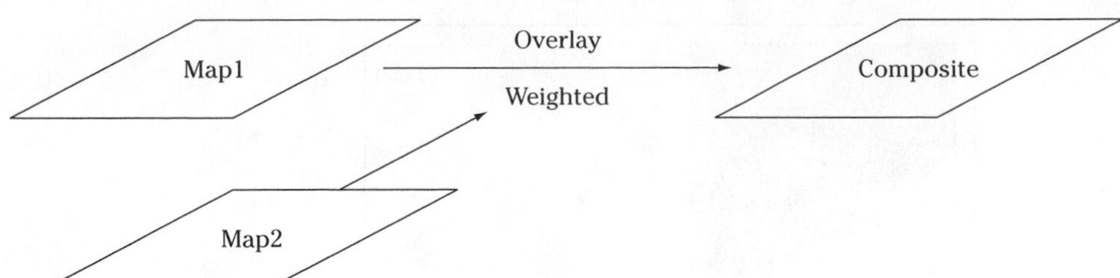

Figure 14.9 Weighted overlay offers an opportunity to vary the importance or contribution each grid provides to the overlay operation.

can also be assigned importance values. This **weighted overlay** approach is an elegant solution to the problem encountered by McHarg in his acetate overlay methodology—that of assumed equal values for each layer. In his environmental sensitivity work he had to assume that each measure of sensitivity was equally as important as any other. While this makes the modeling simple, it is unrealistic. We know, for example, that some factors, like areas that have underground gaslines, are going to be more at risk than areas with high shrink-swell clay soils for constructing large buildings. As such, the gasline layer might be assigned a higher value than areas known to have high shrink-swell clays. In short, we *weight* the layers based on their importance to our model. As with the development of friction surfaces and impedance values, such weights might be difficult to assign and might also require the use of scalar values. However they are implemented, weighted overlay operations are extremely useful and increasingly commonplace. Figure 14.9 provides a sample flowchart of a simple weighted overlay model.

Before leaving the methods of map overlay, there is a special-case type of vector-based overlay that resembles intersection overlay, but which typically is applied for quite different reasons. The method, called **clip**, extracts the input graphics (entities) that overlay those being clipped. The process uses a user-defined template, sometimes called a **clipcover**, that represents an area of interest. This type of overlay can be used so that multiple temporal layers will be identical in spatial extent, or simply to reduce the amount of area that will be analyzed later. The clip process will retain all point, line, area, and surface entity information (Figure 14.10). You might recall that we alluded to this method when creating post-interpolation surface layers.

TYPES OF VECTOR OVERLAYS

You have seen the different map overlay approaches that are applied for GIS analysis. While this provides you with a conceptual framework to decide which approach is most appropriate for particular tasks, you need also to examine how software operationally implements these approaches, particularly with regard to vector data models. If you are familiar with vector GIS data models you might want to skip this section, but if you are new to the discipline, you might want to read on. Particularly important is the idea that some graphics software and many CAC, CAD, and AM/FM systems have the same appearance of functionality of GIS. While some of these are moving in this direction, it is important to know

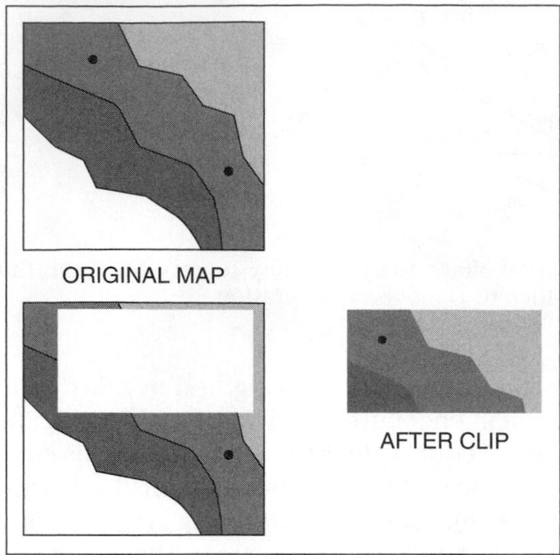

Figure 14.10 The clip operation enables the operator to select an identical study area from a number of other layers, or it can be applied to removing areas of an existing study that are not part of a particular analysis.

the operational levels needed to perform complete vector overlay analysis. The purpose is not to rate these different packages, but merely to provide you with basic guidelines for determining whether or not the software you are choosing will meet your overlay needs.

Graphical Overlay

The first and simplest method of automated vector overlay is very similar to the analog method in that we apply the display symbols of precategorized data to a single Cartesian surface. Such data might include a range-graded choropleth map of population, a single value-by-area map showing the locations of wild plant species, land use polygons, line symbols indicating road networks, or point symbols showing archaeological sites. The process of graphic overlay is akin to the method of thematic combination that is used to produce more general maps, like those in a regional atlas. Such a map allows the reader to see a considerable variety of factors laid out on the same graphic at the same time and can be used to show graphically how these different factors might or might not be spatially associated. To perform this relatively simple operation with the computer requires only that the software be able to keep track of the locations of each point, line, and polygon to ensure that all the coordinates can be represented on the screen.

The result of this operation is a visual graphic, not a GIS layer containing attribute information linked to the entities. The software is not capable of combining the attributes for each object displayed because the attributes are most often merely labels attached to the graphical elements. There are no tables that associate the labels with other attributes, and there isn't any topology. The image is merely a composite artifact of superimposing the individual images to make a single, unified image. Such products can be printed or plotted as hard-copy output or, in some cases, saved as graphical computer files. So why aren't these documents GIS layers?

The question is an important one, and it continues to confuse even some experienced GIS users. Let's use an example from drawing software. Say that we have digitized a map of roads using a drawing package (even older forms of CAD software are good examples). On another layer of the software, we digitize the locations of water bodies from the identical map. Next, we combine the two layers into a composite that shows both roads and lakes. Visually we can locate the roads that would get us to the lake so we can go boating. This is useful information, so we save the composite map for another time. Later, another user wants to ask the software which roads enter the area and run parallel to and within 100 yards of the lakefront. Although the solution to this simple query can readily be observed by a person who has found your map and is looking at it, there is no way for the computer to answer the question because it does not know explicitly what names might be associated with what roads. It has no associated database; instead, the names are also graphical devices placed on or near the entities. The computer also lacks the capability to determine any measure of proximity or connectedness because it does not know how to determine these measurements. It is not a GIS map layer. With a little work, it would be possible to import this map to a functional GIS system, to build the database for road names, to create topological structures, and then to answer the question asked. Thus enhanced, our original composite has become more than a graphical device designed specifically for the communication paradigm: It is now a bona fide map suitable for analysis under the analytical or holistic paradigm.

Graphic overlays are very useful within their analytical limits. It is not always necessary to perform computer-intensive queries to determine neighbors or locate the spatial congruity between two objects. These tasks can easily be done by hand. Much of the utility of this type of overlay lies in its simplicity and the rapidity with which it can be produced. Many older **AM/FM** (automated mapping/facilities management) systems operated on the same structure and provided exactly what the users wanted. That is, they produced maps of selected items reasonably fast for users to take into the field or job site, where the graphics could be read and interpreted manually. It is only when the complexity of the databases or the complexity of the queries is very large that the limitations of manual interpretation of automatically produced maps became a problem.

Let us say, for example, that you have a map of underground gas lines, overlaid with a map of streets (including street names) and housing, and a map of population. Someone in the area of the map reports smelling gas. You look at your map to see where the leak might be located, but the lines are rather dense because of the large number of symbols representing residences. It is also difficult to separate the lines that represent roads from those that represent gas lines. You are aware that there are several above-ground shutoff valves, but the complexity of the map obscures the tiny symbols that show them. Such uses require overlay operations (as well as others) that can keep track of multiple attributes for each entity and can highlight the shutoff valves, as well as the locations of residences that might need to be evacuated. All this could be done by hand, given enough time. It might, however, prove easier to use an individual map for each theme to prevent the graphical symbols from interfering with the map interpretation. In short, you need something more for this emergency situation than a basic graphic overlay is capable of producing.

Topological Vector Overlay

The idea of topological data structures introduced in Chapter 5 outlined how the modern GIS can explicitly relate points, lines, and areas in a single map. You saw how these relationships allow you to determine functional relationships among the entities, such as defining left and right polygons associated with a specific line, defining the connections among line segments to examine traffic flows, or searching for selected entity combinations on the basis of individual or related attributes. It also established a method for overlaying multiple polygon layers to ensure that the attributes associated with each of the entities could be accounted for, and so the resultant polygons with multiple attribute combinations would be maintained. This topological result, known as the least common geographic unit (**LCGU**), was introduced by Chrisman and Peucker (1975) specifically to show how the changes in polygonal entities can eventually reach a point at which no further divisions are possible. Associated with these smallest divisions, then, is a specific set of attributes that also cannot be broken down into any further categories. In addition, the attributes associated with the LCGU would effectively emulate the "darkest" or "most sensitive" area polygons found through using the McHarg (1971) method of manual cartographic overlay. For each of the remaining polygons, there will be a number of attribute combinations with lesser "darkness."

With the advent of object-oriented programming languages, the need for topological data models is lessened. The strength of the object-oriented model is that objects have properties that can be inherited. Such inheritance can be transferred through overlay methods just as the properties were linked in overlays employing topological data models. As graphics software, CAC, CAD, and AM/FM packages migrate to object-oriented data models, the line between them and geographic information systems will become even more indistinct than it is currently. Prior to selecting your software, be sure to examine the degree to which attributes are carried forward during the overlay process.

A Note about Error in Overlay

You have seen in the raster examples that the operational use of cartographic overlay tends to involve numerical data as well as categorical (nominal) data. Overlay data may be ordinal, interval, scalar, or ratio; they may be range-graded to indicate groups of values for a given set of polygons. The values, in turn, place constraints on the use of simple Boolean logic and **identity overlay** operations that are designed more for handling nominal data than for mathematical manipulation. The more precise mathematical values give us other powerful options for overlay. Instead of being restricted to Boolean logic and set theory, we now have available the same set of mathematical operators we used in mathematically based raster overlay. Any fully operational database management system can allow mathematical manipulations of the data in the attribute tables associated or linked with the GIS's graphics. Because the vector GIS normally contains the attributes and their labels and can link these with the entities, the process is even easier than it would be in the simplest types of raster system, and nearly identical to the raster systems that also contain a database management system toolkit.

The operations of mathematically based overlay in vector are performed much the same on the attributes in the database tables and in each individual georeferenced set of grid cells. Except for computational difficulties due to the graphical procedures, the user will see little difference in the results of such operations. This similarity in operational capabilities will prove very important when we consider cartographic modeling in the next chapter. Because of the nature of the output from vector GIS, the visual results of mathematically based overlay will look quite different in vector and in raster. One would expect the cartographic output to resemble a hand-drawn map. The result, however, can be surprising, as many tiny, unexpected polygons of interacting polygons show up, especially around the margins of intersecting polygons. This apparently innocuous visual difference can have quite a profound impact on how we interpret the results of our analysis.

For example, let's compare multiple land use layers representing separate time periods for identical pieces of land, digitized from aerial photographs and classified on the same scale. When we compare a single set of polygons of our urban land use category for just two of these time periods, we notice that the polygons looked identical when mapped separately (Figure 14.11). But when combined, they do not exactly match. Instead, there are a number of sliver polygons at the margins of the overlaid/intersecting polygons (Figure 14.11). Now we must decide: Has there been no change from time 1 to time 2? Have there been slight changes in the land use locations? Was there a problem with the digital input of the data in the first place? Instinctively we know that changes as minor as the slivers in the composite are unlikely, especially because the polygons look so much alike. We can probably eliminate our second possibility as well, again because the polygons are nearly identical in shape and even minor changes in land use would have imposed greater modifications.

Ignoring the possibility of miniscule land use changes, we can conclude that the polygons are identical, and we try to account for variations caused by digitizing. Or if the polygons are essentially the same shape but one is slightly rotated relative to the other, the slivers might be due to a minor rotation or distortion in one of the aerial photographs from which the polygons were digitized. In this case, we might be able to perform a computerized form of rubber-sheeting. This remedy is readily available in GIS software but rarely gives 100 percent co-occurrence even if done with care. The process also requires that we anchor a number of points that we know should be located at exactly the same place on both layers. Although these are held in place, the rest of the map can be "conflated" or rubber-sheeted to move the polygons until

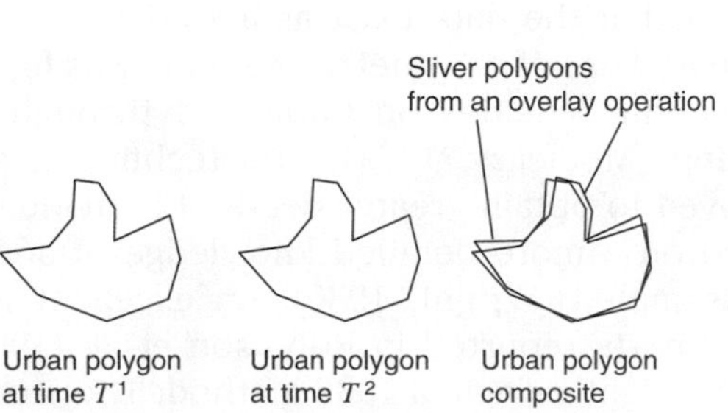

Figure 14.11 Sliver polygons created by vector overlay. These sliver polygons result from minor differences in two marginally intersecting polygons. Do the discrepancies indicate real change or error? Note that most of the slivers occur on the periphery of the two otherwise identical polygons. Care must be taken in interpreting the results of vector overlay, especially if small amounts of change are important.

correspondence is achieved. A common result of this is that other polygons that were originally well aligned move out of register. In short, there is no easy solution to this common problem. Some users select raster over vector overlay to avoid the slivers produced through vector overlay.

Problems are compounded when multiple layers are overlaid, especially if the layers encompass fundamentally different themes. For example, if we overlay a map of soils with one of vegetation, we no longer have the luxury of assuming that there will be a perfect one-to-one relationship between soils and vegetation. In fact, the purpose of the overlay might be to see where the categories deviate. Our dilemma is to separate the error from the actual polygon differences.

If, as in our last example, each map has its own unique amounts, types, and sources of error for both entities and attributes, how do we account for the error in multiple layers? Unfortunately, despite ongoing research attempting to identify how much error is propagated with multiple layers (for a good general survey of this topic consult Burrough and McDonnell, 1998), there are few general principles and even fewer answers, especially if the layers are from widely different sources (Chrisman, 1987). Your decisions must be based on your own knowledge about the data quality, input methods, and field methods to obtain them. Your response to error is heavily dependent on how precise your output must be for the problem at hand. It is logical to assume that if you have many different layers, the results of overlay will be only as good as the worst map employed, but it is very important to note that this is not often the case. This "**weakest link hypothesis**" approach, although appealing, is based on the assumption that all layers are equally important. As you have seen, however, you have the capability, in either raster or vector, of weighting each theme through weighted overlay. In addition, even if you do not explicitly weight each theme, some will ultimately be more important to your model results than others. Before you overlay your layers, you should become familiar enough with your data to be able to decide what types and amounts of error you can accommodate.

DASYMETRIC MAPPING

Before we leave the topic of polygon overlay, we should examine an old cartographic technique for detailing polygonal information on the basis of other variables. The technique, called **dasymetric mapping**, is based on the idea of the choropleth map discussed in Chapter 3, and it gives insights into some of the approaches currently implemented by GIS. There are others that have not yet been considered by the GIS community but might prove useful. Dasymetric mapping requires that the data exist as a statistical surface (see Chapter 3). The first documented use of dasymetric mapping was for improving the categorization of population densities on Cape Cod through a method called **density zone** outlining (McCleary, 1969). The technique, also called density of parts, was employed to obtain greater detail of individual densities of unresampled areas based on a more detailed knowledge of some smaller subareas that had been resampled (Wright, 1936). An excellent example of the application of density of parts reported in Robinson et al. (1995) closely associates the example with Wright's original 1936 method. The student of GIS should become familiar with this technique because of its power to improve the quality of quantitative data contained in polygons by comparison with more detailed data for another map layer.

We have already employed one method of dasymetric mapping without identifying it as such. Pure dasymetry, first used by Hammond (1964) in his work on landforms, entails the delineation or classification of areas of geomorphic type on the basis of a recategorization of topographic data. Topographic surfaces perfectly fit this approach, which calls for continuous distributions of an infinite number of points. When we studied reclassification of continuous surfaces, we created neighborhoods for "intervisibility," "south-facing slopes," "steep slopes," and so on by grouping preselected ranges of our data sets. This was a modified form of pure dasymetry. Once again, we see that the modern techniques we use so often have roots in a time well before the advent of the computer.

Several more forms of dasymetric mapping could prove potentially useful. Among the most powerful is the type called "use of other regions with the assumption of correlation," with its strong implication of a form of cartographic overlay. Thus, we might look at our example of density of parts and suggest that this was also a comparison of variables between or among maps. The assumption-of-correlation approach is strikingly different, however. Instead of isolating portions of an area for which detailed study improves the information content and then using this information to improve what we know about the remainder, we use either **limiting variables** or **related variables** contained in other layers.

We used the limiting variables form of dasymetry when we discussed exclusionary factors within overlay. As an example of how this might be applied to improve the quality of our polygonal data and the models we produce from them, let's say that we have a map of Minnesota that shows the population by county. We also have a map giving as a polygonal value the area in square miles for each county. By overlaying the two layers, having divided the population map by the area map, we obtain a map of population density for each county in the state. However, Minnesota is the "land of 10,000 lakes," which implies a substantial surface on which people do not reside unless they have houseboats. To improve our population density map result, we should have "excluded" from the area map the amount of water contained in each county. If we had used this "limiting variable" to create a map of land area by county and employed it in producing our population density map, the population density of counties with large water bodies would have gone up. Such techniques can be used frequently in conjunction with cartographic overlay to isolate and exclude areas that bias quantitative polygonal data.

A final example of dasymetric mapping, called related variables, corresponds strongly to our use of mathematically based cartographic overlay methods as in both raster and vector GIS. You have seen that variables contained in polygon layers often interact in ways more complex than simple exclusion. Statistical tests such as correlation and regression are often used to show how geographically dispersed variables are related to one another and how these relationships can allow us to predict variations in one based on the changes in the other. This is no less true when using GIS. If we, for example, know that there is a high correlation between percentage of cropland and percentage slope on the land, we can predict the amount of cropland on the basis of this correlation and its associated regression line. With this information, we can develop a detailed predictive map of "% cropland" on the basis of the slope alone. Alternatively, we might have layers of existing percentages of cropland and slope. By cartographically overlaying these, we would create a map showing the true relationships between these two variables in a particular area, or we could overlay our composite on a map of "predicted % cropland." The differences between the two would give a visual display, as well as a quantifiable difference between the actual and the predicted, which could be used to evaluate the predictive model. The areas that do not fit the predictive model could then be compared to other layers to develop hypotheses concerning variations in the model. This is an excellent example of how GIS can be used to develop hypotheses for scientific applications of GIS or to create predictive GIS models for decision making in commercial applications of GISs. We will look at the use of this powerful application in the next chapter.

As you have seen, dasymetric mapping has a great deal of potential to improve the use of GIS in both academic and commercial settings. Still, the subject is relatively untapped within the GIS literature. Raster GIS lends itself readily to many forms of dasymetry, and several methods are being employed every day without the knowledge of the users. Vector GIS is also capable of performing some—perhaps all—of these techniques, but few have attempted in a serious, systematic manner to evaluate its potential (Gerth, 1993). An awareness of dasymetry as a potential set of cartographic overlay methods will likely result in vast improvements in the GIS overlay toolkit currently available to users.

A related statistical approach that is often useful employs the integrated terrain unit mapping approach (Dangermond, 1976). Very much like biophysical mapping units, **integrated terrain units (ITUs)** generally reflect the assumptions that all the data are integrated on the ground and that you employ a method, such as digitizing from aerial photographs, that requires you to extract these interacting variables from the same source of information. This ensures that the variables you include are related spatially, if not in a logical sense. If you do not have the capability of employing ITUs and can't find a logical reason, or at least a statistical reason, that would indicate a relationship among your variables, perhaps you should reevaluate the solution to your problem. This is no different from testing the relationships between the surrogate radiometric responses gathered from a satellite and the land features they are meant to represent.

Terms

AM/FM
areal correspondence
biophysical mapping
clip
clipcover
complement overlay
conditional overlay
dasymetric mapping
density zone
exclusionary variables
graphic overlay
identity overlay
integrated terrain unit (ITU) mapping
intersection overlay
least common geographic unit (LCGU)
limiting variables
line-in-polygon overlay
overlay
point-in-polygon overlay
related variables
rules-of-combination overlay
selective overlay
union overlay
weakest link hypothesis
weighted overlay

Review Questions

1. What are some obvious limitations of the manual overlay process? What advantages have been offered through automation?
2. What are the advantages of raster-based overlay operations over their vector-based counterparts? What are the disadvantages?
3. Why is overlay still among the strongest operations in most modern commercial GISs?
4. From your own studies, give examples of how point-in-polygon and line-in-polygon could be useful to you.
5. What are the limitations of graphic overlay operations? Give an example of how graphic overlays would prohibit subsequent analyses on the product of the overlay process.
6. What is dasymetric mapping? How can it be employed in vector GIS? Raster?
7. What is the difference between related variables and limiting variables in dasymetric mapping? Give an example of the use of each variable type.
8. What problems or difficulties might you encounter when comparing multiple layers with vector overlay techniques? Are there any approaches you might consider to solve these problems? What should you do before you overlay to lessen this?
9. Give some examples of how portions of dasymetric mapping might readily be applied with existing raster or vector GIS. Why do you suppose there has been so little attention paid to the use of dasymetric mapping in GIS overlay?

References

Adams, J.S., 1970. "Residential Structure of Midwestern Cities." *Annals of the Association of American Geographers*, 60(1):37–62.

Burrough, P.A., and R.A. McDonnell, 1998. *Principles of Geographical Information Systems*. New York: Oxford University Press.

Chrisman, N., 1987. "The Accuracy of Map Overlays: A Reassessment." *Landscape and Urban Planning*, 14:427–439.

Chrisman, N., 1996. *Exploring Geographic Information*. New York: John Wiley & Sons.

Chrisman, N., and T.R. Peucker, 1975. "Cartographic Data Structures." *American Cartographer*, 2(1):55–69.

Dangermond, J., 1976. "Integrated Terrain Unit Mapping (ITUM)—An Approach for Automation of Polygon Natural Resource Information." Publications of the Environmental Systems Research Institute, no. 160, 11 pages. Redlands, CA: ESRI.

Gerth, J.D., 1993. "Towards Improved Spatial Analysis with Areal Units: The Use of GIS to Facilitate the Creation of Dasymetric Maps." Unpublished MA paper, Ohio State University, Columbus.

Hammond, E.H., 1964. "Analysis of Properties in Landform Geography: An Application to Broad-Scale Landform Mapping." *Annals of the Association of American Geographers*, 54:11–19.

Hills, G.A., P.H. Lewis, and I. McHarg, 1967. *Three Approaches to Environmental Resource Analysis*. Cambridge, MA: Conservation Foundation.

McCleary, G.F., Jr., 1969. "The Dasymetric Method in Thematic Cartography." Ph.D. dissertation, University of Wisconsin, Madison.

McHarg, I.L., 1971. *Design with Nature*. Garden City, NY: Natural History Press.

Monmonier, M., 1991. *How to Lie with Maps*. Chicago: University of Chicago Press.

Robinson, A.H., J.L. Morrison, P.C. Muehrcke, A.J. Kimerling, and S.C. Guptill, 1995. *Elements of Cartography*, 6th ed. New York: John Wiley & Sons.

Sauer, C.O., 1925. "Morphology of Landscape." In: *Land & Life*, J. Leighly, Ed. Berkeley: University of California Press, 1963, pp. 315–350.

Simpson, J.W., 1989. "A Conceptual and Historical Basis for Spatial Analysis." *Landscape and Urban Planning*, 17:313–321.

Steinitz, C., P. Parker, and L. Jordan, 1976. "Hand-Drawn Overlays: Their History and Prospective Uses." *Landscape Architecture*, 56(4):146–157.

Tyrwhitt, J., 1950, "Surveys for Planning." In *Town and Country Planning Textbook*. London: Architectural Press.

Wikle, T.A., 1991. "Computers, Maps and Geographic Information Systems." *National Forum*, summer, pp. 37–39.

Wright, J.K., 1936, "A Method of Mapping Densities of Population: With Cape Cod as an Example." *Geographical Review*, 26:104–115.

CHAPTER 15

Cartographic Modeling

In this chapter we will view the analysis subsystem as a set of interacting, systematic, ordered map operations that together perform some very complex modeling tasks. These complex models are composed of much smaller, simpler model components that can be solved with relatively few analytical operations. Each one can then be combined with others to create larger modules of more complex models, and these can be combined further until your entire model is constructed. Like any system, a spatial model can almost always be broken down into these component parts as long as you are familiar with how the model works.

Although you have seen some sample flowcharts along the way, this chapter emphasizes their development as an integral part of the modeling process. Flowcharting forces you to think carefully about what data elements you need to obtain the answers you are seeking. It is easier to detect missing themes, or to spot elements that will be used many times, by examining a flowchart during model development than to correct a finished model. Flowcharting of cartographic models should be a required preliminary step to the development of any model. During this process, you may discover that you do not have a clearly defined objective, whereupon you will have to rethink your goals for building the model in the first place. In addition, even if your primary interest is not modeling but database development, you cannot properly develop a database without first specifying the final product that is to be obtained from the modeling process.

When you examine the models and flowcharts given in this book, you will probably find dozens of ways to improve on the approach or to enhance the model by including additional information. The models given should not be considered to be absolutely complete. Nor should you assume that they represent the only solution to the respective problems. Instead, you should envision alternative solutions for each problem—perhaps even creating an alternative flowchart for each. Then make a comparison—it may appear that your approach is more easily accomplished (i.e., a more elegant solution) or more realistic (i.e., more likely to give an accurate representation of what is being modeled). Practice is essential to good flowcharting, as it is to good modeling.

LEARNING OBJECTIVES

When you are finished with this chapter you should be able to:

1. Understand how commands are assembled to produce a cartographic model.
2. Illustrate the cyclical movement from subsystem to subsystem inside a GIS.
3. Know the difference between deductive and inductive cartographic modeling and the advantage of deductive over inductive approaches.
4. Determine the difference between explicitly and implicitly spatial variables and discuss the role of spatial surrogates for nonspatial variables.
5. Produce and explain a simple model flowchart.
6. Explain the problem of cartographic modeling constraints that are too severe or too loose and define methods of circumventing these problems.
7. Describe the idea of weighting of coverages as they impact the outcome of cartographic models, especially with regard to how these weights might affect the error component of the output.
8. List and discuss three basic areas of cartographic model verification.
9. Describe alternatives to cartographic output for GIS results.
10. Discuss the role of prototyping and preplanning in the production of a successful cartographic model.

MODEL COMPONENTS

In the last few chapters you've seen a wide variety of possible individual techniques for working with points, lines, areas, and surfaces. Each of these simple techniques alone can be powerfully applied to the solution of geographic problems. And, of course, each has a wide array of possible options and subtle variations that could radically change the result of an analysis. As a practical matter, the number of combinations could be described as nearly infinite.

Fortunately, there are limitations to the ways in which the options can be applied. To develop a neighborhood function that shows the number of land uses within a specified area, for example, you would use a measure of diversity, ignoring all other options because "diversity" correctly answers your question. This constraint on the numbers and types of options used to answer specific questions greatly simplifies GIS analysis. It also focuses your attention on the techniques or options that are absolutely necessary. You will recall that there are many options for overlaying one thematic map with another. You first decide what result you are looking for and what is the best solution. For example, an overlay operation using the logical complement may produce a map that has the same polygonal entity shapes as a map based on union. One approach requires you to perform a second operation to reclassify the result because the attributes are the opposite of what you want. Both achieve the same answer, but one is more efficient than the other. Efficiency is very important in working with large models, both to prevent you from making mistakes and to simplify the overall model as you begin combining many commands.

As with the options for each command, the potential number of combinations is daunting. Fortunately, the potential for interaction of commands is also limited by the nature of the models you are creating and the types of objects on which you are operating. For example, although it is common to reclassify polygonal data prior to polygon overlay operations, this is not generally done prior to topographic viewshed analysis. So not only are the nature of the command operations linked to the models and their objects, but their sequencing is also driven by them. It would be very useful to be able to create a set of do's and don'ts for which commands will be combined with selected others and which should not. Unfortunately, such a cookbook would be too large to be practical. Instead, you will have to experiment, learning from your own experiences and those of others around you. Formalized procedures for combining GIS functions to analyze complex spatial systems are now available, and we will examine them in some detail.

THE CARTOGRAPHIC MODEL

The term **cartographic modeling** was coined by C. Dana Tomlin and Joseph K. Berry (1979) to designate the process of using combinations of commands to answer questions about spatial phenomena. Their formal definition is that a cartographic model is a set of interacting ordered map operations that act on raw data, as well as derived and intermediate map data, to simulate a spatial decision-making process (Tomlin, 1990). Let's examine this definition in greater detail.

The first condition of the definition is that the map operations must interact with one another. Each individual operation performed on a theme or element must have as its purpose the creation of a result, usually another theme or variable, that can be used by the next operation. The next operation may be a further manipulation of the same output, for example, to isolate certain map variables, or it may operate on a separate input variable or map layer, perhaps as a map overlay operation.

Let us say that you have a "landcover" map and you want to determine whether there are large, contiguous, federally owned rangelands available so you can petition the government for permission to graze your cattle there. You first reclassify the map to create a new layer, called "landuse," to isolate the land uses related to the land covers (e.g., reclassifying grassland polygons as either pasture or natural grasslands). Then you create a map called "whosegrass" to show pasture ownership, and reclassify "whosegrass" to separate out only federal land. Your next map operation is to reclassify that map based on a size operation to create "bigfed," which isolates contiguous (a neighborhood function), federally owned grasslands that are greater than or equal to that 25 hectares (Figure 15.1).

Notice that you operated on both raw data ("landcover") and derived data ("bigfed"). A cartographic model employs each analytical operation on either raw or intermediate map data. Each step is combined to incrementally add information to the data by giving the data a context about which it relates.

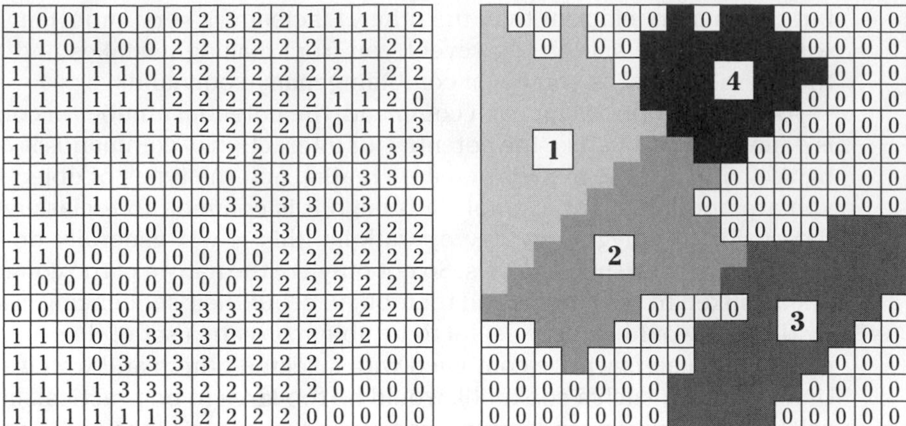

Figure 15.1 Reclassifying neighborhoods by size. Clumping to isolate the polygons of 25 hectares or larger.

Our original layer, "landuse," was not specific enough: It related to all possible land cover in the study area, without regard to what the question might have been. As we manipulated this layer we created value-added layers of specific attributes that could be used to decide whether the layer contained federally owned pastures large enough to graze a certain number of cattle.

This brings us to the last portion of the definition. Our manipulation of raw and intermediate data was designed to simulate the process of deciding whether the coverage in question contained "large enough" contiguous areas owned by the federal government. A negative result will require you to look elsewhere for such grasslands, whereas a positive result will prompt you to begin seeking the permits necessary for your grazing operations. A positive result means simply that there are sites available. But suppose you want the option of increasing the size of your herd. You would then have to perform yet another operation on the derived coverage "bigfed." By asking the computer to rank by size the federally owned pasture polygons that exceed 25 hectares, you will be in a position to determine which combinations would allow you to increase your grazing herd. These steps add information to each preceding data element incrementally.

Now that you have completed your simple decision model, think carefully about the GIS subsystems involved. The first step was to *input* the "landcover" map, which was then *stored* and/or *edited*. When you were satisfied with the accuracy of your landcover map you *retrieved* it as digital *output* for *analysis* and reclassification as a land use map. The product of the reclassification is a *stored* map layer. Next you *retrieve* the new layer to continue its *analysis*. At any point you might want to view "output" of the map to examine your progress. The process can continue to input, retrieve, store, edit, output, and analyze these and different layers at will.

Clearly, the foregoing process is not linear—we did not go directly from input to storage to retrieval to output, as in cartographic systems. Instead, we moved about the different GIS subsystems as necessary. In short, the process of cartographic modeling is cyclical rather than linear (Figure 15.2). The cyclical nature of cartographic modeling allows us the greatest possible level of data transformations to produce our final **spatial information product (SIP)**.

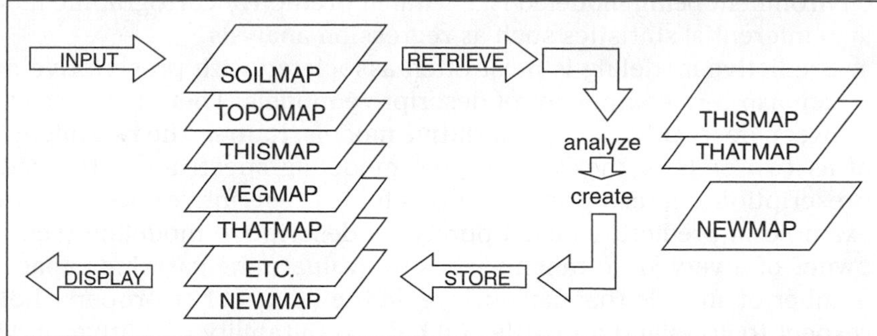

Figure 15.2 Cyclical modeling process. The illustration shows the movement from one geographic information system (GIS) subsystem to another to produce new maps.

TYPES OF CARTOGRAPHIC MODELS

There are strong similarities between the categorization of statistical techniques into descriptive and inferential types and the major types of cartographic models. Getis separates spatial analysis procedures into **exploratory analysis** and **confirmatory analysis**. Exploratory is more descriptive of how spatial layers are spatially correlated, while confirmatory assumes that we know such correlations exist and are trying to authenticate the degree to which these correlations exist. Specific to the development of cartographic modeling, Tomlin (1990) employed loosely synonymous terms. He classifies cartographic models into **descriptive cartographic models** (like descriptive statistics) and **prescriptive cartographic models** (like inferential statistics). Because of their origins and the explicit link to cartographic modeling, we will adopt this terminology for our discussion.

Descriptive cartographic models describe and sometimes explain patterns and pattern associations produced from analysis. The simplest descriptive models portray existing conditions by isolating preselected phenomena and presenting the results unambiguously. Despite their simplicity, descriptive models are among the most common because they offer a relatively straightforward way of producing easily recognized patterns of spatial phenomena. The output can be viewed as a final product or as the precursor to more complex models.

It is a natural progression to go from the description of existing conditions to explanation and eventually to prediction. These predictions allow us to select the best circumstances for locating industry, preventing hazards, siting a dam, or providing waterfowl habitat. A city planner might be able to predict locations of potential urban expansion by using knowledge of where expansion has taken place in the past to identify spatial phenomena that might be used as predictors.

The latter case illustrates the predictive potential of descriptive cartographic models. Such **predictive models** allow the user to determine what factors are important in the functioning of the study area. They also permit the user to determine how these factors are associated with each other spatially. Prediction on the basis of these associations can, of course, be very tricky. It requires that the variables have a clear and verifiable causal relationship. Remember, spatial association does not assume a cause-and-effect relationship. Knowledge of the environment being modeled is as vital in predictive cartographic modeling as it is in inferential statistics such as regression analysis.

Predictive modeling is most often associated with prescriptive models, but it can also be a component of descriptive models. There is no clear separation between descriptive and prescriptive models. Rather, the two might be thought of as two ends of the cartographic modeling spectrum, with prediction and prescription increasing as we approach the prescriptive end. Take the following example of prediction based purely on descriptive modeling techniques. The owner of a very large ranch wishes to evaluate the carrying capacity (i.e., the number of animals that can survive on the land) of her property holdings with respect to grassland for cattle and habitat suitability for native wildlife such as quail. The model requires her to develop a database showing all the vegetation types on the ranch. In addition, she needs to know the aboveground biomass (the amount of vegetation by weight that occurs above the ground level) for the grassland vegetation, as well as the locations of large patches of noxious weeds (that might be harmful to her stock) and significant clumps of shrubs (needed by the quail for shelter and foraging). From these variables, she produces a cartographic model showing the areas that have the minimal habitat needed to support a reasonable quail population. She also produces a thematic map indicating the minimum necessary carrying capacity for the cattle. The model is descriptive in that it shows where the owner could reasonably expect to put her cattle and where quail might survive as well. At this point she could simply herd her cattle into the adequate grazing portions of the ranch and prevent the livestock from moving into the quail habitat by fences or other means.

Suppose, however, that she does not wish her cattle merely to graze successfully but rather wants them to have only the very best grazing lands, to ensure that the animals are fattened as quickly as possible. Such a manipulation of the database could reasonably be called predictive, because in effect it is predicting that the areas with the highest carrying capacity will result in the most efficient use of the land for grazing. This model is only slightly more complex than the original, but it carries with it an element of prescription because it "**prescribes**" the "best" use of the land.

Prescription also shows the model user how manipulating existing attributes can improve an overall situation. For example, some of the best grazing land may contain many shrubs, making it prime quail habitat as well. If the same location also contains poisonous weeds, the rancher may decide to plant (or transplant) shrubs in areas that are less productive for cattle while eliminating the noxious weeds from the most productive grazing sites.

The GIS could be used to "prescribe" the best places for moving the shrubs as well as targeting for other manipulation locations with noxious weeds. Thus, the GIS is now used to suggest the appropriate manipulation of existing holdings to ensure the best solution to the original problem. It is the highest form of predictive model because the user has to determine the interactions of many variables through time, and a much higher degree of predictive capability is called for.

You have probably noticed a common theme of predictive modeling, even if an individual case is not purely prescriptive. All these models require, as a prerequisite, a description of existing conditions. That is, when developing prescriptive models, you describe first and then prescribe. The change from one to the other is a gradual transition from pure description with no action needed to increased predictive power and increasing prescriptive capabilities. The nature and complexity of the problem will dictate the type of model you employ, no matter what you call it.

INDUCTIVE AND DEDUCTIVE MODELING

Whether you are working with descriptive models, prescriptive models, or somewhere along the spectrum, there are generally two logics involved in model formulation. The first, the **inductive model**, begins by examining the placement, patterns, sizes, shapes, and other properties of specific map entities and moves toward a general conclusion concerning those properties. For example, you could examine the spatial characteristics of black bears based on their known locations to build a model of where they should appear. Such approaches sometimes lack design but are useful, especially in a research environment where spatial data have been collected over a long period of time for individual research projects. Under such a setting, the empirical spatial data are often perused for important patterns and generalizations. Different map layers are compared and tested for correspondences among the elements of others to identify similar spatial patterns or associations that might indicate an ongoing process within the study area. A useful caveat here is that the data being used for inductive models should have been sampled within a spatial sampling scheme with the model formulation in mind.

Unlike inductive models, **deductive models**, move from general knowledge to specific answers. Most often you begin with a specific recipe or formulation that addresses very specific questions. Adopting our previous example, if we have been studying and mapping black bears for a long time, we can relate their patterns to other known factors. We might want to examine what particular habitat changes are causing a reduction in population density from what we normally deserve. The primary advantages of the deductive approach to cartographic modeling include a ready ability to perform statistical hypothesis testing, and time savings resulting from a focus on building spatial data oriented specifically on the solution to the working hypotheses.

FACTOR SELECTION

Recall from Chapter 6 that one of the problems we face in database creation and subsequent model building is determining appropriate variables. In some cases the choice is obvious. If we are going to model a transportation net, common sense tells us to include all available road network data. It is just as obvious that we need to know the road types, which are one-way and two-way, which have traffic signals, and so on. All these factors are accessible as attributes of raw input data elements. However, not all data are so easily obtained, nor are all variables explicitly defined.

Suppose you want to model the potential impact of a hazardous chemical spill along a rail line in an urban setting. You have explicit information about the rail line, and possibly about the train manifest. You also have explicit spatial information about the number of housing units located near the rail line. Because the population in your area is highly mobile, however, you don't know exactly how many people live in each household, nor do you have information about any special-needs populations such as elderly or handicapped people. To develop your model, you will need to estimate these variables with surrogates. Perhaps you will sample the households in each neighborhood to get an idea of the average number of people living in each unit and the average number of special-needs individuals. By generalizing these data, you can obtain an estimate for each neighborhood, and then use these estimates in your potential spill impact model.

MODEL FLOWCHARTING

Whether you are using a deductive or an inductive approach, an extremely useful technique to assist you in formulating your model and determining the appropriate data elements is the **flowchart**. Model flowcharting requires that you isolate each item or element (input layer) that is to be used in your model. Each element should have a very specific, unique theme that is representative of a single factor or group of factors in your model. Flowcharting allows you to determine whether you have all the necessary map elements. It also allows you to evaluate whether each is unique. If you have several redundant data elements representing the same theme, you can eliminate one or more of these from your model and the input time will be reduced. There are two types of model flowcharts: **formulation flowcharts**, which are designed to conceptualize your model and isolate important elements, and **implementation flowcharts**, which allow the modeling to take place. Formulation flowcharts start with the desired outcome and decompose the relevant parts until all parts are accounted for. The implementation flowchart moves from the basic map elements, combining them and their intermediate layers until arriving at the final outcome. In short, the two flowcharts are mirror images of each other. We will examine how the flowchart works by dissecting the formulation flowchart first.

Figure 15.3 is a formulation flowchart of a deductive (and prescriptive) cartographic model designed to locate the best site for a mountain cabin. The specific result requires the combination of four major groups of factors. Let's examine the components of infrastructure first. You need to know about the availability of running water, electricity, gas, roads, and other facilities (sewerage, etc.). Next are political or legal factors such as land ownership, zoning, whether the site meets federal standards for fire and/or earthquake hazards; these all place constraints both on where you can build and how much the effort will cost.

A third group of model factors are aesthetic. You are building a mountain cabin to use as a retreat from your usual environment, so you want the cabin to be out of sight of the city while still within driving distance. In addition, you want an unobstructed view of the mountains, and you may also want the cabin to face in a particular direction, perhaps west, so you can view sunsets.

The fourth group of siting factors is physical. You need a relatively level slope in an area that does not have dense tree cover or with a clearing. Soil engineering properties such as soil stability and existence of high shrink-swell

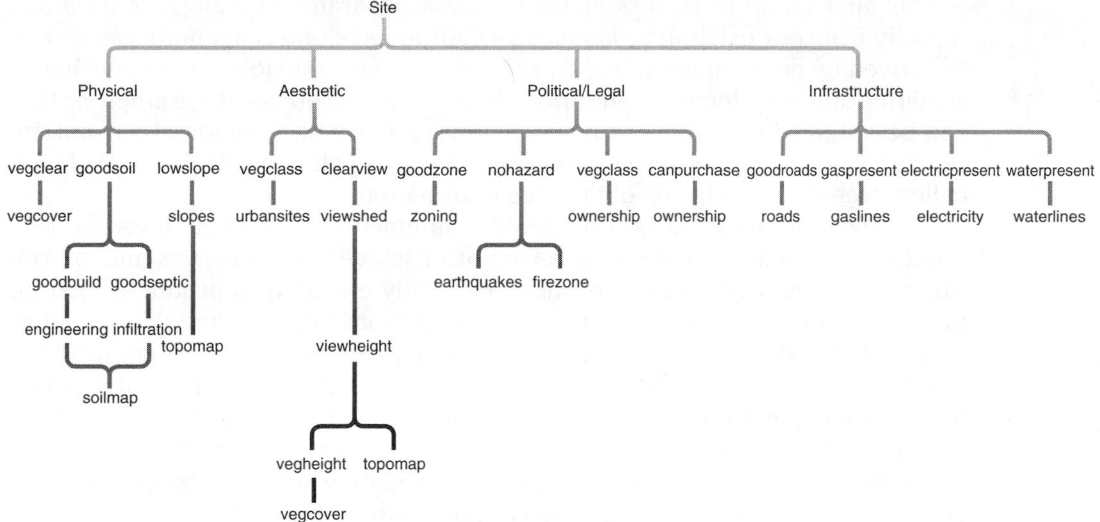

Figure 15.3 Simple mountain cabin model flowchart. This shows each of the elements as well as the interim maps needed to produce the final solution—the site.

clays will influence building costs. You may not have access to sewerage facilities, requiring the construction of a septic system. This means the soils infiltration is also a necessary physical factor.

Let's look closely at how each data element is chosen to represent individual unique factors (later we will examine how these elements are combined to complete the model and indicate the best site for our mountain cabin). We begin by looking at the infrastructure portion of our model. For data regarding the first factor—accessibility to running water—we need a map layer that shows where all the municipal water lines are located. The layer "waterline" would be created by digitizing an analog map of water lines. Likewise, we create layers for electric lines ("electricity"), locations of gas lines ("gaslines"), and all available roads in our study area ("roads").

Our political and legal factors require us first to obtain "ownership" from an analog cadastral map. This layer allows us to determine locations we could purchase. Next, we obtain a zoning map and digitize "zoning," from which we can legally build a single-family dwelling. We also need to know whether there are locations that are earthquake zones or fire hazards. We obtain a map of earthquake-prone areas and digitize these ("earthquakes"), and from forest fire hazard maps we create a theme called "firezone." Although the last two layers are physical parameters rather than political or legal, it is likely that permit restrictions are already in place for such areas.

An evaluation of aesthetic qualities requires maps showing the locations of nearby municipalities ("urbansites"). Topographic maps provide information about obstructions ("topomap"). Maps of forest types tend not to include information about the sizes of trees because of the amount of time required to obtain such information. You might have to create your own forest cover map, either by performing a ground survey of some of your more suitable locations or by using aerial photography of the area and mapping the forest stands yourself to create the needed layer ("vegcover").

Our final group of factors includes physical parameters. Slope is obtained directly from our existing topomap layer. Although slope is a unique element, it is derived by modifying existing data. Thus, we have not violated our rule about selecting unique elements. Another element of our physical parameters that can be obtained from an existing map shows areas with little forest vegetation, or with forest vegetation containing an appropriate clearing. The existing layer, called "vegclass," helps us obtain this information.

Our two soil parameters will likely be obtained from a single layer as well. A detailed soil map will normally have both engineering properties and infiltration properties; either of these will be explicitly encoded to produce separate maps, or they will be associated through a database system and connected to each of the soil-type polygons. We will create these maps by first creating the soil layer ("soilmap"). The methods we use to keep track of the various soil properties depend on our GIS software. Alternatively, we could simply encode the soil polygons and retain a paper copy of the soil survey from which we obtained the soil polygon information. These survey reports associate a large variety of soil factors with each soil-type polygon (Table 15.1).

Working Through the Model

Look closely at the flowchart we used for the mountain cabin siting model: all the appropriate elements are included, and they all flow to the final site for our cabin. There is still something missing, however. Ask yourself how the intermediate layers were created from the initial data elements. To complete our flowchart, we must indicate what operations the GIS used to create these intermediate layers. Not only is this step important in modeling, it also assists us in determining the needed functionality of any GIS for a specific project. Let's consider a revised version of the flowchart that includes this functionality (Figure 15.4).

To avoid using software-specific commands, we will use generic, mnemonic commands that are descriptive of the function of the command. To implement the model on your own software, you need only to change these names to the specific commands you would use. Let's begin by examining the infrastructure parameters: roads, gas lines, water lines, and electric utilities. There is a high probability that none of these will directly serve your cabin site. We need to know whether they are close enough—say, within 500 meters—to make it feasible to extend them to your cabin. To find out, we begin with one of the existing elements. The digital electric utility layer will tell us where they are located, and it will not be too difficult to create a buffer of 500 meters on each side of the existing electric line entities. The result is a map called "electricpresent," which shows potential cabin sites that are reasonably accessible to electric power. Next, we build a buffer around your gas lines layer extending, say, 250 meters, indicating the likely areas where gas lines could be extended to your cabin—a layer called "gaspresent." Roads are not nearly as important because we have a four-wheel-drive vehicle. You are comfortable traveling as much as 750 meters offroad to get to your cabin, so we can build a buffer of 750 meters around the existing roads layer to produce "goodroads."

TABLE 15.1 Typical Page from a Soil Survey Report Showing Soil Polygons and Small Portions of Tabular Information Available for Each Soil Type

Soil Name and Map Symbol	Dwellings Without Basements	Dwellings with Basements	Small Commercial Buildings	Local Roads and Streets	Septic Tank Absorption Fields	Sewage Lagoon Areas	Playgrounds
Ev: Eudora part	Severe: floods	Severe: floods	Severe: floods	Severe: frost action	Moderate: floods	Moderate: seepage	Moderate: floods
Gravelly land: Ge	Moderate: slope	Moderate: slope	Moderate: slope	Slight	Severe: small stones	Severe: seepage, small stones	Severe: small stones
Judson: Ju	Moderate: shrink/swell	Moderate: shrink/swell	Moderate: shrink/swell	Severe: frost action	Slight	Moderate: seepage	Slight
Kennebec: Kb Kc	Severe: floods	Severe: floods	Severe: floods	Severe: floods, frost action, low strength	Severe: floods, wetness	Severe: floods, wetness	Moderate: floods
Kimo: Km	Severe: floods, shrink/swell	Severe: floods, shrink/swell	Severe: floods, shrink/swell	Severe: shrink/swell, low strength	Severe: percolates slowly	Slight	Severe: wetness
Martin: Mb, Mc, Mh	Moderate: shrink/swell	Severe: shrink/swell	Severe: shrink/swell	Severe: low strength, shrink/swell	Severe: percolates slowly	Moderate: slope	Moderate: too clayey, percolates slowly
Morrill: Mr	Moderate: shrink/swell, low strength	Moderate: shrink/swell, low strength	Moderate: shrink/swell, low strength	Moderate: shrink/swell	Severe: percolates slowly	Moderate: slope	Moderate: percolates slowly
Ms	Moderate: shrink/swell, low strength	Moderate: shrink/swell, low strength	Severe: slope	Moderate: shrink/swell	Severe: percolates slowly	Severe: slope	Severe: slope
Oska: Oe	Severe: shrink/swell	Severe: shrink/swell, depth to rock	Severe: shrink/swell	Severe: shrink/swell	Severe: depth to rock, percolates slowly	Severe: depth to rock	Moderate: percolates slowly, too clayey, depth to rock
Pawnee: Pb, Pc, Ph	Severe: shrink/swell	Severe: shrink/swell	Severe: shrink/swell	Severe: shrink/swell	Severe: percolates slowly	Moderate: slope	Moderate: percolates slowly
Sharpsburg: Sc, Sd	Severe: shrink/swell	Severe: shrink/swell	Severe: shrink/swell	Severe: shrink/swell, low strength	Severe: percolates slowly	Moderate: slope	Moderate: percolates slowly
Stony steep land: Sx	Severe: depth to rock, slope	Severe: depth to rock, slope	Severe: depth to rock, slope	Severe: depth to rock, slope	Severe: depth to rock, slope	Severe: depth to rock, slope	Severe: depth to rock, slope
Thurman: Tc	Slight	Slight	Moderate: slope	Slight	Slight	Severe: seepage	Severe: slope
Vm: Vinland part	Moderate: depth to rock, slope	Moderate: depth to rock	Severe: slope	Moderate: depth to rock	Severe: depth to rock	Severe: depth to rock	Severe: depth to rock
Wabash: Wc Wh	Severe: wetness, floods, shrink/swell	Severe: wetness, floods, shrink/swell	Severe: wetness, floods, shrink/swell	Severe: wetness, floods, shrink/swell	Severe: percolates slowly, floods, wetness	Severe: floods, wetness	Severe: wetness, floods, percolates slowly

Source: U.S. Department of Agriculture.

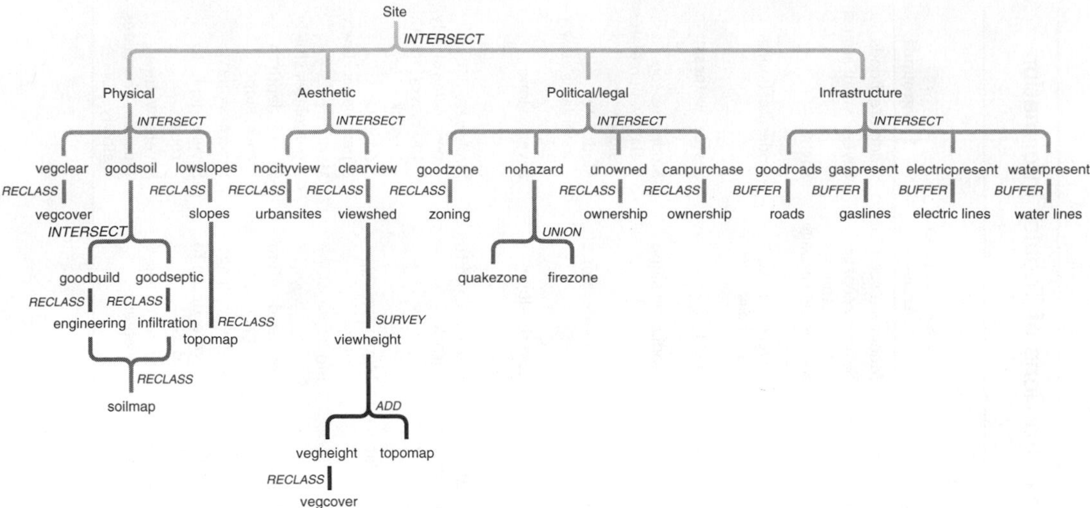

Figure 15.4 Detailed mountain cabin model flowchart. This flowchart shows the processes necessary to move from one interim map to the next to obtain the final solution. *Source:* Modified from S.A. Carlson and H. Fleet. Systems Applications Geographic Information System (SAGIS) and Linked Analytical/Storage Packages. Workshop, annual meeting of the Association of American Geographers, 1986.

We now have three separate, intermediate themes, each representing the limits of your requirements for accessibility to roads and electricity, water, and gas utilities. Now we must decide how to combine these to produce a final map, called "infrastructure," to represent all these factors. Because you need *all* these elements, rather than just one or two, they must be overlaid in a way that results in a map showing where they intersect. Therefore, the overlay procedure must use some form of "and" search, or intersection. The resulting map should display at least one area that satisfies all three criteria.

What happens if no area satisfies all of these infrastructure criteria? This is not an unlikely outcome. If our composite map shows no polygon intersections, we could focus on the factor that is the farthest from intersecting the other two. Let's say that the gas line buffer and the roads buffer intersect, but the electric lines do not. If the distance limits for that factor can be extended—in other words, if the model constraints can be relaxed—we might be able to fulfill the requirements of our infrastructure submodel. For example, we might find that for an additional charge, you could extend the electric lines to 1,000 meters. This approach to improving model performance by constraints relaxation is common in cartographic modeling.

If constraints relaxation is impossible—for example, if the electric company is unwilling or unable to extend powerlines 1,000 meters from their existing facilities—we may have to reduce the importance of one or more of your factors. In creating the buffers, we based their sizes on the absolute maximum limits for the corresponding factor. That is, we assumed that all factors were equally important. This is not always the case—in fact, it is rare. Therefore, unless you are willing to abandon the project, you must weight the four infrastructure factors. Roads are important beyond the 750-meter limit because you will need to get to and from the cabin regularly. This factor must retain a high degree of importance. You could heat your cabin with gas if it is available or even propane if not. What is left is electricity, which was the problem in the first place. The solution to your problem is to downgrade your perceived need for electricity. Thus, you could reduce the importance of electricity to zero, eliminating it as a model constraint. Operationally we perform a weighted overlay rather than an intersection overlay. This shows us how the model itself helps us decide which commands to use and how to use them.

Now that we have satisfied our infrastructure submodel constraints, we proceed to the political and legal factors. Areas not zoned against single-family dwellings ("goodzone") are obtained by reclassifying the zoning layer. Likewise, we create a layer of property without a registered owner ("unowned") and another of properties whose owners may be willing to sell or have offered the land for sale ("canpurchase"). Both these maps use the same initial coverage ("ownership"), and both are produced by reclassification. The remaining political and legal maps, "quakezone" and "firezone," contain the information we need, but your cabin must be located outside both of these areas. We can combine these using an "or" (union) overlay. The resulting output, called "nohazard," shows all areas that do not have seismic and/or fire hazards.

We now have the four intermediate themes needed to complete our political/legal submodel. As before, to satisfy all these constraints, we use an intersect overlay to combine all the politically/legally viable polygons. Here, we must be careful how we intersect the layers. We must be sure that all four layers indicate good areas for locating your cabin. If one of the themes, for example, the nohazard layer, actually shows locations that have hazards, your cabin will be located in a rather warm and possibly shaky location.

The third submodel, aesthetic factors, has fewer components than the other three but is somewhat trickier to calculate. Beginning with the urban sites theme, we may not be able to eliminate all small towns and villages from the study area. In fact, you probably want a nearby community as a source of supplies. You decide that you are willing to live near towns with fewer than 1,000 people. We could reclassify the map to create a "nourban" theme that appears to satisfy that criteria. But we need to know where the large towns and cities are so we can determine whether they can be viewed from your potential location. To do this we reclassify "urbanzone" to show where all the cities are located ("cities"). This layer is then combined with the theme we are about to create to eliminate such unsightly anthropogenic artifacts from your view.

The second portion of the aesthetic submodel requires that you can see around any peaks and any tall vegetation. Remember that the vegetation is growing on the terrain itself. To incorporate this feature, we must find an approximate height for mature forest stands. Then we reclassify our vegcover on the basis of the known vegetation height classes and call the result "vegheight." This theme of estimated tree heights is stored as ratio data rather than nominal data. To show what the combined vegetation and topographic elevations are, we add the topomap data (also in ratio format) with the vegheight data to produce a layer called "trueheight." This is the intermediate theme we need to determine what can and can't be seen. We use trueheight to compute a viewshed that shows what areas are visible from your proposed cabin site ("clearview"). Remember, however, you don't want to see cities. The clearview layer shows areas that are visible from the hillslopes, and the cities map shows all communities exceeding 1,000 people. By combining these two layers through intersection overlay, we produce a map that shows viewable areas with cities and viewable areas without cities. The viewable areas without cities will ultimately be combined with the other three submodels.

Completing the model, we next look at the three major physical parameters: vegetation, soils, and slope. Returning to the vegcover layer, it is easy to reclassify the vegetation categories into those with large, mature tree stands and those with other types of vegetation or with relatively small trees that can be cleared. The resulting layer, called "vegclear," should isolate areas with no mature tree stands. Keep in mind that the scale of most vegetation maps is rather small, offering little detail. Thus, any decision relying on such data should be checked on the ground. Some mature tree stands on very rugged terrain are not very high and hence may not be as difficult to clear as a vegetation map implies. This is another area in which it is very easy to modify constraints to accommodate the decision process. In this case, having mature trees may not be that important because of the ready availability of chainsaws. In addition, a few trees, properly placed, could provide shade without obstructing the view or interfering with the construction of your cabin.

Among the most severe physical constraints in siting a mountain cabin is slope. A severe slope will either prohibit the project altogether or radically increase its cost. A slope of 15 degrees or less would be ideal. To find all the areas that fit that constraint, we begin with the topomap and perform a neighborhood function that reclassifies slope so that low slopes are those of 15 degrees or less ("lowslope").

The two soil characteristics that must be considered are engineering properties and infiltration capacity. The first determines the capacity of the soil to withstand the weight of your mountain cabin ("goodbuild), and the second determines if the soil will support a septic tank ("goodseptic"). These layers are then intersected to create a layer called "goodsoil." This layer completes the three physical factors dictating where you can put your mountain cabin. Through an intersection overlay we create the final submodel layer ("physical").

Our final solution is to intersect physical, aesthetic, political/legal, and infrastructure maps, because all these constraints must be met. If our model works as planned, we should have at least a single, small portion of land that suits all your needs.

Not so fast! Just when we thought we were finished, another problem appears. Earlier we discussed the problem of constraints that were too tight. When we looked at the infrastructure factors, we discovered that insistence on electricity limited your choice so severely that the idea of building a mountain cabin would have to be abandoned. There is also the unlikely possibility that our outcome illustrates that there are very good soils with few very steep locations, lots of good mountain views, no major zoning or ownership problems, and plenty of close infrastructure. If this nearly fairy-tale scenario were true, our model would indicate large portions of the study area in which building conditions were appropriate. Under these circumstances you could build your mountain cabin almost anywhere. Thus, an apparently wonderful outcome shows that the model was not very useful in helping us make a decision.

When the modeling constraints were too tight we loosened them to make the model more flexible. Now the model constraints are too lax so we must tighten them a little, to give the GIS user the very best mountain cabin location from the model. As before, there are two ways to accomplish this. The first is to tighten the criteria. For example, we could require soils of the absolute best quality for building the cabin and installing a septic tank. We could also require slopes of less than 7 degrees, not 15. Another factor that could be modified is the size of town or city that would be acceptable within your viewshed. As you can see, quite a number of changes could be made to the model. In fact, it is generally easier to produce a range of values for each factor to produce ranked answers rather than Boolean "yes" and "no" answers.

As with loosening constraints, there is an alternative approach to tightening them. While physical, aesthetic, political/legal, and infrastructure parameters are all important, they are not all equally important. Physical limitations of the soil and slopes have a substantial impact on your ability to build the cabin. However, if there are no lands zoned for single-family dwellings, and the zoning commission is unwilling to give you a variance, the physical parameters are no longer relevant. What remains is one set of "absolute" constraints and another set that might be overcome by modification of the constraints or by a willingness to invest more in the project. Determining which factors can be weighted and which are absolute and preemptive helps the cartographic modeler adjust the solution to the individual situation. You must be flexible and very aware of the impacts of these weights on the final outcome. There are a few examples in the literature of the use of weighting and reweighting to improve a model and to satisfy the ultimate users (Davis, 1981; DeMers, 1989, 1998. Lucky and DeMers, 1987).

Conflict Resolution

It is important to consider factor weights and the different thematic layers when producing a cartographic model, but this may not be enough to produce results that satisfy all possible conditions. Consider the problem of producing a cartographic model that involves two essentially competing views of the world (e.g., timber harvesting versus protecting the endangered spotted owl). Even without federal mandates, there are essentially two opposing but no less valid views about the best use of large portions of land.

How do we reconcile these two conflicting modeling tasks? Let's look at one readily available technique suggested by Dana Tomlin. Although not the only approach, it shows that the competing portions of a problem can be conceptually isolated and a reasonable compromise accomplished without employing exotic software, alternative logics, or high-priced negotiation teams. The approach, called ORPHEUS, begins by creating two separate models, one for each competing side. After isolating all the competing factors, we begin to negotiate weights for each of the factors for each party (Tomlin and Johnston, 1988). The process involves evaluating the most and the least important factors for each participant and trying to accommodate the most important factors for both parties. With time, the planned result is small concessions on each side, ultimately allowing both competing demands to coexist. This methodology is iterative, with each round resulting in different output maps. Here the cartographic output provides both visualization and a basis for more negotiation. The process takes time to perfect, and it seems that there is no substitute for experience.

Sample Cartographic Models

To conclude our examination of model flowcharting, let's look at three basic examples of models and their flowcharts. Our sample cases have separate themes and use quite different approaches both to modeling and to flowcharting. This should give you an idea of some of the options available.

We begin with a descriptive model of deer habitat quality (Carlson and Fleet, 1986) (Figure 15.5). This simplified model is based on the presence or absence of some factors essential to the survival of the deer: availability of water, vegetation for browsing, and shelter. Availability of water is based on a hydrology layer, whereas browsing and shelter are based on the vegetative cover. Browsers normally obtain their food from woody plants, which have their leaves off the ground. Thus, the necessary forage and shelter factors are based on a reclassification of vegetation cover into woody shrubs and trees as useful and grasses and forbs as nonuseful vegetation types. The model is vague about the types of vegetation needed and whether all types of water can be drunk. Still, it should be easy to imagine how these factors could be obtained from the available maps by employing documented deer food, water, and shelter requirements. This approach is inductive because it begins with empirical data and generalizes to habitat polygons.

A second portion of the model is designed to determine habitat quality on the basis of spatial configuration of the landscape units. The primary considerations here are the size, amount of edge for each landscape compartment or patch (i.e., edginess), and integrity (Euler or Lacunarity measure). This portion recalls our earlier discussion in Chapter 14 about the use of landscape ecological units in cartographic modeling. In fact, this is an excellent, easy-to-understand

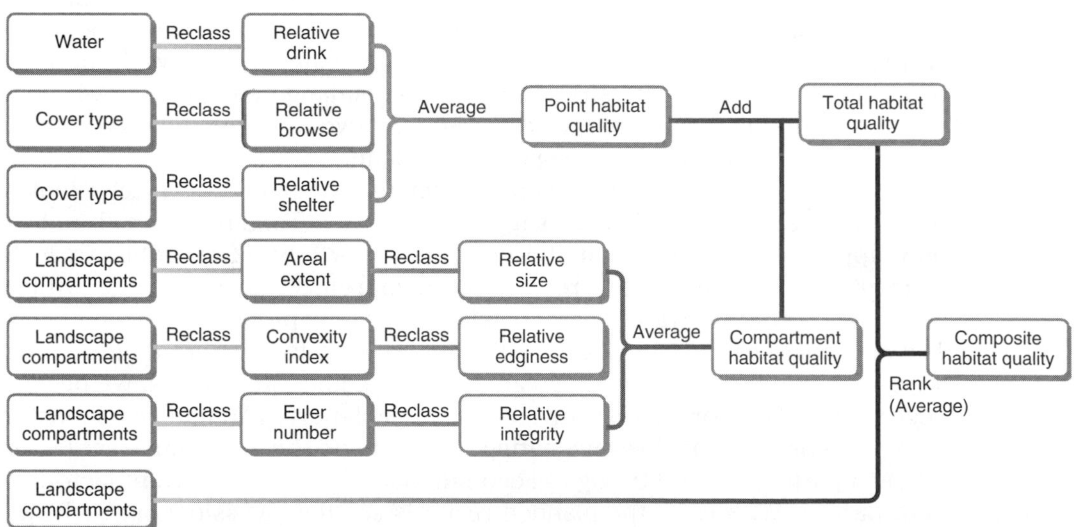

Figure 15.5 Simple deer habitat model flowchart. This flowchart is based on the presence or absence of some primary survival factors: shelter, food and water, and the spatial configuration of the landscape with respect to size, amount of edge habitat, and the integrity or patchiness or the landscape. *Source:* Adapted from Joseph K. Berry (personal teaching notes). Used with permission.

example of landscape spatial variables as they are applied to a single-species habitat.

The final portion of the model totals the habitat quality components for the separate landscape components and scores or ranks them through, in this case, averaging (a mathematical overlay) to obtain the composite habitat quality. Once again, it should be fairly obvious how simple submodels can be combined to achieve a relatively complex model of deer habitat. By compartmentalizing each set of operations and each landscape compartment, we can easily produce the individual models.

Our second model is a prescriptive model that attempts to define appropriate areas to place wind generators (Carlson and Fleet, 1986) (Figure 15.6). There are only two basic layers: topographic elevation data and vegetation communities. The topographic layer is classified to provide for the orientation or aspect of the topographic slopes, which have been assigned ordinal categories ranging from best to worst (for wind generators). Level ground will generally be the most useful because of the absence of obstructions. Northwest is the next best because the prevailing winds come from that direction. As one moves further away from northwest, usefulness declines.

Another necessary factor is distance to obstructions—in this case, trees and slopes—which is easily obtained by using a functional distance operation; this model uses a command called "Radiate." The required generator height, h′, based on the heights of obstructions and the distance just calculated, is a simple difference between the obstruction height and the ratio of the obstruction

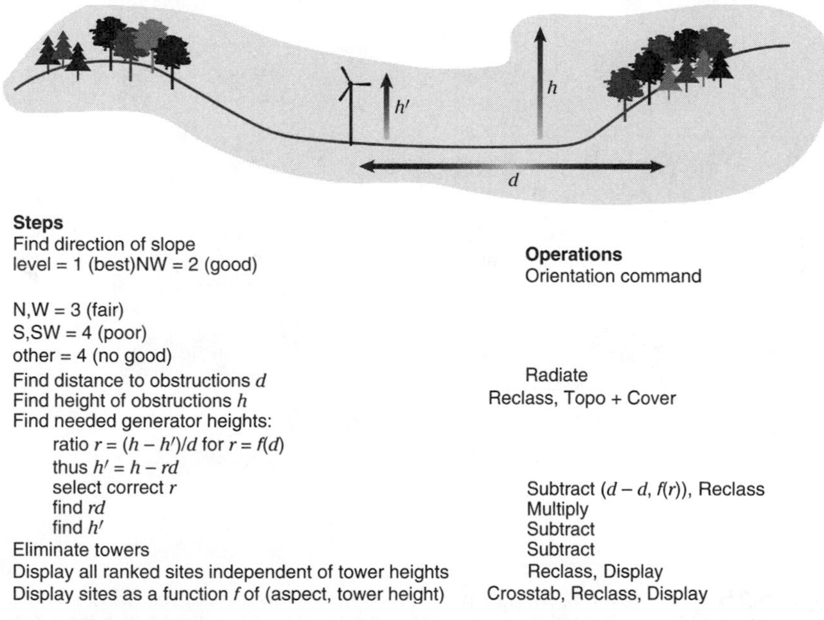

h′
h
d
Steps
Find direction of slope
level = 1 (best)NW = 2 (good)
Operations
Orientation command
N,W = 3 (fair)
S,SW = 4 (poor)
other = 4 (no good)
Find distance to obstructions d
Radiate
Find height of obstructions h
Reclass, Topo + Cover
Find needed generator heights:
ratio r = (h – h′)/d for r = f(d)
thus h′ = h – rd
select correct r
Subtract (d – d, f(r)), Reclass
find rd
Multiply
find h′
Subtract
Eliminate towers
Subtract
Display all ranked sites independent of tower heights
Reclass, Display
Display sites as a function f of (aspect, tower height)
Crosstab, Reclass, Display

Figure 15.6 Wind generator siting flowchart. Model flowchart for the siting of wind generators based on prevailing winds, terrain elevation, and tree heights. This example is designed more as an algorithm than as a flowchart. The mapped data, consisting of elevations and vegetation communities, can be manipulated by means of the commands in the "Operations" column. *Source:* Adapted from C. Data Tomlin (personal teaching notes). Used with permission.

height minus the generator height divided by the distance. The model assumes that towers will be no higher than 80 feet by subtracting that value from the topography, then ranks the sites independent of the tower heights. Finally, the model ranks the sites independent of tower heights, then recombines the aspects to obtain sites available for wind generation. The flowchart for this model is considerably different from what you have seen in this text so far. It is more algorithm than flowchart, but some may feel more comfortable with this recipe approach, although it is relatively easy to transform both the classic flowchart and the algorithmic approaches.

Our last model is based on a predictive statistical method known as regression. It is designed to predict breakage during timber harvesting (Berry and Tomlin, 1984) (Figure 15.7). Such data are valuable to the lumber companies in determining profits. In this model, the independent variables (grids or layers) used are percent slope, tree diameter, tree height, tree volume, and percent tree defect. Each variable has a measurable impact (indicated by the regression coefficient) on the predicted breakage. By multiplying each coefficient by its associated independent variable, the model produces weighted maps that, when added to each other and a constant (a necessary portion of a regression equation), yield the total predicted breakage during tree harvesting. Because

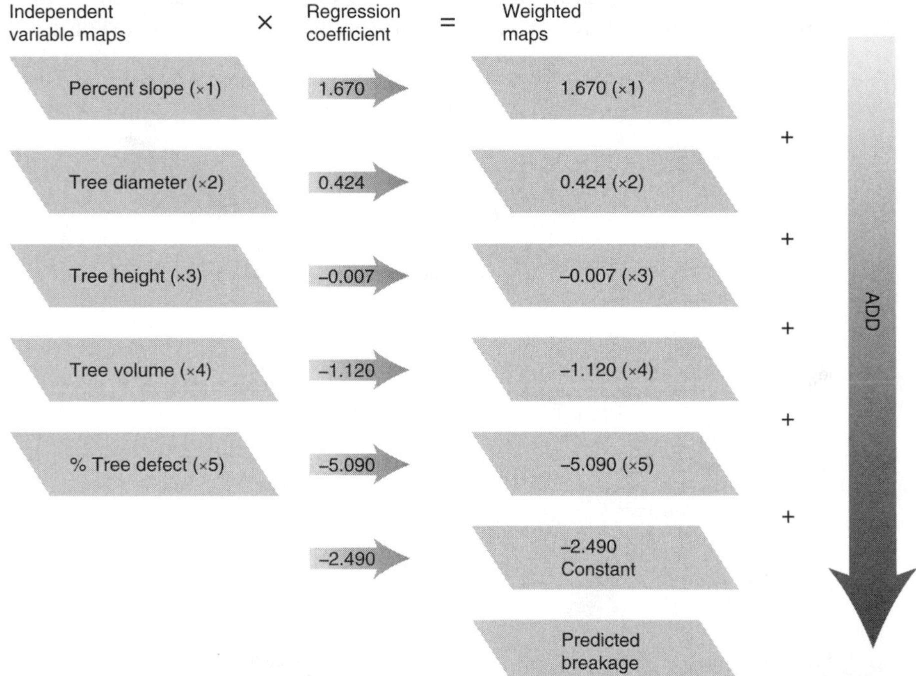

Figure 15.7 Timber breakage predictive model using regression. The flowchart illustrates the use of dasymetric mapping techniques embedded in the following regression equation for conventional felling practices:

$$Y = -2.490 + [1.670 \times (1)] + [0.424 \times (2)] - [0.007 \times (3)] - [1.120 \times (4)] - [5.090 \times 950]$$

Source: Adapted from Joseph K. Berry (personal teaching notes). Used with permission.

this predictive model uses tested, statistically significant predictions for each independent variable layer, the results are highly robust and easily tested for validity, a subject covered later in this chapter. As before, this provides yet another way of flowcharting a model that might be more amenable to the statistically or mathematically inclined modeler.

MODEL IMPLEMENTATION

Thus far, we have assumed that our model is going to work according to plan and that our results will naturally conform to our expectations. Many GIS practitioners have, through trial and error, seen that it is best to take a conservative approach to model implementation. Although input consumes much more of our time than does any other phase of GIS, we can nevertheless find ourselves spending many hours correcting mistakes if our implementation fails. For this reason alone, it is wise to select a small subset of the study area on which to prototype the model, a technique that you will see is very useful in model verification later on.

Using our subset study area and our model flowchart, we begin at the elemental level for each submodel and perform the necessary operations. This activity allows us to examine some of the following basic questions:

1. If we are reclassifying our categories, is our database designed to permit retrieval of the appropriate values?
2. If there are alternatives to a single technique or alternative combinations of techniques, which ones are the most likely to give us the correct response with the fewest steps (**parsimony**)?
3. Are we absolutely sure that the operations selected are representative of how the modeled environment really works?
4. How are we going to deal with missing variables, if we have any, in each layer?

There are many more questions, but this short list makes the point: Be sure you know what you are doing before employing the full database. By prototyping the operations on a small version of the database, you have a better chance to evaluate what is happening at each step than you would if you used the entire database. In addition, a small sample database makes it possible to test some of the results by hand as they are generated, thereby saving time during model verification.

Another important aspect of implementation is determining which intermediate layers will be needed to continue modeling and which should be retained for later model verification. Many GIS practitioners take an extremely conservative approach here, believing that if a layer is created, it probably should be kept, just in case. If you have sufficient disk space on your computer, this is certainly a reasonable approach. However, if your study area is very large or if you have terrabytes of detailed data, the result can be a model that quickly exceeds your capacity to store the intermediate layers.

Some professional GIS software provides a mechanism to create a live implementation flowchart that flows from the basic map elements to the final model results. The storage advantage of such systems is that the intermediate layers do not have to be preserved because their creation is determined by the flowchart itself. These active flowcharts are essentially graphical modeling languages and require only that you populate the elemental layers with existing data. A modeling advantage of such systems is that they allow the user to adjust portions of the model, whether in functions or in data. This allows a more incremental approach to modeling, even allowing for the development of simple models that can be improved on as more data are acquired or more knowledge about factor interactions discovered.

Whether using a graphic modeling language or not, it is important that, at a minimum, the elemental layers be preserved. If a graphic modeling language is not employed, the intermediate layers and the final products should be saved. In this case, during model implementation you will be creating a large variety of useful output. Some of it will be needed to make refinements to the model during implementation, others you will want to keep in case additional models from the same database are required, and still others will be needed for model verification.

Because implementation is an exciting part of GIS, it is tempting to try out different modifications just to see what they do. To avoid this temptation, keep in mind that the GIS is capable of producing all sorts of meaningless output. Stick to your flowchart unless you have good reason to deviate from it. Continued model tweaking will frequently result in many layers that cannot be identified later. If you have too large a collection, it can always be purged; because of the sheer numbers, however, you may wind up discarding needed layers during cleanup operations.

This brings me to a couple of additional cautionary notes. When creating intermediate and output maps, give them mnemonic names, as I have throughout this discussion, to remind you of what they are and/or how they were created. Document every layer you produce. Keep records (metadata) of what you have done and of what each layer represents, how it was derived, and so on. Protect your data. This will reduce the danger of throwing out good layers with bad. It will also help to prevent you from unalterably modifying good layers.

MODEL VERIFICATION

Cartographic output is an appealing method of communication. The map has a visual impact that allows people to accept the document as a matter of course, without questioning either the data from which it was produced or the cartographic or modeling methods by which the product was derived. It should therefore not be surprising that many GIS models are accepted as fact, just as most maps are accepted as fact. However, this is seldom the case.

The solution is **model verification**. There are few examples of model verification either in practical applications or in the literature. Despite the sparse treatment cartographic model verification has received, most would agree that it is extremely important. So how do we begin? First, we need to agree on what we mean by verification. I define the term loosely to describe not just correct models but also useful models. After all, if the results of your analysis are correct but unusable, you have ultimately failed to provide the desired product. Within this loose definition, three fundamental questions should be asked:

1. Do the data used in the model truly represent the conditions we are attempting to model?
2. Are the model factors combined correctly to describe or prescribe the correct decision-making process?
3. Is the final solution acceptable by the users and/or useful to them as a decision-making tool?

The first question might appropriately be asked during input, but this is not always possible. Remember that one important reason for flowcharting a model is to determine missing variables. Another is to determine whether appropriate solutions could be derived with the existing data. At times, however, the final models produced are counterintuitive, or as we begin to put the pieces of the model together, we find that the factors we used don't work. Such unsatisfactory results usually mean that something was missing from the original conceptual model. Obviously, it is impossible to evaluate something that wasn't there, even after a flowchart has been produced. Unusual or counterintuitive results might suggest that the model is incomplete or that the variables used were not adequately representative of model constraints. This is another good reason to build a small subset of a proposed database and test the model in a small study site that is typical of the overall study area (DeMers, 2002).

If the prototype results are negative, we must begin to dissect the model to ascertain where the problem lies. This is another reason that flowcharting the model is so important. Most often, a problem with a model will appear in one or more of the submodels. In our mountain cabin model, for example, we may experience difficulties with one or more of our physical parameters because very positive mountain cabin site characteristics keep cropping up in areas known to have dense forests. The discrepancy could be due to dated information. That is, our information on vegetation might have been compiled when the stands of trees in question were immature. By updating our information on vegetation stands and running the model again, we should obtain a more realistic outcome.

The foregoing example shows one of the fundamental approaches to evaluating whether data are representative of the important factors—namely, testing them against known conditions. An excellent example of this is provided by Duncan et al. (1995), who used bird census data to test whether their Florida scrub-jay habitat model accurately represented the known locations of these birds. The study is particularly useful because it shows that a simple pattern comparison is an insufficient basis for this decision. Instead, these workers used statistical testing to evaluate the results and followed up with a logical discussion of the possible reasons for lack of correspondence.

Another problem that often occurs in determining whether model elements match needed real-world parameters is missing variables. Williams (1985) clearly indicates the possibility of a GIS model with missing variables. Some of these variables may simply be unavailable and hence must be eliminated from the model. In such a scenario, the final model should state which variables are unavailable and acknowledge explicitly that the model is not a complete depiction of the whole. In some cases, the variables are not missing but rather are either too vague or ill defined to be used, or are by definition aspatial variables that cannot properly be placed in an explicitly spatial modeling context in the absence of spatial surrogates (DeMers, 1995).

Our next verification parameters involve the twofold question of whether the factors have been properly combined. The first part asks whether the GIS algorithms are operating properly. Many GIS operate somewhat blindly in their analysis functions because the software vendor fails to provide specific information about how the algorithms work. Research on a variety of GIS software has examined the differences in performance using the same functionality on an identical database (Fisher, 1993). The results suggest that the user might want to test these different software systems in much the same way. This is seldom possible in an operational setting, however, because of the high cost of purchasing and maintaining multiple GIS software and hardware configurations. An alternative is to produce an artificial database with known parameters—a sort of control database—against which to test each analysis operation for the validity of results. This approach, like the former, is time-consuming, and you should consider carefully the importance of embarking on it. If your results, especially for highly complex operations, are extremely sensitive, it might prove useful. There are few guidelines for producing control databases. You must rely on your own knowledge of the available data and your notion of how you would expect the software to perform.

The other part of our factor interaction question assumes the algorithms are accurate, but examines whether all the correct steps were performed in just the right sequence. To test the validity of the modeling in this case, we might use a small portion of the model and its elemental database. You then reverse the model by following the flowchart from the trunk to the branches. This should produce the same map elements we started with (Tomlin, 1990). Any step that does not produce the same map elements will immediately become obvious, and we will also know which portions of the overall model need adjustment. If we use a graphical modeling language, we can make adjustments and run the entire model again. If not, and we have kept our intermediate layers, we can run the model from a point where it is verifiably correct instead of starting the model at the very beginning.

The second part of our model verification task is to determine whether the factors are combined correctly. This is much more difficult. Now we must ask whether the factors themselves interact in a manner that actually simulates or models reality. Because of the nearly infinite number of combinations of factors and model types, there is no simple recipe. Here you must rely on your knowledge of the environment you are attempting to model and your intuition. In most cases the modeler becomes so familiar with both the environment and the data representing the environment that any discrepancies between the data and the model are immediately apparent. This is identical to an evaluation of whether we are using the correct parameters. If the model seems to be counterintuitive, there is a good chance that something is wrong. If you have eliminated bad data as a possibility, and you know that the algorithms were operating correctly, you are left with the possibility that the model itself is incorrect. In some cases incorrect or counterintuitive results may, as in statistical analysis, result in a further investigation of how the environment actually operates. In this case, the GIS becomes a useful tool in advancing science as well as implementing it.

Finally, the third major category of model verification concerns the usefulness of the model as a decision tool. There are many forms of output, not just the cartographic output. In some cases a tabulation of the results might be more useful to the client than a map. If, for example, the client is the publisher of a newspaper who is trying to increase circulation, the analysis might produce a list of local residents who are not current subscribers, together with their telephone numbers. The tabulated results could, of course, be used for telemarketing. In this case a list would be far more useful to the client than a map showing a distribution of circulation.

Or suppose that an analysis yields predicted numbers of people infected with a virus for each county of a state or region. Some clients would prefer to see a tabulation of these results from high to low rather than as choropleth map. This is partly because some users are unfamiliar with the choropleth map and have difficulty interpreting the results. It may also be that some users want to see the raw data, at least prior to cartographic classification, with a view to determining their own critical values for classification.

Still other forms of output are possible. In a few communities, GIS is being used for emergency services. In addition to producing a map showing the shortest route to a fire or other emergency, the system can send an electronic alarm signal directly to the nearest emergency service, thereby shortening response time and, it is hoped, saving lives and property. Although this is perhaps an unusual example of GIS output today, as the technology changes and as more and different clients begin using GIS, such output will become commonplace. More importantly, such output is more useful by the client.

This leads us to another important aspect of model verification—acceptability of the results to the user. Clients may be searching for particular results to coincide with their intuitive or preconceived ideas about what a correct solution is or because they intend it to serve their political or legal mandates. If a client has provided data to be included in the model and has indicated how the model should perform, it might be assumed that the model will meet these expectations. If, for example, we have included values provided by the users (e.g., the quality of a particular parcel of land for a preselected land use), we are limited to using those values. The client will probably also expect that the handpicked values will produce model output that will either favor or discourage using the land for the designated purposes. In such a case, the GIS is being used to validate the client's existing attitude and to document its derivation. If the results are not what the client wanted, we may have to backtrack through the modeling steps, as before, to see where the model deviated from expectations. In some cases, we will be asked to modify how we combined variables (often a form of constraints relaxation), and in other cases you may be able to teach the client that the initial view was incorrect (DeMers, 1995).

In another situation, clients may require us to manipulate the model to obtain a particular answer to support a legal mandate to, for example, protect a particular piece of land. Again, the client is looking for a means of validating a decision that has already been made. In any event, remember that regardless of whether the model exactly coincides with the client's view, the client is not obliged to accept the results as produced. These realities, although not always pleasant, should be considered when selecting your clientele. Such ethical questions are part of an ever-increasing body of research.

Terms

cartographic modeling
confirmatory analysis
deductive models
descriptive models
exploratory analysis
flowchart
formulation flowchart
implementation flowchart
inductive models
model verification
parsimony
predictive models
prescribe
prescriptive models
spatial information product (SIP)

Review Questions

1. With so many commands and so many options to each command available in a GIS, how can anyone possibly decide how to build a model?
2. What do we mean when we say that the process of cartographic modeling is cyclical? Within your own area of interest, create a scenario that illustrates this process.
3. What advantage does deductive cartographic modeling have over inductive cartographic modeling? When might you find inductive cartographic modeling most useful?
4. What is the difference between explicit and implicit variables? What is the difference between implicitly spatial and explicitly spatial variables? What are surrogate variables, and how are they used in cartographic modeling?
5. What is the purpose of flowcharting cartographic models? What advantages does this process have over creating cartographic models without them?
6. What do you do if your cartographic modeling constraints are so severe that the model cannot reach a reasonable solution? What are two possible ways around this problem?
7. What do you do if your cartographic modeling constraints are so loose they give results that are too broad to help you make your decision?
8. Why are weights so important to the creation of cartographic models? Why are some factors preemptive rather than simply weighted higher than others? Can you give examples?
9. What are the three basic areas of cartographic model verification? Suggest some methods of model verification for each.
10. Give some examples of cartographic output that might not be as useful as alternative methods or output from cartographic analysis.
11. Why is it important to prototype models before they are implemented on the entire database?
12. What should we keep in mind as we begin model implementation? Why are experimentation and reliance on serendipity not good practices when we are implementing a model?

References

Berry, J.K., and C.D. Tomlin, 1984. *Geographic Information Analysis Workshop Workbook*. New Haven, CT: Yale School of Forestry.

Carlson, S.A., and H. Fleet, 1986. "Systems Applications Geographic Information Systems (SAGIS) and Linked Analytical/Storage Packages Including Map Analysis Package (MAP), Image Processing Package (IPP) and Manage, A Relational Database Management System." Workshop Materials, Annual Meeting, Association of American Geographers, Minneapolis, MN.

Davis, J.R., 1981. "Weighting and Reweighting in SIRO-PLAN." Canberra: CSIRO, Institute of Earth Resources, Division of Land Use Research, Technical Memorandum 81/2.

DeMers, M.N., 2002. *GIS Modeling in Raster*. New York: John Wiley & Sons.

DeMers, M.N., 1998. "Policy Implications of LESA Factor and Weight Determination in Douglas County, Kansas." *Land Use Policy*, 5(4):408–418.

DeMers, M.N., 1995. "Requirements Analysis for GIS LESA Modeling." In *A Decade with LESA: The Evolution of Land Evaluation and Site Assessment*, F.R. Steiner, J.R. Pease, and R.E. Coughlin, Eds. Ankney, IA: Soil and Water Conservation Society, pp. 243–259.

DeMers, M.N., 1989. "Knowledge Acquisition for GIS Automation of the SCS LESA Model: An Empirical Study." *AI Applications in Natural Resource Management,* 3(4):12–22.

Duncan, B.W., D.R. Breininger, P.A. Schmalzer, and V.L. Larson, 1995. "Validating a Florida Scrub Jay Habitat Suitability Model, Using Demography Data on Kennedy Space Center." *Photogrammetric Engineering and Remote Sensing,* 61(11):1361–1370.

Fisher, P.F., 1993. "Algorithm and Implementation Uncertainty in Viewshed Analysis." *International Journal of Geographical Information Systems,* 7(4):331–347.

Lucky, D., and M.N. DeMers, 1987. "A Comparative Analysis of Land Evaluation Systems for Douglas County." *Journal of Environmental Systems,* 16(4):259–277.

Tomlin, C.D., 1990. *Geographic Information Systems and Cartographic Modeling*. Englewood Cliffs, NJ: Prentice-Hall.

Tomlin, C.D., and J.K. Berry, 1979. "A Mathematical Structure for Cartographic Modeling in Environmental Analysis." In *Proceedings of the 39th Symposium of the American Conference on Surveying and Mapping*, pp. 269–283.

Tomlin, C.D., and K.M. Johnston, 1988. "An Experiment in Land-Use Allocation with a Geographic Information System." In *Technical Papers, ACSM-ASPRS*, St. Louis, MO, pp. 23–34.

Williams, T.H. Lee, 1985. "Implementing LESA on a Geographic Information System—A Case Study." *Photogrammetric Engineering and Remote Sensing,* 51(12):1923–1932.

UNIT 5
GIS OUTPUT
AND DESIGN

CHAPTER 16

Cartography and Visualization

Output is the final product of any analysis. If our output is unintelligible or inadequately communicates the final results of analysis, we have failed in our mission. After all, our purpose is not just to analyze geographical data but also to provide a tool for decision makers. This requires us to effectively and succinctly communicate the results of our analysis. In this chapter we examine the technology of GIS output as well as some basic design criteria for producing good quality, understandable output. Both technology and design are important components of this process because both affect the methods and the effectiveness of our communication. The technological aspects of GIS output are important because the limits of technology impose physical constraints on our ability to produce the kinds of output we need. Beyond the technology, however, we must recognize that users of GIS output have psychological and physiological limits and cultural, gender, age, and environmental biases that affect the way they interpret the results.

Traditionally, the output for GIS analysis has been cartographic, and this still remains the predominant form. Therefore, we focus much of our attention in this chapter on the development of readable, meaningful cartographic output that effectively communicates its intent. Other forms of visualization are also becoming commonplace because of the advances in computer technologies and research in scientific visualization and user interfaces. We examine some typical forms of noncartographic visualization, again with the ultimate use in mind.

LEARNING OBJECTIVES

When you have completed this chapter you should be able to:

1. Explain with examples the fundamental importance of map design in communicating the results of GIS analysis.
2. Describe, with examples, the conflict between intellectual and aesthetic map design considerations and suggest how these can be resolved.
3. Describe the use of the three basic map design principles and be able to use them in designing and organizing maps with your own software.

4. Explain the difference between multiple single theme map displays and a single multivariate map from the perspective of the potential user.
5. Show examples of the possible map design constraints and considerations for two- and three-dimensional GIS output.
6. Suggest and apply different types of linear and area cartograms for GIS output.
7. Provide concrete examples of the appropriate use of animated maps, dynamic displays, and flythroughs for appropriate scientific visualizations.
8. Discuss the special nature of surface data from the perspective of display, particularly with regard to perspective, viewing distance and angle, visual overlay, and flythroughs.
9. Identify and apply some common effective types of noncartographic output and defend their application over cartographic forms (including cartograms).
10. Explain the impact of modern technology on GIS output design.
11. Discuss the role of the Internet on the types, immediacy, and methods of delivering GIS output.
12. Provide specific suggestions for methods of visualizing the temporal dimension within space-time GIS modeling.
13. Suggest the future role of virtual worlds and immersion GIS environments in providing a visualization tool for decision making.

OUTPUT: THE DISPLAY OF ANALYSIS

GIS analysis is of little use unless we can present the results so that the user can quickly and effectively understand the purpose of the analysis and what the results mean. GIS output can be permanent or transitory; visual, tactile, auditory, or electronic in nature; and either audience specific (e.g., tactile maps for the blind) or understandable by a broad general audience. Permanent output refers to hard-copy graphics or tactile displays—and magnetic, laser, flash, or other forms of memory that are designed to preserve the product for extended periods of time. Ephemeral output might include transitory voice communications, electronic signals, video displays, and other forms that are meant to elicit a near-time response and then vanish. This is exemplified by the use of computer monitor displays of intermediate GIS output that suggest the need for further analysis. Each of these types of output presents different opportunities and unique challenges for their creation and for attaining a maximum level of communication effectiveness.

The literature related to nonvisual methods of output communication is growing at much the same rate as the methods themselves. At present, however, the proportion of such output is small, and we will leave their examination for more advanced texts. Output can be produced in human- or machine-compatible form (Burrough, 1986). Machine-compatible forms are often used for storage, while human-compatible output is typically permanent and targets a viewing audience. Machine-compatible forms are important and require decisions about media, data structure, and volume compatibility between and among computer systems and their storage devices because many forms of computer data and storage devises exist. But many of the technical translation problems are readily solved. Human users, however, have various backgrounds and experiences, and often unequal levels of graphic understanding (graphicacy). Therefore, we concentrate primarily on human-compatible output forms. We will examine both permanent and ephemeral types because both can be used to communicate results to human users.

Because traditional GIS input is cartographic, the primary output today still consists of maps. Thus, we will spend a large portion of this chapter on cartographic output, but we'll also look at some alternative methods and discuss reasons for using these alternative output methods instead of, or in addition to, maps. Among the more important topics we examine will be the direct impact of computer technologies not only on the methodology, but also on the potential design options not available in the past.

CARTOGRAPHIC OUTPUT

When we examined the shift from pure communication to a combination of communication and analysis, you saw that analysis without communication is unproductive. Today's GIS software allows us to create maps in many different forms based on a wide variety of analyses on an ever increasing availability of spatial data. In fact, the advent of readily available databases is in large part a response to the increasing availability of GIS software and the burgeoning number of users. These factors make an understanding of the basics of map design more important than ever, especially because the majority of GIS analysts today have limited experience with cartographic production and design. More importantly, because there are now far more users of many different types of maps, attention to the role of audience in selecting the appropriate map type and designing it for effective communication is essential.

The object of creating a map is to elicit from the readers a response that allows them to understand the appearance of the environment as displayed in graphic form (Robinson et al., 1995). This objective is in large part a function of the **purpose** of the map, which is inextricably linked to the intended **audience**. The vast array of data types, map styles, symbols, scales, typography, and other iconographic devices are meant to work together. You have seen that most GIS output maps are not **general reference maps** whose purpose is to display, on a single document, a wide variety of different geographical phenomena. Instead, we will focus on the second and most common category of map employed in GIS: the thematic map. These maps are designed to represent the behavioral relationships of a selected theme. For GIS output we could easily replace "theme" with "solution," because the map will most often answer a predefined question or provide the key to a complex decision-making process.

Thematic Maps and Cartograms

Thematic map design involves the selection, creation, and placement of appropriate symbols and graphic objects to show explicitly the important features and spatial relationships of the objects being studied. Most often, however, the thematic map will also need a spatial reference system as a comparative framework for locating these thematic objects in geographic space (Robinson et al., 1995). In some cases—for example, when satellite data are the primary graphical objects—the representational framework becomes especially important because the reader may be relatively unfamiliar with the color representation of earth features through this medium. In that case they will not be able to interpret the data without a reference system that includes notations of well-known objects. In most other cases, the reference system provides a method of demonstrating the areal extent of the thematic objects and their relationships to other spatial locations.

Beyond the inclusion of a reference framework, perhaps the first rule of thematic map design is to remember that you are designing the map primarily to be read, analyzed, and interpreted (Muehrcke and Muehrcke, 1992), not displayed as artwork. Simply put: eliminate needless objects. Any object placed on the map should communicate something about the spatial distributions and arrangements. An object should not be placed on the map if it has any objective beyond providing a means to communicate your results. There should be no extraneous objects or fancy flourishes that distract rather than clarify the intent.

Even after eliminating extraneous map objects, many decisions still remain to be made. Map design involves selecting, processing, and generalizing the output data purposefully and with forethought, and then using appropriate symbols to portray them in an intuitive manner (Robinson et al., 1995). This entails a sometimes difficult mix of art and science, impression and logic. To make the best use of these components, you should produce a number of preliminary sketches of how the map will look before beginning the actual design. Although your GIS software may provide a graphic environment for map composition, it is a good idea to begin with pencil and paper, and outline the selection and placement of objects that will be represented. This manual operation will save you time and effort later when using the software, because sketching is generally easier and very often faster than placing, moving, and relocating complex graphic objects on screen. It also requires no complex hardware and can be done whenever the ideas are flowing. Doing the work by hand also focuses attention on the choices that have to be made and the look you are trying to achieve for the map, rather than on the commands or operations needed to perform the various tasks.

Map design can be very complex, especially with the large number of options available with graphical software. Additionally, the logical or **intellectual objectives** and the graphic or **visual objectives** of map design will sometimes be in conflict. These conflicting demands on map production indicate quite clearly that design considerations are nearly always resolved by compromise (Robinson et al., 1995). For example, if you are trying to place house symbols on your map, both logic and aesthetics dictate a consistent approach. For example, you might want to place each house symbol in its exact (analog) position. But if

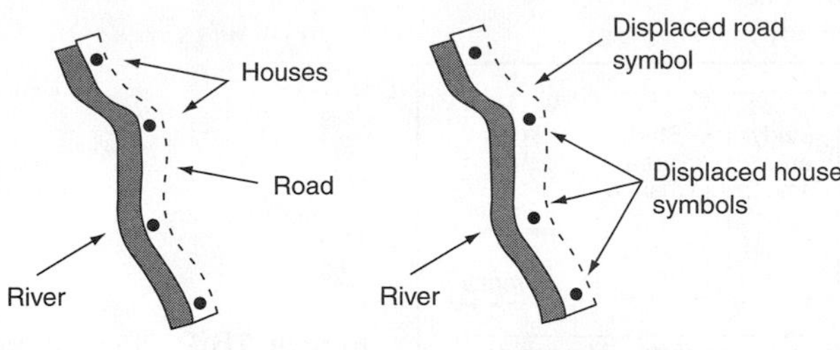

Figure 16.1 Symbol placement compromise. Closeup of a portion of a map that illustrates the compromise necessary to allow placement of the stream symbol adjacent to house symbols. One set of cartographic objects must be physically moved to accommodate the other.

you have a line symbol representing a stream that runs very near the house symbols, either the stream symbol or the house symbols must be adjusted to make room. Under such circumstances, you might prefer to move all the house symbols to one side of their actual locations rather than changing the location of the stream (Figure 16.1).

The choice displayed in Figure 16.1 reflects the premise that the location of a physical landscape object takes precedence over the location of a cultural feature or even its text-based equivalent because they are not necessarily indicating the precise location and footprint of the house. In fact, GIS software can now increase or decrease the size of the symbol graphically without changing its actual configuration inside the database. This technique can be used to emphasize or deemphasize symbols or, as in this case, to resolve a symbol conflict. Cartographers often deal with symbol decisions and symbol conflicts, and have therefore established a rather exhaustive set of **conventions** and traditions. These are a result of both accidental trial-and-error and purposeful testing. The conventions act as compromise guidelines for cartographic representations. Deviations from established norms will almost always result in a map that is less effective than it could be at communicating the results of your analysis. We will now look at the basic processes, perceptual considerations, and design principles needed to produce effective map products. You are encouraged to consult frequently with textbooks on thematic cartography or more specific texts on map design (Brewer, 2005).

The graphic design process begins with knowledge of the basic cartographic elements and symbology, combined with a mental image of how the completed map should appear. The first step is to visualize the type of map, the configuration and amount of available space, the objects to be placed on the map, and a basic layout. This initial step of intuition results in a general graphic map plan. Again, good practice suggests sketching this out freehand on a piece of paper (Figure 16.2) rather than using the computer, but this is no longer common practice.

As your plan develops, you need to consider symbols to use and decide on class limits, colors, line weights, and other graphical elements. The final stage in the design process consists of fine-tuning the work you did in the preceding stage. This stage should employ only minor modifications to the graphical plan. Among the more important considerations here is the development of design prototypes on your monitor prior to hard-copy output. Because hard-copy devices employ a different medium than your monitor, the output never exactly replicates what's on the screen. Colors may be more vibrant, slightly different

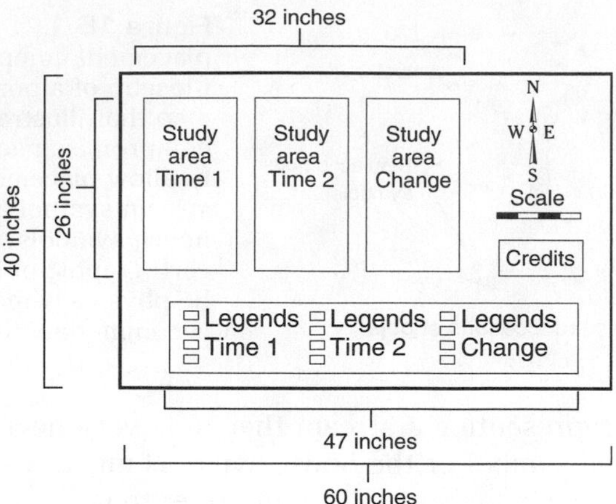

Figure 16.2 Preliminary map composition. Computer-generated preliminary drawing showing the general object types, their approximate placement, and a general view of how the map will be composed.

in tone, or appear more easily separable on your monitor than as hard copy. When you are finished with the design prototype, it is wise to test the output by plotting or printing a small version, or perhaps a small portion of the map, to examine it for these possible pitfalls. Final adjustments can then be made before printing the entire map.

Among the most important aspects of the creation of thematic maps as a result of GIS analysis is providing a map that is easily understood by the user. Depending on the nature of your model, data measurement levels, map skills of your audience, and the capabilities of your software, your result could be a choropleth map or an isarithmic map; it might employ graduated symbols or flow lines; it might be quantitative or qualitative; or it might be an orthographic or perspective view. The output possibilities are varied and each should be considered with a view to communicating results and providing decision support.

Traditionally, thematic maps have been seen as orthographic forms that present the display as if the map reader were observing the map from directly overhead. These maps have also concentrated on the production of graphical forms on the basis of their relative shapes and sizes as seen in the real world. Although these forms of thematic cartography are still among the most frequently used, they are far from the only ones available, especially with the computer graphics formats available with most commercial GIS.

Among the first vector graphics display options used for presenting somewhat nontraditional thematic maps were **fishnet maps** or **wire-frame diagrams**. This now prevalent form of cartographic output produces an impression of three dimensions by changing the viewer's perspective from vertical to oblique. Although these outputs are not often useful as analytical devices, they are extremely powerful at communicating the output of analysis, especially where topography is involved. Three additional design factors must be kept in mind when designing these oblique views: **angle of view**, **viewing azimuth**, and **viewing distance**. There is ample literature from which to obtain details concerning these three aspects (Imhof, 1982; Kraak, 1993), as well as other factors such as sun angle (Moellering and Kimerling, 1990) and the impact of DEM resolution. In general, the importance of each factor depends on the perception of size, vertical relief, and the portion of the study area that is most important to the map reader. Today's graphical software allows us to visually overlay other coregistered map data on these apparent three-dimensional maps. Thus, the viewer can readily identify relationships between topographic surfaces and other factors such as land use or vegetation. Although we most often envision topographic surfaces for viewing in perspective, we can also envision such uses to visualize forest canopy structure (Song, Chen, and Silbernagel, 2004) or urban analysis (Shiode, 2000), as well as any of the nontopographic statistical surfaces we have previously discussed.

A more exotic perspective is available through the use of nontraditional cartographic forms that have been relatively neglected among the community of users. These graphic devices, called **cartograms**, look like maps but the spatial arrangements of the depicted objects have been modified, not by their locations in geographic space, by the changes in value of the variable being measured. Thus, the distances, directions, and other spatial relationships are relative rather than absolute. Cartograms are seen occasionally but are often proven to be quite useful. For example, many urban mass-transit networks are posted inside the municipal buses and trains. These linear cartograms plot the relative sequential locations of each connecting stop on a straight line, as opposed to showing the actual route taken. In addition, the spacing between stops is not necessarily a true depiction of the scaled distance between each stop. These conventions simplify the map and emphasize the most important aspect of the spatial relationship—in this case, sequence rather than distance. While riding in a bus or subway car, your most important concern is where you want to get off, perhaps followed by where the transit line you are traveling on will connect with the next (Figure 16.3). Such **routed line cartograms** are also found in many road atlases that indicate, by using straight lines rather than road networks, both the distances between major cities on the map and the estimated travel times.

Cartograms modify geographic space to produce a distorted but focused view of the reality they represent. Another form of cartogram, the **central point linear cartogram**, could easily be used to modify the output from a functional distance model used by your GIS (Bunge, 1962) (Figure 16.4). And among the most used forms of cartograms are **area cartograms**, which vary the sizes of each mapped study area on the basis of the value being examined. A classic example of the use of this technique was presented by de Blij and Muller (1994), who modified the physical size of each nation to illustrate the amount of population rather than the physical land area. An area cartogram can be either contiguous, with all the areas touching, or noncontiguous, where the areas are not touching (Campbell, 1991) (Figure 16.5). The latter type is sometimes called an exploded cartogram and has the advantage of not requiring the lines to be connecting, thus simplifying the production of the output.

The interpretation of cartograms may require a minor rethinking by users unfamiliar with this form of output. Still, the time spent should be minimal. And once users become really fluent in the reading of cartograms, the results generally are well received. However, many users will want to have traditional cartographic forms against which to compare this radically different device of graphic communication.

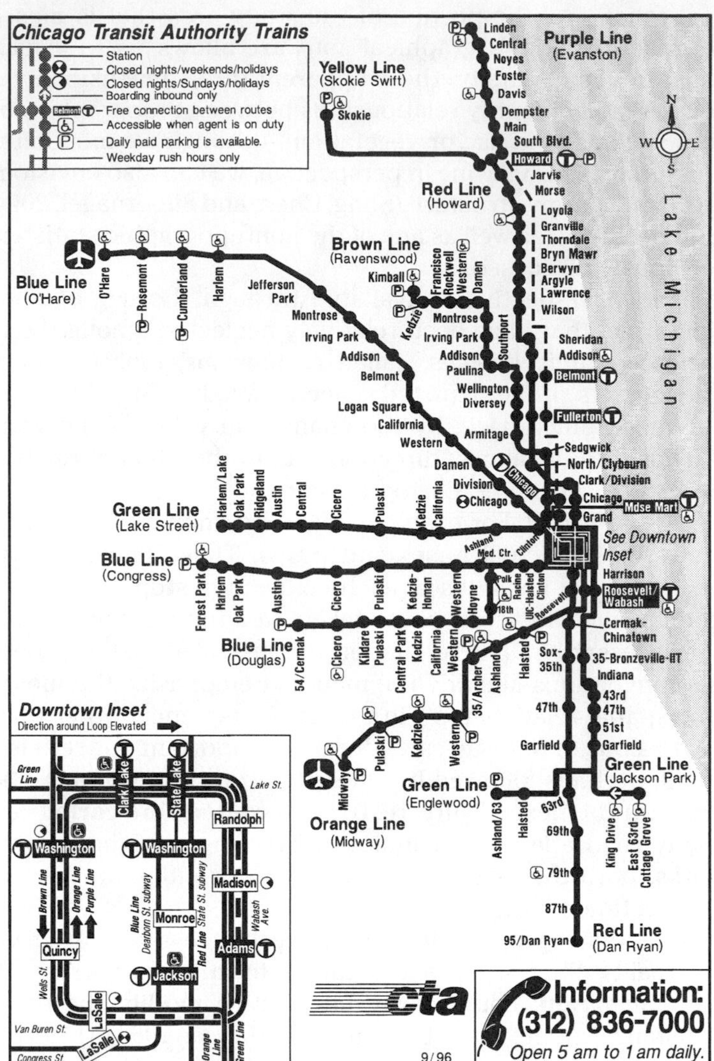

Figure 16.3 Linear cartogram. Linear cartogram for the Chicago transit system. *Source:* Chicago Transit Authority, used with permission.

You may not see many examples of cartograms as you look over the output from GIS analysis at conferences and meetings. This is partly because most of the members of the GIS community either are not familiar with the form or are not used to producing it. A second—and perhaps more important—reason for the relative lack of cartogram output is that few vendors offer software that readily performs this task in an automated environment, despite it's having been developed in the late 1960s (Tobler, 1967). Some have modified these early programs for their own particular needs (Jackel, 1997; Wolf, 2005 [http://gis.esri.com/library/userconf/proc05/abstracts/a1155.html], last visited 2/20/07). In any event, the use of cartograms as an alternative to traditional map output is sure to increase as both GIS operators and users become more familiar with them.

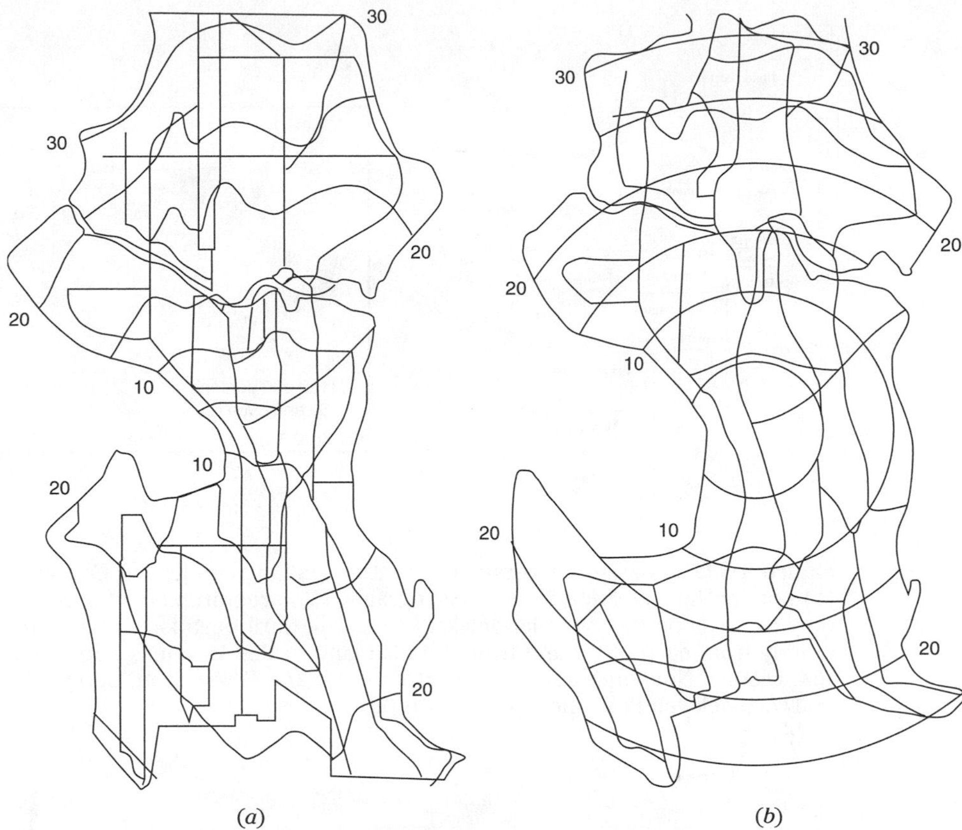

(*a*) (*b*)

Figure 16.4 Central point cartogram. (From Bunge, 1962.) (*a*) The illustration shows a conventional representation of travel times from central Seattle in 5-minute intervals. (*b*) This is a cartogram showing functional distances in 5-minute intervals. *Source:* Adapted from W. Bunge, in *Theoretical Geography*, 1962. Lund Studies in Geography. Series C. General and Mathematical Geography, No. 1 (Lund, Sweden: C.W.K. Gleerup, Publishers for the Royal University of Lund, Sweden, Dept. of Geography).

Multivariate Display

Many circumstances suggest that a number of related variables might need to be visualized at a single glance. One simple solution to this problem is to display a number of these maps as small graphic devices in columns or rows for side-by-side comparison. While this is helpful, it is less than optimal for direct comparison both because of the displacement of the graphics and the scale at which they are displayed. An alternative approach is to use bivariate and multivariate maps to allow two or more variables to be simultaneously displayed on a single map.

The primary purpose of bivariate and multivariate mapping is to cartographically display more than a single thematic variable simultaneously to examine potential spatial colocations for possible functional correspondences. Bivariate mapping displays only two factors. This can be done by using bivariate point symbols, such as graduated pie charts, to compare two different variables located at a single place (Figure 16.6). A weather station model (Figure 16.7) is

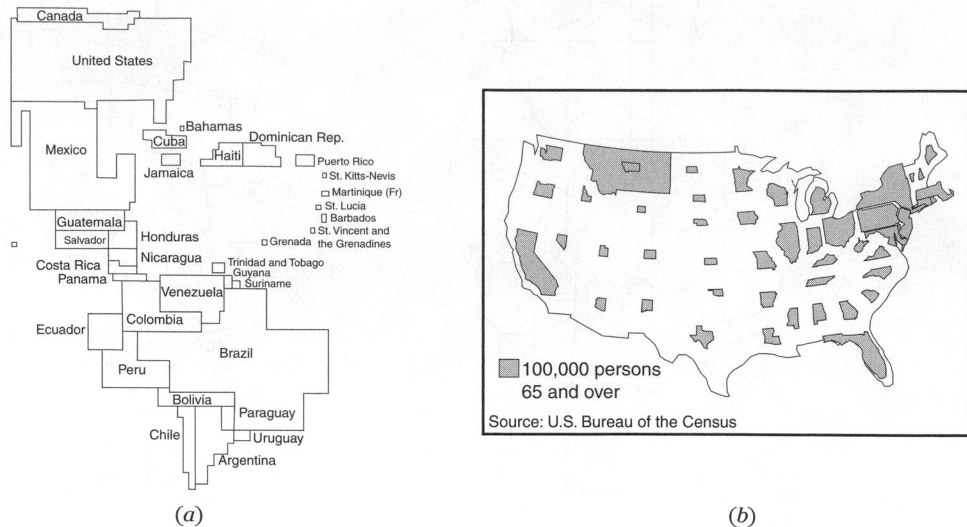

Figure 16.5 Contiguous versus noncontiguous cartograms. (*a*) Contiguous and (*b*) noncontiguous value by area cartograms. (The requirements for contiguity applies only to areas on the same landmass; hence, the Caribbean islands are shown separately from each other and from the mainland of the Americas.) *Source:* Part (*b*) J.M. Olson, "Noncontiguous Area Cartograms," *The Professional Geographer*, 28(4), p. 377. Blackwell Publishers. Used with permission.

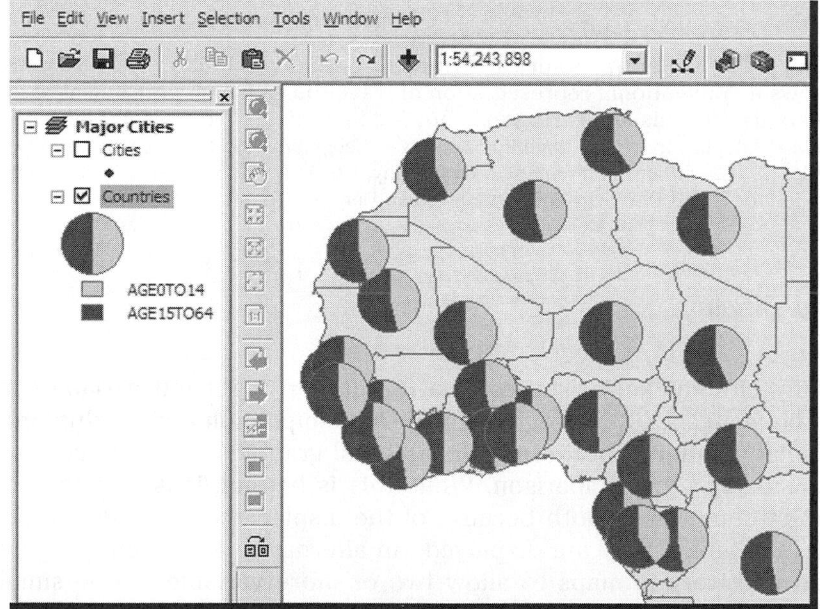

Figure 16.6 Bivariate point map. Here is a preliminary data view of a map of part of Africa that uses graduated circles converted into pie charts so different portions of the population can be displayed separately. In this case the light portions of the pie show the portion of the population of each country that is between 0 and 14 years of age, while the dark portions represent the portion 15 years and older.

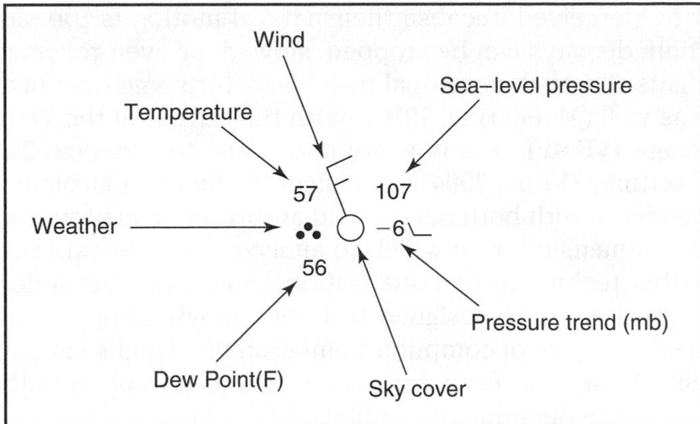

Figure 16.7 Multiple point symbol. The weather station model is a classic example of a single point symbol that displays multiple pieces of information. Such information as percent cloud cover, wind direction and speed, temperature, humidity, and many others are included in a single symbol.

perhaps one of the most common multivariate point symbols, and illustrates how many variables can be displayed at a single location.

Point and line object symbols can also be graphically displayed on top of area symbols, representing any range of data levels from nominal categories (e.g., vegetation types) to ratio data (e.g., elevation values). The physical colocation of the point and line patterns, while not analytical, will provide ample visual clues suggesting possible spatial correspondences. Dot distributions of, for example, a species of bird, visually displayed over a map of potential natural vegetation or current land use might provide some quite useful scientific hypotheses without the actual process of analytical overlay processing. You have also seen that polygons representing different categories (or different time periods for the same category) can be overlaid through a variety of mathematical and set theoretical methods. But this does not, in itself, represent pure geographic visualization. An alternative approach might be to provide a set of color schemes or simultaneous symbol sets that will visually appear on the same document. Unfortunately, despite the power of the computer to create such graphic displays, the resulting visualizations are often less than optimal for interpretation or hypothesis formulation. Instead, current **visualizations** of areal features such as choropleth variables are employing a combination of multiple symbol sets combined with animation and dynamic displays—the subject of the next section.

Dynamic and Interactive Display

Today, an increasing number of vendors offer an alternative to static cartographic displays, whether univariate, bivariate, or multivariate. For two-dimensional maps, animation allows the viewer to inspect dynamic systems as they are seen in real life (Moellering, n.d.) or to illustrate phenomena that would otherwise not be perceived because their natural motion is too slow (Töbler, 1970). In addition, displays can be stopped, slowed, or even reversed to obtain additional insights. Three-dimensional maps have for some time been viewable as animations as well (Moellering, 1980). With the advent of the **Virtual Reality Markup Language (VRML)**, it is now commonplace to visualize 3-D terrain is quite realistic settings (Knox, 2004). The effect is one of an airplane flying over terrain that is covered with both natural and anthropogenic patterns, giving the viewer an added dimension from which to analyze possible data relationships. The power of this technique of visualization should not be underestimated. Because the human eye was designed to notice moving objects more readily than static objects, the use of computer animation of all kinds can only enhance a viewer's ability to see pattern interactions and to formulate both questions and answers about the phenomena studied.

Many of these dynamic maps allow for user input, whereby the value of a single data class is modified when compared to another. This enhances the viewer's ability to examine possible spatial correspondences in preparation for future map overlay operations, because the map variables and their input can be carefully controlled. Time is a common factor that is applied in these types of dynamic mapping displays, where change that occurs slowly can be speeded up to be observed, or where change that occurs quickly might have a critical moment that can only be observed through very slow motion or even stop action methods. This is not unlike the methods used by law enforcement to view video displays in shopping centers and stores to detect criminal activity.

Among the more innovative approaches to this problem is the use of conditioned choropleth maps (Carr et al. 2005), which partition multivariate data into two or more subsets with one dependent variable and two independent variables. By dynamically updating the maps and their related statistics, the result is enhanced hypothesis generation over that possible with static multivariate map display.

Web Mapping and Visualization

Cartographic output, whether traditional or nontraditional, has proven quite useful for many decision-making situations. Its limitation is one of distribution and access. Modern digital technology allows maps to be shared over great distances through the Internet. One of the many examples of this is the Topozone website (www.topozone.com), which allows users to specify specific locations, scales, and formats of data to be retrieved (Figure 16.8). Another example that has proven quite useful is Terrafly (www.Terrafly.com), which lets users not only peruse digital imagery data, but scroll or "fly" through the data from place to place, and change the spatial scale as well. These two commercial examples demonstrate a distributed environment, sometimes called a distributed GIS, consisting of a client (the map user) and a server (the map or spatial data provider). There are many of these applications in development, some of which allow the user to download GIS-compatible data for insertion into existing databases. Websites that allow this ready access are available through university websites, commercial operations like GEOWeb, where clients and data

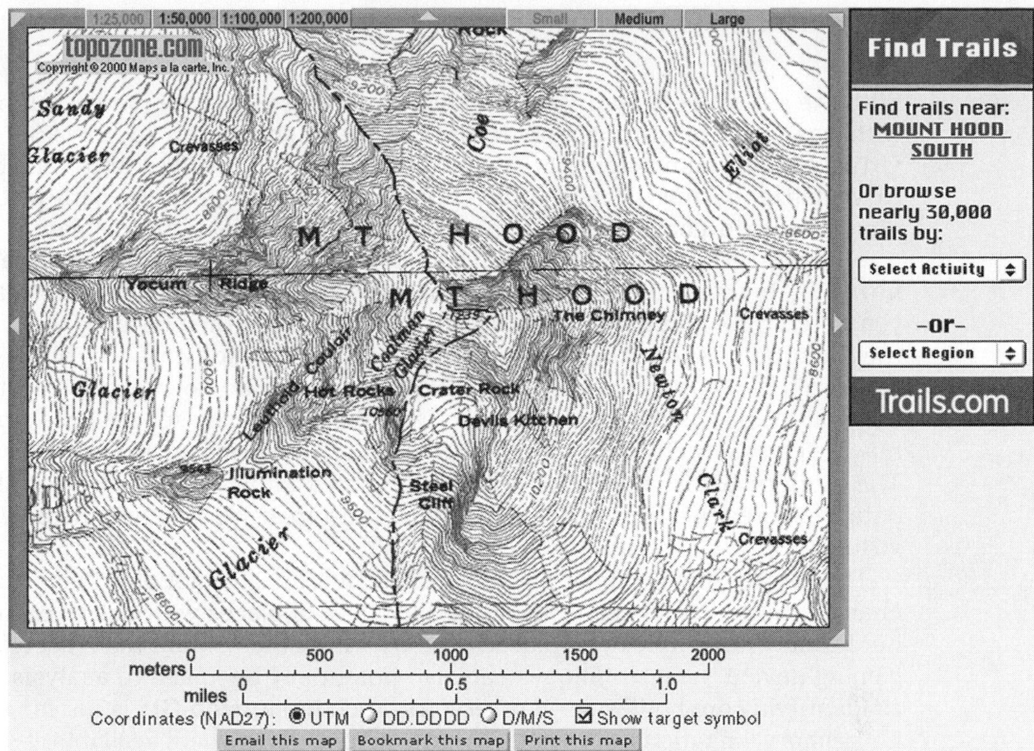

Figure 16.8 Maps on the Internet. A map of Mt. Hood retrieved from the Internet.

distributors share their data, and ESRI's Geographic Data Community, where both academics and commercial vendors provide ready access to their archived data in multiple formats.

Beyond data viewing, browsing, and sharing, the GIS output subsystem has been greatly enhanced with the development of GIS application software capable of not only distributing maps in digital form over the Internet, but also allowing an increasing amount of query and analytical capabilities over the Internet as well. This means the client is no longer required to own the GIS software at all unless substantially more sophisticated analyses are necessary. Such dynamic maps are evident in the MapQuest and Google map searches so commonplace on the Internet. In such simple applications users are able to locate places of interest, addresses, and even produce travel directions from place to place. In this way the output of GIS is not just the maps, but also the added capability to allow the client to search, pan, zoom, and analyze the data contained within them as well, in a dynamic, distributed, interactive environment. As more datasets become available, and as the software becomes more sophisticated, the trend toward distributed GIS will increase. While at present there are few fully functional distributed GIS websites available, there will soon be a time when much of GIS-based analysis will be performed at a considerable distance from the server. This is particularly true with the advent of increasingly fast Internet access routes via DSL and digital cable. As this availability of web-based spatial operations increases, the already fuzzy lines among the four GIS subsystems will become even fuzzier.

Virtual and Immersion Environments

With the advent of modern computing technologies, especially when combined with vast amounts of geographic data, it is no small step to envision both virtual GIS and immersion GIS. The idea of **virtual GIS** involves both the ability to visualize existing data and the capability to envision these data in three dimensions, to change our perspective and move about the display at will. The advantage of such virtual environments goes beyond the visualization, however, because the viewer also has access to the data, whether real or generated through a modeling process. These data can then be analyzed within the traditional GIS environment. In the future, some GIS software may allow both visualization with such languages as VRML and the ability to take one directly to the analytical capabilities of the GIS as well. We are already seeing, for example, how we can use software such as Google Earth to move in and around our earth. The modern GIS is capable of doing this, plus it allows us to select available data from areas we visit and analyze them, thus greatly enhancing the power of virtual GIS.

The new frontier is one of viewing perspective. Imagine not only being able to change the viewing distance, angle, and location of places you wish to examine, but literally being able to be placed directly within them. More than just a fancy gaming device, such techniques can also potentially be linked to analysis such as identifying ore bodies or oil reservoirs. Such **immersion GIS** is not far away. It's primary limitations at present, as with virtual GIS, are data availability rather than technological sophistication.

Mapping the Temporal Dimension

Conceptually, time has always been considered a part of GIS. However, there has been limited capability to analyze time. Most applications employed a simple time slice approach, in which several temporal datasets were created and their categories compared through overlay processing, or with array tracking in the case of changing point locations (e.g., tracking a wild animal with telemetry devices). There are many software packages capable of handling several variables through many time slices, and many that can track and manage multiple variables through space, but they have limited temporal flexibility. Some modern GIS software is now incorporating a powerful ability to perform many operations iteratively. For example, to model the spread of a disease or the movement of fire with time requires multiple changes in state and conditions. Typically we would run a model to demonstrate a single time slice, and then run it again for the next, and the next, and so on. Now we can set up a conditional operation that runs the model, interprets the intermediate results, resets the state variables and their conditions, and continues until the model reaches a preselected number of iterations or time periods. This single innovation is going to have a major positive impact on the modeling capabilities of the GIS, especially where processes are well understood or where massive amounts of data are available for testing.

NONCARTOGRAPHIC OUTPUT

Despite the prevalence of maps as a form of GIS output, certain data are better presented in noncartographic form or in addition to the map product. There are two general reasons for this relatively frequent occurrence: either the map provides output that is less immediately understood because of the audience, or the map is not the desired output from analysis. In the former case, the map usually is supplemented with other forms of output, whereas in the latter, it most often is replaced. Let's look at a few examples of enhanced or alternative output.

Perhaps the classic example of replacing map output with alternative types is the enhancement of emergency services such as the 911 system by means of GIS. In the standard implementation of this system, residents of large metropolitan areas may not get a speedy response when they dial the 911 network to report a fire because there are many fire stations throughout the city that could respond. The 911 operator wants to make certain that the closest station will always be the one to take the call. In the enhanced implementation, information about, say, the location of a fire, is transmitted electronically to an automated GIS system while the operator is still on the phone. The GIS matches the address of the actual road network inside the database. In turn, the program determines which fire station should respond and reacts by immediately sending an alarm signal to that unit, alerting the dispatcher to the current emergency. In this case, the result of the GIS analysis is an electronic response rather than a map.

The program could also be designed to determine the shortest or the quickest path from the fire station to the fire, and this information could be electronically transmitted to an output device at the appropriate fire station. The output could be in the form of a text-based route, perhaps including a map. Although computer networking and GIS implementation today are seldom so sophisticated, the scenario is not beyond current capabilities. It is much more likely that multiple output would be available under less pressing (i.e., nonemergency) circumstances.

Suppose that you are using the network functions of your GIS to route a moving van from one location to another. This is an analytical task commonly performed by auto clubs for individual motorists as well or for Web sites like Expedia for travel planning. If you have used these services, you know that there are generally two types of output. The first is a map that highlights the roads that lead most efficiently from one place to another. The second is a text-based road log that shows, in miles, the distance traveled on each road, the names of the roads, the intersections at which you are to turn and the direction of the turn, as well as the number of interchanges when available. Although the correct route could be discerned by careful attention to any of the individual components, together they provide additional information that makes navigating along the highways a simpler task. Even though many of us are comfortable with a regular map, having the road log along makes us more confident, particularly when the available map is highly generalized or where turns are close together. Thus, road logs offer a good example of how cartographic and noncartographic output can complement each other to provide a more comprehensive solution to a problem.

Some new automobiles are now equipped with GIS units linked to digital maps that allow for onboard navigation. With these systems, it is now possible to interact with the map to determine your current position and to make decisions as to where you go next. Such systems are becoming more commonplace and indicate an effective use of interactive computer output in everyday use.

Even less sophisticated uses of interactive output from GIS could easily be envisioned with the introduction of GIS in, for example, county land records offices or appraisal districts. By keeping current records of landownership (parcel) records, for example, it would be easy to determine who currently owns a single property. In such a case, realtors or neighborhood associations could search for ownership records by selecting individual properties rather than examining endless text-based records. The system might also increase the speed of data entry and the currency of the cadastral data for the records officials as well.

Tables and Charts

There are many other alternative forms of output from GIS (Garson and Biggs, 1992), but among the most common are tables and charts. Earlier in the book we considered a GIS client in the newspaper business and saw that the most useful output would be a tabulation of potential customers. Here, a simple noncartographic output designed to increase circulation could be produced without the use of expensive equipment.

In other situations, the use of tables and charts could greatly improve the understanding of cartographic results. Assume that you are analyzing land use patterns as they change through time in an area that is losing natural habitat and agricultural lands as a result of population pressures. A series of maps might show where the changes from wildlife habitat and farming to urban land uses are most severe. The map legend may include tabulations of total land uses by type for each time period. However, there is much more specific information available in the database. The tabulation could be enhanced by simply producing a bar graph or pie chart under each map to show the percentages of each land class. In addition, you might produce a bar graph showing the percentage change for each land class for each subsequent time period. You could improve this simple graphic technique by creating a matrix showing "from" and "to" changes for each time period. Such a table would prove very useful in understanding which types of land use were responsible for the loss of the largest proportion of natural habitat and which were most responsible for the loss of agricultural lands.

The foregoing example illustrates that for any GIS analysis, there are many options for noncartographic output, both in type and in form. Just as there are design issues in cartographic display, there are also design issues for noncartographic output. One simple example is encountered very often among users of computerized spreadsheets. Nearly all these software programs allow the user to produce a wide array of different types of tabular output, graphs, and diagrams showing the results of analysis. The design questions that arise are numerous and encompass complexities associated with color perceptions, personal preferences, and the capabilities of output devices. Let us look at just a few of the more important considerations.

Tabular output from GIS analysis is of two general types. The first is contained in the textual material of the map legend and can often be treated as part of the map design process because it is used to connect the attribute data to the entities presented on the map. A second type of tabular output employs lists of attribute data, tables showing relationships between raw attribute data and map entities, text showing extended-data dictionary entities for improved understanding of what the legend means, correspondence matrices showing relationships among sets of attributes, and so on. These outputs can be placed inside the exterior map boundary (the neat line) if they are to be presented as a combined finished product, or they can be output separately for the user to peruse at will (i.e., alone or in conjunction with a separate map).

Design Considerations

Design considerations for text-based output involve the basic concepts of purpose, readability, and audience, familiar from your introduction to map design. Before producing text and tabular data, consider the needs of the user. First, will the text-based output effectively replace the map as a communication device? Will it sufficiently enhance the readability of the map output to warrant its production or inclusion? In our example of the newspaper publisher needing lists of potential subscribers together with their telephone numbers, the answer to the first question is an obvious yes. This is exactly what would serve this client best. Many other situations will be far more subtle, and you will have to determine the value of the tables themselves or the value added to the map by including of tables and text.

If you intend to include tables and text, the next item you should consider is readability. Text and tables are fairly commonplace and more easily understood by the general public than maps are; hence, they tend to be easier to design. Still, a few things should be considered to ensure basic readability. First, the font should be simple in style and large enough to be seen at the desired combination of viewing distance and lighting conditions. Fancy fonts may by themselves seem particularly attractive, but simpler is usually better. Keep in mind the need for contrast when you select the color of your text output. If the tables or additional text are to be included inside the map boundary, be sure that their colors contrast sufficiently with the map background. To further ensure separation, add a border, taking care that it is not so heavy that it conflicts with graphic devices on the map proper. Unless specified by your audience, it is best to avoid acronyms and abbreviations whenever possible. The less explanation your tables require, the more useful they will be. For large tables, you may decide to outline major categories, or even all the cells in the table, since these options are familiar to many from word processing or spreadsheet programs. Text or tables lying inside the map boundaries should be separated with a border, which should be functional but not so heavy that it conflicts with other graphic devices on the map. Finally, employ the same common sense in depicting your data that you used in designing your map.

As you produce text and tables, you should also remember the nature of your audience, using terms that will be familiar to the client. In addition, the type of table you create should be useful as well as understandable. Returning to our example of land use change in this section, a set of tables indicating the amount of change in each category for each time period may not contain as much information as a change matrix table or to–from matrix. However, if the client is interested only in the amount of increase or decrease in a select few land use types, the additional information from a matrix may be both uninteresting and generally confusing. The text or table should say something about the phenomena studied, but no more than what the reader needs. Alternatively, if you or your client really need to know both the amounts and the directions of the land use changes, a change matrix might be entirely appropriate. Before you produce your tabular output, be specific about what you or your client needs to know and how best to present it.

Tables can easily become too large or cumbersome to be useful. Monstrosities with 40 rows and 40 columns are not likely to be useful because they require far too much time to examine. An easy alternative is to present the results not as tables but rather as graphic representations of tabular data. Most common computer spreadsheet programs illustrate a wide variety of traditional business graphics devices for the presentation of data: line graphs, pie charts, bar graphs, and the like. There are, however, far more options available to represent the phenomena that can be output from GIS analysis. Among these are triangular graphs representative of, for example, soil textures; climographs indicating the relationships among soil moisture, temperature, and precipitation variables for a region; and wind roses or star diagrams indicating predominant directions of wind, glacial flow, or other phenomena for selected points (Monkhouse and Wilkinson, 1971). It would be futile to try to examine each in detail, or even to consider all the possibilities. Instead we'll look again at some basic considerations for the more general types of line graph, trusting in your own background and experience to select the appropriate specimens for the data with which you are working.

The most common line graphs employ a Cartesian coordinate system with the standard ordinate (vertical scale) and abscissa (horizontal scale). Another, less common type of graph is the polar or circular graph, where the points are plotted on the basis of a combination of angular bearing (the vectorial angle) and distance from a point of origin (the radius vector). In one variation, the points of the line graph are plotted on a triangular grid. Such curves are common when showing the relationships of three variables: in a soil texture triangle, for example, the percentages of sand, silt, and clay are represented on the same graph. For all line graphs, in general, the observed points for the variables, usually classified as dependent and independent, are connected either by individual straight-line segments between pairs of observations or by a smooth curve indicating the general trend along the points. A modification of the standard Cartesian line graph is the histogram or bar graph, which displays the response of each observation by means of a wide band extending either vertically from the abscissa or horizontally from the ordinate. The histogram is more often used when the number of observations is low; it is more generally representative of the noncontinuous nature of many observations. For data that represent continuous variables, such as changes in barometric pressure, the lines themselves replace the points (Monkhouse and Wilkinson, 1971). You are not likely to encounter these true line graphs in GIS analysis, where the spatial dimension is normally stronger than the temporal dimension.

Line graphs can be simple charts displaying a single series of values connected by a line, or they can resemble a polygraph, producing multiple lines with each representing a selected variable. Simple histograms may have a single series of values rising like ribbons, or multiple values each represented by a ribbon of unique color or pattern. These multiple line graphs are produced primarily to allow the direct comparison of numerous variables under the same circumstances. Multiple line graphs can be produced in two general fashions: as multiple sets of adjacent lines or bar symbols, or as stacked graphics. For stacked graphic versions, the values of each observation are physically combined with those of the preceding values to give an impression of area. Such plots are especially useful if the values themselves are meant to indicate an additive relationship. In the case of stacked histograms, for example, the height of the bar is based on a combination of all the different variables at a single observation point. Under this circumstance, if the bar was meant to represent the variables %sand, %silt, and %clay, the total bar would be at 100 percent because each of the values is additive.

Designing line graphs has become relatively simple with the power of the computer. Some software, however, reduces our options for output or preselects the ways in which the output will appear. These default settings can, and often should, be overridden by the user. Good line graph design choices for GIS output can be indicated in general with a few simple comments. Although most software offers very colorful and stylish options, such as the appearance of three dimensions for histograms and ribbons for lines, these tend to be unnecessary and generally are less readable. Try to avoid using large, garish symbols as data points because these outsized symbols often interfere with the lines. Unwieldy data points can easily be replaced by numerical values, which are more informative and less confusing. Avoid radical differences in direction of the internal marks used to produce shading patterns, such as diagonal lines next to crossed lines next to diagonal lines in the other direction. Such inelegant graphical design will often result in unnecessary eyestrain. (The same can be said for adjacent choropleth patterns on maps.)

Keep the important lines prominent. Although most software allows you to place grids on your graph, the fewer you use and the lighter the line shade and line width, the more prominently the important lines will show. Make the size of the vertical and horizontal scale extents as close to the maximum value as possible. This is not always easy with available software, and you may want to ignore this rule if you have multiple graphs that show relationships between or among different places. In this case, all the graphs should carry the same maximum sizes to avoid confusion, or the false impression that small values are larger than they really are. Finally, remember what you are seeking: simplicity and legibility. Simple clear fonts, and lines and symbols with good contrast, should remain the most important criteria.

The sheer numbers of graph types and options can be somewhat overwhelming. There are a number of good references, primarily based on business graphics and graphics presentations. You should consult these or others like them when you intend to include business-type graphics in your work (Cleveland and McGill, 1988; Holmes, 1984; Meilach, 1990; Robertson, 1988; Sutton, 1988; Tufte, 1983; Zelazny, 1985).

A final note should be made about another form of noncartographic output that is becoming more prevalent in GIS output. Many commercial vendors now include the option of incorporating digital photographs showing, for example, buildings, site characteristics, and specimens of plants and animals found in certain areas. These digital images, often explicitly linked to the attribute database, can add a great deal to the understanding of what is represented on the map by giving the reader a set of familiar examples. The images can either be scanned into appropriate software or photographed directly with the use of digital cameras or even videotape. More of these images are likely to appear as output from a GIS in the near future. Like all graphical and nongraphical devices, they should be used sparingly and only to achieve a desired effect. Because of the graphical complexity typical of these images, it is easy to overwhelm the associated map. And, as always, make sure the images are appropriate, useful, and add value to the output.

Terms

angle of view
area cartograms
audience
cartograms
central point linear cartograms
conventions
fishnet maps
general reference map
immersive GIS
intellectual objectives
linear cartogram
purpose
routed line cartograms
viewing azimuth
viewing distance
virtual GIS
Virtual Reality Markup Language (VRML)
visual objectives
visualization
wire-frame diagrams

Review Questions

1. Why is it important to be familiar with cartographic design in GIS? Why is it even more important with GIS than with traditional, hand-drawn cartography?
2. What is our primary concern when producing a map from GIS analysis? How does this matter relate to the map as an art form?
3. Give a concrete example, other than what is in your text, of how intellectual and aesthetic design considerations can conflict. Show how the problem you describe can be solved.
4. Given the ready availability of computer software for designing and organizing graphics, why should you begin the map design process with pencil and paper?
5. Indicate, through concrete examples, the three basic design principles.
6. Show, with examples, what some of the possible design constraints might be in GIS map output.
7. Give an example of how too much data can produce problems in map design especially with regard to multivariate maps or with small-scale maps with large numbers of features.
8. Give an example of map design problems due to insufficient data.

9. What are some design considerations involved in producing good quality, three-dimensional map output?
10. Give as many examples as you can think of to describe how you might find animation useful as a form of cartographic output.
11. Discuss, with examples, some of the unique properties of topographic data that produce difficulties and opportunities for cartographic visualization. In your discussion be specific about potential solutions to each problem and each opportunity relative to a specific application.
12. What are cartograms? What limits the frequency of their application in GIS today? When might they prove useful as output from GIS analysis?
13. Provide some examples of the use of linear cartograms, contiguous area cartograms, and discontiguous area cartograms.
14. List some types of noncartographic output. What are some of the more unique types? Think about how you might decide whether these should replace or augment an existing map.
15. What are the basic design considerations for the production and use of line graphs as GIS output?
16. How does technology impact the design considerations for GIS output?
17. Suggest how virtual environments might be useful for visualization of GIS analysis.
18. Provide several examples of situations where immersive environments might be employed for visualizing GIS analysis.

References

Brewer, C.A., 2005. *Designing Better Maps: A Guide for GIS Users*. Redlands, CA, ESRI Press.

Bunge, William, 1962. *Theoretical Geography*. No. 1 in Lund Studies in Geography, Series C, General and Mathematical Geography. Lund, Sweden: C.W.K. Gleerup, Publishers, for the Royal University of Lund, Sweden, Department of Geography.

Burrough, P.A., 1986. *Geographical Information Systems for Land Resources Assessment*. New York: Oxford University Press.

Campbell, John, 1991. *Map Use and Analysis*. Dubuque, IA: Wm. C. Brown, Publishers.

Carr, D.B., D. White, and A.M. MacEachren, 2005. "Conditioned Choropleth Maps and Hypothesis Generation." *Annals of the Association of American Geographers*, 95(1):32–53.

Cleveland, W.S., and M.E. McGill, 1988. *Dynamic Graphics for Statistics*. Belmont, CA: Wadsworth Publishers.

de Blij, H.J, and P.O. Muller, 1994. *Geography: Realms, Regions and Concepts*, 7th ed. New York: John Wiley & Sons.

Garson, G.D., and R.S. Biggs, 1992. *Quantitative Applications in the Social Sciences*. No. 87 in Analytic Mapping and Geographic Databases Series. Newbury Park, CA: Sage Publications.

Harley J.B. (1988). "Maps, Knowledge and Power." In Cosgrove, D., and S. Daniels, Eds, *Iconography of Landscape: Essays on the Symbolic Representation, Design and Use of Past Environments*, pp. 277–311. Cambridge: Cambridge University Press.

Holmes, N., 1984. *Designer's Guide to Creative Charts and Diagrams*. New York: Watson-Guptill.

Imhof, E., 1982. *Cartographic Relief Presentation*, edited by H.J. Steward. Berlin and New York: Walter de Gruyter, 1982.

Jackel, C.B., 1997. "Using Arc View to Create Contiguous and Noncontiguous Area Cartograms," *Computers and GIS*, 24(2):101–109.

Knox, D., 2004. "3D Terrain Visualization and Virtual Fly-through for Tourism Conservation Using Geographic Information Systems (GIS)." Ph.D. Dissertation, University of South Queensland.

Kraak, M.J., 1993. "Cartographic Terrain Modeling in a Three-Dimensional GIS Environment." *Cartography and Geographic Information Systems,* 20:13–18.

Meilach, D.Z., 1990. *Dynamics of Presentation Graphics*, 2nd ed. Homewood, IL: Dow Jones-Irwin.

Moellering, H., 1980. "The Real-Time Animation of Three-Dimensional Maps." *American Cartographer,* 7:67–75.

Moellering, H., n.d., Traffic Crashes in Washtenaw County "Michigan, 1968–1970," Highway Safety Research Institute, University of Michigan, Ann Arbor.

Moellering, H., and A.J. Kimerling, 1990. "A New Digital Slope–Aspect Display Process." *Cartography and Geographic Information Systems,* 17:151–159.

Monkhouse, F.J., and H.R. Wilkinson, 1971. *Maps and Diagrams: Their Compilation and Construction*, 3rd ed. London: Methuen & Co. Ltd.

Muehrcke, P.C., and J.O. Muehrcke, 1992. *Map Use: Reading, Analysis, and Interpretation.* Madison, WI: J.P. Publications.

Robertson, B., 1988. *How to Draw Charts and Diagrams*. Cincinnati, OH: North Light.

Robinson, A.H., J.L. Morrison, P.C. Muehrcke, A.J. Kimerling, and S.C. Guptill, 1995. *Elements of Cartography*, 6th ed. New York: John Wiley & Sons.

Shiode, N., 2000. "3D Urban Models: Recent Developments in the Digital Modeling of Urban Environments in Three Dimensions," *GeoJournal*, 52(3):263–269.

Song, B., J. Chen, and J. Silbernagel, 2004. "Three-Dimensional Canopy Structure of an Old-Growth Douglas-Fir Forest," *Forest Science* 50(3):376–386.

Sutton, J., 1988. *Lotus Focus on Graphics* (5 vols.). Cambridge, MA: Lotus Development Corporation.

Töbler, W.A., 1967. *Automated Cartograms*. Department of Geography, University of Michigan, Ann Arbor.

Töbler, W., 1970. "A Computer Movie Simulating Urban Growth in the Detroit Region." *Economic Geography*, 46:234–240.

Tufte, E.R., 1983. *The Visual Display of Quantitative Information*. Cheshire, CT: Graphics.

Wolf, E., 2005. Creating Contiguous Cartograms in ArcGIS 9. Proceedings of the ERSI International User Conference. http://gis.esri.com/library/userconf/proc05/papers/pap1155.pdf

Zelazny, G., 1985. *Say It with Charts: The Executive's Guide to Successful Presentations.* Homewood, IL, Dow Jones-Irwin.

CHAPTER 17

GIS Design

Improper or incomplete GIS design is the most common causes of system failures. By system failures we are generally not speaking about the software failing to operate, but rather that the intended goals of the system or its database are not being realized. Some GIS operations are incapable of producing the desired results because of poorly selected or badly organized data, incorrect data models, or software with limited capabilities; in some cases, system managers have underestimated the time involved in developing the databases. Conversely, you also see well-designed GIS that are underutilized, either because system complexity prohibits easy use or because of lack of understanding of the overall capabilities of the software. In still other cases, systems that were once highly useful have lost that utility because of aging software and hardware configurations, aging data, or other shortcomings that rob the system of the flexibility required to adapt to changing demands. Finally, some failed systems will be seen as direct reflections of personnel problems that arise when organizational structures are changed forever by the introduction of new technologies into previously manual settings. Much of what we know of successful systems can be learned by examining failed systems, just as successful businesses have learned from disasters experienced by competitors.

In this chapter we focus on the business side of GIS for two reasons. First, the vast majority of GIS clients are institutional: consulting firms; service and utilities organizations and businesses; and governments. In short, most users provide a product or service requiring the application of GIS. Our second reason for concentrating on the business side of GIS is that the success or failure of the system, regardless of user or application, depends on much the same set of criteria used in the business community: If you can make a system work in the budget-conscious business world, it is likely to be successful for other applications.

Ultimately, our business is the business of digital geography. Although the current subject matter may seem to deviate considerably from what you have studied thus far, success depends on being able to combine business-oriented systems theory and operations research with the basic data models, abstractions, and analytical capabilities of the practicing geographer. The utility of geography is what makes GIS a successful enabling technology. Companies use GIS because they need to answer geographic questions. They employ GIS design experts—liaisons between the organization and the geography they need to solve spatial and spatio-temporal problems.

Design activities come in three interacting classes: analytical model design, spatial and spatio-temporal database design, and system design. Analytical model design attempts to develop effective, realistic mathematical models of spatial settings, situations, and related processes. It endeavors to represent the elements of real-world settings required for decision making. To be effective, we must consider the real world in the context of existing data models and structures. Spatial and spatio-temporal database design seeks to optimize the organization of the necessary spatial data within a computing environment to allow both effective and sustained system functioning. These latter two portions of GIS design are nearly inseparable, while system design, the development or selection of conceptual GIS models that link current geographic information science theory and available technology with spatial problems, more often has an institutional context. This chapter examines all three of these general classes of design activity: (1) application design, (2) analytical design, and (3) database design. It then proceeds to look at the larger institutional issues of system design, and concludes with a focus on the implementation of the GIS with a view toward adapting to changing needs and conditions in an efficient and effective manner to guarantee the long-term functioning of the GIS.

LEARNING OBJECTIVES

When you are finished with this chapter you should be able to:

1. List and define the components of analytical and database design.
2. Explain the close relationship between analytical GIS design and GIS database design.
3. Describe the process of isolating the spatial components of a GIS problem in general terms, and provide a concrete example of how this works within a specific knowledge domain (e.g., designing a wildlife preserve or analyzing a crime pattern).
4. Define system design and discuss why it is important for GIS.
5. List and explain the function of the internal and external players in the GIS institutional setting.
6. Describe the idea of conceptual design as it applies to GIS, and have a feel for how people problems are a primary concern in this area.
7. Discuss what design creep is, enumerate its possible negative effects, and explain how this can be avoided.
8. Define spatial information products (SIPs), discuss how they can drive the design of a GIS, and describe how they are integrated from local views to global views of the GIS in an organization.

9. Discuss the idea of cost-benefit analysis as it applies to GIS in an institutional setting. In particular, describe how costs and benefits are determined.
10. Understand the major database design considerations, especially with a view toward system verification and validation.

APPLICATION DESIGN

GIS applications are numerous and extremely varied. While we can learn by observing successful applications elsewhere, it is still necessary to prepare each application individually to address all of its uniqueness and subtle differences. Some applications are one-time-only and short lived, while others are meant to address long-term spatial data handling. Some applications are aimed at an academic setting where research is the focus of the design; some are aimed at a business environment where the bottom line is the goal; and still others are primarily aimed at government planning and decision making and support. In some applications there is a more enterprise-wide system environment where the GIS is meant to serve many users' needs, whether they be short term or long term. Regardless of the nature of the system, application design is a methodology that aims at ensuring that the scope and intended purpose of the GIS are well defined and that the necessary resources to make that happen are in place and remain so during the life of the application. This implies that the system is also capable of changing as the needs of the application changes. In short, this acknowledges that projects within a larger organization change, grow, and sometimes go away.

Among the first concepts developed in the field of system design was the **project life cycle**. In most operational GIS settings, many projects are going on simultaneously. For each project, we must make basic decisions about what needs to be done, when it needs to be done, and who will be responsible for doing it. Because each project has a beginning, a middle, and an end (we hope), we can say that it has a life cycle, which in turn dictates the operations and organizational structure that will bring the project to a successful conclusion. If we are working on a single project, the project life cycle will focus on that single undertaking. The principles remain the same, but the operational details will differ.

Whether the project is to be performed in house, within a small group, or within a larger organizational structure, there must be methodology to give a framework for finishing the job. You might think of an assigned term paper as a simple example of a project having a life cycle. The paper may require field work, computations and analysis, review of the literature, compilation of results, and writing. You don't (I hope) begin writing the paper until the preparatory operations are complete. There is a need for a structure that enables you to perform the correct tasks in the right order to get to the final result. Because you are the only one involved, defining the tasks and establishing the proper order of execution probably will suffice. In a setting characterized by many interconnected projects designed to meet the needs of management, however, two other objectives of the project life cycle must be met: There is a need to ensure consistency among all the different projects so that all will have the same level of reliability, and there must be built-in points at which management can determine whether it is time for the project to go forward. We can formalize these three objectives of the project life cycle as follows:

1. Define the activities of the project and the order in which they are to be performed.
2. Ensure consistency among many projects in the same organization.
3. Provide points for management decision making regarding starting and stopping individual phases of the larger project.

The project life cycle is only a guide to management, a framework for the project; the manager still makes the fundamental decisions. Many important aspects of the organization (supporting the workers, fighting political battles, ensuring good morale, etc.) are still fundamentally up to the manager. The life cycle provides guidelines to facilitate good, timely decision-making (Yourdon, 1989).

Among the first approaches to the project life cycle was a technique called **bottom-up implementation**. This method, although flexible in terms of the number of steps and the exact steps taken, is highly structured and progresses linearly. The steps cover the definition of user requirements, the specification of functional needs, systems analysis, detailed design, the testing of individual modules, the testing of subsystems, and finally system testing. In the computer industry, this approach is often called the **waterfall model** of system design (Boehm, 1981; Royce, 1970) (Figure 17.1).

The waterfall model was developed to provide a structure for the systematic movement from requirements analysis through testing and final operation of an information system. It generally cascades from conceptual design through

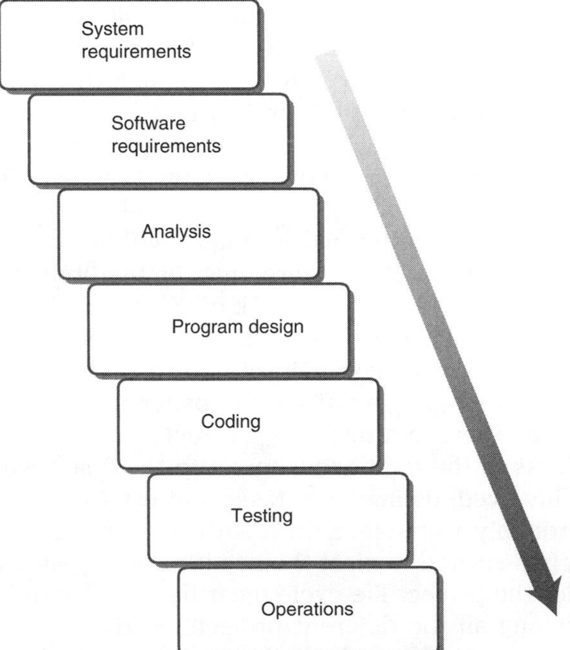

Figure 17.1 Waterfall model of system life cycle.

program detail design, code creation, and code testing to program implementation. There are several problems with the waterfall model as a tool for developing a GIS. First, because the model requires each step to be completed properly before the next proceeds, any delay in a single step will slow the entire system (Yourdon, 1989). Under this model, if we know most but not all of the GIS user requirements, we could not proceed to the input phase until the rest of the requirements were received. Another problem with the waterfall model is its linear structure. Although we may want our GIS system to progress in a linear fashion, nearly every implementation is going to encounter problems. And, as it turns out, people are better at making improvements to an imperfect system than at anticipating all possible problems before the initial implementation. In addition, clients may neglect important details at the outset or discover new uses for the GIS as they see the system implemented. Or a client may find the economic situation changing, resulting in either more uses being included or costly portions of the system being cut. If we followed the waterfall approach to system development, we are likely to have a completed or nearly completed system before we discover that other things need to be added or mistakes need to be fixed.

SOME GENERAL SYSTEMS CHARACTERISTICS

The more specialized a GIS is, the less adaptable it is for new tasks. If, for example, you are developing a GIS system specifically oriented toward digital remote-sensing applications, you are most likely to incorporate data models and techniques designed to enhance and classify the pixels obtained from a satellite system. If you decide later to incorporate a gravity model designed for use with a vector data structure, your tools will likely be inadequate for the new tasks. You will have to add vector data modeling capabilities to the system and then you will have to train the users. This is nearly like creating a new system and placing it on top of the existing one.

The size of the system has much to do with the amount of resources that must be applied. In general, the larger the system, the more resources needed to run it. For example, if you are creating a GIS for a modestly sized region with many possible tasks and hundreds of themes, the input alone is going to consume more time for digitizing, more costs for obtaining data, more data entry, and probably more copies of the software than in a smaller system. Once the system is implemented, you will need more people to analyze data, you will need people to keep the data secure, you will need more output devices, and you will have to hire more managerial staff to keep track of the many parts of the analysis.

Fortunately, systems can be broken into component parts, each of which can be managed separately. This is another basic principle of systems, and one that is of paramount importance for larger systems. Each of the separate analyses can be envisioned as a subsystem that can be operated much like a single project. Strongly related to the ability to break GIS tasks into smaller components is the concept that many systems, perhaps even most, have a tendency to grow. So even if your original GIS operations performed only a few analytical tasks, as the knowledge and skill of the personnel grow, and as the utility of the system becomes better known both to management and to the client base, the system will have to grow to adapt. In fact, the more successful the operation, the more likely it is to grow. But success is often partly a function of communicating success to the institution and its sponsors. Most commercial operations serve many people, so knowledge of the institutional setting of the operation becomes important.

PROJECT DEFINITION

Given the vast number and types of possible GIS projects that can be performed, the scope and definition of the GIS is paramount to its overall functioning, especially for long-term enterprise projects (Tomlinson 2003). The general purpose definition for the GIS might include whether it will take the organization in a new direction or if it will support or expedite the continuation or migration of existing processes, products, and/or legal mandates, as in the case of government organizations. This requires understanding the underlying nature of the specific proposed GIS as (a) one primarily of data archiving and cataloging, (b) one meant to be employed for everyday decision making, or (c) one more aptly suited for a research or academic application. Among the easiest ways to approach such a project definition is to first explicitly specify the mission of the organization as it currently exists without the presence of the GIS. This can often be found in readily available organizational documents such as charters, mission statements, organizational diagrams, and the like.

Once established, the potential of using GIS to accomplish the defined goals involves working backward through the organization's operations and plugging in **spatial information products (SIPs)** derived from the software to replace those spatial products, spatially derived answers and decisions, and spatially related variables developed prior to system implementation. In some cases this type of project definition is performed in-house, but larger organizations normally employ GIS design firms to collect information about the organization, compile it with institutional feedback, and develop appropriate conceptual foundations for its ultimate creation.

Inextricably linked to the definition of a GIS project is an initial determination of the necessity for a system in the first place and, if applicable, its relative merit to support or enhance the organization's goals and objectives. Many assumed applications of GIS might require a less complex computer-assisted cartographic system or a CAD system to store plans and maps for later retrieval rather than a full-function GIS with its many analytical tools. Or an organization may have little need for the spatial or graphic subsystem and rely more heavily on database storage and retrieval than on GIS technology. The first task of the GIS designer is to determine whether or not a simpler, less expensive, and easier-to-manage system might be more applicable and more easily integrated into the organization.

If a decision is made to adopt some level of GIS, the degree to which it is capable of supporting, expediting, or enhancing the existing operations may well determine the appropriate investment in it. Such a determination normally consists of a thorough examination of the organizations work flow and product catalog to determine what the outputs are, which departments have the greatest need for GIS services to enhance their performance, and which needs are critical and which can be addressed later. In many ways this is a form of what computer scientists call **reverse engineering**, because it effectively and systematically decomposes the organization into smaller components that can be analyzed one at a time. Some portions of an organization, for example, the secretarial operations, may have no need for GIS at all. Some groups, for example, a mapping division, might have a pressing and significant need of the GIS tools. Still others, like a motor pool, might be able to function quite nicely without a GIS, but could improve their functioning if it were eventually employed to keep track of the locations of vehicles in the pool.

The scope essentially measures the size of the project in both geographic space and time. The study area is most often defined for government bodies by legal mandate, for business and commercial applications by potential market availability, and for research scientists by the subject and its environmental constraints. Once the geographical boundaries are defined, the life expectancy of the GIS operations will also be necessary for the allocation of human, financial, intellectual, and educational resources. Such resource planning is essential because the setup expenses for GIS integration, particularly with regard to data entry, can be prohibitively large if not done carefully.

ANALYTICAL MODEL DESIGN

There is no single volume that provides a comprehensive set of recipes or best practices for the application of the GIS toolkit to each of the myriad tasks to which it could potentially be applied. Such a book would likely be impossible to write because of all the possibilities. However, Mitchell (1999) has created a set of useful tables outlining some of the more common tasks the GIS practitioner might encounter, their conditions of use, types of output resulting from their implementation, and the trade-offs encountered through their application. These are summarized on the inside covers of this text.

Components and Procedures

Each task requires a specific set of analytics—usually a subset of the full GIS toolkit. In some cases, the components are so limited that application-specific software may be a viable option to full-scale GIS software. Examples might include applications that are related to hydrological modeling, where hydrogeological or surface water modeling software might be appropriate. Alternatively, if the project is part of a larger set of GIS operations, then the inclusion of such hydrological modeling components within the software will be important. Some GIS software comes as a set of modifiable modules that can be pieced together to complete a given set of tasks. Others include the basic operations sets but offer specific modules for subsequent purchase. Modules might include image processing, image segmentation, surface analysis, data visualization, spatial statistics, and the like. Before beginning a project it is important to recognize which modules are needed. This will not only drive data decisions, but possibly software purchase decisions as well.

GIS Tools for Solving Problems

In Chapter 15 we discussed the idea of confirmatory and explanatory problem solving in GIS. There are many considerations which come into play when deciding how to apply the software to either of these primary types of modeling that go well beyond the individual tasks that might be employed (Mitchell 1999). Among the most important of these considerations are the choice of software, designing the database so it allows you to perform the necessary data searches, establishing an effective spatial domain for your study, selecting an appropriate study area that incorporates the necessary spatial domain, and obtaining data that have the appropriate level of detail and necessary coordinate systems and projects. Each of these is explained in some detail in the following pages.

Selecting the Software

Whether for an individual or an organization, it is not always simple to select the correct software system, particularly since many software packages are beginning to look very much alike. This is particularly true of the commercial systems, but several open-source GIS packages are viable solutions even for complex modeling situations because they conform to a rigorous set of standards developed by the Open GIS consortium [http://www.opengeospatial.org (last viewed 11/08/07)]. For individuals and even some organizations, the choice is often an economic one. In others, the choice may require a specific data model, although this is much less common today because of the ability of the software to move from one data model to another almost seamlessly. Finally, the choice of system my be driven by the expected SIPs for the project. The choice of hardware platforms and peripherals is dictated by economic constraints, compatibility with software, accuracy requirements, learning curves, and personal preference.

So how do we decide for ourselves or for our clients which system to purchase? A reasonable approach is to prepare a document indicating analytical needs, cost limitations, accuracy needs, and training needs. Then these specifications are submitted to a large number of vendors for competitive bidding. You may wish to append a request that each bidder provide a list of current customers, both satisfied users and clients who have experienced problems. The responses you receive will tell you about the kinds of organization that use the system (giving you an idea of their modeling needs) and will allow you to learn from firsthand responses about various users' personal satisfaction levels. The final selection of the GIS should be among the last decisions to be made, but is, unfortunately, often one of the first.

Scientific Models and GIS

Much of what drives the GIS analytics is formalized by econometric, geographic, environmental, transportation, and other types of scientific models. Such well-known types of **models** as population dynamics, transportation flow, potentials (gravity) models, soil erosion models, hydrological models, and so on are now commonly incorporated into the GIS software. Some software is specifically designed to address selected topics, and this should have a major impact on the choice of system. Software such as CrimeMapper is specifically designed for criminal hot-spot mapping and includes an appropriate user interface and symbol set. Other software, such as the programs available from RockWare, is designed to work with stratigraphic data. There are packages designed to model transportation systems and traffic flows at different levels of detail, those created specifically for hydrological modeling, and still others whose functions are specific to modeling sound wave generation. Some systems are stronger with raster data types and have large image-processing suites as a central component, while others are more adept at producing quality cartographic output.

The choice of a general, all-purpose GIS versus a specific focused system requires an analysis of some pretty basic system characteristics. First, if the system is a short-term duration, project-specific task that is not likely to be repeated, a focused system might be the most appropriate alternative. If, however, there are many GIS tasks that will need to be addressed over long periods of time, a more robust, expandable system is more likely to fit the need. Beyond the selection of the software based on the scientific models it was designed to address, however, is the need for data compatibility and compatibility with existing models already in place within the organization. Finally, the chosen system will be most effective if the software vendor is familiar with the modeling needs and can provide both system and modeling support, especially in the formative period of GIS adoption.

DATABASE DESIGN

The database is a critical component of the GIS. But a collection of data without structure and without an understanding of the way in which the data are to be used is not a database. Many projects fail because they begin with an extensive set of data collected without focus and without a project in mind. The utility of the database is contingent on selecting the proper data for the proper tasks, organized so that the necessary queries can be performed efficiently and effectively. The following sections describe some important aspects of database design.

Modeling Tools

Fundamental to effective GIS modeling is the recognition of an explicit spatial nature of both the geographic objects and their underlying processes. A failure to do so is a primary reason that high-powered geographic information systems are often underutilized. This is particularly true with regard to processes, but this in turn is often caused by a failure to perceive a spatial pattern of existing conditions in the first place. Once a pattern is perceived, it is at least possible to research the likelihood that the pattern is either purely happenstance or that there is a process or set of processes responsible for it. While this may be simple in some cases, in others it may require substantial research. In both cases, the first step is to perceive or recognize the pattern and then to describe it.

DeMers (2002) provides some detail on the variety of tools that allow us to perceive spatial patterns, but a synopsis is valuable here. Maps, aerial photographs, and satellite imagery that show a planimetric view of our world are obvious tools for acquiring pattern descriptions. These graphical forms explicitly display the spatial locations, associations, and juxtapositions of objects in geographic space. Frequent exposure to such tools sensitizes the viewer to the spatial nature of the world around us. Alternative tools for recognizing spatial patterns include statistical analysis, charts, and graphs. These latter are particularly useful for relating one set of variables to another. The scattergram, for example, is a commonly recognized method of discovering the relationships among variables.

We have already seen how patterns, especially nonrandom patterns, typically both develop from underlying spatial processes and produce changes in these processes as well. Not everybody is aware of this relationship, however. There are many examples of businesses that fail because they do not recognize the power of the spatial proximity of their potential customers. Customers who are nearby are generally far more likely to frequent stores that require less travel time than they are to travel great distances for the same products or services, even when the cost is less at a more distant store. This is particularly true when the distance incurs a cost of both fuel and time or when nearby stores cater to the particular needs of its proximate clientele. If you are a GIS consultant, you may need to explain this to your client.

Establishing the Effective Spatial Domain of the Model

Our last example showed that there is a relationship between distance and clients. This may also be true in the context of the modeling process itself. There are relationships between animal predators and prey in wildlife modeling, between criminals and victims in crime mapping, and many other situations. Depending on your specific need, the relationships among spatial variables may not always be this obvious. Before starting a GIS project it is important to identify which factors are most likely to be important to the system being modeled. In short, you need to be aware of the effective spatial domain for each model you build. An effective and relatively simple way of doing this is to begin with the expected outcome from the GIS and work backward. DeMers (2002) often compares this to envisioning a GIS modeling process as a tree with the trunk as the expected outcome, the branches as the general categories of thematic elements, and the leaves as the more specific data elements that are input to the GIS and from which the model is built. It is precisely for this reason that the development of a GIS flowchart is used to identify possible missing data elements, aspatial components, and poorly understood relationships. The mountain cabin example in Chapter 15 is designed specifically to illustrate the process. While it explicitly illustrates the flows among data elements, the example does not provide much guidance as to what the computational linkages might be. This requires a more detailed look at the role of the powerful GIS analytics in connecting the data elements.

Study Area

Study areas are most often driven by a formal decision about why the area is under study. Thus, for a county-level study, the entire county will be the study area. For a study based on a natural phenomenon, such as pollution in a stream, the entire stream drainage basin may be selected as the study area because the pollution will come from all upland areas that drain into all the stream's tributaries. For a study for a lumber company contemplating the use of a GIS to keep track of its logging operations and land holdings, the landownership will be the deciding factor for selecting the study area. In short, the study area may be selected on the basis of political boundaries, physical boundaries, or landownership, or it may reflect primarily the economic and data constraints that are found in the early stages of the design process; in any event, its choice is project driven.

If there are detailed data for a small portion of a prospective study area but more general data for the rest, the detailed data need not dictate the extent of the study area. Instead, the larger area of interest can be considered to be the overall study area, and the subarea for which there is more detailed information can act as a prototype for a detailed analysis of the larger study when data become available or affordable. In addition, the detailed data for the subarea might be useful for improving the knowledge of the larger area through the use of density of parts or another dasymetric mapping technique.

Whenever any form of interpolation is scheduled, the study area boundary, at least for the elevational coverage, must be extended sufficiently to ensure that the interpolation will produce the correct results. Finally, the larger the study area, the more expensive in time and money will be the production of the database. Often, cost awareness is a major constraint in defining a study area in the first place. In addition, the larger the study area, the greater the chance that there will be a lack of data for all the coverages needed to perform the analysis. Selecting a smaller study as an initial prototype is a good choice if it will allow the organization to demonstrate the utility of the system. Once that has been effectively done, the system sponsor is much more likely to supply more resources to increase the size of the study area.

Scale, Resolution, and Level of Detail

Related to the extent of the study area is the scale at which you want your data input. This is seldom a straightforward matter because of the disparate scales of cartographic data available for building cartographic databases. As you remember, the smaller the scale of the maps, the larger the amount of error and the greater the generalization. A good rule of thumb is to take the largest-scale maps available for the task. The importance of the aerial coverage to the models being built will also dictate what scales will be acceptable. Although there are no formal guidelines for determining an acceptable scale, most GIS professionals apply the "best available data" approach, recognizing that more detail is better than less detail.

When working with raster data models, you will also need to know the acceptable size of grid cells within the larger map area. Again, smaller provides better detail but also drastically increases the data volume (i.e., the smaller the grid cells, the more you will need a single coverage). Although storage may not be a problem for most modern computer systems, functions that require extensive neighborhood searches will be slowed by massive increases in the grid array. In some cases, the grid cell size may be dictated by the smallest item you wish to represent, in others by the requirements of the model (DeMers, 1992), and in still others by issues of compatibility with other digital data, such as input from a satellite.

Classification

With the different sources of data for GIS comes a need to consider the potential classification systems appropriate for the modeling tasks. It might prove useful to examine the kinds of input data that are available prior to deciding on the classification system. The use of a more detailed classification is often preferred for two reasons: it gives the user the largest amount of data, and if one theme can be compared to another using a low level of detail, it is relatively easy to aggregate classes, whereas disaggregation may be difficult or even impossible. Of course, the most detailed classifications also present two possible problems. First, the classifications may be designed for a particular task that has little to do with the project at hand; and second, the more detail, the greater the chance for errors in classification in the first place.

But classification is more than just selecting the correct level of classification detail. It is important to remember that in GIS we frequently are making comparisons among maps. This means that we need to be able to correctly compare the classifications from one coverage to another. We also need to be able to produce consistent classifications within a single coverage, often from different sources. Suppose, for example, that you require a regional coverage of soil classes based on different county-level soil surveys. When surveys have been developed in widely differing time periods, it will be necessary to **crosswalk** them to ensure a uniform classification. Burrough (1986) provided a more detailed discussion of the problems of classification, especially as they related to potential errors in the GIS.

Coordinate System and Projection

Deciding which coordinate systems and projections to use is often dictated by the regional extent of the study area and by the availability of data. Again, there are no hard-and-fast rules, just some basic commonsense guidelines that can be applied to the problem. For study areas covering many states, it is generally advisable to avoid the use of any coordinate system that has a limited geographic range of accuracy. As for projections, you have seen that the properties needing to be preserved (i.e., the properties that will be most important to analysis) are those that will dictate the correct projection. In general, for both coordinate system and projection, spatial and temporal compatibility are vital to the decision-making process.

Conceptual, Logical, and Physical Models

For a database to be effective, you must know how each of the pieces of the data relate and will eventually interact with each other inside the system. This **conceptual data model**, or **conceptual data schema**, uses tools such as entity-relationship (E-R) diagrams to map out how these interactions look. The system designer is looking for data dependencies in that one piece of information depends on another. In short, if one piece of information changes, the dependent pieces of information will also have to change. For example, if a parcel of agricultural land in a GIS database is sold to a developer who then changes the land use to apartments, the ownership information change will also require a change in the land use information.

Once a conceptual schema is established, one converts this to a **logical data model** that is a representation of the data organized based on the particular database management technology to be employed by the GIS (e.g., hierarchical, network, relational, objet-oriented, etc.). For example, with relational systems the data are described in terms of tables and columns, while in object-oriented systems they will be represented as classes, attributes, and associations. Other forms, such as XML, describe the data in terms of tags. The logical data model is sometimes confused with the **physical data model**, which involves intense application of the selected database management technology. This might include which drives on computers contain which portions of the database, and in which parts of the organization or even which parts of the world which they might be located. This might impact such operations as system recovery in cases of hardware crashes.

INSTITUTIONAL/SYSTEM DESIGN

System design is concerned with the interactions of individual people, groups of people, and computers as they function within organizations (Yourdon, 1989). The introduction of such new technology as GIS requires different or additional training for people doing the work, requires that funds be allocated for software and hardware, changes the flow of information within the organization, and therefore modifies the organization's structure. And now that we have large databases to work with, we must be more concerned about the integrity and quality of the data that are accessible by many different individuals. Wide accessibility, in turn, increases the need for security and quality control measures within the organization.

GIS **system design** can be subdivided into two highly interactive parts: **technical design** (internal) issues and **institutional design** (external) issues. The internal issues deal most often with the system functionality and the database. Will the system work the way we need it to? Can we answer the questions we need answered? Have we got the correct data in the right format? Do we have people with the right training to run the system? Can our system adapt to changing demands? These are some of the general questions that accompany the process of internal or technical design. But although we need to be sure that our system functions properly, we also need to understand the relationship between the GIS operation and the organizational setting. Do we have the funding necessary to permit continued operations? Can we obtain data at reasonable cost? Do we need to employ application developers to customize the software? Will we have adequate software support from our GIS vendor? Will we be legally responsible for errors made through our analyses? And are we meeting the larger goals of the organization beyond the immediate end of performing GIS analyses? All these questions are important institutional considerations.

The technical design issues cannot be separated from the institutional issues. Even GIS operations that are brilliantly successful from the technical design standpoint will fail if we lose the support of our organization or external sponsor. For example, even if your GIS-based efforts to enhance delivery of newspapers within *The New York Times* organization are successful, if upper management sees the added expense of operating the GIS as excessive, your operation will likely be suspended. Or if your government organization fails to obtain funding for the use of GIS, you will have to abandon such efforts. And, of course, if your system fails to provide the needed answers to the questions at hand, you are far more likely to lose your institutional support. Internal and external forces impact the overall system in which the GIS operates. And, as with any system, proper planning can make all the parts work together more effectively.

GIS INFORMATION PRODUCTS

In large part, GIS information products are the result of analysis within the software. What the specific information products are depends on the nature of the organization, its goals, and its experience with the system. Although many organizations have a fairly structured set of products—for example, Rand McNally Corporation makes a relatively definable number of products (Calkins and Marble, 1987)—others, especially scientific research organizations, have a large and ever-changing set of potential products that may be difficult to define (Rhind and Green, 1988; Wellar, 1994). But even in scientific organizations, there are often only general ideas of what the GIS is capable of doing and what the technololgy might offer the individual institution.

How Information Products Drive the GIS

Organizations that have a particular goal in mind when they consider implementing a GIS are quite likely to have some ideas about what the output of analysis might look like, even if they have only a vague sense of how it might be derived. In addition, they most likely know roughly what general data might be part of the system. In fact, many organizations recognize the value of the data in an automated form even before they are able to see the potential utility for analysis. To extract particular spatial information products from the users, we must recognize the close linkage between what goes into the system (data availability) and what comes out.

But because members of the user community cannot be expected to be instant GIS experts, it is our job to *help* them define their needs. You should note that I did not say it is our job to define their needs! We are present not to tell them what they need, but instead to help them tell us what they need. Thus, the systems analyst must play the role of educator to elicit descriptions of needed products from the users. Perhaps this will clarify my insistence on the need to understand the analytical capabilities of the GIS as well as the database creation and management tasks.

We first determine how a GIS could best serve each individual user within the organization. This is best done in person, rather than through a review of questionnaires, because interactive question-and-answer sessions are most likely to lead to an understanding of the tasks each user normally performs and to enhance explanations of functions the GIS might be able to perform to match those tasks. Any documentation indicating what the user produces (sample maps, written responses to questions or reports of decisions, applicable legal mandates, etc.) will prove useful in making a good match between the existing products and the SIPs obtainable through GIS. As you perform this review, keep in mind that the users will be more interested in what can be obtained than in how it is derived. As the process moves toward implementation, the how questions can be addressed.

Organizing the Local Views

Because we most often will be working with many potential users of differing individual needs, it can be useful to keep track of the relationships between each user and each SIP. Tomlinson (n.d.) suggested using a **decision system matrix**, listing the users along the side and the products along the top; Figure 17.2 shows a highly simplified version. Not only will this device allow us to keep track of the SIPs for each user, but later, when we must integrate each of the local (individual user) views into a larger global view, data from this matrix will provide much of what we need. An alternative to the decision matrix is an **organizational diagram** such as that shown in Figure 17.3. The global view at the bottom of the diagram illustrates the overall organization's needs, the next tier isolates the local views of individual users, and the products—differentiated by user—appear at the top. This type of diagram, like the matrix format, is useful for integrating the local views into a more general **global view**. Both can also be used to decide which products are the most important, based on which are most often called for by individual users.

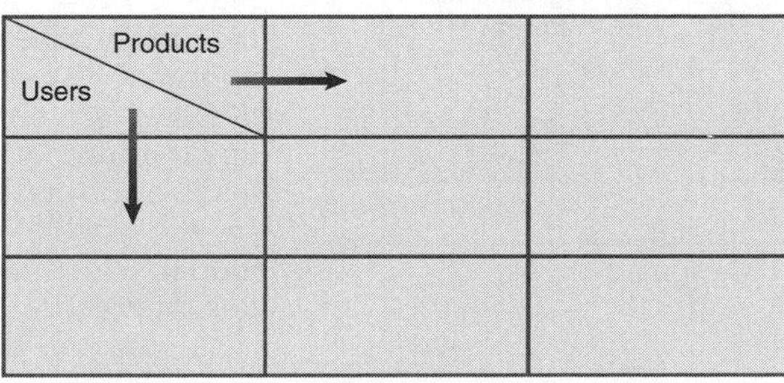

Figure 17.2 Decision system matrix. Simplified decision system matrix used to organize individual user views of the geographic information system (GIS).

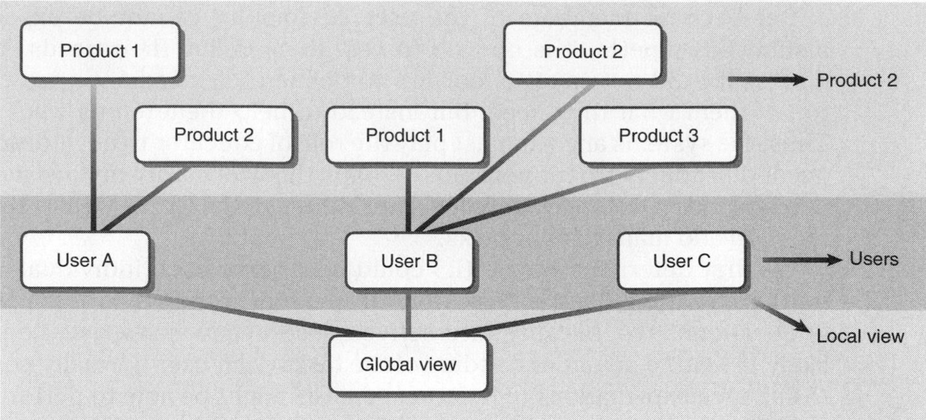

Figure 17.3 View integration. Organizational chart used to display individual user views of the geographic information system (GIS). *Source:* Duane F. Marble, Department of Geography, Ohio State University, Columbus, Ohio. Used with permission.

Avoiding Design Creep

Design creep, which occurs in the absence of organizational learning, can be defined in terms of its results: at one extreme is a system with more functionality than is necessary, and at the other is a system having an incorrect functionality. Figure 17.4a shows a model for a rather structured approach to GIS design, proceeding from the feasibility study at the left through the design phase and finally into the implementation phase. The two curves indicate the relationship

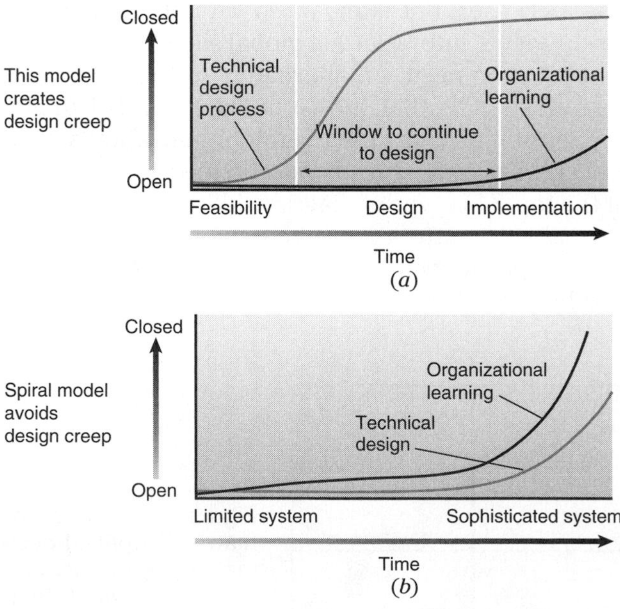

Figure 17.4 Avoiding design creep. Two approaches to system design: (*a*) a structured, linear approach that leads to design creep, and (*b*) the Marble spiral model, which avoids design creep.

between the technical design process (light curve) and organizational learning. As you can see, organizational learning begins late in the process, when the system is almost completely designed and is nearly ready for implementation. Thus, users are forced to learn and operate a system that may not meet their needs.

Figure 17.4b shows the more flexible spiral model (Marble, 1994), in which organizational learning drives system development. As users become familiar with what a GIS is capable of, they can describe their system needs at a point well in advance of the run-up to implementation. You can also see that the more sophisticated the organizational learning, the more sophisticated the system becomes, but the learning curve is always above the design curve. In short, the users are driving the design of the GIS, rather than being driven to learn by an in-place system. The latter approach avoids design creep and allows the system to grow in complexity as the organization's needs increase.

VIEW INTEGRATION

As you have seen, most organizations will have many GIS users, and it is vital that the needs of each be ascertained and addressed. The individual sets of requirements are called **local views** (of what the GIS is going to do and how it is to do it), and as a preintegration task we often find it useful to list a set of preferences for each of these local views, based on management needs. We also need to decide on a methodology—perhaps a decision matrix or an organizational chart—for later integration of the local views (called **view integration**). One useful technique is to produce a pairwise (dyadic) grouping of similar individuals or groups. This can be performed easily with either the decision matrix or the organizational chart approach. The most common needs, those being met for the most people, will provide us with one method of identifying a set of priorities that can be compared to the larger organizational goals.

When conflicts arise among local views, additional discussion will be necessary to pinpoint the problems. Conflict resolution may require arbitration, modification, and even redefinition of the local views to achieve correspondence to the overall organizational goals. If isolated needs are found to be incompatible with the global view, several iterations of intermediate views may be required to combine individual local views into larger groups. This merger of local views will ensure that the vast majority of individual user requirements are being met. Final decisions for determining the global view will often rest with management, where individual user needs are considered as part of the overall needs.

SYSTEM IMPLEMENTATION

Once a GIS is implemented, its impact on an organization can be profound. Tasks that once took enormous effort might now be performed quickly and efficiently, thus freeing up time for other tasks. Other operations, such as data input, data editing, and computer- and software-related tasks, may all be new to the organization and will require substantial input of time and resources. The changing nature of the software and the hardware will also require additional and continuous training and upgrading. Among the most profound influences such new technologies have on organizations is that their impact are often permanent and usually result in substantial changes in the day-to-day operations. Some organizations are more adaptable to such changes than others.

The Institutional Setting for GIS Operations

Whether a GIS is operating in a university environment, a business, or a government organization, it is not operating in a vacuum. If you are doing your own work, it is likely that you will need access to the hardware and software, but you should be sensitive to the requirements of other users so that your project does not interfere with theirs. If you are working on a larger project, you will need to coordinate your activities with the other users while keeping the goals and objectives of the overall project in mind. In turn, a project funded by external sources will require the system manager to meet the goals and objectives of that outside sponsor. The hardware and software most often will come from outside sources as well. As you can plainly see, there are many players in most GIS operations, and these players are both internal and external.

The System and the Outside World

Figure 17.5 shows a simplified model of the relation between most GIS operations (the system) and the world at large. Inside the system there are interactions with the people responsible for day-to-day operations of the GIS, with those who operate the system, and with those responsible for project management and oversight. This is illustrated by the two-way internal arrow in Figure 17.5. The other arrows show inputs and outputs to and from the outside world. This is the larger framework within which the GIS must operate. Now let's identify the internal (system) players and the external (world) players and see how they interact.

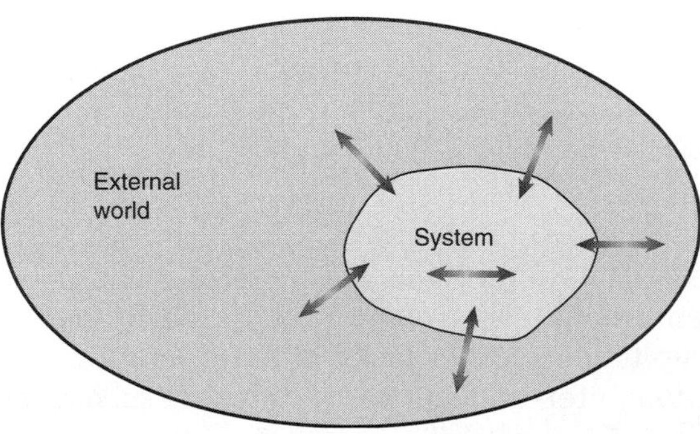

Figure 17.5 Internal and external players. The geographic information system (GIS) in relation to the outside world.

Internal Players

The system is comprised of three basic sets of players, each with different objectives. The **system users**—those who will use the GIS to solve spatial problems—are most often people who are well trained in GIS, perhaps in a specific GIS. Digitizing, error checking, editing, analysis of the raw data, and output of the final solutions to queries are the primary tasks of the system users. Whether trained in university courses or on the job, the system users will require additional training because of the constant changes both in software and in the demands of the system for new analytical techniques. In many GIS operations, this group is likely to be the most transitory, with new employees replacing those who have moved on to different positions.

System operators are responsible for day-to-day system operations, more often performing tasks that allow the system users to function efficiently. Their work involves troubleshooting when programs hang or the complexity of the analysis requires additional insights. They are also responsible, in most cases, for training the users. This suggests that in many cases system operators have experience as system users. They also know enough about the hardware and software configurations to be able to make adjustments for upgrades. In addition to acting as system administrators, they frequently act as database administrators, keeping track of the security and integrity of the database to prevent possible loss or corruption of data.

The security and integrity of GIS databases are integral to the long-term operability of a GIS, particularly within large organizations. System operators are responsible for ensuring that the most current themes are not corrupted by system users who are not familiar with the overall database contents, structure, and possible uses. They are responsible for granting access to the databases through password protection, workspace allocation, and other security measures. This is especially important where the database is to be modified through data input, editing, or analysis, where old themes are to be replaced by new. Beyond providing secured access, it is often necessary for the system operator to archive one or more copies of the database so that mistakes do not corrupt the existing data and so that computer malfunctions do not destroy valuable information.

Beyond providing for system and database integrity, the system operator is also responsible for data security involving sensitive GIS data. Such situations might arise where absolute locations of, for example, endangered species should be hidden through data aggregation to avoid exploitation of that information. Another might arise in defense operations where only portions of a database might be provided to each individual system user whose security clearance might limit their access to the overall database, or even in research situations where field researchers have shared valuable data with GIS scientists on a cooperative basis. The system operator should examine the nature of each security issue and outline specific, formal procedures prior to starting a project. In most cases, these procedures should be explained to both the internal and external players.

Both system users and system operators function within a larger organization often called the **system sponsor**. The system sponsor is the institutional parent that provides the funding for the software, hardware, operating expenses, and salaries. It also provides the political viability of the system. In other words, if there is no larger organization, there is probably no GIS. The system sponsor can be a university research group that obtained grant funding for a project, a government body whose work requires the use of GIS, or a commercial organization that performs GIS operations for paying clients. Funding and political viability of a GIS system also require that the system sponsor engage in the long-term planning needed to ensure a steady supply of projects and continued funding.

External Players

Because most GIS operations today do not use homemade software, they rely on the first of our external players, the **GIS supplier**, a company that develops and markets GIS software systems of one or more types. Suppliers also are responsible for providing software support and updates of the software as new and improved methods are put into the system; they may work in concert with hardware development companies to provide bundled packages of software and hardware. In many cases, training is provided for GIS users either onsite or at this supplier's facility, through contacts with the system administrator.

The data suppliers, the second external player in the GIS world, tend to be either private or public. The private company may provide internally generated data or data obtained from public agencies, modified to better fit needs expressed by the user community. Public agencies, primarily federal government agencies, provide data for large portions of the country. In many cases, these data are designed for use by the agencies themselves, but the data are available to external users as well. Private companies are most often for-profit organizations that supply data at a cost-plus-profit level. In some cases, the data can be provided for specific regions and in the formats needed for a specific use. Some public agencies offer data at cost, or in some cases at no cost at all, provided the use is for nonprofit GIS operations. Appendix A lists some public sources of data.

Another group of external players, a group that is becoming increasingly important as GIS software becomes more sophisticated and more object oriented, is the **application developers**. Application developers provide user interfaces to reduce the reliance on specialized GIS professionals to perform common tasks. In many cases the programming is done in macro languages provided by the GIS supplier to support applications development without the need for interfacing with traditional computer languages. Although in the past most application developers were internal personnel who had years of experience with the analytical capabilities of the GIS software, today there are many small consulting firms specializing in the development of applications for third-party clients. Application programmers can limit the training costs incurred as new GIS users enter the organization. They provide a "point and click" environment that allows the system to be used by analysts having only a limited amount of knowledge of the internal functionality of the GIS software.

Finally, we have the **GIS systems analysts**. Members of this group of external players specialize in the study of systems design. Most often systems analysts are part of a team of professionals responsible for determining the goals and objectives of the GIS within an organization, fine-tuning the system so that it provides the appropriate analytical techniques and ensuring the successful integration of the system into the organizational framework. In short, the systems analysts act as navigators for organizations using GIS. Commonly, the systems analysts are part of a larger consulting firm specializing in the implementation of GIS for third-party users, but in some cases they are employees of the GIS suppliers.

Terms

application developers
bottom-up implementation
conceptual data model
conceptual data schema
crosswalk
decision system matrix
design creep
GIS supplier
GIS systems analysts
global view
institutional design
local views
logical data model
models
Marble spiral model
organizational diagram
physical data model
project life cycle
reverse engineering
software design
spatial information product (SIP)
system design
system operators
system sponsors
system users
technical design
view integration
waterfall model

Review Questions

1. What is the concept of a system life cycle? What are its objectives? Why is the waterfall model of the system life cycle inappropriate for designing and implementing GIS in organizations?
2. What is a system sponsor? Why is it important in the overall, long-term functioning of a GIS in the institutional context?
3. Who are some internal players in the institutional setting? What important tasks do they perform?
4. Who are some external players in the institutional setting? What role do they play in organizational implementations of GIS?
5. What is conceptual GIS design? Why is it not enough for designing a GIS for an organization? What are some of the people problems of GIS?
6. What is a cost–benefit analysis and what role does it play in designing a GIS? Discuss some of the methods used to measure GIS benefits.
7. What is design creep? Explain how it can be avoided.
8. Describe the importance of a flexible, open-ended GIS design process that emphasizes educating the users prior to implementation. How does this help eliminate design creep?
9. List and describe the potential benefits of GIS in an organizational framework. Create a fictitious organization and discuss the potential benefits for it.
10. What are SIPs? How are they related to the input to a GIS? How do they drive the implementation of a GIS?

11. Describe and diagram the decision system matrix approach to organizing local views of a GIS. Do the same for the organizational chart approach. Suggest how these methods might be used for later view integration.

12. What major database design considerations do we need to look at for GIS design? Define some general rules for each.

References

Boehm, B., 1981. *Software Engineering Economics*. Englewood Cliffs, NJ: Prentice-Hall.

Burrough, P.A., 1986. *Geographical Information Systems for Natural Resources Assessment*. New York: Oxford University Press.

Calkins, H.W., and D.F. Marble, 1987. "The Transition to Automated Production Cartography: Design of the Master Cartographic Database." *American Cartographer*, 14(2):105–119.

DeMers, M.N., 1992. "Resolution Tolerance in a Forest Land Evaluation System." *Computers, Environment and Urban Systems*, 16:389–401.

DeMers, M.N., 2002. *GIS Modeling in Raster*. New York: John Wiley & Sons, Inc.

Marble, D.F., 1994. *An Introduction to the Structured Design of Geographic Information Systems* (Source Book, Association for Geographic Information). London: John Wiley & Sons.

Mitchell, A., 1999. *The ESRI Guide to GIS Analysis: Volume I: Geographic Patterns & Relationships*. Redlands, CA: ESRI Press.

Rhind, D.W., and N.P.A. Green, 1988. "Design of a Geographical Information System for a Heterogeneous Scientific Community." *International Journal of Geographical Information Systems*, 2(2):171–189.

Royce, W.W., 1970. "Managing the Development of Large Software Systems," *Proceedings, IEEE Wescon*, pp. 1–9.

Tomlinson, R.F., n.d. *The GIS Planning Process*. Ottawa, Canada: Tomlinson Associates Ltd., Consulting Geographers.

Tomlinson, R.F., 2003. *Thinking About GIS*. Redlands, CA: ESRI Press.

Wellar, B., 1994. "Progress in Building Linkages Between GIS and Methods and Techniques of Scientific Inquiry." *Computers, Environment and Urban Systems*, 18(2):67–80.

Yourdon, E., 1989. *Modern Structured Analysis*. Englewood Cliffs, NJ: Prentice-Hall.

APPENDIX A

Software and Data Sources

This appendix is a sample of available GIS software and sources of data for use with the software. Although this list is not complete, it should be enough to get you started either in a learning setting or a commercial operation. Additional data sources and vendors can be found by searching through some of the new GIS, remote sensing, and surveying trade journals, as well as through your instructor.

GIS SOFTWARE VENDORS

Arc/Info, Arc/View, MapObjects, Arc/GIS and others
Environmental Systems Research Institute
380 New York Street
Redlands, CA 92372
(714) 793–2853

Spans GIS
Intera TYDAC Technologies Inc.
1500 Carling Avenue, Ottawa, Ontario K1Z 8R7
Canada
(613) 722-7508

GeoMedia
Integraph Corporation
Huntsville, AL 35894
(205) 730–2000

ERDAS IMAGINE
ERDAS, Inc.
2801 Buford Highway
Atlanta, GA 30329
(404) 248-9000

GRASS

GRASS Information Center
USACERL
Attn: CECER-ECA
P.O. Box 9005
Champaign, IL 61826-9005

Atlas GIS

Strategic Mapping Inc.
4030 Moorpark Avenue
San Jose, CA 95117-4103
(408) 985-7400

GisPlus

Caliper Corporation
1172 Beacon Street
Newton, MA 02161
(617) 527-4700

MapInfo

MapInfo Corporation
200 Broadway
Troy, NY 12180-3289
(518) 274-6000

IDRISI

The Clark Labs for Cartographic Technology and Geographic Analysis
Clark University
950 Main Street
Worcester, MA 01610
(508) 793-7526

MapGrafix

ComGraphix, Inc.
620 E Street
Clearwater, FL 34616
(813) 443-6807

OSU Map-for-the-PC

Duane Marble
Department of Geography
The Ohio State University
Columbus, OH 43210

DATA SOURCES

U.S. Department Of Agriculture (USDA)
Natural Resources Conservation Service (NRCS)

The following information is available from the National Cartographic and Geographic Information Systems Center

USDA-NRCS
P.O. Box 6567
Fort Worth, TX 76115
(817) 334-5559
Fax: (817) 334-5290

Soil Survey Geographic Database (SSURGO). This is the most detailed soil mapping for the United States, ranging in scale from 1:12,000 to 1:31,680. It duplicates the original soil survey maps found in county soil survey documents.

State Soil Geographic Database (STATSGO). Generalized SSURGO data at a scale of 1:250,000, which covers state, regional, and multicounty areas.

National Soil Geographic Database (NATSGO). Data based on generalized state-level soils maps and formed from the major land resource area (MLRA) and land resource region (LRR) boundaries. Digitized at 1:7,500,000.

U.S. Department Of Commerce
Bureau of the Census

Census data in the form of TIGER files and Summary Tape File content data are available from

Customer Services
Bureau of the Census
Washington Plaza, Room 326
Washington, D.C. 20233
(301) 763-4100
Fax: (301) 763-4794

National Oceanic and Atmospheric Administration (NOAA)
National Environmental Satellite, Data, and Information Service (NESDIS)

U.S. weather records, collected by the National Climatic Data Center (NCDC), and international weather records, collected by the World Data Center A (WDC-A), are available from:

NOAA
National Climatic Data Center
Climate Services Division
NOAA/NESDIS E/CC3
Federal Building
Asheville, NC 28801-2696
(704) 259-0682
Fax: (704) 259-2876

National Geophysical Data Center (NGDC)

Solid earth, solar-terrestrial physics, and snow and ice data from worldwide sources are available from

National Geophysical Data Center
Information Services Division
NOAA/NESDIS E/GC4
325 Broadway
Boulder, CO 80303-3328
(303) 497-6958
Fax: (303) 497-6513

National Oceanic Data Center (NODC)

Oceanographic station data, bathythermograph, current, biological, sea surface, and GEOSAT data are available from

National Oceanographic Data Center
User Services Branch
NOAA/NESDIS E/OC21
1825 Connecticut Avenue, NW
Washington, D.C. 20235
(202) 606-4549
Fax: (202) 606-4586

National Ocean Service (NOS)

Sailing charts, coast charts, and harbor charts are available from NOS. For information, contact

Coast and Geodetic Survey
National Ocean Service, NOAA
1315 East-West Highway, Station 8620
Silver Spring, MD 20910-3282
(301) 713-2780

U.S. Department of The Interior
U.S. Geological Survey (USGS)

Digital line graphs, digital elevation models, side-looking airborne radar (SLAR), advanced very high resolution radiometer (AVHRR), digital orthophotoquad, land satellite (LANDSAT) data, système probatoire d'observation de la terre (SPOT) data, defense mapping agency (DMA) data, national wetlands inventory (NWI) data, and national biological service (NBS) data are all accessible from

Earth Science Information Center
U.S. Geological Survey
507 National Center
Reston, VA 22092
(703) 648-5920
Fax: (703) 648-5548
Toll free: 1-800-USA-MAPs

or

Sioux Falls ESIC
U.S. Geological Survey
EROS Data Center
Sioux Falls, SD 57189
(605) 594-6151
Fax: (605) 594-6589

Canadian Data

National Digital Topographic Data are available from

Topographic Surveys Division
Surveys and Mapping Branch
Energy, Mines and Resources Canada
615 Booth Street
Ottawa, Ontario K1A 0E9, Canada
(613) 992-0924

Data from the Canada Land Data System (including the Canada Geographic Information System) are available from

Environmental Information Systems Division
State of the Environment Reporting Branch
Environment Canada
Ottawa, Ontario K1A 0H3, Canada
(613) 997-2510

Canadian soils information system data are available from

CanSIS Project Leader
Land Resource Research Centre
Agriculture Canada, Research Branch
K.W. Neatby Building
Ottawa, Ontario K1A 0C6, Canada
(613) 995-5011

Landsat and SPOT data in Canada are available from

Canada Centre for Remote Sensing
2464 Sheffield Road
Ottawa, Ontario K1A 0Y7, Canada
(613) 952-2171

or

Canada Centre for Remote Sensing
Prince Albert Receiving Station
Prince Albert, Saskatchewan S6V 5S7
(306) 764-3602

Census data for Canada are available from

User Summary Tapes
Electronic Data Dissemination Division
Statistics Canada
9th Floor, R.H. Coates Building
Ottawa, Ontario K1A 0T6, Canada
(613) 951-8200

APPENDIX B

Using the World Wide Web to Find Data and GIS Examples

In Appendix A, I described a number of locations for data available through the mail. Many of these sources of data are now accessible using the World Wide Web (WWW). In fact, even a cursory perusal of the net using the key words geographic information system will provide a wide array of data sources, listserves, interest groups, class materials, active GIS research agendas, GIS output, sources of GIS employment, software and hardware vendors, and the like. A major problem with the Web is that you can spend an enormous amount of time surfing before you find the object of your search. I have compiled a list of sources that have come to my attention while searching. My intent is to shorten the amount of time you spend surfing and increase your productivity. This list is available directly from the New Mexico State University geography department Web page. The universal resource locator (URL) is http://www.nmsu.edu/~geoweb/gisrc.html.

Among the more useful Websites for finding and sharing data is the GIS Data Depot at http://www.gisdatadepot.com. They call themselves the "... largest geo-spatial data repository on the Web!" The site provides sources for both free data and data for purchase. In many cases the data for purchase are bundled to reduce the cost. Additionally, data can be downloaded directly from the Web in some cases, or even created as a custom CD.

Another recent source of GIS data, often for free or in-kind sharing is the geography network at http://geographynetwork.com. Ranging from academic sources, to professional data providers and from ready-to-use datasets to software, this growing site is sure to be an important resource for both teaching and research.

GLOSSARY

Absolute barriers: Barriers that prevent movement through them.

Absolute location: Location on the earth or on a map that has associated with it a specific set of locational coordinates.

Accessibility: A measure of arrangement that focuses on the connectedness of an area object to other area objects.

Accumulation threshold: Low points within the calculation of accumulated distance within a statistical surface (usually a topographic surface) that result in a downward movement or pouring of a surface substance.

Address matching: The process of defining exact addresses and linking them to specific locations along linear objects.

Affective objective: One of two parts of the purpose of a map. This one relates to the overall appearance of the map and controls how information contained in a map is to be portrayed. *See also* intellectual objective.

Affine transformation: A geometric transformation that scales, rotates, skews, or translates images or coordinates within or between coordinate spaces.

Aggregation: One of a number of numerical processes that place individual data elements into classes.

Alber's equal area projections: Widely used equal area conic projections that contain two standard parallels. Good for midlatitude areas of greater east–west than north–south extent.

Alidade: A telescopic survey device that uses magnetic bearings for determining locations on the earth. It is commonly used with an associated traverse table or plane table, so scaled values can be recorded directly on paper or mylar.

Allocate: The process of selecting a portion of geographic space to satisfy a predefined set of locational criteria. This process is usually performed on a vector data structure for the purpose of selecting a portion of a network.

Allocation: A process, often performed in network systems, that assigns entities, edges or junctions to features until the capacity or impedance limit is reached.

Alpha index: A measure of network complexity that examines network circuitry.

AM/FM: Automated mapping/facilities management. Computer-assisted cartography, particularly applied to the display and subsequent analysis of facilities associated with the function of urban and rural areas.

Analytical paradigm: A conceptual model, sometimes called the holistic paradigm, where the map is viewed both as a means of graphic communication and as a means of numerical spatial analysis.

Angle of view: The angular distance above the ground from which a perspective map can be shown.

Angular conformity: The property of some map projections that retain all angular relations on the map.

Application developers: GIS professionals specializing in the development of specific databases, analytical functional capabilities, and appropriate graphical user interfaces to allow specific applications to be performed by non-GIS specialists.

Application requirements model: A database design model focusing on the application needs as a means of properly completing the design.

Applications development: The process employed by a group of GIS professionals to allow non-GIS users easy access to the technology for specific tasks.

Arbitrary buffers: Buffers whose distances are selected without regard for any known criteria.

Area cartograms: Cartograms that vary the size of the polygons on the basis of the statistical value represented, rather than on their actual areal dimensions.

Areal correspondence: The percent overlap of one feature with another.

Arrangements: 1. A number of measurements indicating how spatial data are organized through such attributes as nearness, connectivity, and others. 2. The spatial distribution of internal markings for cartographic symbols to allow visual separation.

Array: A data structure in which a variable, with multiple, sequentially indexed cells that can each store a value, usually a variable, of a similar type.

Aspect: The azimuthal direction of surface features.

Association: Spatial relationship that exists between different elements of the earth that occur at the same locations. An example would be a particular vegetation type occurring on north-facing slopes.

Attribute error: Incorrect or missing attributes.

Attribute pseudo nodes: Spatial pseudo nodes that are a result of explicit attribute changes along a line.

Attributes: Nongraphic descriptors of point, line, and area entities in a GIS.

Audience: Map design consideration based on the background and map reading skills of the potential map user.

Available data: Map design limitation based on the quality, quantity, and reliability of the data used for mapping.

Azimuthal equidistant projections: Types of azimuthal map projections in which the linear scale is uniform along the radiating straight lines through the center. Allows for the entire sphere to be shown on a single projection.

Azimuthal projections: A family of map projections resulting from conceptually transferring the earth's coordinates onto a flat surface placed perpendicular to the sphere.

Barrier: An object whose attributes either stop or impede movement through the rest of the coverage.

Baselines: In the U.S. Public Land Survey, a line passing east west through an origin, used to establish Township and Range corners.

Bifurcation ratio: A method of classifying streams based on the relative positions of trunk streams and tributaries and connectivity within the network.

Binary maps: Maps that contain only two possible values for the same characteristic or attribute, either present or absent.

Biophysical mapping: A traditional analog technique used by environmental planners to combine relevant mapped biological and physical data into meaningful combinations for decision making.

Block codes: A compact method of storing raster data as multifaceted blocks of homogeneous grid cells.

Block function: GIS modeling function that uses a manipulation of the properties of a defined, nonoverlapping block of cells to produce an output block of cells showing the new values.

Boolean overlay: A type of overlay operation that relies on Boolean algebra.

Borders: Higher-order groups of line objects that exemplify some functional or political demarcation from one region to another.

Bottom-up implementation: One of the major weaknesses of the classical (waterfall model) project life-cycle design methodology: final system testing is left until last, thus requiring large chunks of uninterrupted computer time. Because such large chunks of computer time are often hard to obtain, the development often falls behind schedule.

Buffering: The process of creating areas of calculated distance from a point, line, or area object.

CAC: *See* computer-assisted cartography.

CAD: *See* computer-assisted drafting (CAD).

Cadastral: Having to do with cadastre, which involves interests in landownership and management.

Canada Geographic Information System: The first major operational geographic information system. It was designed in the early 1960's to document and manage the vast resources of much of Canada.

Cartesian coordinate system: A mathematical construct defined by an origin and a unit distance in the X and Y direction from that origin.

Cartograms: Maps that emphasize the communication paradigm to a degree that often modifies actual geographic space in deference to the phenomena being displayed.

Cartographic database: A digital database developed from an existing cartographic document. This process results in the appearance of increased accuracy where it does not necessarily exist.

Cartographic modeling: The process of combining individual GIS analytics to create complex models used for decision making.

Cartographic process: The steps involved in producing a map, beginning with data collection and ending with the final map product.

Causative buffers: Buffers whose distances are selected on the basis of some physical phenomena that dictate their location.

Census: A method of collecting data that normally involves tabulating data about an entire population rather than a sample of the population.

Center of gravity: Center of a polygon based on the arrangements of objects within it rather than simply as a measure of the area geometry.

Central place: A node that exhibits a high degree of network linkage intensity.

Central point linear cartograms: Cartograms that show distances from a central point as concentric lines, thus requiring the underlying graphic objects to be modified to conform to those lines.

Centroid-of-cell method: A method of raster encoding where a grid cell is encoded based on whether or not the object in question is located at the exact centroid of the geographic space occupied by each cell.

Centroids: For graphical objects, the exact geometric center.

Chains: In the POLYVRT vector data model, collections of line segments that begin and end with specific nodes that indicate topological information such as to and from directions as well as left and right polygons.

Changes of dimensionality: The process of either increasing or decreasing the dimension of cartographic objects, often on the basis of the scale at which the map is produced. For example, a line (1 dimension) at one scale may be represented as an area at a substantially larger scale.

Chi-square: A statistical test that examines the relationship between observed and expected distributions of objects.

Chroma: The perceived amount of white in a hue compared with a gray tone of the same value level.

Circuitry: The degree to which nodes are connected by circuits of alternative routes. Defined by the alpha index.

Circuits: Type of network that allows movement to and from the initial point.

Circular variance: The circular analog-to-standard deviation. A measure of the degree of variation from a mean resultant length.

Class interval selection: The process of using a variety of statistical techniques to group cartographic data into classes for cartographic display. This data aggregation results in loss of data for analysis, but it is essential for many map production purposes to allow readability of the map.

Classed choropleth map: The traditional method of value by area mapping that first aggregates the data into classes before the polygons are assigned color or shading patterns.

Classless choropleth mapping: A proposed mapping technique that assigns a color or shading pattern to each polygon on the basis of its unit value rather than grouping polygons into classes.

Clip: The process of cutting a portion of one map based on a template (clip cover) that is often a portion of another map.

Clip cover: A template composed of the outside boundary of part or all of one map that is used to select the same geographic area during the process called clip.

Closed cartographic form: Polygonal object whose entire extent is contained within a single map.

Clustered: Spatial arrangement of objects where they occur as groups located close to one another, leaving large empty spaces between. This type of arrangement indicates that the processes operating on the objects are different near the clusters than in the intervening space.

Clustering: The process that produces a spatial arrangement where some objects are located very close to one another while others are widely separated.

Color video: A passive remote-sensing device designed to produce color images of the ground surface and store it on videotape for later retrieval and analysis.

Communication paradigm: A conceptual model where the map is viewed primarily as a means of graphic communication.

Communities: Higher-order groups of objects that either occur in a particular pattern that differs from their surrounding objects or that are linked as a functional region.

Complement overlay: A set theoretic overlay method that results in those areas from the first map that are not included in the second map.

Computer file structures: Any of a number of methods of storing computer data on disk or in virtual memory.

Computer-assisted cartography (CAC) systems: A computerized system designed primarily to assist in the production of maps.

Computer-assisted drafting (CAD): A set of computer programs designed to assist in the process of drafting. It is normally used for architectural purposes but can also be used for drafting maps and as an input to GIS.

Concavity: A measure of the degree to which an area object produces concave shapes along its perimeter.

Conceptual data model: An abstract way of representing computer data whether they be entity or attribute data.

Conceptual data schema: An abstract way of representing computer data within a database or database management system.

Conditional overlay: A method of map overlay that selectively determines which numbers will compare and how those numeric comparisons will be coded in the output map.

Conditions of use: Map design consideration based on such factors as viewing distance and lighting that limit the readability of the document as it is used.

Confirmatory analysis: A type of analysis that attempts to confirm or reject a formal belief or working hypothesis.

Conflation: The computational equivalent of stretching a map until its internal components can be rectified. *See also* rubber sheeting.

Conformal: A type of map projection that attempts to preserve the shapes of objects for small areas.

Conformal projection: A type of map projection that maintains all angular relations normally found on the reference globe.

Conformal stereographic projections: Azimuthal projections that exhibit symmetrical distortion around the center point. Useful when the shape of the area to be mapped is relatively compact.

Conformality: The property of preserving shapes of objects on small areas of the earth when they are represented by projections.

Conical projections: A family of map projections resulting from conceptually transferring the earth's coordinates onto a cone.

Connectivity: The degree to which lines in a network are linked to one another.

Constrained math: Any of a number of mathematical functions applied to cartographic coverages to reclassify their attributes.

Contiguity: A measure of the degree of wholeness within a region or of the degree to which polygons are in contact with one another.

Contiguous regions: Regions that are defined on the basis both of category homogeneity and of all parts being in direct contact with one another.

Continuous: Data that occur everywhere on the surface of the earth, for example, temperature.

Continuous surface: Any statistical surface whose values occur at an infinite number of possible locations.

Contour interval: Class interval illustrating the difference in elevation between contour lines.

Contour line: An isarithm connecting points of identical elevation.

Control points: Any points on a cartographic document for which the geographic coordinates are well known and reasonably accurate. These points are often used during the process of coregistration of two or more coverages or for adjusting the locations of other objects on the same coverage through conflation.

Controls: The limitations on the map design process.

Conventions: For maps, agreed-upon rules that have been carefully selected and are generally accepted by the mapping community.

Convexity: A measure of the degree to which an area object produces convex shapes along its perimeter.

Coregistered: The result of the process of precisely locating the coordinates for two or more GIS coverages so that their spatial locations match each other.

Coverage: The common terminology signifying a single thematic map in a multimap GIS database. Sometimes referred to as a data layer or an overlay.

Cross-sectional profile: A method of visualizing statistical surfaces by examining one transect through the surface viewed from the side.

Crosswalk: The process of matching disparate categories between maps or GIS coverages.

Cumulative distance: Usually associated with cost surfaces. The accumulation of cost in terms of time, energy, and so on as one travels from place to place.

Cut-and-fill analysis: A GIS analysis that attempts to determine the volume of selected portions of the earth.

Cylindrical projections: A family of map projections resulting from conceptually transferring the earth's coordinates onto a cylinder.

Dangling node: A node located at the end of an undershoot.

Dasymetric mapping: A variety of techniques designed to improve area homogeneity for choropleth maps.

Dasymetry: *See* Dasymetric mapping.

Data dictionary: Detailed description of the data contents of a database, with particular attention being paid to explanations of categories.

Data requirements model: A database design model focusing on the required data needs as a means of properly completing the design.

Database: A collection of many files associated with a single general category.

Database management system (DBMS): Any of a variety of computer organizational structures that allow search and retrieval of individual files or items in the database. These systems are generally of three primary types: hierarchical, network, and relational.

Database structures: A set of approaches to organizing large collections of computer files commonly relating to a single major subject. Database structures are designed to enable the user to store, edit, search, and retrieve files or individual pieces of data within those files.

Datum: A reliable starting point from which accurate measurements of the earth's size can be made.

DD: *See* Decimal degrees.

Dead reckoning: A method of surveying that measures distance and direction from one point to the next.

Decimal degrees (DD): Decimal degrees method of designation for geocoded objects.

Decision system matrix: An aid to determining system design requirements, especially for linking spatial information products to necessary data elements.

Deductive models: Cartographic models that move from a general goal to the selection of individual components needed to achieve it.

Degrees, minutes, and seconds (DMS): Degrees, minutes, seconds method of designation for geocoded objects.

DEMs: *See* Digital elevation models.

Density: Measure of the number of objects per unit area.

Density of parts: A type of dasymetric mapping that uses improved quantitative information about subarea density on maps to calculate improved information about the remaining polygons.

Density zone: Polygons that contain a selected range of density of occurrence of objects.

Density zone outlining: A type of dasymetric mapping that creates neighborhoods of uniform density of cartographic objects that are then mapped as individual polygons. Also known as density of parts.

Depressionless surface: A statistical surface that assumes the absence of small depressions or interruptions to allow for general surface analysis.

Descriptive model: A type of cartographic model that describes an outcome based on a set of conditions.

Design creep: Process that occurs when the design process exceeds the organizational learning curve. Usually results in a system that is overdesigned and offers more than the client needs.

Differential rectification: Process used to adjust aerial photographs for planimetric errors due to changes in aircraft altitude and elevational features on the ground.

Diffusion: A spatiotemporal process of objects moving from one area to another through time, or of additional objects appearing where they had not previously existed.

Digital elevation model (DEM): Digital model of landform data represented as point elevation values. Also called digital terrain model (DTM).

Digital line graphs: Digital representations of the graphics contained on USGS topographic maps. These do not include topographic data as one might find in a digital elevation model.

Digital orthophotoquads: Digital version of aerial photographs that are constructed to eliminate image displacement due to changes in aircraft tilt and topographic relief.

Digital photography: A passive remote-sensing device that produces still digital images similar to that from video, but as single frames.

Digitizers: Electronic devices designed to transfer analog cartographic data to digital form.

Direct files: Within an indexed file structure, a condition in which the data items themselves provide the primary order of the files.

Directed network: Type of network that places restrictions on the direction of movement (e.g., a one-way street).

Directed sampling: Sampling where certain attribute qualities or quantities dictate that they must be sampled to properly represent the overall population.

Directional filter: A high-pass filter that enhances linear objects that lie in a particular direction (e.g., NE–SW).

Dirichlet diagrams: *See* Thiessen polygons.

Discrete: Data that occur only in selected places on the earth's surface. An example is human population.

Discrete altitude matrix: A lattice of point values representing elevation that is used to model topography in a GIS.

Discrete surface: Any statistical surface whose values are limited to selected locations.

Dispersed: A spatial arrangement that exhibits the greatest possible distance between objects in a confined space.

Dispersion: A measure of arrangement that focuses on the relationship between one area and its neighbors, especially regarding the average distances and average density of patches in the map.

Distributions: The frequencies or numbers of occurrences of objects within a particular area.

DMS: *See* Degrees, minutes, and seconds.

Dominant type method: A method of raster encoding where a grid cell is encoded based on whether or not the object in question occupies greater than 50 percent of the geographic space occupied by each cell.

Dot distribution map: *See* Dot mapping.

Dot mapping: The cartographic technique of representing one or more objects as points on a map, and where the exact geographical locations are not precisely recorded.

Doughnut buffer: A series of buffers of varying distance produced one inside the other.

Drift: The random, spatially correlated elevational component for a Kriging model. The general trend in elevational values.

Dumpy level: A telescopic survey device that has an elongated base with a movable joint that allows vertical movement so that elevations can be measured easily in steep terrain.

Easting: On some grid systems, a measurement of distance east of a preselected standard starting meridian.

Edge enhancement: A type of filter, often called a high-pass filter, which enhances values that change rapidly from place to place so that these changes can easily be observed.

Edge matching: The process of aligning the edges of two or more cartographic documents.

Edginess: Measure of the amount of edge in a polygonal cartographic form.

Ellipsoid: Hypothetical, nonspherical shape of the earth that resembles a flattening of the earth at the poles due to rotation. Basis for datums.

Encode: To place analog graphic data into a form that is compatible with computer cartographic data structures.

Entities: Points, lines, and polygons as they are represented by computerized cartographic data structures.

Entity–attribute agreement error: Related to attribute error. A mismatch between entity objects and the attributes assigned to them.

Entity error: Error of position for cartographic objects.

Equal area (equivalent) projections: A group of map projections that maintain the property of equal area for cartographic objects. These projections are useful for small-scale general reference and instructional maps.

Equidistant projections: A group of map projections that maintain the property of equal distance along straight lines radiating through a central starting point from which the projection was made.

Equivalent projections: *See* equal area projections.

Euclidean distance: Distance measured simply as a function of Euclidean geometric space.

Euler function: Mathematical measure of spatial integrity that compares the number of regions to the number of perforations within all the regions.

Euler number: Number resulting from the implementation of the Euler function.

Exclusionary variables: Coverage variables whose importance outweighs the results of normal Boolean overlay operations for the purposes of decision making.

Exploratory analysis: A form of analysis in which a data rich environment encourages exploration without initial hypotheses to direct that search.

Extended neighborhoods: Regions that extend beyond those immediately adjacent to the focal point of analysis.

Extrapolation: The process of numerically predicting missing values by using existing values that occur on only one side of the point in question.

False eastings: On some coordinate systems, arbitrary, large values given to the *Y* axis of the false origin to allow only positive distance measured east of that point.

False northings: On some coordinate systems, arbitrary, large values given to the *X* axis of the false origin to allow only positive distance measured north of that point.

False origins: Arbitrary starting points on a rectangular coordinate system designed to allow only positive distances east and north.

Federated databases: Databases whose component parts are derived from disparate sources, each with its own data quality standards.

Field: The place in a database or graphical user interface (GUI) in which data can be entered. In relational database management systems such places are the columns in the tables that store data for a single variable.

Filter: In raster GIS and digital remote sensing, a matrix of numbers used to modify the grid cell or pixel values of the original through a variety of mathematical procedures.

First normal form: Based on the theory of normal forms, the first requirement that all tables must contain rows and columns, and that the column values cannot contain repeating groups of data.

Fishnet maps: A term often applied to wireframe diagrams but representing cartographic statistical surfaces.

Flow accumulation: The additive buildup of surface data with distance.

Flowchart: A graphical device that illustrates the exact coverages and data elements, the operations to be applied to each, and the order in which the operations are applied to produce a cartographic model.

Focal function: GIS modeling function that uses properties of an immediate neighborhood for characterizing the target cell.

Foreign key: In relational database management systems, a column in a secondary table linked by a primary key in the primary table that is being used to join two tables.

Formulation flowchart: A type of GIS flowchart that begins with the expected output and works backward to determine all the elements needed to populate the database.

Fragmented regions: Regions based on category homogeneity but with parts spatially separate from each other.

Free sampling: In the join count statistic, the method of testing that assumes that we can determine the expected frequency of, within, and between category joins based on theory or known patterns.

Freeman–Hoffman chain codes: Compact raster data models that use eight unique directional vectors to indicate the directional orientation or change of linear features.

Friction surface: Some assigned attribute on a surface representation of a portion of the earth that acts to impede movement.

Functional distance: Distance measured as a function of difficulty rather than as a simple geometric distance.

Fuzzy tolerance: A user-defined distance of error to allow for minute digitizing mistakes.

Gamma index: A measure of complexity of a network that compares the number of links on a given network to the maximum possible number of links.

GBF/DIME: Geographic base file/dual independent map encoding system. A topological vector data model created by the U.S. Census Bureau and based on graph theory.

General reference map: A map, usually at small scale, whose primary objective is to show locations of different features on a single document.

Geocaching: An outdoor treasure hunt where participants use GPS and other navigational methods to hide and seek containers (geocaches) throughout the globe.

Geocoding: The process of inputting spatial data into a GIS database by assigning geographical coordinates to each point, line, and area entity.

Geocomputation: The use of computer technology to solve spatial problems. Often called quantitative geography.

Geodatabase: A collection of geographic objects used within the object oriented ArcGIS© software.

Geodetic framework: A carefully measured system of ground-based coordinates designed to ensure accurate locations for cartographic documents.

Geodetic reference system: *See* GRS80.

Geographic data measurement: The processes and data levels that combine to characterize data observed on the surface of the earth.

Geographic database: A digital database developed from field-based spatial data. This process results in decreased accuracy and increased generalization because of the sampling necessary to produce it.

Geographic Grid: A reference system that uses latitude and longitude to determine the locations of objects on the surface of the earth.

Geographic visualization (GVIS): A spatial approach to scientific visualization whereby the cartographic output is designed to elicit a response from the map reader that results in the formulation of new scientific hypotheses.

Geographically referenced: The condition where objects are linked explicitly to geographic locations on the surface of the earth.

geoid: Generalized physical model of the earth based on compilations of all variations from the ellipsoid.

Geostatistics: A class of statistical techniques employed to analyze and predict values associated with spatial or spatio-temporal phenomena.

GIS supplier: A source, (often either an agency or business), of geospatial data used within a geographic information system.

GIS systems analysts: GIS professionals specializing in the proper design of the overall GIS operations.

Global positioning system (GPS): A satellite-based device that records locational (X, Y, Z) and ancillary data for portions of the earth.

Global view: Organizational view of a GIS database requiring the integration of multiple local views.

Gnomonic projections: Azimuthal projections where all great circles are represented as straight lines. Useful for marine navigation.

GPS: *See* Global positioning system (GPS).

Graph theory: A mathematical theory that examines the relationships among linear graphs.

Graphic data structure: A method of storage of analog graphical data into a computer that enables the user to reconstruct a close approximation of the analog graphic through some output procedure.

Graphic overlay: A type of overlay in which the graphics entities are clearly shown in the output but where a new, unique set of data is not produced.

Graphic scale: One method of representing map scale where a preselected, real-world distance is shown as a bar or line on the map document. Sometimes called a bar scale.

Graphicacy: The level of understanding of graphic devices of communication, especially maps, charts, and diagrams.

Gravity model: A measure of the interaction of nodes based on their distance and some functional measure of their individual importance.

Grid: A network of parallel and perpendicular lines used for determining location on the surface of the earth.

Grid cells: Raster data structures in which the geographic space is quantized into rectangular shapes of equal size and shape.

Grid system: A system of horizontal and vertical lines on a globe or projected map that allow for locating objects in geographic space and making measurements among objects.

GRID/LUNR/MAGI: Raster GIS data model where each grid cell is referenced or addressed individually and organized in columnar fashion.

Ground control points (GCPs): Points of known geographic location used to register satellite imagery and other coverage data to the geodetic framework.

GRS80: Standard geographic reference system based on NAD83.

Heterogeneity: Opposite of homogeneity. The amount of diversity within a selected region.

Hierarchical data structures: Computer database structures employing parent child or one-to-many relationship that requires direct linkages among items for a search to be successful.

Hierarchical structure: The graphic appearance of some cartographic objects appearing to be more important than others, resulting from one or more cartographic techniques to produce this result.

High-pass filter: *See* Edge enhancement.

Holistic paradigm: *See* Analytical paradigm.

Homogeneity: The degree to which attributes in a region are similar. If all attributes are identical within a selected region, they are said to be homogeneous.

Homogeneous sampling: A process of sampling where individual polygonal areas containing some homogeneous mix of attributes are selected for sampling.

Hue: The color of a cartographic symbol based on the wavelength of electromagnetic radiation reaching the viewer.

Hybrid systems: GIS systems whose entity and attribute tables are separate and are linked through a series of pointers and identification codes.

Identity overlay: The use of tabulations of data from two or more coverages to make decisions about how attribute values are to be combined. This technique can be employed to enhance a number of other methods of overlay by providing a decision tool for their use.

IMGRID: Raster GIS data model where grid cells are referenced as part of a two-dimensional array and the thematic attributes are coded in Boolean fashion (i.e., 1 or 0 only).

Immediate neighborhoods: Regions limited to those immediately adjacent to the focal point of analysis.

Immersive GIS: A virtual world within a computer system in which the user of the software is placed within the world they are analyzing.

Impedance value: A user-specified numerical value used to simulate the effects of barriers or friction surfaces. These are also employed in the same way for network modeling, where the impedance values indicate the degree to which a network allows travel.

Implementation flowchart: A type of flowchart that moves from individual data elements toward the final expected outcome.

Incorrect attribute values: Attribute error resulting from the assignment of incorrect or nonexistent attribute values to entities.

Incremental distance: The simple addition of distance at each successive step.

Indexed files: Computer file structures that allow faster search than ordered sequential because each entry is assigned an index location through the use of software pointers.

Inductive model: A method of cartographic model that begins with individual observations (data elements) and proceeds to develop general patterns.

Information theory: A field of study that deals with information content. It is used in GIS primarily as an estimate of the amount of sampling necessary to encode a graphic object.

Innovation diffusion: A set of geographic models that examines the movement of ideas, innovations, or strategies through geographic space.

Inscribe: A geometric shape that is enclosed by and "fits snugly" inside another geometric shape.

Institutional design: Design issues concerning the role of institutional considerations such as training and personnel, data security, organizational functioning, and many others.

Institutional setting: The size, type, organizational hierarchy, goals, and overall objectives of each organization likely to adopt GIS.

Integrated systems: GIS systems whose entity and attribute tables are shared.

Integrated terrain unit (ITU) mapping: A unit of geographic space defined by the explicit collection of various attributes at one time, usually from aerial photography.

Intellectual objectives: Those aspects of cartographic design that attempt to preserve, as much as possible, the accuracy of representation.

Interactions: A measure of arrangement that focuses on the proximity, sizes, and amount of edge between neighboring areas, especially regarding the edges of these areas.

International date line: Meridian drawn at 180 degrees east and west of the prime meridian.

Interpolation: A process of predicting unknown elevational values by using known values occurring on multiple locations around the unknown value.

Interval: Comparative data with a relatively high degree of accuracy, but with an arbitrary starting point (e.g., degrees Centigrade).

Intervisibility analysis: Analytical technique that allows for the determination of visibility from one object to another and back.

Inverse map projection: The process of converting from two-dimensional (projected) map coordinates to geographical coordinates.

Inverted files: Within an indexed file structure, a condition in which the data items are organized by a second or topic file that provides the primary order of the files. In such a system, the topic or inverted file is searched rather than the data items themselves.

Irregular lattice: A series of point locations whose interspatial distances are not identical.

Isarithm: General term for a line drawn on a map to connect points of known continuous statistical surface values.

Isarithmic map: The result of the process of isarithmic mapping.

Isarithmic mapping: The process of using line symbols to estimate and display continuous statistical surfaces.

Island pseudo nodes: Spatial pseudo nodes that result when a single line connects with itself.

Isolated State Model: An early model to explain the arrangement of agricultural activities on the basis of their distance to a single market.

Isolation: A measure of arrangement that focuses on the distance between objects in geographic space.

Isoline: A line connecting points of known or predicted equal statistical surface value.

Isometric: A form of data point selection within isoplethic map production in which absolute locations are used for sampling data points.

Isometric map: A map composed of isolines whose known Z values are sampled at point locations.

Isoplethic map: A map composed of isolines whose known Z values are recorded for polygonal areas rather than at specific point locations.

Isotropic surface: A measure of simple Euclidean distance from some central point outward throughout a coverage where no obstructions or frictional changes exist.

ITUs: Integrated terrain units.

Kriging: An exact interpolation routine that depends on the probabilistic nature of surface changes with distance.

Lacunarity: One counterpart to the fractal dimension that describes the texture of a map. It deals with the size and distribution of holes within an area whereby a map with large holes has high lacunarity.

Lag: On a semivariogram, the distance between sample locations, plotted on the horizontal axis.

Lambert's conformal conic projections: Conformal map projections that have concentric parallels and equally spaced, straight meridians meeting the parallels at right angles. These provide good directional and shape relationships for east–west midlatitudinal zones.

Lambert's equal area projections: Azimuthal equivalent projections with symmetrical distortion around a central point. Useful for areas that have nearly equal east–west and north–south dimensions.

Land information systems: Subset of geographic information systems that pay special attention to land-related data.

Latitude: Angular measurement north and south from the equator.

Least common geographic unit (LCGU): A computer graphics construct that provides topological information for vector overlay operations to determine the ability or lack of ability of polygons to be further divided.

Least convex hull: Straight-line segments connecting all outside points within a distribution in such a fashion that the smallest possible polygon is produced.

Least-cost distance: A distance measure based on the minimization of friction or cost for a single path.

Least-cost surface: A distance measure based on the minimization of friction or cost for an entire surface.

Legibility: The ability of a cartographic object to be visually observed and recognized by the map user.

Limiting variables: Variables that result in the absence of one or more attributes because of how they interact with one another. A method of dasymetric mapping.

Line dissolve: In vector GIS, the process of eliminating line segments between polygons so that they become larger polygons with identical attributes.

Line-in-polygon overlay: A method of overlay that combines a coverage with linear objects with a polygon coverage to determine the numbers and extent of linear objects that fall within selected polygons.

Line intersect methods: A group of techniques for analyzing the spatial distribution of linear objects by drawing one or more sample lines at random and noting where they intersect the coverage lines.

Line-of-sight analysis: A technique in which point to point visibility is determined along a topographic surface.

Linear cartogram: A form of cartogram in which the length of the line is directly proportion to the quantity being measured (e.g. time) rather than its actual linear distance.

Linear directional mean: The average directional value of a number of linear objects.

Linear interpolation: A type of interpolation linearity measure of the ability of a digitizer to be within a specified distance or tolerance of the correct value as the puck is moved over large distances.

Linearity: A measure of the ability of the digitizer to be within a specified distance (tolerance) of the correct value as the puck is moved over large distances.

Local: A type of raster GIS function in which the output values of grid cell values at particular locations are functions of the input values of grid cells at those same locations.

Local operator models: Methods of interpolation that rely heavily on existing elevational data that are located within close proximity to the points being predicted.

Local views: Individual users' views of a GIS database.

Locate: Determining the best location for some form of geographic activity, most often used for economic activities.

Location–allocation: A group of models designed to determine the best locations for activities in geographic space and to assign a portion of geographic space to existing facilities based on demand and location.

Locational information: Any information about the absolute or relative coordinates of points, lines, or areas. This information is often used to reclassify coverage values.

Logical data model: A representation of an organization's data, organized in terms of a particular data management system or system technology.

Longitude: Angular measurement east and west from the prime meridian.

Low-pass filter: *See* smoothing.

Mandated buffer: Buffers whose distances are dictated by legal mandate (e.g., frontage along homes dictated by zoning ordinances).

Map algebra: A mathematical language, based on a locational specific modification of matrix algebra, designed to work with raster GIS datasets and software.

Map analysis system (MAP): Raster GIS data model where grid cells are referenced as part of a two-dimensional array and the thematic attributes for each coverage are referenced by separate number codes or labels, thus allowing a range of values for each category in a single theme.

Map legend: The portion of the map document that describes the symbols. Conceptually, the map legend ties the entities and attributes together in an analog map document.

Map projections: Any of a number of approaches to transfer the spherical earth onto a two-dimensional surface. Each projection is an approximation and imposes its own limitations on the utility of the map.

Marble model: A spiral structured GIS design model that uses an increasing level of detail to ensure design flexibility.

Mathematical overlay: A method of overlay, usually associated with raster GIS, that uses mathematical and algebraic expressions to create new coverage attribute variables.

Mean center: Center of a group of point objects located in space derived by calculating the average *X* and *Y* coordinate distances.

Mean resultant length: The result of averaging the resultant length by the number of observed component vectors.

Measurable buffers: Buffers whose distances are selected based on exact measurements of existing phenomena as they occur in space.

Mercator projections: Conformal map projections introduced specifically for nautical navigation. It maintains the property that rhumb lines always appear as straight lines.

Meridians: Lines of longitude drawn north and south on the globe and converging at the poles.

Metadata: Data about data. An overall description of the contents of a database.

Method of ordinates: A simplified method for calculating volumes.

Missing attributes: Attribute error resulting from a failure to assign labels to point, line, or polygonal features or to grid cells.

Missing labels: Entity–attribute agreement error resulting from failure to label a polygon during the input process.

Mixed pixels: *See* mixels.

Mixels: Mixed pixels—pixels whose spatial extent is coincident with more than one category of objects on the earth. Mixels often result in category confusion during the classification process.

Model verification: The process of comparing expected results to observed results of a cartographic model to ensure it is performing as it should.

Models: In GIS an ordered set of map operations designed to simulate some spatiotemporal system in the real world.

Morton Matrix: A set of tiles that result from the ordered coding methodology known as Morton Sequencing.

Morton Sequencing: A four-fold clockwise coding system designed to provide a searchable filing system for keeping track of spatial tiles.

Most important type method: A nonsystematic form of raster encoding in which, when classification conflicts occur, the user determines which of the conflicting categories is the most important and will, therefore, be retained.

Mountain: A form of topographic feature that requires specification of the minimum height and some formal definition of the height to distance ratio to be classified as such. Definitions vary but usually contain these typical attributes.

Multiresolution Seamless Image Database (MrSID): It is a proprietary lossless data compacting methodology, that purports to provide rapid display of very large graphic datasets.

Mythical man-month: The erroneous concept that if a problem arises, the simple addition of more workers to the problem will result in its solution. This idea ignores the necessary learning curve for people asked to perform tasks with which they are unfamiliar.

Nearest neighbor analysis: A statistical test to compare the distance between each point object and its nearest neighbor to an average "between neighbor" distance.

Neighborhood functions: GIS analytical functions that operate on regions of the database within proximity of some starting point or grid cell.

Network complexity: Overall network patterning that combines the number of links and the degree of connectivity.

Network systems: Computer database structures employing a series of software pointers from one data item to another. Unlike hierarchical data structures, network systems are not restricted to paths up and down hierarchical pathways.

Networks: 1. Two or more interconnected computers often used to allow communications between them. 2. Higher order linear objects. Interconnected lines (arcs, chains, strings, etc.) defining the boundaries of polygons or a linear object that allows movement (e.g., road network).

Node: In vector data structures, a point that acts as the intersection of two or more lines (links) and explicitly identifies either the beginning or ending of each line to which it is attached.

Nominal: Level of data measurement that is noncomparative, usually representing a description or name.

Nondirected sampling: Data sampling that does not consider specific factors existing within a potential study area to make a determination of whether or not it should be sampled.

Nonfree sampling: In the join-count statistic, the sampling procedure that compares the number of joins of a random pattern of joins to those measured.

Nonlinear interpolation: A method of interpolation that accounts for the nonlinear nature of elevational change with distance.

Nonstationarity: In Kriging, the concept that statistical variables, in this case elevation values, are not stationary (do not stay the same) with horizontal distance.

Normal forms: In relational database management systems, a theory of table design that specifies what types of values columns may contain and how the table columns are to be dependent on the primary key.

North American datum (NAD27): Accepted U.S. datum as of 1927 based on a flatness ratio of the earth of 1/294.9787.

North American datum (NAD83): Datum established for the United States as of 1983, based on estimates of the center of the earth rather than surface measurements.

Northing: On some grid systems, a measurement of distance north of a preselected standard starting parallel.

Nugget variance: Spatially uncorrelated noise associated with a Kriging model.

Object-oriented database management systems: GIS systems based on object-oriented programming methods and demonstrating object inheritance.

Ontology: In GIS the precise utilization of words as descriptors of entities and attributes as they are encoded inside the computer and employed during analysis.

Ordered sequential files: Computer file structures that are ordered based on some form of alphanumeric scheme much like alphabetizing a mailing list. This file structure makes it more difficult for data input, but allows searches and retrieval with greater speed than with simple lists.

Ordinal: Ranked data (e.g., good, better, best) that are comparable only within a given spectrum.

Organizational diagram: Hierarchical diagram showing the structure of an organization from management to workers.

Orientation: 1. The azimuthal directions in which linear objects are placed. For area objects, this can also be applied to the major axis of the object. 2. The azimuthal directions of internal symbol markings used to separate one category from another.

Orthogonal: The normal point of view for viewing a cartographic document directly from above.

Orthographic projections: Azimuthal projections that look like perspective views of the globe at a distance. Useful for illustrations where the sphericity of the globe must be maintained.

Orthomorphic projections: *See* conformal projection.

Orthomorphic: A form of cartographic projection that preserves shape.

Orthophotoquads: Aerial photographs that are constructed to eliminate image displacement due to changes in aircraft tilt and topographic relief.

Overlay: The operation of comparing variables among multiple coverages.

Overshoot: An arc that extends beyond the arc with which it was meant to connect.

Parallels: Lines of latitude drawn east and west around the globe parallel to the equator.

Parsimony: The degree of simplicity. In GIS modeling the simplicity of the models produced to answer a particular question.

Passive remote sensing: A method of remote sensing that uses ambient energy from the object being sensed, rather than sending out a signal for later sensing.

Pattern: The regular arrangement of cartographic objects or internal symbol markings.

Percent occurrence method: A method of raster encoding where a grid cell is encoded so that it includes all categories falling within the geographic space occupied by each cell, as well as the percentage of each cell it occupies.

Perforated regions: Regions that demonstrate category homogeneity, but within which other dissimilar polygonal forms exist (e.g., islands within a lake).

Perimeter/area ratio: Method of measuring polygons that compares the perimeter of each polygon to its area as a ratio.

Physical data model: The actual way in which logical data models are explicitly implemented within the specified database management technology.

Pixels: In remote sensing, picture elements, each of which contains a preselected amount of geographic space and a set of electromagnetic measurements assigned to each. Pixels most often resemble grid cells when displayed at small scale.

Planar projections: Those map projections that are based on the idea that the spherical earth is transformed onto a tangent or secant plane.

Plane coordinates: *See* rectangular coordinates.

Plane table and alidade: A common traditional field device used for surveying and consisting of a table upon which a draft survey document would be placed, a telescopic device for determining line of site at a distance, and a tripod upon which they both rest.

Planimeters: Mechanical devices designed to measure lengths and areas on analog maps.

Point-in-polygon overlay: The process of overlaying a point coverage with a polygon coverage to determine which polygons contain which points.

Polygon: A multifaceted vector graphic figure that represents an area.

Polynomials: Mathematical equations with two or more terms, used for approximating surface trends. The more complex the polynomial, the more complex the surface estimate.

POLYVRT: Polygon converter model. A vector data model that uses explicit topological data stored separately for each type of entity.

Position measures: Analytical techniques that characterize cartographic objects on the basis of their absolute positions or of their positions and arrangements relative to other cartographic objects.

Pour points: *See* accumulation threshold.

Predictive models: A group of cartographic models (either descriptive or prescriptive) that rely on the prediction of outcomes given a set of conditions.

Prescribe: In GIS models, prescriptive models attempt to determine the best solution to a spatio-temporal problem rather than just an acceptable one.

Prescriptive model: A type of cartographic model that determines the best possible solution from a set of conditions.

Presence/absence method: A method of raster encoding where a grid cell is encoded based on whether or not the object in question is either present or absent from that portion of geographic space occupied by each cell.

Primary data acquisition: A form of data acquisition that normally involves direct contact or high precision instrumentation-based measurement of phenomena rather than using secondary sources or surrugates.

Primary key: A set of attributes in a relational database that are designed as the primary search criteria for the database. For example, a database of students may have as a primary key the student's name.

Prime meridian: Arbitrary starting point for lines of longitude located at Greenwich, England.

Principal meridians: Vertical lines or meridians, used for the U.S. Public Land Survey, to determine the longitudinal bounds of portions of the earth called ranges.

Principal scale: A given representative fraction for a reference globe derived by dividing the earth's radius by the radius of the globe.

Project life cycle: A highly structured model developed to ensure proper software development that proceeds from module testing to subsystem testing and finally system testing.

Projection family: A group of map projections based on a conceptual model of transferring the spherical earth onto a cylinder, a cone, or a flat surface. The conceptual model is visualized as if the spherical coordinates are physically transferred onto these individual surfaces by projecting a light source located at the center of the sphere toward the projected surface.

Proximal region: The area of influence defined for each point object when subjected to Thiessen polygonal analysis.

Proximity: The closeness of one spatial object to another.

Pseudo nodes: A node where two, and only two, arcs intersect, or where a single arc connects with itself.

Public Land Survey System (PLSS): Common grid system used in the United States to divide the land into square mile sections, each equal to 640 acres.

Purpose: A primary limitation on the design of a cartographic document relating to what the map is to be used for.

Quadrat analysis: A standard method of testing point distribution patterns by comparing the numbers of objects from one subarea (quadrat) to another.

Quadrats: Uniform subareas designed for sampling point objects.

Quadtrees: Compact data structures that quantize geographic space into variably sized quarters, each of which exhibits attribute homogeneity.

Quantize: The process of dividing data into quanta or packets. In GIS the process is most commonly associated with the encoding of geographic space into some form of raster data structure.

Radio telemetry: A method of tracking large animals with the use of radio collars that transmit a signal that can be received either on the ground or through low-flying aircraft.

Random: Spatial arrangement of objects where their locations relative to one another are unpredictable. This type of arrangement is indicative of processes that operate through pure chance occurrences.

Random noise: The nonspatially correlated error or residual component for a Kriging model.

Random walk: A line intersect method for analyzing the distribution of linear objects by drawing a zigzag path and recording its points of intersection with lines on the coverage.

Range: On a semivariogram, the critical value where the variance levels off or stays flat.

Range lines: A set of vertical lines used to determine longitudinal position within U.S. Public Land Survey ranges.

Range-graded classifications: The result of a cartographic classing technique that groups ranges of numerical values into single classes.

Raster: A form of GIS graphic data structure that quantizes space into a series of uniformly shaped cells.

Raster chain codes: A compact method of storing raster data as chains or grid cells bordering homogeneous polygonal areas.

Ratio: The highest level of data measurement; it includes an absolute starting point and allows ratios of values to be produced (e.g., salaries).

Ray: A line based on optical geometry representing the uppermost bound of visible objects from a viewer location.

Ray tracing: Analysis of locations on a map that are visible from a viewer location by using a series of rays.

Reality: A map design constraint resulting from the often irregular shapes of real geographic objects.

Record: An individual piece of computer information.

Rectangular coordinates: Coordinate system based on a two-dimensional planar surface, using *X* and *Y* coordinate locations from an origin or starting point. The origin is normally based on an arbitrary Cartesian coordinate system. This is the common coordinate system used by digitizer tablets.

Reductions: In mapping, the idea that the cartographic document is normally smaller than the area it represents.

Reference globe: A hypothetical reduction of the earth and its coordinates mapped onto a sphere that is reduced in size to that chosen for the scale of the flat map being produced.

Region growing: In remote sensing, the procedure that is most often used in unsupervised classification as it attempts to define groups of pixels (regions) with similar spectral signatures.

Regions: Areas of relative uniformity of content within a coverage.

Registration points: Digitizer locations that locate the corners of the map document so that the user can digitize different parts of the document at different sessions.

Regular: Spatial arrangement of objects where all objects are the same distance away from their immediate neighbors. Such an arrangement implies some form of process that places these objects this way.

Regular grid: A series of locational lines whose interspatial distances are identical.

Regular lattice: A series of point locations whose interspatial distances are identical.

Regular sampling procedure: A sampling procedure in which each data point is evenly spaced from its neighbors.

Related variables: Variables from two coverages whose interactions can be defined by mathematical equations. A method of dasymetric mapping.

Relational database structures: Computer database structures employing an ordered set of attribute values or records known as tuples grouped into two-dimensional tables called relations.

Relational join: In relational database management systems, the process of creating functional links between two or more tables sharing some common attribute.

Relations: Two-dimensional tables of relational database attribute records.

Relative barriers: Barriers that, like friction surfaces but at discrete locations, act to impede movement.

Relative location: Location determined in relation to a second object rather than by its own coordinates.

Remote sensing: The observation of objects or groups of objects, normally at a distance, most often with the use of some form of mechanical or electronic device. The data can either produce an image or be stored for later retrieval.

Repeatability: Deals with how close the digitizer will read to an original value if it is recorded more than once.

Representational fraction (RF): One method of representing map scale where both map units and earth units are represented by the same measurement units and are shown as a fraction (e.g., 1:24,000).

Resolution: 1. The amount of earth surface represented by a single grid cell in a raster GIS. 2. The ability of a digitizer to record increments in space. The smaller the units it can handle, the better its resolution.

Resultant length: The length of the resultant vector calculated through the use of the Pythagorean theorem on component vectors.

Resultant vector: A sum of vectors that occur at varying angles and possess different lengths.

Retrieval monitor: The name given to the database management system in the original Canada Geographic Information System that was designed to keep track of both the graphics database and its associated entities by separating them into unique databases and linking them together.

Reverse engineering: In GIS design, the process of examining the nature of an organization and deconstructing it to see how it operates, with the ultimate goal to determine how GIS can be applied to most benefit the organization.

Root mean square (RMS) error: A measure of the difference between known locations and those that have been interpolated. The calculation involves summing the squared differences between the known and unknown points, dividing that by the number of data points and finding the square root of that result.

Rose diagram: A circular graph that displays the orientation of linear objects or phenomena by starting at the center of a circle and drawing each observation as a single line outward from that point.

Rotation: The process of moving part or all of a cartographic object in some azimuthal direction around some focal point.

Rough surface: A statistical surface whose values change rapidly and often unpredictably with changes in horizontal distance.

Routed line cartograms: Cartograms that show the order of objects located along a line but do not show the actual distances between them.

Routing: The process of using networks to define travel paths from one place to another.

Roving window neighborhood functions: A group of analytical functions that employ a moving filter to reclassify the values of the coverage.

Rubber sheeting: A method to adjust coverage features in a nonuniform manner. Used to rectify objects within a coverage.

Rules-of-combination overlay: A method of overlay that allows the user to define which categories will be combined, which will take precedence, and which may be ignored.

Run-length codes: A compact method of storing raster data as strings of homogeneous grid cells.

Runs test: A simple statistical technique for analyzing the pattern of sequences of data, such as those derived through line intersect methods.

Saddle-point problem: For interpolation, the problem that arises when one pair of diagonally opposite *Z* values forming the corners of a rectangle is located above and the second pair below the value the algorithm is attempting to solve.

Sampled area: The portion of a larger area that is to be sampled to draw inferences about the whole.

Sampling frame: A complete list of all members of a population to be sampled, or the entire area within which a sample is to be selected.

Scalar: A level of data measurement that is based on a specific problem set and an estimate of the level of compliance within it. Scalars are often estimated by interpolating based on a known minimum and a known maximum but with no measureable numbers for internal numbers.

Scale: In mapping, the size relationship or ratio between the map document and the portion of the earth it represents.

Scale change: The process of changing the size of part or all of a cartographic object.

Scale-dependent error: Spatial data error that is primarily a function of the scale of the input map document.

Scale factor (SF): On a reference globe, the actual scale divided by the principal scale. By definition, the scale factor for a reference globe should be 1.0 because the actual scale and the principal scale should be the same everywhere.

Scan lines: Lines of pixels resulting from the process of scanning used by the remote sensing device.

Scanning radiometers: Passive remote sensing devices that receive electromagnetic radiation in groups of wavelengths.

Second normal form: Based on the theory of normal forms, the second requirement that every column that is not part of the primary key must be completely dependent on the primary key.

Secondary data acquisition: The use of nondirect, often surrogate sources of data to take the place of data that are not normally obtainable through direct observation.

Section: A portion of land, specifically $1/36$ of a township, determined by the U.S. Public Land Survey to occupy one square mile and consisting of 640 acres.

Selective overlay: *See* rules-of-combination overlay.

Semivariance: One-half the square of the standard deviation between each elevational value and its neighbors.

Semivariogram: A graphical device used in the process of Kriging that compares the variance of the difference in value of elevational points to their distances.

Shape measures: Analytical techniques that characterize cartographic objects based on comparisons of their own dimensional attributes (e.g., perimeter area ratios), comparisons to known geometric shapes (e.g., circles), or to an alternative geometry (e.g., fractal geometry).

Shortest path: A coverage created by determining the shortest Euclidean distance from one place to another. This can be determined for a surface or for networks.

Shortest-path surface: A coverage created by determining the distance from a point, line, or area to all other locations on the coverage. This coverage includes the shortest path as part of the result.

Side-looking airborne radar (SLAR): A system mounted at the base of an aircraft that sends out and receives radar signals in a sweeping motion perpendicular to the flight path of the aircraft.

Simple list: The simplest computer file structure that stores data as it is input, but does not organize it in any sequence or order. This type of file structure allows very easy input, but makes searches and retrieval operations very difficult.

Sinuosity: The relationship between straight-line distance and actual distance of linear objects.

Sinuous: Wandering wildly in a lateral direction along its length.

Size measures: Analytical techniques that characterize cartographic objects based on their absolute sizes or comparisons of their sizes with other cartographic objects.

Skew: Measure of the squareness of the results of digitizing four corners of a document on a digitizing table.

Slicing: A group of GIS functions that selectively change the class interval for statistical surfaces.

Sliver polygons: Small polygons, often without attributes, that result either from digitizing the same line twice or following the overlay of two or more coverages.

Slope: Measure of the amount of rise divided by the amount of horizontal distance traveled.

Smooth surface: A statistical surface whose values change at a relatively constant rate with changes in horizontal distance.

Smoothing: A type of filter, often called a low-pass filter, that reduces the value of extraordinarily high cell or pixel values through some process of averaging.

Software design: The structured process of transforming software functional requirements to specific code.

Software engineering: A subfield of computer science specializing in the systematic approaches to software code generation.

Soundshed: A set of polygons that display areas of differential acoustic penetration.

Spacing: The attribute of spatial objects that shows their distances from one another.

Spaghetti model: The simplest vector data structure representing a one-for-one translation of the graphic image. Often used as a data structure for vector input.

Spatial: Anything dealing with the concept of space. In the geographic context, primarily dealing with the distribution of things on the surface of the earth.

Spatial arrangement: The placement, ordering, concentration, connectedness, or dispersion of multiple objects within a confined geographic space.

Spatial cognition: The ability to recognize the spatial properties or nature of a problem.

Spatial information product (SIP): The desired output from a set of GIS analysis operations.

Spatial integrity: The degree of perforation of a perforated region.

Spatial pseudo nodes: Pseudo nodes that represent island polygons or where some attribute along a line changes attributes. Both of these are acceptable pseudo nodes.

Spatial surrogates: Spatially explicit data used to replace data that do not exhibit spatial properties.

Spatially correlated: Sets of objects that are spatially associated may also show statistically significant relationships that may indicate some connection between the processes acting on both objects or even the processes acting on one set of objects influencing those of another set of objects in the same area.

Splines: Mathematical methods of smoothing linear cartographic objects. Often used in interpolation methods.

Stability: The property of a digitizer that deals with the propensity of the device to change the readings it provides as it warms up.

Stack: A computer data storage structure based on the last in, first out (LIFO) protocol. The items on the top must be removed before the objects below may be manipulated.

Standard parallels: Lines on a reference globe that maintain a scale factor of 1.0.

Static neighborhood functions: Analytical functions that modify the attributes of neighboring cells without the use of a moving filter.

Statistical surface: Any ordinal, interval, or ratio data that can be represented as and operated on as a surface.

Step rate: The incremental distance covered by an output device during the process of drawing lines. The smaller the step rate, the finer the resolution of the output document.

Stereographic projections: *See* conformal stereographic projections.

Stratified sampling: A sampling strategy in which the total study area is divided into smaller, relatively homogeneous areas (strata), within which sampling takes place at a rate specific to that smaller strata.

Stream magnitude: Based on the Shreve method of stream channel analysis, the intersection of first and second order streams results in a third order link and the intersection of a second and third order stream results in a fifth order link. These order numbers are called stream magnitudes.

Stream order: Based on the Strahler method of stream channel analysis, the stream orders increase as stream segments of the same order value intersect. So as two first order streams come together they form "second order" streams, two second order streams linking form "third order" streams, and so on.

Structure: *See* drift.

Substantive objective: One of two parts of the purpose of a map. This one relates to the information the map must include and controls what is to be mapped.

Summit: In topographic analysis the top of a hill or mountain, indicated by the highest elevational value occupying the feature.

Supervised classification: In remote sensing, the process of classifying digital data with the interaction of a user. The user normally selects pixels that are of a known category and uses these to train the software about the electromagnetic properties likely to be associated with each category.

Supplier for GIS (GIS supplier): Third-party vendor for GIS software, and sometimes hardware, used by the GIS community.

Surface-fitting models: Interpolation methods that attempt to fit the existing data points into a mathematical surface equation.

System design: The process whereby the output from system analysis is transformed from technology-independent statements of user requirements to the system that is going to be used to implement the GIS.

System operators: GIS professionals responsible for ensuring that the system functions correctly and for maintaining the hardware and software in a condition that allows the users to perform their tasks.

System sponsor: The organizational leadership responsible for making funding and logistical decisions and for selecting priorities for resource allocation within the organization.

System users: GIS professionals who interact most closely with the GIS through data input, storage, editing, and performing the actual analytical functions necessary to provide answers to queries.

Systematic sampling: In statistical methods, the selection of every k^{th} element of a sampling frame (whether spatial or aspatial).

Target population: The portion of a whole population that is likely to contain attributes for which a sample is being taken.

Targeted analysis: An analysis that is directed toward reclassifying a target cell or target groups of cells as a function of neighboring values.

Technical design: GIS design issues that deal with system requirements, hardware and software needs, and other noninstitutional issues.

Technical limits: A set of limitations on map design based on the physical limitations of the equipment to produce readable results.

Template: A user-selected area on a map that is employed as the outline for subsequent coverages to compare their attributes.

Terrain analysis: A set of geospatial analytical techniques specifically designed for operating on the topographic form of the statistical surface.

Tessellation: Any of a number of subdivisions of a two-dimensional plane (or three-dimensional volume) into polygonal tiles (polyhedral blocks) that completely cover a plane or volume. Planes or volumes can be either divided into regular or irregular tessellations.

Texture: The visual appearance of a symbol resulting from the spacing and placement of internal markings.

Thematic maps: Maps whose primary purpose is to display the locations of a single attribute or the relationships among several selected attributes.

Theodolite: A modern optical survey device that includes digital readouts and improved accuracy over its predecessors. This device is still limited in its ability to provide survey data for large areas.

Thiessen polygons: A method of creating polygons or proximal regions around point objects by defining them mathematically, dividing the space between each point, and connecting these distances with straight lines.

Third normal form: Based on the theory of normal forms, the requirement that every nonprimary key column must be nontransitively dependent on the primary key.

Tic marks: On the cartographic document, known point locations designed to address the registration (latitude/longitude) of the document during input.

TIGER: *See* Topologically Integrated Geographic Encoding and Referencing system.

Tiling: Process of separating large databases into predefined subsections for archival purposes.

Too many labels: Entity-attribute agreement error resulting from placing more than one label point within a single polygon during input.

Topological models: A vector data structure that incorporates explicit spatial information about the relative locations of objects in the database. These are necessary to allow advanced analysis in a GIS.

Topologically Integrated Geographic Encoding and Referencing (TIGER) system: A topological vector data model created by the U.S. Census

Bureau as an improvement over GBF/DIME. Points, lines, and areas are explicitly addressed, allowing direct retrieval of census block data, and real-world objects are portrayed in their true geographic shape.

Topology: The mathematically explicit rules defining the linkages of geographical elements.

Total analysis of neighborhood: A function that analyzes the entire neighborhood and returns new values for that entire neighborhood.

Township: A unit of area determined by the U.S. Public Land Survey to occupy 36 square miles and bounded by range lines and township lines.

Township lines: Lines in the U.S. Public Land Survey, drawn east and west, and measured north and south that determine the latitudinal extent of portions of the land.

Translation: The process of moving part or all of a cartographic object to a new location in Cartesian space.

Transverse projections: A type of Mercator projection that does not exhibit the property that all rhumb lines are straight. As a result, scale exaggeration increases away from the standard meridian, limiting the usefulness of this type of projection to a small zone along the central meridian.

Trapezoidal rule: A methodology for calculating the centroid of a highly irregular polygonal shape.

Traveling salesman problem: A complex computer problem in which a traveling salesman is expected to visit a large number of spatially separated points by doing the least amount of traveling.

Travelsheds: Surfaces indicating the results of cumulative distance analysis across an entire surface. Such surfaces typically show all possible travel distances (or other non-Euclidean distance measure such as time) from a single point or area to all other possible points.

Trend surface: The result of interpolation based on the fit of a polynomial surface by least squares regression through the set of sample points. This calculation minimizes the variance of the surface relative to the input data points. Trend surface is used to determine general trends rather than detailed accurate solutions.

Trend surface analysis: An interpolation routine that simplifies the surface representation to allow visualization of general or overall changes in elevation value with distance.

Triangulated irregular network (TIN): A vector data model that uses triangular facets as a means of explicitly storing surface information.

Triangulation: A method of surveying that measures distance and direction from a baseline outward to the selected points.

Trilateration: A method of surveying that measures both from and to a baseline and each point to be located.

Tuples: Individual records (rows) in a relational database structure.

Undershoot: An arc that does not extend far enough to connect with another arc. Sometimes called dangling arcs or just dangles.

Undirected networks: Type of network that does not place restrictions on the direction of movement.

Uniform: A distribution of objects that exhibits the same density in one small subarea as that in each additional subarea.

Union overlay: A form of logical overlay in which polygonal attributes common to the two or more layers being combined will result.

Unit value: The actual numerical value assigned to individual polygons in a choropleth map.

Universal polar stereographic (UPS) grid: A grid system used to cover the polar areas of the globe. Each circular polar zone is divided in half at 0 and 180 degrees meridian.

Universal Transverse Mercator (UTM) grid: A projected map coordinate system that divides the earth into 60 north-south zones, each 6 degrees wide.

Unsupervised classification: In remote sensing, the process of classifying digital data without the interaction of a user. The process, most often statistical, randomly selects pixels and attempts to find spectral similarities between the selected pixels and the remainder of the pixels in the image.

Value: The lightness or darkness of a cartographic object compared to a standard black and white value.

Value-by-area mapping: A method of mapping whereby each polygon is assigned a specific ordinal, interval, or ratio attribute value.

Variance–mean ratio (VMR): An index showing the relationship between the frequency of subarea variability to the average number of points per quadrat.

Vector: A graphic data structure that represents the points, lines, and areas of geographical space by exact *X* and *Y* coordinates.

Vector chain codes: A method of storing lines as a series of straight line segments defined as a set of directional codes, with each code following the last like links in a chain.

Vector resultant: The result of combining the magnitudes and directions of component vectors.

Verbal scale: One way of representing map scale, where both map units and earth units are stated as text (e.g., one inch equals 100 feet).

Vertical exaggeration: A technique used for production of cross-sectional profiles that increases the vertical dimension for easier viewing of surface changes.

View integration: The process of satisfying multiple user needs for a GIS database and its applications by incorporating each local view into the global view.

Viewing azimuth: The geographic direction from which a perspective map can be displayed so as to observe its different sides.

Viewing distance: One of a number of factors that can be modified to effectively display perspective maps.

Viewshed: A polygonal map resulting from analysis of topographic surfaces that portrays all locations visible from a preselected viewpoint.

Viewshed analysis: An analytical technique that calculates the locations visible from one or more observation locations.

Virtual GIS: A GIS environment in which the objects are wholly or partially artificial constructs that represent either past environments or predicted environments depending on the application.

Virtual Reality Markup Language (VRML): a file format specifically designed to represent 3-dimensional interactive vector graphics. VRML was developed with the World Wide Web in mind.

Visibility analysis: An analysis of points or areas visible from one or more observation points that results in either a set of visible points or a viewshed.

Visual contrast: The relationship between cartographic symbols and the underlying map background, often responsible for legibility.

Visual objectives: Those objectives of cartographic design that attempt to preserve, as much as possible, the visual aesthetic aspects of the document.

Visualization: A set of techniques designed to assist in the understanding of data through often unique visual display methodologies.

Voronoi diagrams: *See* Thiessen polygons.

Waterfall model: *See* project life cycle.

Weakest link hypothesis: The concept (generally rejected) that suggests that in map overlay operations, the layer that is of the poorest quality controls the quality of the output. The reasons for the poor quality can be error, scale, representation, or class limit method; however, the incorrect assumption of this hypothesis is that all layers are of equal weight and/or equal importance.

Weighted mean center: Center of a group of point objects located in space and modified by a weighting factor applied to each point. The weight is representative of some physical, economic, or cultural factor.

Weighted overlay: A method of map overlay that allows the user to define the importance of each map layer. Normally weighted overlay is based on percentage where the total must equal 100%.

Weighting methods: Nonlinear interpolation methods that weight the predicted elevation values by the distance to its nearest neighbor elevation values.

Weird polygons: Graphical artifacts resembling real polygons but lacking nodes. Usually the result of digitizing points for polygonal objects out of sequence.

Wire-frame diagrams: A perspective view of a set of statistical data displayed as a skeleton form.

Witness trees: Trees left behind by surveyors during pioneer periods in the United States. These trees were left as witnesses for the metes and bounds survey method and have also been used more recently as indicators of previous forest vegetation.

World geodetic system (WGS84): Modified version of GRS80, developed for the United States military in 1984.

Zonal function: GIS modeling function that uses properties of a defined region of cells (including fragmented regions) to characterize the target cell.

Zones: On the UTM grid system, the earth is divided into columns 6 degrees longitude wide. Each of these is called a zone and is numbered from 1 to 60 eastward beginning at the 180th meridian.

Zoom transfer scope: A mechanical, optical device designed for accurately transferring information from aerial photographs to projected map documents.

INDEX

PHOTO CREDITS

Chapter 0
Page 5: Jill Toyoshiba/Kansas City Star/NewsCom. Page 7: Age fotostock/SUPERSTOCK. Page 9: Brand X/SUPERSTOCK.

Chapter 2
Page 38 (left and right): Courtesy Del Webb Corporation. Page 46: Courtesy of Robert Barron, www.texashiking.com. Page 48: Courtesy NAVSTAR/UNAVCO. Page 51: From "Soil Survey Manual," published in 1993, United States Department of Agriculture.

Chapter 6
Page 133: Courtesy GTO Calcomp, Inc. Page 135: Courtesy Agfa Corp. Page 136: Photo courtesy The International Boundary and Water Commission. Graphics courtesy Dona Ana County, New Mexico.

Chapter 8
Page 188: Courtesy GeoHealth, Inc. Pages 190 and 192: Courtesy Michael N. DeMers.

Chapter 11
Page 255: Courtesy USGS.

Chapter 14
Page 373: Courtesy Topozone.